Texts and Monographs in Physics

Series Editors:

R. Balian, Gif-sur-Yvette, France
W. Beiglböck, Heidelberg, Germany
H. Grosse, Wien, Austria
E. H. Lieb, Princeton, NJ, USA
N. Reshetikhin, Berkeley, CA, USA
H. Spohn, München, Germany
W. Thirring, Wien, Austria

Springer
Berlin
Heidelberg
New York
Hong Kong
London
Milan
Paris
Tokyo

Physics and Astronomy ONLINE LIBRARY

http://www.springer.de/phys/

A. Bohm A. Mostafazadeh
H. Koizumi Q. Niu J. Zwanziger

The Geometric Phase in Quantum Systems

Foundations, Mathematical Concepts,
and Applications in Molecular
and Condensed Matter Physics

With 56 Figures

Springer

Professor Arno Bohm
University of Texas at Austin
Physics Department
1 University Station C1600
Austin, TX 78712-0264, USA
E-mail: bohm@physics.utexas.edu

Dr. Hiroyasu Koizumi
Himeji Institute of Technology
Faculty of Science
Hyogo 678-1297, Japan
E-mail: koizumi@sci.himeji-tech.ac.jp

Professor Joseph Zwanziger
Indiana University
Department of Chemistry
Bloomington, IN 47405, USA
E-mail: jzwanzig@indiana.edu

Professor Ali Mostafazadeh
Koç University
Department of Mathematics
34450 Istanbul, Turkey
E-mail: amostafazadeh@ku.edu.tr

Professor Qian Niu
University of Texas at Austin
Department of Physics
1 University Station C1600
Austin, TX 78712-0264, USA
E-mail: niu@physics.utexas.edu

ISSN 0172-5998
ISBN 3-540-00031-3 Springer-Verlag Berlin Heidelberg New York

Library of Congress Cataloging-in-Publication Data
The geometric phase in quantum systems : foundations, mathematical concepts, and applications in molecular and condensed matter physics / A. Bohm ... [et al.].
p.cm. – (Texts and monographs in physics, ISSN 0172-5998)
Includes bibliographical references and index.
ISBN 3540000313 (alk. paper)
1. Geometric quantum phases. 2. Adiabatic invariants. 3. Quantum theory. I. Böhm, Arno, 1936- II. Series.
QC20.7.G44G457 2003   530.12–dc21   2002036470

This work is subject to copyright. All rights are reserved, whether the whole or part of the material is concerned, specifically the rights of translation, reprinting, reuse of illustrations, recitation, broadcasting, reproduction on microfilm or in any other way, and storage in data banks. Duplication of this publication or parts thereof is permitted only under the provisions of the German Copyright Law of September 9, 1965, in its current version, and permission for use must always be obtained from Springer-Verlag. Violations are liable for prosecution under the German Copyright Law.

Springer-Verlag Berlin Heidelberg New York
a member of BertelsmannSpringer Science+Business Media GmbH

http://www.springer.de

© Springer-Verlag Berlin Heidelberg 2003
Printed in Germany

The use of general descriptive names, registered names, trademarks, etc. in this publication does not imply, even in the absence of a specific statement, that such names are exempt from the relevant protective laws and regulations and therefore free for general use.

Typesetting by the authors using a TEX macro package
Data conversion: K. Mattes, Heidelberg
Cover design: *design & production* GmbH, Heidelberg
Printed on acid-free paper   SPIN: 10467482   55/3141/di - 5 4 3 2 1 0

To our wives

Preface

Since Berry's introduction of the adiabatic geometrical phase, a large number of articles have appeared on the theoretical foundations, physical applications, and experimental manifestations of geometric phases. Although there are by now several review articles on geometric phases, there have been no comprehensive books or monographs on the subject. The present volume is intended to fill this gap in the literature. It is aimed at a diverse audience of advanced undergraduate as well as graduate students of physics and chemistry.

Due to their general nature, geometric phases have found applications in several different areas of physics and chemistry. Their theoretical basis has been shown to be related to the most basic concepts of modern mathematics. These make a complete treatment of the subject in a single volume a quite impossible task. We have included in this book an introductory part which offers an elementary discussion of the basic concepts and is based on our graduate level courses and summer school lectures. In the later part of the book we present more advanced subjects on the mathematical foundations of the geometric phase and the applications of the geometric phase in molecular and condensed matter physics. In the preparation of this book priority was given to the clarity of the exposition. We have also made every attempt to make the book as self-contained as possible.

A student with a good understanding of basic quantum mechanics should be able to learn the contents of the book at a reasonable pace. Although we have not assumed a knowledge of differential geometry, familiarity with manifolds and differential forms will certainly facilitate a quick reading. Readers with limited mathematical background should consult Appendix A. Here we offer a discussion of the most basic mathematical concepts together with worked examples. Appendix B provides an overview of the point group theory needed to understand many of the molecular examples of geometric phases.

Chapter 1 includes an introduction to the importance of geometric phases as well as a short historical survey of the developments which led to their discovery. Chapter 2 introduces Berry's adiabatic geometrical phase. This is followed by a discussion of the topological phase of Aharonov and Bohm. Chapter 3 is devoted to a detailed treatment of the quantum dynamics of a magnetic dipole in a precessing magnetic field. This is used as the motivation for the introduction of the non-adiabatic geometric phase of Aharonov and Anandan in Chap. 4. This chapter also discusses the connections between the geometric phase and the theory of fiber bundles. Chapter 5 offers a more detailed introduction to fiber bundles and gauge theories. Chapter 6 includes a thorough discussion of different holonomy interpretations of the geometric phase and their relation to universal classifying bundles and connections. Chapter 7 treats the non-Abelian generalization of the ordinary geometrical phase. Chapters 8 and 9 discuss the emergence and importance of the Abelian and non-Abelian geometric phases in molecular physics. Chapters 10 and 11 provide a wealth of experimental examples in which the geometric phase has been detected and for which knowledge of the geometric phase greatly enhances understanding. The final three chapters survey various manifestations and applications of the geometric phase in condensed matter systems.

Acknowledgments

A. B. would like to express his gratitude to J. Anandan, M. V. Berry, B. K. Kendrick, L. Loewe and C. A. Mead for discussions on the subject of various chapters of this book. He is particularly grateful to C. A. Mead and A. Mondragon for reading parts of the earlier manuscript and for their comments and suggestions. He also thanks the participants and organizers of the NATO Advanced Study Institute and GIFT International Seminar Salamanca where some of the material of this book was first presented.

A. M. would like to thank his PhD advisor Bryce DeWitt and his teachers Can Delale, Metin Arik, Cecile DeWitt-Morette, and Gary Hamrick for their guidance and encouragement, his friends and colleagues Ertuğrul Demircan, Teoman Turgut, Bahman Darian, Rouzbeh Allahverdi, Luis Boya, Don Page, Mohsen Razavy, and graduate students of his mathematical physics course (taught at the Sharif University of Technology, Tehran, in the Spring of 1995) for fruitful discussions and comments. He would also like to acknowledge the financial support of the Killam Foundation of Canada and the hospitality of the Theoretical Physics Institute of the University of Alberta during the preparation of the first draft of the book.

Q. N. wishes to acknowledge his PhD and post-doctoral advisers David Thouless, Walter Kohn and Eduardo Fradkin for their mentorship, and his former students M. C. Chang and G. Sundaram and colleagues Ping Ao, L. Kleinman, and A. H. MacDonald for their close collaborations. He especially thanks Julie Zhu for her love, understanding, and care.

H. K. would like to thank I. B. Bersuker, J. E. Boggs, P. J. Rossky, J. F. Stanton, and R. E. Wyatt for their hospitality during his stay at the Institute of Theoretical Chemistry, the University of Texas at Austin, where a part of the manuscript was prepared. He also thanks K. Makoshi and N. Shima for making this extended stay possible, G. C. Schatz, V. Z. Polinger and Y. Takada for helpful comments and discussions, and B. Kendrick for providing some of the figures.

J. Z. acknowledges many instructive interactions on the subject of geometric phases with E. R. Grant, A. Pines, and M. Bausch-Koenig. He also thanks A. Bohm for inviting him to participate in this project.

The quantum geometric phase has been a research topic in many areas of science: theoretical amd mathematical physics, condensed matter theory, theoretical and experimental chemistry. This book is a collaboration of these diverse areas of physics and chemistry.

We gratefully acknowledge support from the Welch Foundation.

Austin,	*Arno Bohm, Qian Niu*
Istanbul,	*Ali Mostafazadeh*
Kamigori,	*Hiroyasu Koizumi*
Bloomington,	*Josef Zwanziger*
March 2003	

Table of Contents

1. **Introduction** 1

2. **Quantal Phase Factors for Adiabatic Changes** 5
   - 2.1 Introduction 5
   - 2.2 Adiabatic Approximation 10
   - 2.3 Berry's Adiabatic Phase 14
   - 2.4 Topological Phases and the Aharonov–Bohm Effect 22
   - Problems 29

3. **Spinning Quantum System in an External Magnetic Field** 31
   - 3.1 Introduction 31
   - 3.2 The Parameterization of the Basis Vectors 31
   - 3.3 Mead–Berry Connection and Berry Phase
         for Adiabatic Evolutions – Magnetic Monopole Potentials.... 36
   - 3.4 The Exact Solution of the Schrödinger Equation 42
   - 3.5 Dynamical and Geometrical Phase Factors
         for Non-Adiabatic Evolution 48
   - Problems 52

4. **Quantal Phases for General Cyclic Evolution** 53
   - 4.1 Introduction 53
   - 4.2 Aharonov–Anandan Phase 53
   - 4.3 Exact Cyclic Evolution for Periodic Hamiltonians 60
   - Problems 64

5. **Fiber Bundles and Gauge Theories** 65
   - 5.1 Introduction 65
   - 5.2 From Quantal Phases to Fiber Bundles 65
   - 5.3 An Elementary Introduction to Fiber Bundles 67
   - 5.4 Geometry of Principal Bundles and the Concept of Holonomy 76
   - 5.5 Gauge Theories 87
   - 5.6 Mathematical Foundations of Gauge Theories and Geometry
         of Vector Bundles 95
   - Problems 102

Table of Contents

6. **Mathematical Structure of the Geometric Phase I: The Abelian Phase** 107
   - 6.1 Introduction 107
   - 6.2 Holonomy Interpretations of the Geometric Phase 107
   - 6.3 Classification of $U(1)$ Principal Bundles and the Relation Between the Berry–Simon and Aharonov–Anandan Interpretations of the Adiabatic Phase 113
   - 6.4 Holonomy Interpretation of the Non-Adiabatic Phase Using a Bundle over the Parameter Space 118
   - 6.5 Spinning Quantum System and Topological Aspects of the Geometric Phase 123
   - Problems 126

7. **Mathematical Structure of the Geometric Phase II: The Non-Abelian Phase** 129
   - 7.1 Introduction 129
   - 7.2 The Non-Abelian Adiabatic Phase 129
   - 7.3 The Non-Abelian Geometric Phase 136
   - 7.4 Holonomy Interpretations of the Non-Abelian Phase 139
   - 7.5 Classification of $U(\mathcal{N})$ Principal Bundles and the Relation Between the Berry–Simon and Aharonov–Anandan Interpretations of Non-Abelian Phase 141
   - Problems 145

8. **A Quantum Physical System in a Quantum Environment – The Gauge Theory of Molecular Physics** 147
   - 8.1 Introduction 147
   - 8.2 The Hamiltonian of Molecular Systems 148
   - 8.3 The Born–Oppenheimer Method 157
   - 8.4 The Gauge Theory of Molecular Physics 166
   - 8.5 The Electronic States of Diatomic Molecule 174
   - 8.6 The Monopole of the Diatomic Molecule 176
   - Problems 191

9. **Crossing of Potential Energy Surfaces and the Molecular Aharonov–Bohm Effect** 195
   - 9.1 Introduction 195
   - 9.2 Crossing of Potential Energy Surfaces 196
   - 9.3 Conical Intersections and Sign-Change of Wave Functions ... 198
   - 9.4 Conical Intersections in Jahn–Teller Systems 209
   - 9.5 Symmetry of the Ground State in Jahn–Teller Systems 213
   - 9.6 Geometric Phase in Two Kramers Doublet Systems 219
   - 9.7 Adiabatic–Diabatic Transformation 222

10. Experimental Detection of Geometric Phases I: Quantum Systems in Classical Environments 225
10.1 Introduction 225
10.2 The Spin Berry Phase Controlled by Magnetic Fields 225
    10.2.1 Spins in Magnetic Fields: The Laboratory Frame 225
    10.2.2 Spins in Magnetic Fields: The Rotating Frame 231
    10.2.3 Adiabatic Reorientation in Zero Field 237
10.3 Observation of the Aharonov–Anandan Phase
    Through the Cyclic Evolution of Quantum States........... 248
Problems 252

11. Experimental Detection of Geometric Phases II: Quantum Systems in Quantum Environments............. 255
11.1 Introduction 255
11.2 Internal Rotors Coupled to External Rotors................ 256
11.3 Electronic–Rotational Coupling 259
11.4 Vibronic Problems in Jahn–Teller Systems 260
    11.4.1 Transition Metal Ions in Crystals 261
    11.4.2 Hydrocarbon Radicals 264
    11.4.3 Alkali Metal Trimers.............................. 265
11.5 The Geometric Phase in Chemical Reactions 270

12. Geometric Phase in Condensed Matter I: Bloch Bands ... 277
12.1 Introduction 277
12.2 Bloch Theory 278
    12.2.1 One-Dimensional Case 278
    12.2.2 Three-Dimensional Case........................... 280
    12.2.3 Band Structure Calculation 281
12.3 Semiclassical Dynamics 283
    12.3.1 Equations of Motion 283
    12.3.2 Symmetry Analysis 285
    12.3.3 Derivation of the Semiclassical Formulas 286
    12.3.4 Time-Dependent Bands 287
12.4 Applications of Semiclassical Dynamics.................... 288
    12.4.1 Uniform DC Electric Field......................... 288
    12.4.2 Uniform and Constant Magnetic Field 289
    12.4.3 Perpendicular Electric and Magnetic Fields 290
    12.4.4 Transport 290
12.5 Wannier Functions...................................... 292
    12.5.1 General Properties 292
    12.5.2 Localization Properties............................ 293
12.6 Some Issues on Band Insulators 295
    12.6.1 Quantized Adiabatic Particle Transport 295
    12.6.2 Polarization 297
Problems 299

13. Geometric Phase in Condensed Matter II: The Quantum Hall Effect 301
   13.1 Introduction 301
   13.2 Basics of the Quantum Hall Effect 302
       13.2.1 The Hall Effect 302
       13.2.2 The Quantum Hall Effect 302
       13.2.3 The Ideal Model 304
       13.2.4 Corrections to Quantization 305
   13.3 Magnetic Bands in Periodic Potentials 307
       13.3.1 Single-Band Approximation in a Weak Magnetic Field 307
       13.3.2 Harper's Equation and Hofstadter's Butterfly 309
       13.3.3 Magnetic Translations 311
       13.3.4 Quantized Hall Conductivity 314
       13.3.5 Evaluation of the Chern Number 316
       13.3.6 Semiclassical Dynamics and Quantization 318
       13.3.7 Structure of Magnetic Bands and Hyperorbit Levels .. 321
       13.3.8 Hierarchical Structure of the Butterfly 325
       13.3.9 Quantization of Hyperorbits
              and Rule of Band Splitting 327
   13.4 Quantization of Hall Conductance in Disordered Systems 329
       13.4.1 Spectrum and Wave Functions 329
       13.4.2 Perturbation and Scattering Theory 331
       13.4.3 Laughlin's Gauge Argument 332
       13.4.4 Hall Conductance as a Topological Invariant 333

14. Geometric Phase in Condensed Matter III: Many-Body Systems 337
   14.1 Introduction 337
   14.2 Fractional Quantum Hall Systems 337
       14.2.1 Laughlin Wave Function 337
       14.2.2 Fractional Charged Excitations.................... 340
       14.2.3 Fractional Statistics 341
       14.2.4 Degeneracy and Fractional Quantization 344
   14.3 Spin-Wave Dynamics in Itinerant Magnets 346
       14.3.1 General Formulation 346
       14.3.2 Tight-Binding Limit and Beyond 348
       14.3.3 Spin Wave Spectrum................................ 350
   14.4 Geometric Phase in Doubly-Degenerate Electronic Bands 353
   Problem 359

A. An Elementary Introduction to Manifolds and Lie Groups 361
   A.1 Introduction 361
   A.2 Differentiable Manifolds 371
   A.3 Lie Groups 388

B. A Brief Review of Point Groups of Molecules
   with Application to Jahn–Teller Systems 407

References 429

Index 437

Conventions

Throughout the text, the Plank constant $\hbar$ is set to unity and Einstein's summation convention for repeated indices is used (the exceptions are obvious from the context). Space-time coordinates are labeled by Greek letters: $x^\mu, x^\nu, \cdots$ where the superscripts take the values $0, 1, 2, 3$, and $x^0 = ct$. Space coordinates are labeled by letters from the beginning of the Latin alphabet, i.e., $a, b, \cdots$. The Minkowski metric $\eta = (\eta_{\mu\nu})$ is chosen to have signature -2, i.e., $\eta = \text{diag}(1, -1, -1, -1)$ which means $\eta_{00} = +1$, $\eta_{0a} = \eta_{a0} = 0$, and $\eta_{ab} = -\delta_{ab}$, where δ_{ab} is the Kronecker delta function. Furthermore, letters from the middle of the Latin alphabet $i, j, \cdots$ are used to label the coordinates of arbitrary parameter spaces where a choice of a metric is not made. The shorthand notation $\partial_i = \partial/\partial x^i$ is used for derivatives. The symbol $:=$ is used to define the left-hand side in terms of the right-hand side.

1. Introduction

Today quantum mechanics forms an important part of our understanding of physical phenomena. Its consequences both at the fundamental and practical levels have intrigued mathematicians, physicists, chemists, and even philosophers for the past seven decades. A quantum system is usually described in terms of certain vector spaces and linear operators acting on these spaces. The vector spaces and their operators represent the states and the observables of the quantum system. The dynamics of a quantum system is determined by dynamical differential equations, the Schrödinger or the Heisenberg equations, which involve a linear operator called the Hamiltonian.

The Hamiltonian operator yields the energy levels and more importantly describes the evolution of the states of the physical system in time. Standard textbooks on quantum mechanics discuss almost exclusively the properties of quantum systems whose Hamiltonian does not depend on time. In many practical situations, however, the physical parameters which occur in the expression for the Hamiltonian are determined by time-dependent external or environmental factors. The study of time-dependent Hamiltonians is therefore very important in modeling real physical systems. One of the most interesting aspects of a quantum system with a time-dependent Hamiltonian is the occurrence of the geometric phase.

The geometric phase had been ignored in quantum physics for more than half a century. It had not been forgotten, but was thought to be unimportant. In 1928, Fock [82] showed that such a phase could be set to unity by a redefinition of the phase of the initial wave function. Although Fock's proof was limited to non-cyclic evolutions only, his conclusion was generally accepted until around 1980 when Mead and Truhlar [167] and Berry [31] reconsidered cyclic evolutions.

A cyclic evolution is an evolution in which the initial quantum state evolves periodically in time. For a pure cyclic state, this means that the state operator returns to the initial operator after each period while the corresponding state vector evolves into a vector which agrees with the initial vector only up to a phase factor. This phase factor contains, in addition to the usual dynamical phase, a purely geometric part which does not depend on the duration of the evolution.

Cyclic evolutions play an important role in the description of quantum systems in a periodically changing environment. The environment can be either classical such as a magnetic dipole in a precessing external magnetic field, or quantal such as an electron in the changing quantal environment of the collective motion of a molecule.

In 1956 Pancharatnam [208] discovered an analog of the quantum geometric phase in polarization optics. Three years later Y. Aharonov and D. Bohm published their findings on the significance of the electromagnetic vector potential in quantum mechanics [8]. They showed how the presence of a vector potential that did not produce an electric or magnetic field in the configuration space of free electrons could influence their interference pattern. The change in the interference pattern is due to the so-called Aharonov–Bohm phases which are special examples of the geometric phase. The Aharonov–Bohm phases received much attention in the 1960s, but it was not until the 1980s that the importance of the geometric phase was fully recognized.

The geometric phase in molecular systems appeared first in an implicit manner in the study of the $E \otimes e$ Jahn–Teller problem by Longuet-Higgins et al. [159] and by Herzberg and Longuet-Higgins [105]. They noticed that an electronic wave function that could be taken as real in all nuclear configurations behaved as a double-valued function that changed sign when the nuclear coordinates traversed a loop encircling a crossing point of the energy levels (potential energy surfaces) in the nuclear coordinate space.

The first concrete derivation of a geometric phase and the corresponding gauge potential was carried out in 1978 by Mead and Truhlar [167]. They considered the chemical reaction $H + H_2 \rightarrow H_2 + H$, which could be viewed as a wave packet motion from one minimum of the potential energy surface of the H_3 system to another. H_3 is an example of an $E \otimes e$ Jahn–Teller system, and the electronic wave function undergoes the sign changes found by Longuet-Higgins and his collaborators. Mead and Truhlar argued that the double-valuedness of the wave function caused by these sign changes could be avoided by including a vector potential in the electronic Hamiltonian.

This amounts to an improvement of the standard molecular Born–Oppenheimer approximation [48]. The latter is based on the observation that one can divide the motion of the constituents of a molecule into two "parts": the fast motion of the electrons and the slow collective rotations and vibrations of the molecule as a whole. One first investigates the dynamics of the fast variables while keeping the slow variables fixed, and then determines the dynamics of the slow variables. This means that in the Born–Oppenheimer approximation one treats the fast and the slow motions as two separate parts that do not influence each other. If, on the other hand, one does not consider the nuclear coordinates as fixed parameters but as quantum observables whose values change slowly in time, then the gauge potential underlying the geometric phase emerges naturally from the Born–Oppenheimer method.

Conceptually simpler than the gauge theory of the Born–Oppenheimer method is the investigation of quantum systems whose Hamiltonian depends on a set of slowly changing parameters. This was carried out in 1984 in a beautiful paper by Berry [31] who considered quite general quantum systems in a slowly changing classical environment. In this paper Berry derived the same gauge potential and the geometric phase that Mead and Truhlar had obtained from the Born–Oppenheimer method for the molecule. He further showed that indeed the celebrated Aharonov–Bohm phase was a special case of a geometric phase.

Berry's derivation of the adiabatic geometric phase – also known as the Berry phase – made use of the quantum adiabatic approximation which was only relevant for slowly changing Hamiltonians. However, it is easy to show that for a Hamiltonian with changing eigenvectors the adiabatic approximation of the dynamics of a cyclic evolution cannot be exact. Therefore Berry's phase could only be an approximation of the true quantum geometric phase. The latter was introduced for general unitary cyclic evolutions by Aharonov and Anandan in 1987 [7] and subsequently generalized to arbitrary (not necessarily unitary or cyclic) evolutions by Samuel and Bhandari [224].

Soon after the publication of Berry's paper, a number of experiments were performed to observe geometric phases. Among these are the nuclear magnetic resonance experiment by Suter *et al.* [239] and the nuclear quadrupole resonance experiment by Tycko [254]. A manifestation of the geometric phase in polarization optics was also observed in an experiment by Tomita and Chiao [250]. Today, there are many publications on various experimental studies of geometric phases in molecular physics. In particular, the geometric phase effect in the $E \otimes e$ problem was recently verified in a very convincing way by high-resolution spectroscopy of Na_3 and Li_3 [122, 259].

Geometric phases also play an important role in the study of condensed matter systems. One of the earliest results in this direction is due to Zak [281] who noticed that certain non-integrable phases of the Bloch wave function could be identified as a geometric phase. This was later related to the polarization of crystal insulators [124] and used to develop a practical method of calculating piezoelectric and ferroelectric properties [219]. Geometric phases in Bloch waves can also affect the semiclassical dynamics of electrons in metals and semiconductors [141, 237] and have important applications in the theory of the anomalous Hall effect [119]. More spectacularly, the quantized Hall conductance discovered in two-dimensional electron systems can be identified as a manifestation of certain geometric phases [23, 201, 247]. Adiabatic particle transport in Bloch bands and mesoscopic systems [245] may be most directly understood in terms of geometric phases as well. Some other applications of the geometric phase in condensed matter physics include a first-principles calculation of spin waves [198, 202], the dynamics of quantized vortices [249], and fractional statistics [18].

The fact that the geometric phase has important observable consequences in many areas of physics and chemistry is not the only reason why it has attracted so much attention. The geometric phase is also one of the most beautiful examples of what Wigner once called "the unreasonable effectiveness of mathematics in the natural sciences."

Immediately after Berry's introduction of the adiabatic geometric phase, Simon [230] noticed that it could be interpreted as the holonomy of a fiber bundle and that Berry's gauge potential played the role of a connection on this fiber bundle. It was this relation to the beautiful mathematics of fiber bundles that caused the geometric phase to become a fashion in mathematical physics. When the theory of fiber bundles was established and when the mathematics of the universal classifying bundles and connections was developed, no one could imagine that these would be directly related to a quantum mechanical phase factor which could be measured in say an interference experiment. The universal connection is the "natural" mathematical object which classifies the geometric structures on arbitrary (finite-dimensional) principal fiber bundles. It is incredible that this mathematical entity is exactly the gauge potential whose integral over a closed path of states gives the Aharonov–Anandan phase with Berry's gauge potential as its limiting case [44, 184], and that this connection is related to Mead's vector potential which was discovered in the study of molecular structure.

2. Quantal Phase Factors for Adiabatic Changes

2.1 Introduction

In the quantum mechanical description of a physical system, one has a finite or infinite dimensional Hilbert space of state vectors and a set of linear operators acting on these state vectors. The linear operators are interpreted as the observables. If a quantum system is not isolated from its environment, the observables are described by operators that depend on a set of parameters, $R = (R^1, R^2, \cdots)$. Each value of R characterizes a particular configuration of the environment. In particular, a changing environment is described by time-dependent parameters, $R = R(t)$.

For a quantum system in a classical environment, the parameters R label the points of a smooth manifold M. Every change of the environment is then described by a curve $\mathbf{C} : [0, T] \to M$, with points $R(t) \in \mathbf{C}$. The manifold M is called the *parameter space* of the quantum system. The geometric properties of the parameter space depend on the specifics of the system. In general, it can be a complicated manifold consisting of several coordinate patches.[1] In particular, the Hamiltonian operator $h(R)$ is a "smooth" and single-valued function of $R \in M$. Here by the smoothness of the Hamiltonian we mean that its eigenvalues and eigenvectors are smooth functions of R.

If the physical system is in a quantal environment, then the parameters are (generalized) eigenvalues of the observables of the quantal environment. For example in a diatomic molecule, the quantum system consists of the electrons, the environment is identified with the nucleus, and the components of the position operator of the internuclear axis are the parameters describing the environment.

Another example of a quantum physical system whose Hamiltonian depends upon environmental parameters is a (quantum) magnetic moment $\mathbf{m}$ in a rotating (classical) magnetic field $\mathbf{B}$ of constant magnitude $B := |\mathbf{B}|$. The parameter-dependent Hamiltonian is given by[2]

$$H = H_0 - \mathbf{m} \cdot \mathbf{B} = H_0 - Bg\left(\frac{e}{2mc}\right)\hat{\mathbf{R}} \cdot \mathbf{J} = H_0 + b\,\hat{\mathbf{R}} \cdot \mathbf{J}. \quad (2.1)$$

---
[1] A review of smooth manifolds is provided in Appendix A.2.
[2] For a derivation of this Hamiltonian see [43, §IX.3].

Here H_0 is the part of the Hamiltonian which is independent of the magnetic field and will therefore be ignored;[3] $\hat{\mathbf{R}}(t) = \mathbf{B}/B$ is the unit vector pointing in the direction of the magnetic field; $\mathbf{J}$ is the angular momentum of the quantum system and $b := -Bg(\frac{e}{2mc})$ is a constant. In (2.1), we have further used the familiar relation between the magnetic dipole moment and angular momentum

$$\mathbf{m} = \frac{e}{2mc} g \mathbf{J}. \tag{2.2}$$

The parameter-dependent part of the Hamiltonian is

$$h(\hat{\mathbf{R}}) = b\,\hat{\mathbf{R}} \cdot \mathbf{J}, \tag{2.3}$$

and the parameter space of the environment is the unit sphere

$$S^2 = \{\hat{\mathbf{R}} \in \mathbb{R}^3 \;:\; |\hat{\mathbf{R}}| = 1\}.$$

We can use as the parameters the polar and azimuthal angles (θ, ϕ) corresponding to the unit vector $\hat{\mathbf{R}} \in S^2$.

The evolution of the states of the quantum system in the external environment is described by the Schrödinger equation

$$i\frac{d\psi(t)}{dt} = h(R)\psi(t). \tag{2.4}$$

Here $\psi(t)$ denotes a state vector which belongs to the Hilbert space $\mathcal{H}$ and represents a pure state of the system. The general (mixed) states are described by the statistical operators $W(t)$ whose evolution is given by the Liouville–von Neumann equation

$$i\frac{dW(t)}{dt} = [h(R), W(t)]. \tag{2.5}$$

This is the generalization of a basic postulate of quantum mechanics for conservative systems [43, §XII]. The evolution of states of the non-conservative system is determined by the dynamical equation (2.4) or (2.5) and the environmental process. The latter is described by the way in which the environmental parameters R change in time. For a given physical situation, the change in these parameters, i.e., the path in the environment's parameter space $R(t) = (R^1(t), R^2(t), \cdots)$, must be specified.

In our example where the Hamiltonian is given by (2.1), the environmental processes are described by how the direction $\hat{\mathbf{R}}$ of the magnetic field changes. A particularly interesting situation arises if the direction of the magnetic field changes periodically; e.g., it may perform rotations in the 1-2 plane with angular velocity ω, $\hat{\mathbf{R}}(t) = \mathbf{e}_1 \cos \omega t - \mathbf{e}_2 \sin \omega t$, or it may run through other

---

[3] We shall ignore H_0 in our analysis, for its presence has no consequences as far as the phenomenon of the geometric phase is concerned.

closed paths **C** in the parameter space. In this case the Hamiltonian returns to its original form as time progresses from $t = 0$ to the period $t = T = 2\pi/\omega$.

We postulate that the space of physical states does not only contain the solutions of (2.4) for one given fixed value of the parameters R or for one given environmental process $t \to R(t)$, but for all $R \in M$. This means that there is a single space of physical state vectors $\mathcal{H}$ for all values of R. For any given value of R, one may choose an orthonormal basis of eigenvectors $|n; R\rangle$ of the parameter-dependent Hamiltonian[4]

$$h(R)|n; R\rangle = E_n(R)|n; R\rangle, \qquad (2.6)$$
$$\langle m; R|n; R\rangle = \delta_{mn},$$

and write the Hamiltonian according to its spectral resolution

$$h(R) = \sum_n E_n(R)|n; R\rangle\langle n; R|. \qquad (2.7)$$

Here as well as in the rest of this book we assume that the Hamiltonian has a discrete spectrum.[5]

Given an environmental process along with a time parameterization $R(t)$, one obtains from the parameter-dependent Hamiltonian $h(R)$ a time-dependent Hamiltonian $h(t) := h(R(t))$. In view of (2.7), the spectral resolution of $h(t)$ is given by

$$h(t) := h(R(t)) = \sum_n E_n(R(t))|n; R(t)\rangle\langle n; R(t)|. \qquad (2.8)$$

It is important to note that in general the projection operators

$$\Lambda_n(R(t)) := |n; R(t)\rangle\langle n; R(t)|, \qquad (2.9)$$

corresponding to the eigenstates of $h(t) = h(R(t))$ change with time.

We have assumed that the observables are single-valued functions of R over the whole parameter space of the environment. Single-valuedness of the observables means that if the same value of R occurs more than once (i.e., at different times) during a process, then the observables are the same at each occurrence. In particular, if the environmental process is periodic, i.e., if the environmental parameters $R(t)$ traverse a closed path **C** and return, after some period T, to their original values,

$$\mathbf{C}: \; R(0) \to R(t) \to R(T) = R(0), \qquad (2.10)$$

---

[4] If in addition to $h(R)$ other observables $A_i(R)$, $i = 1, \cdots, N-1$ are needed to form a complete set of commuting observables for any given value of R, then their eigenvalues $a_i(R)$ will also be needed to label the basis vectors and $|n; R\rangle = |n, a_1, \cdots, a_{N-1}; R\rangle$.
[5] The case of systems with continuous energy spectra is considered in [195].

then the Hamiltonian, its eigenvalues and the projection operators (2.9), which are uniquely defined by (2.7), are the same at $R(T)$ as they are at $R(0)$, i.e.,

$$h(R(T)) = h(R(0)), \qquad (2.11)$$
$$E_n(R(T)) = E_n(R(0)), \qquad (2.12)$$
$$|n; R(T)\rangle\langle n; R(T)| = |n; R(0)\rangle\langle n; R(0)|. \qquad (2.13)$$

Though the observables are single-valued functions of R the basis vectors $|n; R\rangle$ themselves will in general not be single valued over the whole parameter space. In other words, in general it is impossible to define smooth single-valued $|n; R\rangle$ for all $R \in M$. Therefore it is usually necessary to use different parameterizations over different patches of the parameter space. For example for $h(\mathbf{R}) = b\mathbf{R}(\theta, \varphi) \cdot \mathbf{J}$, this means that the vectors $|n; \mathbf{R}(\theta, \varphi)\rangle$ are different functions of the polar and azimuthal angles (θ, φ) for different patches of the unit sphere.[6] Thus in general (2.13) will not imply

$$|n; R(T)\rangle = |n; R(0)\rangle \quad \text{for} \quad R(T) = R(0), \qquad (2.14)$$

but only

$$|n; R(T)\rangle = e^{i\zeta_n}|n; R(0)\rangle \quad \text{for} \quad R(T) = R(0), \qquad (2.15)$$

where $e^{i\zeta_n}$ is a phase factor. This is because the eigenvectors $|n; R\rangle$ are determined by (2.6) only up to a phase factor. We can define a new set of eigenvectors $|n; R\rangle'$ by performing the phase transformations

$$|n; R\rangle \quad \rightarrow \quad |n; R\rangle' = e^{i\zeta_n(R)}|n; R\rangle, \qquad (2.16)$$

where $\zeta_n(R)$ are arbitrary real phase angles. The basis $\{|n; R\rangle'\}$ constitutes just as valid a basis of eigenvectors of the Hamiltonian as $\{|n; R\rangle\}$.

We will restrict ourselves to phase transformations (2.16) for which the phase factors $e^{i\zeta_n(R(t))}$ are single-valued functions. These are also called *gauge transformations*.

In general, if we go from one patch $O_1 \subset M$ of the parameter space to a neighboring patch $O_2 \subset M$ with a different parameterization, then eigenvectors of $h(R)$ in the overlap region $R \in O_1 \cap O_2$ are related by phase transformations of the form (2.16). In order to simplify our analysis, however, we shall assume for the moment that the closed path $\mathbf{C}$ of (2.10) can be placed into one single patch $O \subset M$ and the basis vectors can be chosen to be smooth and single-valued functions.[7] We shall then choose for the basis vectors, the smooth and single-valued functions $|n; R\rangle$ which are defined

---

[6] A thorough discussion of the manifold structure of the sphere S^2 is provided in Appendix A.2. The necessity for different choices of the basis vectors $|n; R\rangle$ on different patches of S^2 is explained in detail in Chap. 3, (cf. (3.17) and (3.25)).

[7] This assumption is made solely to simplify our analysis. In fact, our treatment can be easily extended to the general case where $\mathbf{C}$ lies in two or more patches without any serious difficulty. In this case, one has to simply perform the necessary phase transformations in passing from one patch to the other.

over the patch O of the parameter space and satisfy (2.14). In each patch $O \subseteq M$, gauge transformations (2.16) transform single-valued basis vectors into vectors which are also single valued.

So far we have considered the observables[8] which change in time due to the change of the environmental parameters. Next we wish to consider the states $W(t)$ whose evolution is governed by the Schrödinger (2.4) or Liouville–von Neumann equation (2.5). Specifically, we shall consider a pure state $|\psi(t)\rangle\langle\psi(t)|$. The state $|\psi(t)\rangle\langle\psi(t)|$ can be described by the state vector $\psi(t)$ which satisfies the Schrödinger equation (2.4).

The time evolution of $\psi(t)$ can also be described by an operator $U^\dagger(t)$,

$$\psi(t) = U^\dagger(t)\psi(0), \tag{2.17}$$

which satisfies, as a consequence of (2.4), the integral equation

$$U^\dagger(t) = I - i\int_0^t h(t')U^\dagger(t')dt'. \tag{2.18}$$

This operator is called the *time-evolution operator*. Since the Hamiltonian $h(t)$ is assumed to be Hermitian, the time-evolution operator $U^\dagger(t)$ is necessarily unitary.

An expression for $U^\dagger(t)$ in terms of $h(t)$ can be given by successively substituting the right-hand side of (2.18) for the $U^\dagger(t)$ that appears in the integrand. This yields

$$\begin{aligned} U^\dagger(t) &= I + \frac{1}{i}\int_0^t dt' h(t') + \left(\frac{1}{i}\right)^2 \int_0^t dt' \int_0^{t'} dt'' h(t')h(t'')U^\dagger(t'') \\ &= I + \frac{1}{i}\int_0^t dt' h(t') + \left(\frac{1}{i}\right)^2 \int_0^t dt' \int_0^{t'} dt'' h(t')h(t'') \\ &\quad + \cdots + \left(\frac{1}{i}\right)^n \int_0^t dt_1 \int_0^{t_1} dt_2 \cdots \int_0^{t_{n-1}} dt_n h(t_1)h(t_2)\cdots h(t_n) \\ &\quad + \cdots. \end{aligned} \tag{2.19}$$

Note that the operators appearing in the integrands have decreasing time arguments reading from left to right, $t \geq t' \geq t'' \geq 0$; $t \geq t_1 \geq t_2 \geq \ldots \geq t_n \geq 0$, and that their order is important since they do not generally commute

$$[h(t), h(t')] \neq 0 \qquad \text{for } t' \neq t. \tag{2.20}$$

For the case that $h(t)$ commutes at different times,

$$[h(t), h(t')] = 0 \qquad \text{for all } t, t', \tag{2.21}$$

the order of the operators in (2.19) is not important, and it can be shown (Problem 2.1) that

---
[8] The basis vectors $|n; R(t)\rangle$ are also observables.

$$U^\dagger(t) = \sum_{n=0}^\infty \frac{1}{n!}\left(\frac{1}{i}\right)^n \int_0^t dt_1 \ldots \int_0^t dt_n h(t_1)\ldots h(t_n) = e^{\frac{1}{i}\int_0^t dt' h(t')}. \quad (2.22)$$

Equation (2.21) means that the projection operators $\Lambda_n(R(t)) = |n; R(t)\rangle \langle n; R(t)|$ commute (Problem 2.2):

$$[\Lambda_n(R(t)), \Lambda_{n'}(R(t'))] = 0 \quad \text{for all } n, n' \text{ and } t, t'. \quad (2.23)$$

In view of the assumed continuity of the $\Lambda_n(R(t))$, this means that the projection operators and eigenspaces of $h(t)$ are in fact time independent.[9] Therefore the projection operators may be written as $\Lambda_n(R(t)) = \Lambda_n = |n\rangle\langle n|$, where $|n\rangle$ are time-independent vectors. The resulting spectral resolution (2.8) of $h(t)$ may then be used in (2.22) to give the following spectral resolution of $U^\dagger(t)$:

$$U^\dagger(t) = \sum_n e^{\frac{1}{i}\int_0^t dt' E_n(t')}|n\rangle\langle n|. \quad (2.24)$$

(If the eigenvalues are also independent of time, i.e., if the Hamiltonian is time independent, $h(t) = h$, then (2.22) goes into the standard expression for conservative systems $U^\dagger(t) = e^{-ith}$.) But for general time-dependent Hamiltonians, where (2.20) holds, (2.18) cannot be integrated. In general, the time-evolution operator $U^\dagger(t)$ is written as a time-ordered product[10]

$$U^\dagger(t) = \sum_{n=0}^\infty \frac{1}{n!}\left(\frac{1}{i}\right)^n \int_0^t dt_1 \cdots \int_0^t dt_n \mathcal{T}(h(t_1)\cdots h(t_n))$$

$$=: \mathcal{T}\left(e^{\frac{1}{i}\int_0^t dt' h(t')}\right), \quad (2.25)$$

where $\mathcal{T}$ is the *time-ordering operator* which rearranges a product of time-labeled operators such that the operators whose time labels take smaller values appear to the right, e.g.,

$$\mathcal{T}(h(t_1)h(t_2)h(t_3)) = h(t_3)h(t_1)h(t_2) \quad \text{if } t_3 \geq t_1 \geq t_2.$$

We should like to remark that (2.25) is only a formal solution. In fact, an expression for the spectral resolution of $U^\dagger(t)$ which is valid in this most general case is not known.

2.2 Adiabatic Approximation

In this section we solve (2.4–2.5) under various approximations. We use the initial condition that at $t = 0$ the state is an eigenstate of $h(R(0))$ with eigenvalue $E_n(R(0))$ which means

---

[9] The eigenvalues $E_n(t)$ may, however, depend on time.
[10] For more details see for example [115].

$$|\psi(0)\rangle\langle\psi(0)| = |n; R(0)\rangle\langle n; R(0)|, \quad \text{or} \quad \psi(0) = |n; R(0)\rangle. \tag{2.26}$$

The second form of this initial condition (the second equation in (2.26)) follows after fixing an arbitrary phase factor to unity. First we consider the case where R is a *fixed* parameter.[11] Then the solution of (2.4), with the initial condition (2.26), is

$$\psi(t) = e^{-iE_n(R)t}|n; R\rangle = e^{-iE_n(R)t}\psi(0). \tag{2.27}$$

This is essentially the same result we obtain if $R(t)$ changes, but only in such a way that (2.21) is fulfilled. In this case, (2.24) with the initial condition (2.26) yields

$$\psi(t) = e^{-i\int_0^t dt' E_n(R(t'))}|n; R(0)\rangle = e^{-i\int_0^t dt' E_n(R(t'))}\psi(0). \tag{2.28}$$

We next consider the less drastic assumption, called the *adiabatic approximation* [46, 82, 121].

For a Hamiltonian $h(R(t))$ whose parameters change in time, the interaction with the environment can cause the physical system to jump from the n-th eigenstate $|n, R(0)\rangle\langle n, R(0)|$ at $t = 0$ into any other eigenstate $|m, R(t)\rangle\langle m, R(t)|$, $m \neq n$ at a later time t. A very particular situation arises if this does not happen, i.e., when the state remains an eigenstate of $h(R(t))$ at all times t with the same energy quantum number n. This means that $|\psi(t)\rangle\langle\psi(t)|$ changes in such a way that at all times t

$$|\psi(t)\rangle\langle\psi(t)| \stackrel{\text{adiabatic}}{=} |n; R(t)\rangle\langle n; R(t)| = \Lambda_n(R(t)). \tag{2.29}$$

This time development is called the *adiabatic time development*. We shall instantly show that an adiabatic time development (2.29) can at best be an *approximation*, unless the evolving state does not change in time.

States which do not change in time, i.e.,

$$|\psi(t)\rangle\langle\psi(t)| = |\psi(0)\rangle\langle\psi(0)| \quad \text{for all} \quad t, \tag{2.30}$$

are called *stationary states*. Clearly the solutions (2.27) and (2.28) describe stationary states

$$|\psi(t)\rangle\langle\psi(t)| = |n; R(0)\rangle\langle n; R(0)|. \tag{2.31}$$

One can show that the stationary states of conservative systems are always energy eigenstates (Problem 2.3). For general time-dependent Hamiltonians this is not the case. The assumption of adiabaticity (2.29) is a generalization of the assumption that the state is stationary (2.30). Since $\Lambda_n(R(t))$ changes in time, the adiabatic state $W(t) = |\psi(t)\rangle\langle\psi(t)|$ of (2.29) changes in time, whereas the stationary state (2.30) does not change. The state (2.29) will

---

[11] This is the assumption that has been used for the old Born–Oppenheimer approximation in molecular physics.

always be the n-th eigenstate of $h(R(t))$ but its eigenvalue $E_n(R(t))$ may change in time; and even if that happens not to be the case, $|n; R(t)\rangle\langle n; R(t)|$ can still change. Equation (2.29) describes a curve $t \to |\psi(t)\rangle\langle\psi(t)|$ in the set of projection operators whereas the path associated with (2.30) consists of a single point in this set.

The set of one-dimensional projection operators $W(t) = |\psi(t)\rangle\langle\psi(t)|$ of the Hilbert space $\mathcal{H}$ is called the *projective Hilbert space* or simply the *state space* and denoted by $\mathcal{P}(\mathcal{H})$. It is the projective space of (pure) physical states.

In terms of $\mathcal{P}(\mathcal{H})$, the adiabaticity assumption (2.29) means that if $R(t)$ traverses a path in the parameter space M then the pure state $|\psi(t)\rangle\langle\psi(t)|$ traverses a path of energy eigenstates $t \to \Lambda_n(R(t))$ in $\mathcal{P}(\mathcal{H})$. In particular if $R(t)$ traverses the closed path $\mathbf{C}$ of (2.10) in M, then because of (2.13) the pure energy eigenstate $\Lambda_n(R(t))$ traverses a closed path

$$\mathcal{C} : \Lambda_n(R(0)) =$$
$$W(0) \to W(t) \to W(T) \stackrel{\text{adiabatic}}{=} \Lambda_n(R(T)) = \Lambda_n(R(0)) \quad (2.32)$$

in $\mathcal{P}(\mathcal{H})$. If the eigenprojectors traverse a closed path $\mathcal{C}$ in $\mathcal{P}(\mathcal{H})$ then the eigenvectors $|n; R(t)\rangle$ traverse a closed path in $\mathcal{H}$:

$$|n; R(0)\rangle \to |n; R(t)\rangle \to |n; R(T)\rangle = |n; R(0)\rangle. \quad (2.33)$$

This is because these basis vectors had been chosen to be single-valued functions in the patch that contained the closed curve $\mathbf{C}$ of (2.10). Their time dependence is determined by the time dependence of the environmental parameters $R = R(t)$, not the Schrödinger equation.

The dynamical equation (2.4–2.5) and the adiabaticity assumption (2.29) are two separate conditions on the state $|\psi(t)\rangle\langle\psi(t)|$ and may, therefore, not be generally compatible with each other. In order to show that this is indeed the case, let us substitute (2.29) in (2.5). This yields

$$i\frac{dW(t)}{dt} = [h(R(t)), W(t)] = [h(R(t)), \Lambda_n(R(t))] = 0. \quad (2.34)$$

Equation (2.34) means that $W(t)$ does not change in time,

$$W(t) = |\psi(t)\rangle\langle\psi(t)| = |\psi(0)\rangle\langle\psi(0)| = W(0), \quad \text{for all } t. \quad (2.35)$$

Hence, any adiabatically evolving state (2.29) (which obeys (2.5)) must be stationary. In particular, it cannot have a non-trivial *cyclic evolution*[12]

$$W(0) \to W(t) \to W(T) = W(0), \quad (2.36)$$

---

[12] In this book we shall use the term *cyclic evolution* to mean that the evolving state $W(t)$ satisfies (2.5) and (2.36), but is not stationary, i.e., it changes in time. The initial state $W(0)$ of a cyclic evolution is said to be a *cyclic state*.

2.2 Adiabatic Approximation

in which $W(t)$ changes in time. Therefore, *exact adiabatic cyclic evolutions do not exist*. The adiabatic equality $\stackrel{\text{adiabatic}}{=}$ in (2.29) can only be an approximation.

We shall next examine the condition of the validity of the *adiabatic approximation*.

First we express the evolving state vector $\psi(t)$ in the basis $\{|n; R(t)\rangle\}$,

$$\psi(t) = \sum_m c_m(t)|m; R(t)\rangle. \tag{2.37}$$

Now suppose that initially ψ is an eigenvector of the initial Hamiltonian, so that (2.26) hold. Then the adiabatic approximation is a valid approximation if and only if we can ignore all the coefficients $c_m(t)$ in (2.37) except $c_n(t)$, i.e.,

$$\psi(t) \stackrel{\text{adiabatic}}{=} c_n(t)|n, R(t)\rangle. \tag{2.38}$$

Note that because of (2.26)

$$c_n(0) = 1. \tag{2.39}$$

Substituting the expression (2.38) for $\psi(t)$ in the Schrödinger equation (2.4) and using (2.6), we find

$$\left(\frac{d}{dt}c_n(t) + iE_n(R(t))c_n(t)\right)|n; R(t)\rangle \stackrel{\text{adiabatic}}{=} -c_n(t)\frac{d}{dt}|n\ R(t)\rangle. \tag{2.40}$$

Next let us compute the inner product of both sides of (2.40) with $|m, R(t)\rangle$ for $m \neq n$. Then in view of the orthogonality of basis vectors, in particular $\langle m; R|n; R\rangle = 0$ for $m \neq n$, we find

$$\langle m, R(t)|\frac{d}{dt}|n, R(t)\rangle \stackrel{\text{adiabatic}}{=} 0, \quad \text{for all} \quad m \neq n. \tag{2.41}$$

This is the necessary and sufficient condition for the validity of the adiabatic approximation. We can express the left-hand side of this equation in terms of the matrix elements of the time derivative of the Hamiltonian. In order to show this, we take the differential (exterior derivative[13]) of both sides of (2.6),

$$dh(R)|n; R\rangle + h(R)d|n; R\rangle = dE_n(R)|n; R\rangle + E_n(R)d|n; R\rangle,$$

and compute the inner product of both sides of this equation with $|m; R\rangle$ for $m \neq n$. This yields

$$\langle m; R|dh(R)|n; R\rangle + E_m(R)\langle m; R|d|n; R\rangle = E_n(R)\langle m; R|d|n; R\rangle, \tag{2.42}$$

where we have used (2.6) and $\langle m; R|n; R\rangle = 0$. We can express (2.42) in the form

---
[13] A review of exterior differentiation is provided in Appendix A.2.

$$\langle m;R|d|n;R\rangle = \frac{\langle m;R|[d\,h(R)]\,|n,R\rangle}{E_n(R)-E_m(R)}, \quad \text{for all} \quad m\neq n. \tag{2.43}$$

For a given environmental process, $R=R(t)$, we can replace the differentials (exterior derivatives) appearing in (2.43) with the ordinary time-derivatives by dividing both sides of this equation by dt. This leads to

$$\langle m;R(t)|\frac{d}{dt}|n;R(t)\rangle = \frac{\langle m;R(t)|[\frac{d}{dt}h(t)]\,|n,R(t)\rangle}{E_n(R(t))-E_m(R(t))}, \quad \text{for all} \quad m\neq n. \tag{2.44}$$

In view of (2.44), the condition (2.41) for the validity of the adiabatic approximation is expressed in the form

$$\frac{\langle m;R(t)|[\frac{d}{dt}h(t)]\,|n,R(t)\rangle}{E_n(R(t))-E_m(R(t))} \stackrel{\text{adiabatic}}{=} 0, \quad \text{for all} \quad m\neq n. \tag{2.45}$$

The adiabatic approximation is a valid approximation if and only if the left-hand side of (2.45) can be neglected. The appearance of the time derivative of the Hamiltonian in (2.45) is the reason why this approximation is called the adiabatic approximation. It is important to note that the left-hand side of (2.45) has the dimension of frequency. This means that in order to decide whether it can be neglected, one must have an intrinsic frequency (or energy) scale for the quantum system.[14] For our example of a magnetic moment interacting with a rotating magnetic field (2.3) the intrinsic frequency scale is given by b.

2.3 Berry's Adiabatic Phase

In the preceding section we showed that for a periodic Hamiltonian $h(t) = h(R(t))$, an adiabatically evolving state $W(t) = |\psi(t)\rangle\langle\psi(t)|$ with the initial condition $W(0) = \Lambda_n(R(0))$ traverses a closed path $\mathcal{C}$ in the state space $\mathcal{P}(\mathcal{H})$. This does not necessarily mean that the normalized state vector $\psi(t)$, which fulfills the Schrödinger equation (2.4), also traverses a closed path in $\mathcal{H}$. In general the path

$$C:[0,T]\ni t \longrightarrow \psi(t)\in\mathcal{H}, \quad \text{with} \quad \langle\psi(t)|\psi(t)\rangle=1, \tag{2.46}$$

is not closed in $\mathcal{H}$, but satisfies

$$C(T)=\psi(T)=e^{-i\alpha_\psi}\psi(0). \tag{2.47}$$

For the case of a time-independent Hamiltonian (2.27) or for (2.28), the phase factor is

---

[14] Of course, if the left-hand side of (2.45) vanishes identically, then such an energy scale is not required.

2.3 Berry's Adiabatic Phase

$$e^{-i\alpha_\psi} = e^{-iE_n T}, \quad \text{or} \quad e^{-i\alpha_\psi} = e^{-i\int_0^T dt' E_n(t')}. \tag{2.48}$$

This is called the *dynamical phase factor*. As we shall now show for a general time-dependent Hamiltonian $h(R(t))$, there is an additional phase factor which is called the *geometric phase* or *Berry phase*. We shall in general use the term *Berry phase* for the geometrical phase obtained in the adiabatic approximation (2.29).

If the adiabatic approximation is valid, we can express the evolving state vector according to (2.38), where the coefficient c_n satisfies (2.40). We can obtain the explicit formula for c_n by calculating the inner product of both sides of this equation with $|n; R(t)\rangle$. In view of the orthonormality of the eigenvectors $|n; R(t)\rangle$, this yields

$$\frac{d}{dt} c_n(t) = -c_n \left(iE_n(t) + \langle n; R(t)| \frac{d}{dt} |n; R(t)\rangle \right), \tag{2.49}$$

where $E_n(t) := E_n(R(t))$. We can easily integrate this equation and obtain

$$\int_{c_n(0)}^{c_n(t)} \frac{dc_n}{c_n} = -i \int_0^t E_n(t')dt' - \int_0^t \langle n; R(t')| \frac{d}{dt'} |n; R(t')\rangle dt'. \tag{2.50}$$

In view of (2.39), this leads to

$$c_n(t) = e^{-i\int_0^t E_n(t')dt'} e^{i\gamma_n(t)}, \tag{2.51}$$

where

$$e^{i\gamma_n(t)} := e^{i\int_0^t i\langle n;R(t')| \frac{d}{dt'} |n;R(t')\rangle dt'}. \tag{2.52}$$

The coefficient $c_n(t)$ in (2.38) is a phase factor. This is because $\psi(t)$ is obtained from $\psi(0)$ of (2.26) by a unitary time development (2.17) and is therefore a normalized vector. Therefore $\gamma_n(t)$ is a real phase angle. It is important to note that $\gamma_n(t)$ is only defined up to an integer multiple of 2π.

A remarkable property of the phase angle $\gamma_n(t)$ is that it does not depend on the time dependence of the integrand in (2.52). In fact, it can be directly defined in terms of a (curve) integral

$$\begin{aligned}\gamma_n(t) &= \int_0^t i\langle n; R(t)| \frac{d}{dt} |n; R(t)\rangle dt \\ &= \int_{R(0)}^{R(t)} i\langle n; R(t)| \frac{\partial}{\partial R^i} |n; R(t)\rangle dR^i \\ &= \int_{R(0)}^{R(t)} \mathbf{A}_i^n(R) dR^i,\end{aligned} \tag{2.53}$$

over a vector-valued function

$$\mathbf{A}^n(R) := i\langle n; R|\nabla|n; R\rangle. \tag{2.54}$$

This vector-valued function is called the *Mead–Berry vector potential* [31, 165, 167]. As (2.54) indicates, the Mead–Berry vector potential $\mathbf{A}^n$ is defined only with the use of the single-valued basis eigenvectors $|n;R\rangle$ of $h(R)$. Since a smooth single-valued basis vector may in general not be found on the whole parameter space M, but only on its patches, the same is true for $\mathbf{A}^n$. We will discuss this problem in detail for the example (2.1) in the next chapter. Here we will assume that the curve $\mathbf{C}$ lies in a single patch over which a complete set of smooth and single-valued basis vectors $|n;R\rangle$ exists.

The Mead–Berry vector potential may also be expressed as a (local) differential one-form,[15]

$$A^n = A_i^n \, dR^i := i\langle n;R|\frac{\partial}{\partial R^i}|n;R\rangle dR^i = i\langle n;R|d|n;R\rangle, \tag{2.55}$$

which is defined on the same patch as $|n;R\rangle$. Here the differentials dR^i with $i = 1, 2, \cdots, \mathcal{M} = \dim(M)$ are the basis differential one-forms (covariant vectors), and "d" is the exterior derivative operator. The one-form A^n is called the *Mead–Berry connection one-from*. It can be used to yield the following expression for the phase angle $\gamma_n(t)$ of (2.53),

$$\gamma_n(t) = \int_{R(0)}^{R(t)} i\langle n;R(t)|d|n;R(t)\rangle = \int_\mathbf{C} A^n, \tag{2.56}$$

where $\mathbf{C}$ is the curve traced by the parameters R in the parameter space M.

Returning to our calculation of $\psi(t)$ we obtain by inserting (2.51) into (2.38),

$$\psi(t) \stackrel{\text{adiabatic}}{=} e^{\frac{1}{i}\int_0^t E_n(t')dt'} e^{i\gamma_n(t)}|n;R(t)\rangle. \tag{2.57}$$

This equation shows that in addition to the dynamical phase factor (2.48), there exists another phase factor $e^{i\gamma_n(t)}$ which is given in terms of the eigenvectors $|n;R\rangle$ of $h(R(t))$ according to (2.56).

In view of (2.16) the basis eigenvectors $|n;R(t)\rangle$ are only determined up to a phase factor. Thus, the additional phase factor in (2.57) can be transformed away by a phase transformation. From the requirement that the state vectors $\psi(t)$ have the least possible oscillatory behavior, $\|\frac{d\psi(t)}{dt}\|$ = minimum, Fock [82] derived the condition $\langle n;R(t)|\frac{d}{dt}|n;R(t)\rangle = 0$ which – according to (2.53) – leads to the phase choice of unity for this extra phase factor. Though Fock did not consider the case of a cyclic change as in (2.10) or (2.32) and (2.36), his phase choice was universally accepted and the extra phase factor was ignored for half a century. In the following we shall first attempt to eliminate γ_n and then show that in general this is not possible.

From the definition of the Mead–Berry vector potential (2.54), it follows that under a gauge transformation (2.16), $\mathbf{A}^n(R)$ transforms according to

---

[15] A review of differential forms is provided in Appendix A.2.

2.3 Berry's Adiabatic Phase

$$\begin{aligned}
\mathbf{A}^n(R) \to \mathbf{A}'^n(R) &= i\,\langle n; R|'(\nabla|n, R\rangle') \\
&= i\,\langle n, R|e^{-i\zeta_n(R)}(\nabla\,e^{i\zeta_n(R)}|n; R\rangle) \\
&= i\,\langle n; R|\nabla|n; R\rangle + i\,e^{-i\zeta_n(R)}(\nabla\,e^{i\zeta_n(R)}), \\
&= \mathbf{A}^n(R) - \nabla\zeta_n(R).
\end{aligned} \qquad (2.58)$$

Alternatively, we have for the connection one-form

$$A^n(R) \to A'^n(R) = A^n(R) - d\zeta_n(R). \qquad (2.59)$$

In view of (2.58), under a gauge transformation the phase angle $\gamma_n(t)$ transforms according to

$$\begin{aligned}
\gamma_n(t) \to \gamma'_n(t) &= \int_{R(0)}^{R(t)} \mathbf{A}'^n(R)\,dR \\
&= \gamma_n(t) - \zeta_n(R(t)) + \zeta_n(R(0)).
\end{aligned} \qquad (2.60)$$

If we do the calculations that led to (2.57) using $|n; R\rangle'$ in place of $|n; R\rangle$, we obtain (2.57) with the primed quantities on the right-hand side. Using (2.16) we obtain for the primed quantities

$$e^{i\gamma'_n(t)}|n, R(t)\rangle' = e^{i\gamma'_n(t)}e^{i\zeta_n(R(t))}|n; R(t)\rangle.$$

If $\zeta_n(R(t))$ is an arbitrary single-valued function modulo 2π, we can choose it so that the phase factor $e^{i\gamma'_n(t)}e^{i\zeta_n(R(t))}$ becomes unity and we find in place of (2.57)

$$\psi(t) = e^{-i\int_0^t E_n(t')dt'}|n, R(t)\rangle. \qquad (2.61)$$

This also fulfills the initial condition (2.26). Since $|n, R\rangle'$ is as valid a basis eigenvector as $|n, R\rangle$, we can use it to describe the time development of the state vector, i.e., use (2.61) which involves the dynamical phase factor only. This is Fock's result [82].

The above arguments made use of the fact that $\zeta_n(R(t))$ was arbitrary. If after some period T the environmental parameters return to their original values as described by the closed path $\mathbf{C}$ of (2.10), then one cannot choose $\zeta_n(R(T))$ freely to remove $\gamma_n(T)$.

For $R(T) = R(0)$, the single-valuedness of $e^{i\zeta_n(R)}$ implies

$$e^{i\zeta_n(R(T))} = e^{i\zeta_n(R(0))}, \quad \text{or} \quad \zeta_n(R(T)) = \zeta_n(R(0)) + 2\pi \times \text{integer}. \qquad (2.62)$$

Therefore according to (2.60)

$$\begin{aligned}
\gamma_n(T) \to \gamma'_n(T) &:= \oint_\mathbf{C} A'^n_i(R)\,dR^i \\
&= \oint_\mathbf{C} A^n_i(R)\,dR^i - 2\pi \times \text{integer} \\
&= \gamma_n(T) - 2\pi \times \text{integer}.
\end{aligned} \qquad (2.63)$$

Thus $\gamma_n(T)$ – which is only defined modulo 2π – is invariant under the gauge transformation (2.16) and cannot be removed. We therefore have

$$\psi(T) = e^{-i\int_0^T dt E_n(t) + i\gamma_n(T)} |n; R(T)\rangle \qquad (2.64)$$

with $\gamma_n(T)$ given by the loop integral over the closed path **C** of (2.10),

$$\gamma_n(\mathbf{C}) := \gamma_n(T) = \oint_{\mathbf{C}} A_i^n(R) dR^i \qquad \text{modulo } 2\pi,$$

$$= \oint_{\mathbf{C}} A^n \qquad \text{modulo } 2\pi. \qquad (2.65)$$

If we insert the initial condition (2.26) and (2.14) into (2.64) we obtain

$$\psi(T) = e^{-i\int_0^T dt E_n(t)} e^{i\gamma_n(\mathbf{C})} \psi(0). \qquad (2.66)$$

The phase angle $\gamma_n(T)$ is called the *Berry phase angle*, and $e^{i\gamma_n(T)}$ is called the *Berry phase factor*.

We have shown that for a closed path the extra phase factor cannot be transformed away. This does not mean that $\gamma_n(C)$ could not be zero. Indeed this will turn out to be the case for many systems. In this case the vector potential $\mathbf{A}^n(R)$ (and the connection one-form $A^n(R)$) need not be zero but will be "trivial", which means

$$\mathbf{A}^n(R) = \nabla\zeta(R) = -ie^{-i\zeta(R)}\nabla e^{i\zeta(R)}, \qquad (2.67)$$
$$A^n(R) = d\zeta(R). \qquad (2.68)$$

where $\zeta(R)$ is a well-defined function of R.[16] Then, according to (2.58), the Berry phase angle can be transformed to zero by the gauge transformation with $e^{-i\zeta(R)}$. The cases that we are interested in are those for which the Berry phase angle is different from zero (or a multiple of 2π). The Hamiltonian (2.1) provides the standard example of such a case. We shall present a detailed discussion of this system in the next chapter.

From (2.58) we see that the Mead–Berry vector potential satisfies the same gauge (phase) transformation rule, (2.58), as the vector potential of electromagnetism. The set of phase factors $e^{i\zeta_n(R)}$ form the group $U(1)$. We therefore have here a gauge theory with gauge (symmetry) group $U(1)$ and gauge potential $\mathbf{A}^n(R)$. This is the reason for which we call the phase transformation (2.16) a gauge transformation. Whereas $\mathbf{A}^n(R)$ is not an invariant quantity with respect to a gauge transformation (it transforms according to (2.58)), the Berry phase is gauge invariant. If the parameter space is three dimensional

---

[16] Note that we still require the curve **C** to lie on a single patch of the parameter space. If this is not the case, then (2.67) may be satisfied on individual patches but since the curve **C** does not in general lie in one patch, the corresponding Berry phase may still be non-trivial. Such a phase is called a *topological phase*. We shall describe topological phases in the next section.

and the parameter R is a three-dimensional vector $\mathbf{R} = (R^1, R^2, R^3)$, then we have a complete analogy with electrodynamics. However, the physical meaning of the quantities associated with the Berry phase is different. The gauge potential (2.54) is defined in terms of the eigenvectors of the Hamiltonian and has nothing to do with electromagnetism. For an $\mathcal{M}$-dimensional parameter space we again have a $U(1)$-gauge theory, but the gauge transformations and gauge potentials now depend on $\mathcal{M}$ parameters $R = (R^1, \cdots, R^\mathcal{M})$ and $\mathbf{A}^n(R)$ consists of $\mathcal{M}$ components $A_i^n(R)$; $i = 1, \cdots, \mathcal{M}$. In this case we have a $U(1)$ gauge theory over an $\mathcal{M}$-dimensional parameter space.

In analogy to electrodynamics we can define a gauge field strength tensor F^n with the components

$$F_{ij}^n := \frac{\partial}{\partial R^i} A_j^n - \frac{\partial}{\partial R^j} A_i^n, \quad i, j = 1, 2, \cdots \mathcal{M}. \tag{2.69}$$

This is an antisymmetric covariant tensor field of rank two, i.e., it is a differential two-form

$$F^n = \frac{1}{2} F_{ij}^n \, dR^i \wedge dR^j = \frac{\partial A_j^n}{\partial R^i} \, dR^i \wedge dR^j = dA^n. \tag{2.70}$$

Here we have used the antisymmetry of the wedge product $\wedge$, and dA^n stands for the exterior derivative of the Mead–Berry connection one-form A^n. The two-form F^n, with A^n given by (2.55), is also called the *Mead–Berry curvature two-form*. It is given by

$$F^n = d(i\langle n; R|d|n; R\rangle) = i(d\langle n; R|) \wedge d|n; R\rangle, \tag{2.71}$$

where we have used the simple identity[17] $d^2 = 0$.

An important property of the curvature two-form is that unlike the connection one-form, it is a gauge-invariant quantity. This follows directly from (2.59), (2.70), and the identity $d^2 = 0$,

$$F^n \to F'^n = dA'^n = dA^n - d^2\zeta = dA^n = F^n. \tag{2.72}$$

The gauge invariance of F^n has two important consequences: F^n is a globally defined object over M, and it may be used to yield a direct formula for the Berry phase.

In order to obtain this formula, we assume that the curve $\mathbf{C}$ bounds a surface $S \subset M$, and use Stokes' theorem to convert the loop integral in (2.65) to a surface integral over S. The result is

$$\gamma_n(\mathbf{C}) = \oint_C A^n = \int_S dA^n = \int_S F^n \quad \text{mod } 2\pi. \tag{2.73}$$

The surface S can be arbitrarily chosen as long as it is bounded by the closed curve $\mathbf{C}$. We denote this by $\partial S = \mathbf{C}$.

---

[17] See Appendix A.2 for a description of this property of the exterior derivative operator.

Next we wish to use the manifestly gauge-invariant expression of the Berry phase angle, i.e., (2.73), to investigate the consequences of a possible degeneracy of the energy eigenvalues. In order to see what happens when two energy levels become degenerate for some values of the parameter, we express the curvature two-form F^n in terms of the eigenvalues $E_n(R)$. Using the completeness of the basis $\{|m;R\rangle\}$,

$$1 = \sum_m |m;R\rangle \langle m;R|, \tag{2.74}$$

we can express (2.71) in the form

$$\begin{aligned} F^n &= i \sum_m [(d\langle n;R|)|m;R\rangle] \wedge [\langle m;R|d|n;R\rangle] \\ &= -i \sum_m [\langle n;R|d|m;R\rangle] \wedge [\langle m;R|d|n;R\rangle] \\ &= i \sum_{m \neq n} [\langle m;R|d|n;R\rangle] \wedge [\langle n;R|d|m;R\rangle], \end{aligned} \tag{2.75}$$

Here, in the second equality, we have used the orthonormality of the basis vectors, i.e., $\langle n;R|m;R\rangle = \delta_{nm}$, which upon differentiation implies

$$(d\langle n;R|)|m;R\rangle = -\langle n;R|d|m;R\rangle.$$

The third equality in (2.75) follows from the antisymmetry of the wedge product of two one-forms. In particular, note that due to this antisymmetry the $m = n$ term in the sum is identically zero.

In order to show the dependence of F^n on the difference of the energy eigenvalues, we substitute (2.43) in the right-hand side of (2.75). This yields

$$F^n = i \sum_{m \neq n} \frac{\langle n;R| \, [d\, h(R)] \, |m,R\rangle \wedge \langle m;R| \, [dh(R)] \, |n;R\rangle}{[E_n(R) - E_m(R)]^2}. \tag{2.76}$$

The Berry phase angle is then given in terms of this quantity according to (2.73), i.e.

$$\gamma_n(\mathbf{C}) = \int_S F^n \quad \mathrm{mod}\ 2\pi. \tag{2.77}$$

The form (2.76) does not depend on the phase factor of the basis vectors $|m;R\rangle$. Hence unlike the connection one-form (2.54) which is expressed in terms of the smooth single-valued $|n,R\rangle$ and can therefore be defined only on a patch in the parameter space M where the latter exist, the curvature two-form (2.76) is globally defined on M (Problem 2.5). Therefore, (2.76) and (2.77) can be used to compute the geometric phase angle even for the cases where the curve $\mathbf{C}$ lies in a region in the parameter space where smooth single-valued $|n;R\rangle$ do not exist.

The formula (2.76) also shows that the singularities of F^n occur at those values of $R = R_0$ where the eigenvalues are degenerate $E_n(R_0) = E_m(R_0)$.

If $R = \mathbf{R}$ is a 3-vector, i.e., $M \subseteq \mathbb{R}^3$, then one can define the dual 3-vector to F^n according to[18]

$$\mathbf{F}_i^n := \frac{1}{2}\epsilon_{ijk} F_{jk}^n,$$

where ϵ_{ijk} are the components of the totally antisymmetric Levi Civita symbol (with $\epsilon_{123} = 1$). In vector notation we have

$$\mathbf{F}^n = \nabla \times \mathbf{A}^n, \tag{2.78}$$

where the right-hand side stands for the curl of $\mathbf{A}^n$. In this case (2.71) is expressed in the form

$$\mathbf{F}^n = i(\nabla\langle n; \mathbf{R}|) \times \nabla|n; \mathbf{R}\rangle, \tag{2.79}$$

and the gauge invariance of $\mathbf{F}^n$ is shown by

$$\mathbf{F}^n \to \mathbf{F}'^n = \nabla \times \mathbf{A}'^n = \nabla \times \mathbf{A}^n - \nabla \times \nabla\zeta = \mathbf{F}^n, \tag{2.80}$$

where we have used (2.58) and the fact that the curl of a divergence vanishes. Similarly we can give the formulas for the Berry phase angle

$$\gamma_n(T) = \oint_C \mathbf{A}^n(R).d\mathbf{R} = \int_S d\mathbf{S}.(\nabla \times \mathbf{A}^n) \mod 2\pi \tag{2.81}$$

$$= \int_S d\mathbf{S} \cdot \mathbf{F} = \int_S d\mathbf{S}. [i \nabla \times (\langle n; \mathbf{R}|\nabla|n; \mathbf{R}\rangle)] \mod 2\pi \tag{2.82}$$

where the surface element $d\mathbf{S}$ is directed normal to the surface S and the line element $d\mathbf{R}$ of C is traversed in the right-hand sense with respect to the direction of $d\mathbf{S}$.

We conclude this section with the following remark: The Berry phase (2.65) has been calculated with single-valued basis vectors $|n; R\rangle$ fulfilling (2.14). If we use instead vectors fulfilling (2.15), which may have their origin in non-single-valued phase transformations $e^{i\zeta_n(R(T))} = e^{i\zeta_n}e^{i\zeta_n(R(0))}$:

$$|n; \widetilde{R(T)}\rangle = e^{i\zeta_n(R(T))}|n; R(T)\rangle = e^{i\zeta_n}|n; R(0)\rangle,$$

then there will be again an extra phase factor for $\psi(T)$ which is now given by

---

[18] This can also be done if M is (a submanifold of) a three-dimensional (pseudo) Riemannian manifold. In this case, however, the coordinates $R = (R^1, R^2, R^3)$ cannot be compared with the dual contravariant vector which is now defined by $(F^n)^i = \frac{1}{2}\epsilon^{ijk} F_{jk}^n$. In this expression $\epsilon^{ijk} = g^{ip}g^{jq}g^{kr}\epsilon_{pqr}$ are components of the contravariant dual of the covariant Levi Civita tensor (ϵ_{pqr}) and (g^{pq}) is the inverse of the (pseudo)Riemannian metric (g_{qr}) on M. The dual curvature vector $((F^n)^i(R))$, therefore, lives on the tangent space $T_R M$ of M at R.

$$\psi(T) = e^{-i\oint_0^T dt' E_n(t')} e^{i\tilde{\gamma}_n(C)} e^{i\zeta_n} \psi(0).$$

Here $\tilde{\gamma}_n(T)$ is the phase calculated from (2.65) with the $|n; R\rangle$ replaced by the $\widetilde{|n; R\rangle}$. This phase angle, calculated with the non-single-valued basis vectors, is not the Berry phase angle; the Berry phase angle is now given by $\tilde{\gamma}_n + \zeta_n$. The Berry phase is thus only defined by (2.65), if the basis vectors are single-valued.

From the above formulas we see that the Berry phase angle $\gamma_n(\mathbf{C})$ is independent of how the closed loop $\mathbf{C}$ is traversed (provided that the condition (2.45) for the validity of the adiabatic approximation is satisfied). This means that it is not sensitive to the details of the dynamics of the quantum system. It only depends upon the path $\mathbf{C}$ (with the provision of adiabaticity) and is thus a geometric quantity. The corresponding phase factor – the Berry phase $e^{i\gamma_n(\mathbf{C})}$ – is therefore called the *adiabatic geometric phase*. Later we will also consider non-adiabatic cyclic evolutions and shall reserve the name *geometric phase* for the non-adiabatic generalization of the Berry phase.

2.4 Topological Phases and the Aharonov–Bohm Effect

In this section we shall consider a class of geometric phases which are topological in nature. The best-known example of a topological phase is the so-called Aharonov–Bohm (A-B) phase. We shall, in particular, show that the A-B phase is a particular case of the geometric phases of Sect. 2.3.

We start our analysis by considering the non-relativistic quantum mechanics of a spinless particle of mass m and electric charge e. The coordinate $\hat{x}^i$ and momentum $\hat{p}_i$ operators satisfy the canonical commutation relations of the Weyl–Heisenberg algebra

$$[\hat{x}^i, \hat{x}^j] = 0, \tag{2.83}$$
$$[\hat{p}_i, \hat{p}_j] = 0, \tag{2.84}$$
$$[\hat{x}^i, \hat{p}_j] = i\delta^i_j \hat{1}, \tag{2.85}$$

where the hats are used to distinguish operators from numbers and $\hat{1}$ is the identity operator acting on the Hilbert space $\mathcal{H}$.

In the position (coordinate) representation, for every state vector $\psi \in \mathcal{H}$ we have

$$\langle x|\hat{x}^i|\psi\rangle = x^i \langle x|\psi\rangle = x^i \psi(x), \tag{2.86}$$
$$\langle x|\hat{p}_i|\psi\rangle = \left(-i\frac{\partial}{\partial x^i} + \omega_i(x)\right)\psi(x). \tag{2.87}$$

Here $i, j = 1, \cdots \mathcal{M}$, $x = (x^1, \cdots, x^{\mathcal{M}})$ are coordinates of the configuration space $\mathcal{M}$ and $\omega_i = \omega_i(x)$ are functions of x satisfying

2.4 Topological Phases and the Aharonov–Bohm Effect

$$\partial_i \omega_j - \partial_j \omega_i = 0. \tag{2.88}$$

Equation (2.88) is a direct consequence of (2.84) and (2.87). It is usually referred to as an "integrability condition".

The functions ω_i are usually neglected since they have no significance in usual quantum mechanics where the configuration space M is a Euclidean space $\mathbb{R}^M$. They become physically important whenever M has a non-trivial topology. Before discussing this issue, however, we shall express (2.88) in terms of differential forms.

Let us first note that besides the integrability condition (2.88), (2.87) also implies that under a coordinate transformation ω_i transform like the tangent vectors[19] $\frac{\partial}{\partial x^i}$. This means that ω_i may be viewed as components of a differential one-form on M, namely

$$\omega := \omega_i \, dx^i. \tag{2.89}$$

In terms of the one-form ω, (2.88) may be stated as the requirement that ω is a *closed* one-form, i.e., its exterior derivative vanishes

$$d\omega = \partial_i \omega_j \, dx^i \wedge dx^j = \frac{1}{2}(\partial_i \omega_j - \partial_j \omega_i) \, dx^i \wedge dx^j = 0. \tag{2.90}$$

Here we have used the antisymmetry of the wedge product of one-forms and (2.88).

Dirac was probably first to notice the possibility of the presence of ω_i in the position representation of the momentum operators $\hat{p}_i$. However, he only considered the case $M = \mathbb{R}^M$ and in view of (2.88) concluded that there always existed a function $f = f(x)$ such that

$$\omega_i(x) = \partial_i f(x). \tag{2.91}$$

The latter relation may be expressed in terms of the one-form ω as the requirement that ω is an *exact* one-form, i.e.,

$$\omega = df. \tag{2.92}$$

The argument used by Dirac [75] to establish (2.91) or alternatively (2.92) is a special case of the application of one of the most celebrated results of modern mathematics, known as the *Poincaré lemma*. This result which historically marks the beginning of the development of algebraic topology, indicates that every closed form is exact provided that the manifold M on which the differential form is defined is (topologically equivalent to) a star-shaped region of some Euclidean space $\mathbb{R}^M$. Loosely speaking, by M being a star-shaped region, we mean that M should not involve any holes or cuts.[20]

---

[19] For a review of the tangent vectors of a smooth manifold, see Appendix A.2.
[20] The precise mathematical condition under which every closed one-form is exact is that the first De Rham cohomology group of M is trivial. If M is simply

In particular if M is $\mathbb{R}^{\mathcal{M}}$, the lemma applies and (2.92) is valid. In this case, as argued by Dirac, the functions w_i may be gauged away by an appropriate gauge (phase) transformation of the state vectors (Problem 2.6). However, when M does involve topological peculiarities, then the hypothesis of the Poincaré lemma is violated and one must not neglect w_i in (2.87). As we shall see in the following there are physically interesting situations where the configuration space is not $\mathbb{R}^{\mathcal{M}}$, but just a region of $\mathbb{R}^{\mathcal{M}}$ which is not star-shaped.

The physical interpretation of the momentum operators $\hat{p}_i$ is that they generate displacements of quantum states in configuration space [75, 222]. This follows directly from (2.86) and (2.87). Let us first examine an infinitesimal displacement $\epsilon = (\epsilon^1, \cdots, \epsilon^{\mathcal{M}})$:

$$\begin{aligned}\langle x|\psi\rangle \rightarrow \langle x|e^{i\epsilon^i \hat{p}_i}|\psi\rangle &= (1 + \epsilon^i \partial_i + i\epsilon^i w_j)\psi(x) \\ &= [\psi(x) + \epsilon^i \partial_i \psi(x)](1 + i\epsilon^i w_i) \\ &= \psi(x+\epsilon)e^{i\epsilon^i w_i},\end{aligned} \quad (2.93)$$

where we have neglected the second-order terms in ϵ^i. Equation (2.93) may directly be generalized to large displacements, simply because the $\hat{p}_j$ commute among themselves.

We have demonstrated that in fact the momentum operators generate displacements of the states $\Lambda = |\psi\rangle\langle\psi|$. For the state vectors, however, they are accompanied by phase factors, unless $w_i = 0$.

Next let us consider displacing the state vector ψ along a smooth curve

$$\mathbf{C}: [0, T] \rightarrow M.$$

Under such a displacement the state vector undergoes the following transformation:

$$\langle x|\psi\rangle \rightarrow \langle x|e^{i\int_{\mathbf{C}} dx^i \hat{p}_i}|\psi\rangle = \psi(x + \Delta x)e^{i\int_{\mathbf{C}} \omega}, \quad (2.94)$$

where $\Delta x = \mathbf{C}(T) - \mathbf{C}(0)$ and ω is the one-form defined by (2.89).[21] For a closed curve ($\mathbf{C}(T) = \mathbf{C}(0)$) this reduces to

$$\langle x|\psi\rangle \rightarrow \langle x|e^{i\oint_{\mathbf{C}} dx^i \hat{p}_i}|\psi\rangle = \psi(x)e^{i\oint_{\mathbf{C}} \omega}. \quad (2.95)$$

Let us now consider the case where M has trivial topology. In this case Dirac's hypothesis is valid, namely there exists a function $f = f(x)$ fulfilling

---

connected, i.e. every closed curve in M can be continuously deformed to a single point, then this requirement is fulfilled. An example of a simply connected space which is topologically different from $\mathbb{R}^{\mathcal{M}}$ is the $\mathcal{M}$-dimensional sphere $S^{\mathcal{M}}$ with $\mathcal{M} > 1$. Simple examples of non-simply-connected spaces (with nontrivial first De Rham cohomology) are the punctured $\mathbb{R}^2$, the circle S^1 and the $\mathcal{M}$-dimensional torus $T^{\mathcal{M}}$. Note that a punctured $\mathbb{R}^{\mathcal{M}}$ with $\mathcal{M} > 2$ is simply connected!

[21] Note that here we assume the initial and the end point of the curve $\mathbf{C}$ lie in a single patch. This is necessary for the quantity Δx to make sense.

2.4 Topological Phases and the Aharonov–Bohm Effect

(2.92). Using f, we can easily show that the phase angle appearing in (2.95) vanishes:

$$\oint_C \omega = \oint_C df = f(\mathbf{C}(T)) - f(\mathbf{C}(0)) = 0. \tag{2.96}$$

An alternative way to arrive at this conclusion is to use Stokes' theorem:

$$\oint_C \omega = \int_S d\omega = 0. \tag{2.97}$$

Here S is any surface bounded by the curve $\mathbf{C}$, $\mathbf{C} = \partial S$, and the second equality is established using (2.90). The latter equation is always valid. Hence one might try to generalize (2.97) to the case where M is not topologically trivial. The fallacy of this argument is that in general there might not exist any S which would only be bounded by $\mathbf{C}$. To demonstrate this point, let us consider a quantum system whose configuration space M consists of a two-dimensional plane $\mathbb{R}^2$ with a disk D of radius ρ removed. For any loop $\mathbf{C}$ in M encircling D, there does not exist any surface S belonging to M and bounded by $\mathbf{C}$. The situation is illustrated in Fig. 2.1.

Although a surface S with $\partial S = \mathbf{C}$ does not exist for this example, we may pursue applying Stokes' theorem using a surface S' whose boundary consists of the loop $\mathbf{C}$ and the boundary ∂D of the disk D. This leads to the following expression for the phase angle.

$$\oint_C \omega = \int_{S'} d\omega + \oint_{\partial D} \omega = \oint_{\partial D} \omega, \tag{2.98}$$

where we have used (2.90) in the second equality. Equation (2.98) shows that the phase does not depend on the particular closed curve used for the

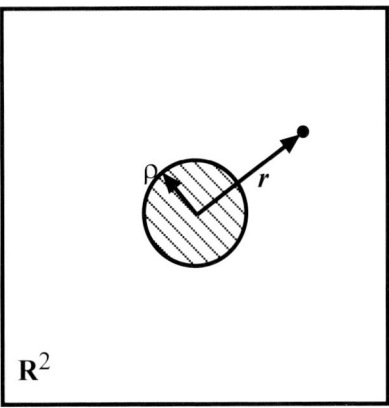

Fig. 2.1. The configuration space of a particle moving in a plane with a disk of radius ρ removed.

displacement, as long as it encircles D once. In fact, in general $\mathbf{C}$ may encircle D, n_+ times counterclockwise and n_- times clockwise. Then we have

$$\oint_{\mathbf{C}} \omega = (n_+ - n_-) \oint_{\partial D} \omega. \tag{2.99}$$

The integer $(n_+ - n_-)$ is known as the *winding number* of the curve $\mathbf{C}$. Equation (2.99) shows that the phase angle remains unchanged under continuous (smooth) deformations of the curve $\mathbf{C}$. This means that it is a *topological property* of the curve, hence the name *topological phase*.

A simple but crucial observation regarding the topological phase

$$e^{i \oint_{\mathbf{C}} \omega} = \left[e^{i \oint_{\partial D} \omega} \right]^{n_+ - n_-}, \tag{2.100}$$

is that it is non-trivial provided that the winding number of the curve is non-zero and that the fundamental phase factor is non-trivial, i.e.,

$$\oint_{\partial D} \omega \neq \text{multiple of } 2\pi.$$

The latter depends on the particular physical processes involved.

The standard physical example which signifies the importance of the topological phases is the Aharonov–Bohm effect. Aharonov and Bohm [8] considered the quantum mechanics of an electron moving around but not actually penetrating a magnetic flux line. It is not difficult to see that the configuration space of this problem is essentially the same as that of Fig. 2.1. This is most obvious if we only consider the planar motion of the electron around a magnetic field confined within an infinitely long tube of cross-section D.

We shall first consider the simpler system of an electron traversing a closed loop $\mathbf{C}$ in the configuration space $\mathbb{R}^3$ and subject to a magnetic field

$$\mathbf{B}^{(\text{el})} = \nabla \times \mathbf{A}^{(\text{el})}, \tag{2.101}$$

where $\mathbf{A}^{(\text{el})} = (A^1, A^2, A^3)$ is the electromagnetic vector potential. The vector potential and the magnetic field may be expressed in terms of differential forms according to

$$A^{(\text{el})} = A_a dx^a, \qquad F^{(\text{el})} = dA^{(\text{el})} = -\frac{1}{2} \epsilon_{abc} \left(\mathbf{B}^{(\text{el})} \right)^a dx^b \wedge dx^c, \tag{2.102}$$

where $A_a := \eta_{ab} A^b = -A^a$ and $\eta_{ab} = -\delta_{ab}$ with $a, b = 1, 2, 3$ are the space components of the Minkowski metric $\eta := \text{diag}(1, -1, -1, -1)$.

The dynamics of the electron is determined by replacing the momentum operator $\hat{\mathbf{p}}$ by its covariant generalization

$$\hat{\mathbf{p}} \rightarrow \hat{\pi} := \hat{\mathbf{p}} - \frac{e}{c} \mathbf{A}^{(\text{el})}, \tag{2.103}$$

2.4 Topological Phases and the Aharonov–Bohm Effect

in the Schrödinger equation for the free electron. As the effect of the magnetic field is incorporated in the definition of the covariant momentum operator $\hat{\pi}$, circulation of the electron along **C** may be performed by using the covariant momentum operator as the generator of the appropriate space displacements. Representing $\hat{\pi}$ in the coordinate representation we have

$$\langle x|\hat{\pi}_a|\psi\rangle = \left(-i\frac{\partial}{\partial x^a} + \frac{e}{c}A_a^{(\text{el})}(x) + \tilde{\omega}_a(x)\right)\psi(x), \quad (2.104)$$

where $\tilde{\omega}_a$ are components of a closed form $\tilde{\omega}$ on $\mathbb{R}^3$. Since the configuration space is $\mathbb{R}^3$, $\tilde{\omega}$ must also be exact. We know however that the one-form $A^{(\text{el})}$ is defined up to gauge transformations, i.e. up to addition of exact one-forms df,

$$A^{(\text{el})} \to A'^{(\text{el})} = A^{(\text{el})} + df.$$

Hence we can absorb $\tilde{\omega}$ in the definition of $A^{(\text{el})}$ and effectively disregard it. This leads to the representation

$$\langle x|\hat{\pi}_a|\psi\rangle = \left(-i\frac{\partial}{\partial x^a} + \frac{e}{c}A_a^{(\text{el})}(x)\right)\psi(x). \quad (2.105)$$

Performing the displacement of the electron, we find

$$\langle x|\psi\rangle \to \langle x|e^{i\oint_\mathbf{C} dx^a \hat{\pi}_a}|\psi\rangle = \psi(x)e^{i\frac{e}{c}\oint_\mathbf{C} A^{(\text{el})}}. \quad (2.106)$$

The *electromagnetic* or *Dirac phase* $e^{i\frac{e}{c}\oint_\mathbf{C} A^{(\text{el})}}$ so obtained is therefore *not* a topological phase. If the magnetic field vanishes, then $A^{(\text{el})}$ becomes also a closed form. Since the configuration space is $\mathbb{R}^3$, the closedness of $A^{(\text{el})}$ implies its exactness and the phase angle vanishes. This is the familiar statement that if the magnetic field vanishes, the vector (gauge) potential is a pure gauge.

Now let us consider the case where the configuration space M is just a portion of $\mathbb{R}^3$ in which the magnetic field is zero. In this case, M may in general have a non-trivial topology. This renders the above argument inconclusive. The vanishing of the magnetic field in this case, still implies $A^{(\text{el})}$ to be closed. This does not however mean that it is exact. In view of this observation, a comparison of (2.87) and (2.105) suggests the identification of the closed one-form ω of (2.89) with the electromagnetic one-form $A^{(\text{el})}$. More precisely, we have

$$\omega = \frac{e}{c}A^{(\text{el})}. \quad (2.107)$$

This provides the physical interpretation of the one-form ω for the system under consideration.

Now we are in a position to analyze the A-B effect. In the A-B system the magnetic field is confined to a tube and the electron is circulating around it. The configuration space M of the electron is the region in space $\mathbb{R}^3$ which lies outside the tube. In view of (2.107), the topological (A-B) phase associated with a closed curve $\mathbf{C}$ is given by

$$e^{\frac{ie}{c} \oint_{\mathbf{C}} A^{(\text{el})}}. \tag{2.108}$$

It is determined by the winding number of $\mathbf{C}$ and the fundamental topological phase:

$$e^{\frac{ie}{c} \oint_{\partial D} A^{(\text{el})}}.$$

Now demanding $A^{(\text{el})}$ to be continuous at the boundary ∂D, we may compute the latter using the information about the interior of the tube. Applying Stokes' theorem, we have

$$\begin{aligned} e^{\frac{ie}{c} \oint_{\partial D} A^{(\text{el})}} &= e^{\frac{ie}{c} \int_D dA^{(\text{el})}} = e^{\frac{ie}{c} \int_D F^{(\text{el})}} \\ &= e^{-\frac{ie}{c} \oint_D \mathbf{B}^{(\text{el})} \cdot d\mathbf{S}} = e^{-\frac{ie}{c} \Phi_B}, \end{aligned} \tag{2.109}$$

where we have employed (2.102) and Φ_B is the magnetic flux through D. The phase so derived is called the Aharonov–Bohm (A-B) phase. Its significance is that although the electron is classically confined to a region where the magnetic field vanishes, it is influenced by the field quantum mechanically.[22]

In the remainder of this section we shall explain how the topological phase of Aharonov and Bohm is viewed as a special case of Berry's geometric phase. Our analysis follows that of [31].

Consider the A-B system and let the electron be confined to a box situated at a point $\mathbf{R}$ of space, with $|\mathbf{R}| \gg \rho$, where ρ is the radius of the tube centered at the origin. We may alternatively suppose that $\rho \approx 0$. The Hamiltonian of the system is then given by

$$\hat{h} = \hat{h}\left(\hat{\mathbf{p}} - \frac{e}{c}\mathbf{A}^{(\text{el})}(\hat{\mathbf{x}}), \hat{\mathbf{x}} - \mathbf{R}\right) \tag{2.110}$$

where we have switched to a coordinate frame centered inside the box.

Now the location of the box $\mathbf{R}$ plays the role of a set of parameters which parameterize the Hamiltonian. Following the analysis of Sect. 2.1, the instantaneous energy eigenvectors $|n; \mathbf{R}\rangle$ are defined by:

$$\hat{h}\left(\hat{\mathbf{p}} - \frac{e}{c}\mathbf{A}^{(\text{el})}(\hat{\mathbf{x}}), \hat{\mathbf{x}} - \mathbf{R}\right) |n; \mathbf{R}\rangle = E_n(\mathbf{R})|n; \mathbf{R}\rangle. \tag{2.111}$$

They may be expressed in terms of the energy eigenvectors ψ_n of the free Hamiltonian,

$$\hat{h}_0 = \hat{h}_0\left(\hat{\mathbf{p}}, \hat{\mathbf{x}} - \mathbf{R}\right), \tag{2.112}$$

according to (Problem 2.7)

$$\langle x|n; \mathbf{R}\rangle = e^{\frac{ie}{c} \int_{\mathbf{R}}^{\mathbf{x}} A^{(\text{el})}} \psi_n(\mathbf{x} - \mathbf{R}). \tag{2.113}$$

---

[22] For a detailed discussion of the theoretical and experimental aspects of the A-B phase see [211] and [101].

Note that the energy eigenvalues $E_n(\mathbf{R})$ are the same for both the Hamiltonians (Problem 2.8) and that the integration in (2.113) is performed along any path joining $\mathbf{R}$ to $\mathbf{x}$ but not intersecting or encircling the tube (Problem 2.9).

Next we compute the analog of the Mead–Berry connection one-form (2.55),

$$\begin{aligned} A^n &:= i\langle n; R|d|n; R\rangle \\ &= i\int dx^3 \psi_n^*(\mathbf{x}-\mathbf{R})\left[-\frac{ie}{c}A^{(\mathrm{el})}(\mathbf{R})\psi_n(\mathbf{x}-\mathbf{R})+d\psi_n(\mathbf{x}-\mathbf{R})\right] \\ &= \frac{e}{c}A^{(\mathrm{el})}, \end{aligned} \qquad (2.114)$$

where the exterior derivative is taken with respect to R and ψ_n are assumed to be normalized. Equation (2.114) together with the definition of Berry's phase (2.56) lead to

$$e^{i\gamma_n} = e^{\frac{ie}{c}\oint_\mathbf{C} A^{(\mathrm{el})}}. \qquad (2.115)$$

Here $\mathbf{C}$ is the closed path in M along which the box containing the electron and therefore the electron itself are transported. Note that the right-hand side of (2.115) is independent of n. Therefore the formula for the phase is valid for any superposition of the energy eigenvectors and hence any wave packet.

Equation (2.115) shows that the A-B phase (2.108) is a particular example of Berry's geometric phase (2.56). However unlike the usual geometric phase the A-B phase is topological in nature. Not only is it independent of the details of the dynamics, but it is even insensitive to the shape of the curve $\mathbf{C}$. The latter is definitely not the case for the geometric phase in general.[23]

Our derivation of the A-B phase as a special case of the geometric phase also applies to the case where the electron is subject to motion in the magnetic field in $\mathbb{R}^3$. In this case it acquires an electromagnetic analog of the geometric phase, the Dirac phase, which is truly geometric. This means that it depends on the geometry of the closed path $\mathbf{C}$ and changes under its continuous deformations.

Finally we would also like to point out that for the system considered here $\langle m; R(t)|\frac{d}{dt}|n; R(t)\rangle = 0$ for $m \neq n$. Hence the condition (2.41) for the validity of the adiabatic approximation is satisfied regardless of the speed with which the electron is circulated about the flux tube.

Problems

2.1) Show that (2.22) is obtained from (2.19), if (2.21) holds.

2.2) Show that (2.21) implies (2.23).

---

[23] This is the main feature which distinguishes topological phases from the more general geometric phases.

2. Quantal Phase Factors for Adiabatic Changes

2.3) Show that the stationary states of a conservative quantum system are always energy eigenstates.

2.4) Show that $\langle n; R|d|n; R\rangle$ is purely imaginary.

2.5) Show that the Mead–Berry curvature two-form F^n, given by (2.71) or (2.76), is invariant under gauge transformations (2.16).

2.6) Show that if (2.92) holds, then there exist a gauge transformation that removes the one-form ω from (2.87).

2.7) Show that $\langle x|n; \mathbf{R}\rangle$ as given by (2.113) satisfies (2.111), provided that ψ_n are eigenvectors of the free Hamiltonian (2.112).

2.8) Show that the introduction of the magnetic field (magnetic vector potential) does not affect the energy eigenvalues E_n.

2.9) Show that (2.113) is independent of the choice of a path along which the line integral is evaluated, as long as the chosen path does not intersect or encircle the flux tube.

3. Spinning Quantum System in an External Magnetic Field

3.1 Introduction

In this chapter we will discuss in detail a quantum particle with magnetic moment $\mathbf{m} = \mu_B g \mathbf{J}$ in an external magnetic field $\mathbf{B}(t) = B\hat{\mathbf{R}}(t)$ whose direction $\hat{\mathbf{R}}(t)$ is changing periodically. In particular we will consider the case in which the direction of the magnetic field precesses around a fixed axis which we take as the 3-axis (z-axis) of our (laboratory) coordinate frame in space ($\mathbb{R}^3$). If the direction rotates slowly ("adiabatically") this system provides an application of the general ideas developed in Chap. 2. The Schrödinger equation for a magnetic moment in a precessing magnetic field has been solved exactly in [216]. Therefore we need not restrict ourselves to the adiabatic approximation and will obtain the non-adiabatic geometric phase. The latter is also known as the *Aharonov–Anandan phase* [7, 9] for their pioneering work. With the help of this example we will then introduce in Sect. 3.5 the non-adiabatic geometric phase for a general cyclic evolution.

The Hamiltonian is given according to (2.1) by

$$h(\hat{\mathbf{R}}(t)) = -\frac{Bge}{2mc}\hat{\mathbf{R}}(t)\cdot\mathbf{J} = b\hat{\mathbf{R}}(t)\cdot\mathbf{J} \tag{3.1}$$

where $\mathbf{J}$ is the angular momentum operator of the quantum physical system, $\hat{\mathbf{R}}(t)$ is the parameter that describes the changing environment, and $b = -\frac{Bge}{2mc}$ is a constant. ($\frac{e\hbar}{2mc} = \mu_B$ is the Bohr magneton and g is the Landé factor of the quantum particle. b is a frequency and $b/(-g) \equiv \omega_L$ is called the Larmor frequency.)

3.2 The Parameterization of the Basis Vectors

For the quantum system described by the Hamiltonian (3.1), the parameter space of the environment is the set of all unit vectors $\hat{\mathbf{R}}$ in the three-dimensional Euclidean space $\mathbb{R}^3$. Each unit vector in $\mathbb{R}^3$ determines a point on the unit sphere, S^2, centered at the origin. Therefore, the parameter space of this system may be identified with the unit sphere S^2 embedded in $\mathbb{R}^3$. We parameterize the points of S^2 by the polar and azimuthal angles (θ, φ) according to

3. Spinning Quantum System in an External Magnetic Field

$$\hat{\mathbf{R}} = \hat{\mathbf{R}}(\theta, \varphi) = \begin{pmatrix} \sin\theta \cos\varphi \\ \sin\theta \sin\varphi \\ \cos\theta \end{pmatrix}, \qquad (3.2)$$

where

$$0 \leq \theta \leq \pi, \text{ and } 0 \leq \varphi < 2\pi.$$

This parameterization associates unique values of the pair (θ, φ) to each unit vector $\hat{\mathbf{R}}$ except for the unit vector

$$\mathbf{e}_3 := \begin{pmatrix} 0 \\ 0 \\ 1 \end{pmatrix} \qquad (3.3)$$

of the north pole $\mathcal{N}$ and the unit vector $-\mathbf{e}_3$ of the south pole $\mathcal{S}$. $\mathbf{e}_3$ and $-\mathbf{e}_3$ are given, respectively, by $\theta = 0$ and $\theta = \pi$ for all values of φ in the range $0 \leq \varphi < 2\pi$; the value of φ is not determined when $\hat{\mathbf{R}} = \pm\mathbf{e}_3$. For a more detailed discussion of the parameterization of S^2 see Appendix A.

The special case in which the magnetic field precesses uniformly about the 3-axis is described by

$$\mathbf{B}(t) = B(\sin\theta \cos\omega t, \sin\theta \sin\omega t, \cos\theta) = B\hat{\mathbf{R}}(\theta, \omega t), \qquad (3.4)$$

with B, θ and $\omega = \varphi/t$ being constants.

The precession of the magnetic field is shown in Fig. 3.1. For $t = 0$ the magnetic field lies in the 1-3 plane

$$\mathbf{B}(0) = B(\sin\theta, 0, \cos\theta) = (B_{12}, 0, B_3) = B\,\hat{\mathbf{R}}(\theta, 0).$$

Figure 3.1 shows the angles θ, φ, and a new angle $\tilde{\theta}$ which will be defined later.

As time progresses the component $B_{12} = B\sin\theta$ rotates in the 1-2 plane

$$\mathbf{B}_{12} := B\sin\theta\,\mathbf{e}_{12} := B\sin\theta\,(\mathbf{e}_1 \cos\omega t + \mathbf{e}_2 \sin\omega t), \qquad (3.5)$$

so that

$$\begin{aligned}\mathbf{B}(t) &= B_{12}\mathbf{e}_{12}(t) + B_3\mathbf{e}_3 \\ &= B[\sin\theta\,(\mathbf{e}_1 \cos\omega t + \mathbf{e}_2 \sin\omega t) + \cos\theta\,\mathbf{e}_3] \\ &= \frac{b}{(-\frac{eg}{2mc})}\hat{\mathbf{R}}(\theta, \omega t). \end{aligned} \qquad (3.6)$$

The Hamiltonian (3.1) takes the form

$$\begin{aligned} h(t) &= b\hat{\mathbf{R}}(t).\mathbf{J} = b(\sin\theta\,\mathbf{e}_{12}(t) + \cos\theta\,\mathbf{e}_3).\mathbf{J} \\ &= b\cos\theta\,[J_3 + \tan\theta\,\mathbf{e}_{12}(t).\mathbf{J}]. \end{aligned} \qquad (3.7)$$

We can rewrite the time-dependent term of the Hamiltonian in the following form

3.2 The Parameterization of the Basis Vectors

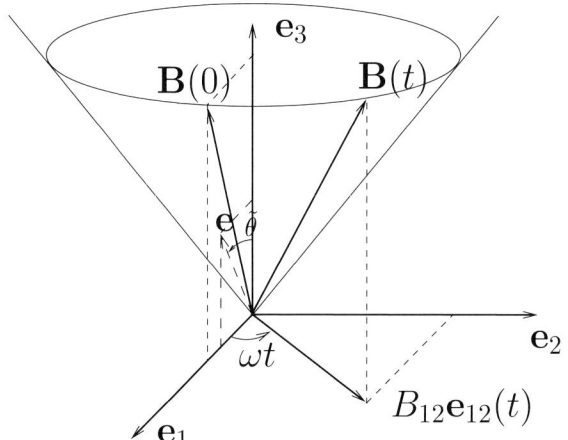

Fig. 3.1. A quantal magnetic moment in an external magnetic field precesses uniformly around a cone of semiangle θ with $\mathbf{e}_3$. (θ, φ) are the polar and azimuthal angles for the rotation of the external magnetic field. $(\tilde{\theta}, \varphi)$ are the "polar and azimuthal angles" for the evolution of the state of the magnetic moment.

$$\mathbf{e}_{12}(t).\mathbf{J} = J_1 \cos \omega t + J_2 \sin \omega t = e^{-i\omega t J_3} J_1 e^{i\omega t J_3}.$$

Thus we obtain from (3.7)

$$h(t) = b \cos \theta J_3 + b \sin \theta \, e^{-i\omega t J_3} J_1 e^{i\omega t J_3} = e^{-i\omega t J_3} h_0 e^{i\omega t J_3}, \qquad (3.8)$$

with

$$h_0 = b\hat{\mathbf{R}}(\theta, 0).\mathbf{J} = b e^{-i\theta J_2} J_3 e^{i\theta J_2}. \qquad (3.9)$$

As seen from (3.8) the time-dependent Hamiltonian $h(t)$ is periodic,

$$h(T) = h(0) = h_0 \quad \text{with period} \quad T = 2\pi/\omega. \qquad (3.10)$$

The eigenvectors (2.6) for this example are defined by

$$h(R)|k; R\rangle = b\hat{\mathbf{R}} \cdot \mathbf{J}|k; R\rangle = bk|k; R\rangle. \qquad (3.11)$$

They are eigenvectors of the operator $\hat{\mathbf{R}}(t) \cdot \mathbf{J}$. The place of the energy quantum number n in (2.6) is taken here by k which is the quantum number for the component of angular momentum along the changing direction of the external magnetic field.

For this example, the eigenvalues $E_k(R(t)) = bk$ of the time-dependent Hamiltonian $h(R(t))$ are constant, but the eigenvectors $|k; R(t)\rangle$ and the eigenprojectors $\Lambda_k(R(t)) = |k; R\rangle\langle k; R|$ change in time (for $\omega \neq 0$). The eigenvalue k is a constant but its physical interpretation changes in time; it is the eigenvalue of the observable $\hat{\mathbf{R}}(t) \cdot \mathbf{J}$ where the direction $\hat{\mathbf{R}}(t)$ changes with respect to the laboratory frame.

If the state vectors $|k; R\rangle$ are also eigenvectors of $\mathbf{J}^2$, then the possible values of k are

$$k = -j,\, -j+1,\, \ldots,\, j-1,\, j, \quad \text{where} \quad j \in \frac{\mathbb{Z}}{2}, \tag{3.12}$$

i.e., j and therefore k take integer or half-integer values, and $j(j+1)$ is the eigenvalue of the square of the angular momentum $\mathbf{J}^2$. We assume here that j and any other additional quantum numbers, which may be connected with the eigenvalues of the rotationally invariant part H_0 of the total Hamiltonian $H^{\text{total}} = H_0 + h(\hat{\mathbf{R}})$, have fixed values and we suppress them throughout the current discussion as labels of the state vectors $|k; R\rangle$.

On the other hand, it is possible that $|k; R\rangle$ are not eigenvectors of $\mathbf{J}^2$ and thus do not have a definite value of j associated to them. This kind of vector will be needed in the example of a diatomic molecule where k is the eigenvalue of the quantum mechanical observable $\mathbf{X} \cdot \mathbf{J}$. Here $\mathbf{X}$ is the position operator of the (direction of the) internuclear axis of the molecule, which rotates in space. This observable describes the slowly changing quantal environment for the electron. The electron is the fast quantum system that follows instantaneously the motion of the molecule as a whole. The electronic angular momentum is not a good quantum number, because one does not have spherical symmetry but only axial symmetry about the internuclear axis $\mathbf{X}$. Therefore the total Hamiltonian commutes with $\mathbf{X} \cdot \mathbf{J}$, but not with $\mathbf{J}^2$.[1] Consequently the molecular states are not eigenstates of $\mathbf{J}^2$ and the argument used to justify (3.12) does not apply. We shall however show, also in this case, that k takes only integer or half-integer values.

The vectors of (3.11) are parameterized by the unit vector $\hat{\mathbf{R}}$, or by the polar and azimuthal angles (θ, φ). They can be obtained by applying (θ, φ)-dependent rotations to an eigenvector $|k; \hat{\mathbf{e}}_3\rangle$ of the component of angular momentum in the direction of the north pole, $J_3 = \mathbf{e}_3 \cdot \mathbf{J}$.

There are many (θ, φ)-dependent rotations $\mathcal{R}(\theta, \varphi) \in SO(3)$ which when applied to $\mathbf{e}_3$ give the unit vector $\hat{\mathbf{R}}(\theta, \varphi)$. We choose the following product of rotations (Problem 3.1 and 3.2).

$$\begin{aligned}\mathcal{R}(\theta, \varphi)\mathbf{e}_3 &= \mathcal{R}_3(\varphi)\mathcal{R}_2(\theta)\mathcal{R}_3(-\varphi)\mathbf{e}_3 = \mathcal{R}_3(\varphi)\mathcal{R}_2(\theta)\mathbf{e}_3 \\ &= \mathcal{R}_3(\varphi)\hat{\mathbf{R}}(\theta, 0) = \hat{\mathbf{R}}(\theta, \varphi),\end{aligned} \tag{3.13}$$

where $\mathcal{R}_3(-\varphi)$ does nothing to the unit vector $\mathbf{e}_3$ and where $\mathcal{R}_2(\theta)$ produces the unit vector $\hat{\mathbf{R}}(\theta, 0)$ which lies in the 1-3 plane at an angle θ with respect to the $\mathbf{e}_3$-axis. The rotation $\mathcal{R}_3(-\varphi)$ has been included in the definition of $\mathcal{R}(\theta, \varphi)$ in order for the rotation $\mathcal{R}(0, \varphi)$ to be independent of φ.

Rotations, like any other continuous transformations, are represented in the space of quantum physical states by unitary operators (representing the group $SU(2)$). The unitary operators that represent the rotations $\mathcal{R}_3(\varphi)$ and $\mathcal{R}_2(\theta)$ are given by

---

[1] Note that in this case $\mathbf{J}$ is the electronic angular momentum or electronic spin.

3.2 The Parameterization of the Basis Vectors

$$U_3(\varphi) = e^{-i\varphi J_3} := I + \frac{\varphi}{i} J_3 + \frac{1}{2!} \left(\frac{\varphi}{i} J_3\right)^2 + \cdots, \tag{3.14}$$

$$U_2(\theta) = e^{-i\theta J_2}, \tag{3.15}$$

and an analogous expression holds for rotations about the $\mathbf{e}_1$-axis. The product of two or more rotations is represented by the product of the corresponding operators. The rotation $\mathcal{R}(\theta, \varphi)$ of (3.13) is thus represented by the operator

$$U(\theta, \varphi) = U_3(\varphi) U_2(\theta) U_3(-\varphi) = e^{-i\varphi J_3} e^{-i\theta J_2} e^{i\varphi J_3}. \tag{3.16}$$

We now choose a fixed normalized eigenvector $|k; \mathbf{e}_3\rangle$ of $J_3 = \mathbf{e}_3 \cdot \mathbf{J} = \hat{\mathbf{R}}(0, 0) \cdot \mathbf{J}$ and transform it using the unitary operator $U(\theta, \varphi)$. The resulting state vector,

$$|k; \theta, \varphi\rangle := U(\theta, \varphi) |k; \hat{\mathbf{e}}_3\rangle = e^{-i\varphi J_3} e^{-i\theta J_2} e^{i\varphi J_3} |k; \hat{\mathbf{e}}_3\rangle, \tag{3.17}$$

is an eigenvector of the operator $\hat{\mathbf{R}}(\theta, \varphi) \cdot \mathbf{J}$ with eigenvalue k,

$$\hat{\mathbf{R}}(\theta, \varphi) \cdot \mathbf{J} \; |k; \theta, \varphi\rangle = k \; |k; \theta, \varphi\rangle. \tag{3.18}$$

In order to prove (3.18), one uses the following transformation property of the angular momentum operators J_i which follows from their commutation relations[2]

$$e^{-i\theta J_2} J_3 e^{i\theta J_2} = J_3 \cos\theta + J_1 \sin\theta, \tag{3.19}$$

$$e^{-i\varphi J_3} J_1 e^{i\varphi J_3} = J_1 \cos\varphi + J_2 \sin\varphi, \tag{3.20}$$

$$e^{-i\varphi J_3} J_2 e^{i\varphi J_3} = J_1(-\sin\varphi) + J_2 \cos\varphi. \tag{3.21}$$

The state vector (3.17) is a smooth vector-valued function of (θ, φ). This vector-valued function gives a unique state vector for all $\hat{\mathbf{R}}$ except for the south pole $\hat{\mathbf{R}} = -\mathbf{e}_3$, where $\theta = \pi$ and (3.19) becomes

$$e^{-i\pi J_2} J_3 e^{i\pi J_2} = -J_3. \tag{3.22}$$

This implies

$$e^{-i\pi J_2} e^{-i\varphi J_3} e^{i\pi J_2} = e^{i\varphi J_3}, \tag{3.23}$$

which may be used to obtain

$$\begin{aligned} |k; \pi, \varphi\rangle &= e^{-i\varphi J_3} e^{-i\pi J_2} e^{i\varphi J_3} |k; \hat{\mathbf{e}}_3\rangle \\ &= e^{-i\pi J_2} e^{2i\varphi J_3} |k; \hat{\mathbf{e}}_3\rangle \\ &= e^{-i\pi J_2} e^{2ik\varphi} |k; \hat{\mathbf{e}}_3\rangle. \end{aligned} \tag{3.24}$$

---

[2] These properties can be easily verified using the Backer–Campbell–Hausdorff formula, $e^A B e^{-A} = B + \sum_{n=1}^{\infty} B_n/n!$, $B_0 := B$, $B_n := [A, B_{n-1}]$, where A and B are operators. See also [35, Equation (e.40)].

This shows that at the south pole different normalized state vectors are obtained as φ varies in the range $0 \leq \varphi < 2\pi$. φ also varies in the range $0 \leq \varphi < 2\pi$ at the north pole $\mathbf{e}_3 = \hat{\mathbf{R}}(0, \varphi)$ but $|k; 0, \varphi\rangle$ is single-valued at the north pole because as a result of the inclusion of the rotation $\mathcal{R}_3(-\varphi)$ in the definition of $\mathcal{R}(\theta, \varphi)$, $|k; 0, \varphi\rangle$ does not depend on φ. It is therefore a smooth single-valued vector function everywhere on S^2 except at the south pole.

A smooth vector-valued function which is well defined at the south pole but not at the north pole is obtained by a gauge transformation (2.16), namely

$$|k; \theta, \varphi\rangle' = e^{-i2k\varphi}|k; \theta, \varphi\rangle = e^{-i\varphi J_3} e^{-i\theta J_2} e^{-i\varphi J_3}|k; 0, 0\rangle. \quad (3.25)$$

The new state vector $|k; \theta, \varphi\rangle'$ differs from $|k; \theta, \varphi\rangle$ by the phase factor $e^{i\zeta(\theta,\varphi)} = e^{-i2k\varphi}$. At the south pole $|k; \theta, \varphi\rangle'$ evaluates to a single vector

$$|k; \pi, \varphi\rangle' = e^{-i\pi J_2}|k; \hat{\mathbf{e}}_3\rangle \quad (3.26)$$

but at the north pole it evaluates to many vectors

$$|k; 0, \varphi\rangle' = e^{-2ik\varphi}|k; \hat{\mathbf{e}}_3\rangle. \quad (3.27)$$

Hence $|k; \theta, \varphi\rangle'$ can be used everywhere on S^2 except at the north pole. Either vector (3.17) or (3.25) can be used in the overlap region $O_1 \cap O_2$ of the two open patches

$$O_1 := S^2 - \{\mathcal{S}\} \quad \text{and} \quad O_2 := S^2 - \{\mathcal{N}\},$$

of S^2, where $\mathcal{N}$ and $\mathcal{S}$ denote the north and south poles of S^2 respectively.

The state vector $|k; \pi, \varphi\rangle'$ is an eigenvector of J_3 with eigenvalue $-k$ and an eigenvector of $\hat{\mathbf{R}}(\pi, \varphi) \cdot \mathbf{J} = -\mathbf{e}_3 \cdot \mathbf{J} = -J_3$ with eigenvalue k (Problem 3.3).

We thus see that two different parameterizations of $|k; \hat{\mathbf{R}}\rangle$ are needed. However, since $|k; \theta, \varphi\rangle$ and $|k; \theta, \varphi\rangle'$ differ only by the phase factor $e^{-i2k\varphi}$ the projection operators and the corresponding subspaces (rays) in the Hilbert space coincide:

$$|k; \theta, \varphi\rangle\langle k; \theta, \varphi| = |k; \theta, \varphi\rangle' \,'\langle k; \theta, \varphi|. \quad (3.28)$$

3.3 Mead–Berry Connection and Berry Phase for Adiabatic Evolutions – Magnetic Monopole Potentials

We next calculate the Mead–Berry connection one-form A^k for the adiabatic evolution of the Hamiltonian (3.1). By definition

$$A^k(R) = A_i^k dR^i = i\langle k; R| \frac{\partial}{\partial R^i}|k; R\rangle \, dR^i, \quad (3.29)$$

where $i = 1, 2$ correspond to the coordinates θ and φ of either of the patches O_1 or O_2 of the sphere S^2. We shall in fact compute the one-forms

3.3 Mead–Berry Connection and Berry Phase for Adiabatic Evolutions

$$A^{k'k}(R^i) = A^{k'k}_i dR^i = i\langle k'; R|\frac{\partial}{\partial R^i}|k; R\rangle dR^i. \tag{3.30}$$

Clearly the Mead–Berry connection one-form A^k is given by the diagonal elements, A^{kk}, of the matrix of one-forms $(A^{k'k})$.

On the patch O_1 in which (3.17) holds, we have

$$\begin{aligned}A^{k'k}_\theta &= i\langle k'; \theta, \varphi|\frac{\partial}{\partial \theta}|k; \theta, \varphi\rangle \\ &= \langle k'\hat{\mathbf{e}}_3|iU^\dagger(\theta, \varphi)\frac{\partial}{\partial \theta}U(\theta, \varphi)|k; \hat{\mathbf{e}}_3\rangle =: r\hat{A}_\theta,\end{aligned} \tag{3.31}$$

$$\begin{aligned}A^{k'k}_\varphi &= i\langle k'\theta, \varphi|\frac{\partial}{\partial \varphi}|k; \theta, \varphi\rangle \\ &= \langle k'; \hat{\mathbf{e}}_3|iU^\dagger(\theta, \varphi)\frac{\partial}{\partial \varphi}U(\theta, \varphi)|k; \hat{\mathbf{e}}_3\rangle =: r\sin\theta\,\hat{A}_\varphi,\end{aligned} \tag{3.32}$$

where $\hat{A}_\theta$ and $\hat{A}_\varphi$ are defined for future use. Next we compute

$$\begin{aligned}iU^\dagger(\theta,\varphi)\frac{\partial U(\theta,\varphi)}{\partial \theta} &= ie^{-i\varphi J_3}e^{i\theta J_2}e^{i\varphi J_3}\frac{\partial}{\partial\theta}e^{-i\varphi J_3}e^{-i\theta J_2}e^{i\varphi J_3} \\ &= e^{-i\varphi J_3}J_2 e^{i\varphi J_3} \\ &= (J_2\cos\varphi - J_1\sin\varphi),\end{aligned} \tag{3.33}$$

$$\begin{aligned}iU^\dagger(\theta,\varphi)\frac{\partial U(\theta,\varphi)}{\partial \varphi} &= U^\dagger(\theta,\varphi)J_3 U(\theta,\varphi) - U^\dagger(\theta,\varphi)U(\theta,\varphi)J_3 \\ &= -(J_1\cos\varphi + J_2\sin\varphi)\sin\theta + J_3(\cos\theta - 1).\end{aligned} \tag{3.34}$$

Substituting (3.33) and (3.34) in (3.31) and (3.32), we find

$$A^{k'k}_\theta(\theta,\varphi) = \langle k'; \hat{\mathbf{e}}_3|(J_2\cos\varphi - J_1\sin\varphi)|k;\hat{\mathbf{e}}_3\rangle, \tag{3.35}$$

$$\begin{aligned}A^{k'k}_\varphi(\theta,\varphi) = \langle k'; \hat{\mathbf{e}}_3|\,[-(J_1\cos\varphi + J_2\sin\varphi)\sin\theta \\ + J_3(\cos\theta - 1)]\,|k;\hat{\mathbf{e}}_3\rangle\,.\end{aligned} \tag{3.36}$$

Since J_1 and J_2 do not contribute to the diagonal matrix elements, we have

$$A^k_\theta(\theta,\varphi) = 0, \tag{3.37}$$

$$\begin{aligned}A^k_\varphi(\theta,\varphi) &= \langle k; \hat{\mathbf{e}}_3|J_3(\cos\theta - 1)|k;\hat{\mathbf{e}}_3\rangle \\ &= -k(1 - \cos\theta), \quad \theta \neq \pi.\end{aligned} \tag{3.38}$$

Next we repeat the same calculation in the patch O_2 in which the basis vectors $|k; \theta, \varphi\rangle'$ of (3.25) are single valued. This leads to

$$A'^{k'k}_\theta(\theta,\varphi) = e^{2i(k'-k)\varphi}\langle k'; \hat{\mathbf{e}}_3|(J_2\cos\varphi - J_1\sin\varphi)|k;\hat{\mathbf{e}}_3\rangle, \tag{3.39}$$

$$\begin{aligned}A'^{k'k}_\varphi(\theta,\varphi) = e^{2i(k'-k)\varphi}\langle k';\hat{\mathbf{e}}_3|\,[-(J_1\cos\varphi + J_2\sin\varphi)\sin\theta \\ + J_3(\cos\theta - 1) + 2\,k]\,|k;\hat{\mathbf{e}}_3\rangle\,.\end{aligned} \tag{3.40}$$

The diagonal matrix elements, i.e., the components of the Mead–Berry connection one-form, are then given by

$$A_\theta'^k(\theta,\varphi) = 0, \qquad (3.41)$$
$$A_\varphi'^k(\theta,\varphi) = \langle k;\hat{\mathbf{e}}_3|\,[J_3(\cos\theta-1)+2k]\,|k;\hat{\mathbf{e}}_3\rangle'$$
$$= k(\cos\theta+1), \qquad \theta \neq 0 \qquad (3.42)$$

According to the general theory of Chap. 2 we expect from (2.58) and $\zeta = -2k\varphi$,

$$A'^k - A^k = -d\zeta(\theta,\varphi) = d(2k\varphi) = 2k\,d\varphi, \qquad (3.43)$$

Comparing (3.42) with (3.38) we have indeed $A_\varphi'^k - A_\varphi^k = 2k$ in agreement with (3.43).

Having obtained the expression for the Mead–Berry connection one-form, we next compute the corresponding curvature two-form. According to (2.70), this is given by

$$F^k = dA^k = \frac{\partial A_\theta^k}{\partial \varphi}\,d\varphi \wedge d\theta + \frac{\partial A_\varphi^k}{\partial \theta}\,d\theta \wedge d\varphi. \qquad (3.44)$$

In view of (3.37) and (3.38) we then obtain

$$F^k = F_{\theta\varphi}^k\,d\theta \wedge d\varphi = -k\sin\theta\,d\theta \wedge d\varphi. \qquad (3.45)$$

We are now in a position to use (2.73) to calculate the Berry phase angle for a closed path $\mathbf{C}$,

$$\gamma_k(\mathbf{C}) = \int_S F^k = -k\int_S \sin\theta\,d\theta \wedge d\varphi = -k\int_S d\Omega \quad \text{mod } 2\pi, \qquad (3.46)$$

where S is a surface, in fact any surface, which has the closed curve $\mathbf{C}$ as its boundary ($\mathbf{C} = \partial S$), and where $d\Omega$ is the element of solid angle (area element of the unit sphere). Here we take the direction in which $\mathbf{C}$ is traversed along the fingers and the direction normal to S, i.e., the direction of $d\theta \wedge d\varphi$, along the thumb of the right hand; cf. Fig. 3.2.

Denoting by $\Omega(\mathbf{C})$ the solid angle subtended by $\mathbf{C}$, i.e.,

$$\Omega(\mathbf{C}) := \int_S \sin\theta\,d\theta \wedge d\varphi, \qquad (3.47)$$

we can write (3.46) in the form

$$\gamma_k(\mathbf{C}) = -k\,\Omega(\mathbf{C}) \quad \text{mod } 2\pi. \qquad (3.48)$$

As we explained in Sect. 2.3, unlike the single-valued basis vectors $|k;\theta,\varphi\rangle$ and the connection one-form A^k, the curvature two-form F^k is independent of the choice of local coordinates. Therefore in (3.46) we can use a surface

3.3 Mead–Berry Connection and Berry Phase for Adiabatic Evolutions

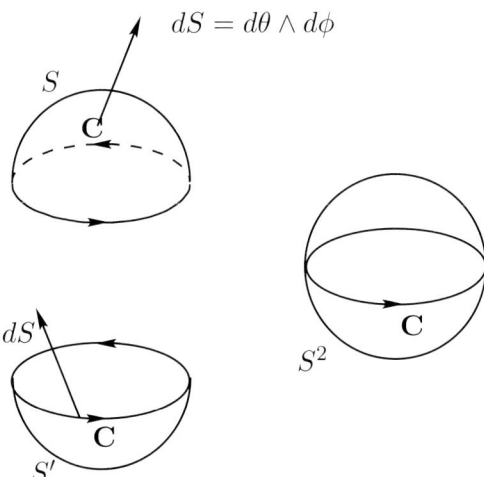

Fig. 3.2. The difference of the line integrals of A and A' can be transformed, using Stokes' theorem, into an integral over the closed 2-surface $S \cup S'$.

$S \subset O_1$ or a surface $S' \subset O_2$, as long as both S and S' have $\mathbf{C}$ as their boundaries, i.e., $\partial S = \mathbf{C} = \partial S'$. In order to investigate the consequences of this property of the curvature two-form, we next compute the Berry phase angle using S'. This yields

$$\gamma_k(\mathbf{C}) = \int_{S'} F^k = -k \int_{S'} \sin\theta \, d\theta \wedge d\varphi = k \int_{S^2 \setminus S} \sin\theta \, d\theta \wedge d\varphi, \qquad (3.49)$$

where the direction of the normal of S' and the direction in which $\mathbf{C}$ is traversed are again given by the right-hand rule. This means that the normal of S' points into the sphere and the normal of S points out of the sphere. In (3.49) we denote by $S^2 \setminus S$ the surface with the same area as S' but whose normal points out of the sphere S^2. We rewrite (3.49) in the form

$$\gamma_k(\mathbf{C}) = k \left(\int_{S^2} \sin\theta \, d\theta \wedge d\varphi - \int_S \sin\theta \, d\theta \wedge d\varphi \right).$$

As the integral of $d\Omega$ over the whole unit sphere is 4π and the integral over S is given by (3.47) we obtain

$$\gamma_k(\mathbf{C}) = +k \, (4\pi - \Omega(\mathbf{C})) \qquad \text{modulo } 2\pi. \qquad (3.50)$$

Comparing (3.50) and (3.48), we find that

$$-k\Omega(\mathbf{C}) = 4\pi k - k\Omega(\mathbf{C}) \quad \mod 2\pi, \qquad (3.51)$$

which can only be satisfied if

3. Spinning Quantum System in an External Magnetic Field

$$k = 0, \pm\frac{1}{2}, \pm 1, \pm\frac{3}{2}, \pm 2, \cdots. \tag{3.52}$$

Thus we conclude that k must be an integer or a half-integer.

If the basis vectors (3.17) and (3.25) are also eigenvectors of $\mathbf{J}^2$ then the possible values of k are already obtained from (3.12). But if these vectors are not $\mathbf{J}^2$-eigenvectors – as is the case for a molecule – then (3.52) is a new result.[3]

So far $\mathbf{C}$ could have been any closed path on the unit sphere. For the special path $\mathbf{C}_1 : [0, T] \to S^2$ given by (3.4), and shown in Fig. 3.1, namely

$$\mathbf{C}_1(t) := R(\theta(t), \varphi(t)) = R(\theta = \text{const.}, \varphi = \omega t), \tag{3.53}$$

the Berry phase (3.48) is given by

$$\gamma_k(\mathbf{C}_1) = -k \int_0^{2\pi} \int_0^\theta \sin\theta' d\theta' d\varphi = -2\pi k(1 - \cos\theta), \tag{3.54}$$

where θ is the constant angle shown in Fig. 3.1.

In the remainder of this section we use the fact that S^2 is a submanifold of $\mathbb{R}^3$ to re-express our results in terms of the Mead–Berry (three-)vector potential $\mathbf{A}^k$ (2.54) and the curvature (three-)vector or the field strength $\mathbf{F}^k$ (2.78). To compute $\mathbf{A}^k$, we may view the connection one-form A^k as a one-form in $\mathbb{R}^3$, i.e.,

$$A^k = A_r^k\, dr + A_\theta^k\, d\theta + A_\varphi^k\, d\varphi,$$

where A_θ^k and A_φ^k are given by (3.37) and (3.38) and $A_r^k = 0$. Next we write A^k in a new (Cartesian) basis $(dr, r\, d\theta, r\sin\theta\, d\varphi)$:

$$A^k = \hat{A}_r^k\, dr + \hat{A}_\theta^k\, r\, d\theta + \hat{A}_\varphi^k\, r\sin\theta\, d\varphi.$$

The new components $\hat{A}_r^k = A_r^k = 0$, $\hat{A}_\theta^k = 0$, and $\hat{A}_\varphi^k$ are called the spherical components of A^k. They yield the components of the Mead–Berry vector potential in the spherical coordinates,

$$\mathbf{A}^k = \hat{A}_r^k\, \hat{\mathbf{e}}_r + \hat{A}_\theta^k\, \hat{\mathbf{e}}_\theta + \hat{A}_\varphi^k\, \hat{\mathbf{e}}_\varphi, \tag{3.55}$$

where $\hat{\mathbf{e}}_r$, $\hat{\mathbf{e}}_\theta$, and $\hat{\mathbf{e}}_\varphi$ are the unit vectors in $\mathbb{R}^3$ along the r, θ, and φ directions, respectively. Alternatively, we could have used the defining relation (2.54) to compute

$$\mathbf{A}^k = i\langle k; \theta, \varphi | \nabla | k; \theta, \varphi \rangle \tag{3.56}$$

$$= \langle k; \hat{\mathbf{e}}_3 | iU^\dagger(\theta, \varphi) \left(\hat{\mathbf{e}}_r \frac{\partial}{\partial r} + \hat{\mathbf{e}}_\theta \frac{1}{r} \frac{\partial}{\partial \theta} + \hat{\mathbf{e}}_\varphi \frac{1}{\sin\theta} \frac{\partial}{\partial \varphi} \right) U(\theta, \varphi) | k; \hat{\mathbf{e}}_3 \rangle.$$

---

[3] See Chap. 6 for a discussion of the topological interpretation of (3.52).

3.3 Mead–Berry Connection and Berry Phase for Adiabatic Evolutions

Both approaches yield the same result, namely

$$\mathbf{A}^k(\theta, \varphi) = \frac{k(\cos\theta - 1)}{r\sin\theta} \hat{\mathbf{e}}_\varphi, \quad \theta \neq \pi. \tag{3.57}$$

Repeating the same computations in the coordinate patch O_2 (which includes the south pole), we find

$$\mathbf{A}'^k(\theta, \varphi) = \frac{k(\cos\theta + 1)}{r\sin\theta} \hat{\mathbf{e}}_\varphi, \quad \theta \neq 0. \tag{3.58}$$

$\mathbf{A}'^k$ is related to $\mathbf{A}^k$ by the gauge transformation

$$\mathbf{A}'^k - \mathbf{A}^k = -\nabla\zeta = \frac{2k}{r\sin\theta} \hat{\mathbf{e}}_\varphi, \tag{3.59}$$

where $\zeta = -2k\varphi$.

More interestingly, we can compute the curvature three-vector (2.78),

$$\mathbf{F}^k = -\frac{k}{r^2} \hat{\mathbf{e}}_r = -\frac{k}{r^2} \hat{\mathbf{R}}(\theta, \varphi). \tag{3.60}$$

This can be obtained either by taking the curl of the vector potential or using the identity

$$F^k = F^k_{\theta\varphi} d\theta \wedge d\varphi = -k\sin\theta d\theta \wedge d\varphi = \hat{F}^k_r (rd\theta)(r\sin\theta d\varphi) = \mathbf{F}^k \cdot d\mathbf{S},$$

to read off $\mathbf{F}^k$ directly.

As seen from (3.60), the curvature three-vector (field strength) is directed along the radial direction $\hat{\mathbf{e}}_r = \hat{\mathbf{R}}(\theta, \varphi)$. It has a very familiar form, namely the form of the magnetic field of a monopole.

An imaginary electromagnetic system, with the vector potential given by (3.57) or (3.58) and the magnetic field given by (3.60), had been envisioned by Dirac about 70 years ago [74] and has since intrigued several generations of theoreticians and experimentalists [28, 35, 68, 87]. It consists of an electric charge e and a magnetic monopole with magnetic monopole strength g with the vector $\hat{\mathbf{R}}$ pointing from g to e. The magnetic field $\mathbf{B}^{\text{mon}}$ of this charge–monopole system is given by

$$e\mathbf{B}^{\text{mon}} = e\frac{g}{4\pi} \frac{\hat{\mathbf{R}}}{r^2}$$

and the electromagnetic vector potential $e\mathbf{A}^{\text{mon}}$ is given by (3.57) and (3.58) with $k := -\frac{eg}{4\pi}$. For this magnetic monopole the field and the vector potential are supposed to be electromagnetic.

If such a magnetic monopole with magnetic charge g exists, then all electric charges must be multiples of $2\pi/g$ because k is integer or half-integer according to (3.52). This would be a remarkable explanation of one of the basic facts in nature. Unfortunately, in spite of all the efforts, this electromagnetic type of monopole has never been found.

The field strength $\mathbf{F}^k$ of (3.60) and the vector potential $\mathbf{A}^k$ of (3.57) and (3.58) are not electromagnetic in nature. They have their origin in a quantum system spinning with fixed angular momentum component k around the axis $\hat{\mathbf{R}}(t)$. The analogy between classical mechanical systems of this type and the charge–monopole system has been known for some time [153]. The question is whether these vector potentials, which are induced by the rapid rotation about the slowly changing axis $\hat{\mathbf{R}}(t)$, lead to any physical consequences for the environment. We will see in Chap. 8 that when the classical environment (represented by the external magnetic field $\mathbf{B}(t)$) is replaced by a quantum environment, then the dynamics of the quantum system will be modified by these vector potentials in the same way as the motion of a charged particle is modified by the electromagnetic monopole potential.[4]

3.4 The Exact Solution of the Schrödinger Equation

In the preceding section we employed the results of Sect. 2.3 to compute the Berry phase. Hence we have implicitly used the adiabatic approximation in our computations. In this section we shall explore the exact time-evolution of the state vector $\psi(t)$ which satisfies the Schrödinger equation

$$i\frac{\partial}{\partial t}\psi(t) = h(t)\psi(t), \quad \text{with} \quad h(t) = b\hat{\mathbf{R}}(\theta,\omega t) \cdot \mathbf{J}. \tag{3.61}$$

In particular we are interested in the solutions of (3.61) which describe cyclic evolutions (2.36):

$$|\psi(\tau)\rangle\langle\psi(\tau)| = |\psi(0)\rangle\langle\psi(0)|. \tag{3.62}$$

The period τ of the pure cyclic state (3.62) can in general be any time period. In the adiabatic approximation, however, it is given by the period of the precession of the magnetic field, $\tau = T = \frac{2\pi}{\omega}$.

We shall first consider the case where the adiabatic approximation is valid. Then substituting (3.54) and (3.11) in (2.57) we obtain the expression for the evolving state vector,

$$\psi(t) \stackrel{\text{adiabatic}}{=} e^{-ibtk}\, e^{i\gamma_k(t)}\, |k;\,\theta,\,\omega t\rangle, \tag{3.63}$$

$$\omega = \frac{2\pi}{T}, \quad \gamma_k(t) = \frac{t}{T}\gamma_k(\mathbf{C}).$$

In particular for $t = T$, we obtain from (2.66)

$$\psi(T) \stackrel{\text{adiabatic}}{=} e^{-i2\pi\frac{b}{\omega}k}\, e^{i\gamma_k(\mathbf{C})}\psi(0), \quad \psi(0) = |k;\theta,\,0\rangle. \tag{3.64}$$

---

[4] See Chap. 8 and [33, 41, 45, 116, 152, 166, 174, 207, 287] for more details.

3.4 The Exact Solution of the Schrödinger Equation

These expressions for the state vector are, as we showed in general in Sect. 2.2, incompatible with the Schrödinger equation (3.61) and can only be approximations. For the example considered here, (3.61) can be exactly solved. In particular the cyclic solutions are known.

In order to find the exact solution of the Schrödinger equation (3.61), we proceed in the following way. The magnetic field (3.4) rotates counterclockwise by an angle ωt about the $\mathbf{e}_3$ axis. Instead of rotating the observables (Hamiltonian, etc.) counterclockwise and keeping the state vector the same (in the laboratory frame) we can rotate the state vector $\psi(t)$ clockwise (by an angle $-\omega t$) about the $\mathbf{e}_3$ axis and keep the observables the same (view the magnetic field from a frame which rotates with $\mathbf{e}_{12}(t)$). For the observable quantities (expectation values) these two points of view give the same result, e.g.,

$$\begin{aligned}\langle \psi(t)|h(t)|\psi(t)\rangle &= \langle \psi(t)|e^{-i\omega t J_3} h_0\, e^{i\omega t J_3}|\psi(t)\rangle \\ &= \langle e^{i\omega t J_3}\psi(t)|h_0|e^{i\omega t J_3}\psi(t)\rangle.\end{aligned} \quad (3.65)$$

This suggests the following unitary transformation of the state vector $\psi(t)$.

$$\psi(t) \to \psi'(t) := \tilde{U}(t)\psi(t) = e^{i\omega t J_3}\psi(t), \quad (3.66)$$

where

$$\tilde{U}(t) := U_3^\dagger(\omega t) = e^{i\omega t J_3}.$$

Inserting $\psi(t) = \tilde{U}^\dagger \psi'(t)$ into the Schrödinger equation (3.61) for $\psi(t)$, we obtain the Schrödinger equation for $\psi'(t)$,

$$i\frac{\partial}{\partial t}\psi'(t) = \left(-\omega J_3 + e^{i\omega t J_3} h(t) e^{-i\omega t J_3}\right)\psi'(t) =: h'(t)\psi'(t). \quad (3.67)$$

The state vector $\psi'(t)$ evolves in time according to the transformed Hamiltonian

$$h'(t) := \tilde{U} h \tilde{U}^\dagger - i\hbar \tilde{U}\frac{\partial \tilde{U}^\dagger}{\partial t} = e^{i\omega t J_3} h(t) e^{-i\omega t J_3} - \omega J_3. \quad (3.68)$$

However, in view of (3.8), the operator h' does not depend on time, for

$$h' = h_0 - \omega J_3 = b\left(\cos\theta J_3 + \sin\theta J_1 - \frac{\omega}{b} J_3\right) =: H. \quad (3.69)$$

This was to be expected since in the rotating frame $\mathbf{B}$ does not change. Because $h' = H$ is time-independent, (3.67) can be immediately integrated. The result is

$$\psi'(t) = e^{-itH}\psi'(0). \quad (3.70)$$

The operator H can also be related to the angular momentum operator (Problem 3.4),

$$H = \Omega \mathbf{e} \cdot \mathbf{J}, \quad (3.71)$$

where
$$\mathbf{e} := \frac{b}{\Omega}\left(\cos\theta - \frac{\omega}{b}\right)\mathbf{e}_3 + \frac{b}{\Omega}\sin\theta\,\mathbf{e}_1, \tag{3.72}$$

$$\Omega := b\sqrt{1 + \frac{\omega}{b}(-2\cos\theta + \frac{\omega}{b})}. \tag{3.73}$$

The unit vector $\mathbf{e}$ lies in the 1-3 plane. Therefore it can be written in the form
$$\mathbf{e} = \cos\tilde{\theta}\,\mathbf{e}_3 + \sin\tilde{\theta}\,\mathbf{e}_1 = \hat{\mathbf{R}}(\tilde{\theta},0), \tag{3.74}$$

where $\tilde{\theta}$ is the angle between $\mathbf{e}$ and $\mathbf{e}_3$, i.e.
$$\cos\tilde{\theta} := \frac{b}{\Omega}\left(\cos\theta - \frac{\omega}{b}\right) = \frac{\cos\theta - \nu}{\sqrt{1 - 2\nu\cos\theta + \nu^2}}, \tag{3.75}$$

$$\sin\tilde{\theta} = \frac{b}{\Omega}\sin\theta = \frac{\sin\theta}{\sqrt{1 - 2\nu\cos\theta + \nu^2}},$$

and ν is a dimensionless parameter given by
$$\nu := \frac{\omega}{b}. \tag{3.76}$$

In view of (3.71), the time development given by (3.70) represents a 'rotation' of the state vector $\psi'(0)$ by an angle Ωt about the $\mathbf{e}$-axis[5],
$$\psi'(t) = e^{-i\Omega t\,\mathbf{e}\cdot\mathbf{J}}\psi'(0). \tag{3.78}$$

---
[5] In the same way
$$\psi(t) = e^{-itbJ_3}\psi(0) \tag{3.77}$$
means a rotation of the state vector $\psi(0)$ by an angle bt about $\mathbf{e}_3$. In order to understand this statement let us assume that the magnetic field is kept constant $\omega = 0$ and $\psi(t)$ represents a pure state for which at $t = 0$ all magnetic moments are aligned along $\mathbf{e}_1 = \hat{\mathbf{R}}(\frac{\pi}{2},0)$. Clearly, the time evolution of this state is given by (3.77). For this state the probability of measuring the observable $|k;\frac{\pi}{2},0\rangle\langle k;\frac{\pi}{2},0|$ at $t = 0$ is one,
$$\mathcal{P}_k^{\mathbf{e}_1}(t=0) = \langle\psi(0)|k;\frac{\pi}{2},0\rangle\langle k;\frac{\pi}{2},0|\psi(0)\rangle = 1.$$

Let us now compare this with the probability that for this state all magnetic moments are aligned along $\hat{\mathbf{R}}(\frac{\pi}{2},bt)$, at time t. In view of (3.17) and (3.77), this probability is given by
$$\mathcal{P}_k^{\hat{\mathbf{R}}(\frac{\pi}{2},bt)}(t) = |\langle\psi(t)|k;\frac{\pi}{2},bt\rangle|^2$$
$$= |(\langle\psi(0)|e^{itbJ_3})(e^{-itbJ_3}|k;\frac{\pi}{2},0\rangle)|^2 = \mathcal{P}_k^{\mathbf{e}_1}(t=0) = 1.$$

This means that the state $\psi(0) = |k;\frac{\pi}{2},0\rangle$ with all the magnetic moments aligned along $\mathbf{e}_1 = \hat{\mathbf{R}}(\frac{\pi}{2},0)$ has developed into a state $\psi(t) = |k;\frac{\pi}{2},bt\rangle$ with all magnetic moments rotated by the angle $\varphi = bt$.

3.4 The Exact Solution of the Schrödinger Equation

Having obtained the exact expression for $\psi'(t)$ we now transform back to the laboratory frame. Using (3.66) and (3.70), we find

$$\psi(t) = e^{-i\omega t J_3} e^{-itH} \psi(0). \tag{3.79}$$

Therefore the solution of the Schrödinger equation (3.61) for arbitrary initial state vector $\psi(0)$ is given by

$$\psi(t) = U^\dagger(t)\psi(0) = e^{-i\omega t J_3} e^{-i\Omega t \mathbf{e} \cdot \mathbf{J}} \psi(0). \tag{3.80}$$

This means that the evolving state $\psi(t)$ is obtained from $\psi(0)$ by transforming $\psi(0)$ by a unitary operator representing a rotation by Ωt along $\mathbf{e}$ followed by a unitary transformation representing a rotation by ωt along $\mathbf{e}_3$.

The relation between (θ, φ) and $(\tilde{\theta}, \varphi)$, i.e., (3.75), involves the physical parameter ν. This relation can be expressed in terms of a function $F_\nu : M = S^2 \to S^2$ which (on the patch O_1 including the north pole) is given by

$$F_\nu(\hat{\mathbf{R}}(\theta, \varphi)) = F_\nu(\theta, \varphi) = \left(\cos^{-1}\left(\frac{\cos\theta - \nu}{\sqrt{\nu^2 - 2\cos\theta\nu + 1}}\right), \varphi\right)$$
$$= (\tilde{\theta}, \varphi) = \hat{\mathbf{R}}(\tilde{\theta}, \varphi). \tag{3.81}$$

A global expression for F_ν may be obtained by viewing the sphere S^2 as embedded in $\mathbb{R}^3$. It is given by

$$F_\nu(x^1, x^2, x^3) = \frac{(x^1, x^2, x^3 - \nu)}{\sqrt{(x^1)^2 + (x^2)^2 + (x^3 - \nu)^2}}, \tag{3.82}$$

where $(x^1, x^2, x^3) \in \mathbb{R}^3$. For $\nu < 1$, the function F_ν is a smooth diffeomorphism[6] of S^2 and consists of a translation in the $\mathbf{e}_3$ direction followed by a projection along $\mathbf{R}$ onto S^2.

Let us next examine the condition (2.45) of the validity of the adiabatic approximation. It is not difficult to see that for our example (3.1), the left-hand side of (2.45) is proportional to the frequency ω of the precession of the magnetic field. Since the intrinsic frequency scale of the system is given by b, the adiabatic approximation is valid if and only if $\nu = \omega/b \ll 1$.

The frequency b is the angular velocity with which the state vector $\psi(t)$ rotates about the direction of the magnetic field $\hat{\mathbf{R}}(\theta, 0)$. For $\omega := \frac{2\pi}{T} = 0$ ($\nu = 0$),

$$\psi(t) = e^{-itb\hat{\mathbf{R}}(\theta,0)\cdot\mathbf{J}} \psi(0). \tag{3.83}$$

This is most evident if we use (3.72) and (3.76) to rewrite (3.80) in the form

$$\psi(t) = e^{-it\omega J_3} e^{-it\Omega \hat{\mathbf{R}}(\tilde{\theta},0)\cdot\mathbf{J}} \psi(0), \tag{3.84}$$
$$= e^{-itb\nu J_3} e^{-itb(\mathbf{R}(\theta,0)\cdot\mathbf{J} - \nu J_3)} \psi(0), \tag{3.85}$$

---

[6] For $\nu = 1$, F_ν becomes ill defined at the poles. We shall re-examine the function F_ν in detail in Chap. 6 and discuss the meaning of its behavior for $\nu = 1$.

and then specialize to the case $\nu = 0$. Equation (3.83) gives the time-evolution of a state vector $\psi(0)$ for the time-independent Hamiltonian

$$h = b\hat{\mathbf{R}}(\theta, 0) \cdot \mathbf{J}, \qquad \theta = \text{constant}. \tag{3.86}$$

Having obtained the exact solution (3.84) of the Schrödinger equation (3.61) we can next identify the cyclic solutions, i.e., the solutions which satisfy (3.62). By definition the cyclic solutions with period τ are obtained by choosing the initial state vector $\psi(0)$ to be an eigenvector of the evolution operator

$$U^\dagger(\tau) = e^{-i\tau\omega J_3} e^{-i\tau\Omega \hat{\mathbf{R}}(\tilde{\theta},0)\cdot \mathbf{J}}. \tag{3.87}$$

The corresponding eigenvalue $e^{-i\alpha_\psi}$ is the total phase factor, for

$$\psi(\tau) = U^\dagger(\tau)\psi(0) = e^{-i\alpha_\psi}\psi(0). \tag{3.88}$$

In general, in order to find the cyclic states for arbitrary τ, one must find the (non-stationary) eigenstates of the evolution operator (3.87). This is quite easy if τ is chosen in such a way that the initial state is an eigenstate of both of the operators appearing on the right-hand side of (3.87). In this case

$$\left[e^{-i\tau\omega J_3},\ e^{-i\tau\Omega\hat{\mathbf{R}}(\tilde{\theta},0)\cdot \mathbf{J}} \right] |\psi(0)\rangle = 0. \tag{3.89}$$

It is not difficult to see that this condition is satisfied for the following values of τ:

$$\text{CLASS A}: \quad \tau = \frac{2\pi}{\omega} = T, \tag{3.90}$$

$$\text{CLASS B}: \quad \tau = \frac{2\pi}{\omega}. \tag{3.91}$$

For both classes there is a discrete set of cyclic solutions. The class A cyclic solutions have the same period as the Hamiltonian and the adiabatic evolution. The Class B cyclic solutions have a period which is given by the intrinsic properties of the quantum system (the Larmor frequency). A particular case of the class B cyclic solutions are the cyclic solutions of the time-independent Hamiltonian (3.86) which satisfy (3.83). The class A and B solutions are not however the only possible cyclic solutions of the Schrödinger equation (3.61). They are merely typical special cases.

In the remainder of our analysis of system (3.1), we shall only discuss the Class A cyclic evolutions. This is mainly because they are directly related with the adiabatic cyclic evolutions. We leave the analysis of the Class B solutions as an exercise (Problem 3.5).

The Class A cyclic solutions are obtained by finding the simultaneous eigenvectors of the operators $e^{-i2\pi J_3}$, $e^{-iT\Omega\hat{\mathbf{R}}(\tilde{\theta},0)\cdot \mathbf{J}}$, and therefore the evolution operator $U^\dagger_A(T) = e^{-i2\pi J_3}e^{-iT\Omega\hat{\mathbf{R}}(\tilde{\theta},0)\cdot \mathbf{J}}$. These eigenvectors which we shall denote by ϕ_k are consequently also the eigenvectors of the operator

3.4 The Exact Solution of the Schrödinger Equation

$$h(\tilde{\theta}, 0) := b\,\hat{\mathbf{R}}(\tilde{\theta}, 0) \cdot \mathbf{J}, \tag{3.92}$$

i.e., they satisfy

$$h(\tilde{\theta}, 0)\,|\phi_k\rangle = bk\,|\phi_k\rangle; \qquad \mathbf{e} \cdot \mathbf{J}\,|\phi_k\rangle = k\,|\phi_k\rangle. \tag{3.93}$$

From (3.17) and (3.18) with θ replaced by $\tilde{\theta}$ and φ by 0 it follows that ϕ_κ is (up to a phase factor) given by[7]

$$|\phi_k\rangle = |k;\tilde{\theta}, 0\rangle = U(\tilde{\theta}, 0)|k;\mathbf{e}_3\rangle = e^{-i\tilde{\theta} J_2}|k;\mathbf{e}_3\rangle. \tag{3.94}$$

To obtain eigenvalues of $e^{-i2\pi J_3}$ we proceed using (3.21) with $\varphi = 2\pi$. This yield

$$\begin{aligned} e^{-i2\pi J_3}\phi_k &= e^{-i2\pi J_3}|k;\tilde{\theta},0\rangle = e^{-i2\pi J_3}e^{-i\tilde{\theta} J_2}\,|k;0,0\rangle \\ &= e^{-i\tilde{\theta} J_2}e^{-i2\pi J_3}\,|k;0,0\rangle \\ &= e^{-i2\pi k}|k;\tilde{\theta},0\rangle. \end{aligned} \tag{3.95}$$

The initial state vectors which lead to a cyclic evolution with period $T = \frac{2\pi}{\omega}$ are thus labeled by integers or half-integers k and given by (3.94),

$$|\psi(0)\rangle = |\phi_k\rangle = |k;\tilde{\theta},0\rangle, \tag{3.96}$$

(after an arbitrary phase factor has been fixed). The total phase of these cyclic evolutions, i.e., the eigenvalues $e^{-i\alpha_\psi}$ of (3.88) are also labeled by k and given by

$$\psi(T) = U^\dagger(T)\psi(0) = e^{-i\alpha_k}\psi(0) = e^{-i2\pi k}e^{-i2\pi\frac{\omega}{\Omega}k}\psi(0). \tag{3.97}$$

Thus, in this case, the possible initial states of a cyclic evolution are very similar to the initial states of an adiabatic evolution. The only difference is that the initial states of an adiabatic evolution are given by the eigenvectors $|k;\theta,0\rangle$ of the initial Hamiltonian $h(R(0)) = b\hat{\mathbf{R}}(\theta,0) \cdot \mathbf{J}$, whereas the initial states of the exact cyclic evolution (3.90) are given by the eigenvectors (3.94) of the operator

$$\tilde{h}(\theta, 0) := h(\tilde{\theta}, 0) = b\,\hat{\mathbf{R}}(\tilde{\theta},0) \cdot \mathbf{J}, \tag{3.98}$$

which is different from $h(R(0))$.

The exact cyclic states are the states with the component k of angular momentum along the direction $\mathbf{e} = \hat{\mathbf{R}}(\tilde{\theta},0)$ rather than $\hat{\mathbf{R}}(\theta,0)$. For large values of ω ($\omega \approx b$) these two directions can be very different, but for $\nu = \frac{\omega}{b} \to 0$ they coincide.

We will next determine the analog of the single-valued eigenvectors $|k; R(t)\rangle = |k; \theta(t), \varphi(t)\rangle$ which are used in the calculation of the Mead–Berry connection one-form and Berry phase. These vectors $\phi_k(t)$ lie on a curve "above" the closed curve

---

[7] We assume that ϕ_κ is normalized.

3. Spinning Quantum System in an External Magnetic Field

$$\mathcal{C}: t \to |\psi(t)\rangle\langle\psi(t)|; \qquad |\psi(T)\rangle\langle\psi(T)| = |\psi(0)\rangle\langle\psi(0)|. \tag{3.99}$$

in the state space $\mathcal{P}(\mathcal{H})$. This means that they fulfill

$$\phi_k(t) = \text{phase factor} \times \psi(t), \quad \text{and} \quad \phi_k(T) = \phi_k(0). \tag{3.100}$$

There are many such vectors all differing by a phase transformation,

$$\phi_k(t) \to \phi'_k(t) = e^{i\zeta_k(t)}\phi_k(t), \quad \zeta_k(0) = \zeta_k(T) \quad \text{mod } 2\pi. \tag{3.101}$$

One and by far the most obvious choice for $\phi_k(t)$ is

$$\begin{aligned}\phi_k(t) &:= U(\tilde{\theta}, \omega t)|k; \mathbf{e}_3\rangle \\ &= e^{-i\omega t J_3} e^{-i\tilde{\theta} J_2} e^{i\omega t k}|k, \mathbf{e}_3\rangle =: |k; \tilde{\theta}, \omega t\rangle.\end{aligned} \tag{3.102}$$

These single-valued basis vectors $|k; \tilde{\theta}, \varphi = \omega t\rangle$ are eigenvectors of the operator

$$\tilde{h}(\theta, \varphi) := h(\tilde{\theta}, \varphi) = h(F_\nu(\theta, \varphi)) = b\,\hat{\mathbf{R}}(\tilde{\theta}, \varphi) \cdot \mathbf{J}, \tag{3.103}$$

where F_ν is the function given by (3.81). The operator $\tilde{h}(R) = h(F_\nu(R))$ is not the Hamiltonian but another parameter-dependent operator which can as well serve to define an orthonormal basis of the Hilbert space $\mathcal{H}$. In terms of these known vectors $|k; \tilde{\theta}, \varphi\rangle$, the cyclic solution of the Schrödinger equation (3.61) is given by

$$\psi(t) = e^{-i\omega t k} e^{-i\Omega t k}|k; \tilde{\theta}, \omega t\rangle. \tag{3.104}$$

This relation expresses the cyclically evolving state vector $\psi(t)$ in terms of the known quantities ω, Ω and $|k; \tilde{\theta}, \omega t\rangle$. It is the generalization of the "adiabatic equality" (3.63). Similarly, (3.97) is the generalization of (3.64). However the phase factor in (3.104) and (3.97) is not in the form of a product of the dynamical and the geometrical parts. Thus, in order to obtain an expression for each part, we must calculate at least one of them independently.

3.5 Dynamical and Geometrical Phase Factors for Non-Adiabatic Evolution

The splitting of the phase factor into a *dynamical* and a *geometrical* part can be performed in two different ways. Either one gives an argument why a certain part of the total phase angle is geometrical and obtains the dynamical part as the difference between the total and the geometrical part, or one defines the dynamical part and obtains the geometrical part as the difference between the total and the dynamical part. We will pursue the latter approach, namely define the dynamical phase angle α_k^{dyn} and obtain the geometrical phase angle α_k^{geom} as a derived quantity,

$$\alpha_k^{\text{geom}} = \alpha_k - \alpha_k^{\text{dyn}}. \tag{3.105}$$

3.5 Non-Adiabatic Geometric Phase

The dynamical phase (angle) for general cyclic evolution was defined by Aharonov and Anandan according to [7, 9]

$$\alpha_k^{\text{dyn}} := \int_0^T \langle \psi(t)|h(t)|\psi(t)\rangle dt. \tag{3.106}$$

This is a reasonable choice, because one can show that the phase angle for the evolution of a stationary state (2.30) is given by (3.106). Furthermore, for an adiabatic cyclic evolution the right-hand side of (3.106) yields the expression for the dynamical phase angle (2.48).

We can easily calculate the dynamical phase (3.106) for the Hamiltonian (3.1),

$$\alpha_k^{\text{dyn}} = k 2\pi \left(\frac{\omega}{\Omega} + \cos\tilde{\theta} \right). \tag{3.107}$$

Using (3.105) and (3.97), we then find

$$\alpha_k^{\text{geom}} = k 2\pi (1 - \cos\tilde{\theta}) = k 2\pi \left(1 + \frac{\omega}{\Omega} - \frac{b}{\Omega} \cos\theta \right). \tag{3.108}$$

This has the same form as (3.54) for the adiabatic case except that θ is replaced by $\tilde{\theta}$ of (3.75).

In the adiabatic approximation, $\nu = \frac{\omega}{b} \ll 1$, we can expand the expressions in (3.73) and (3.75) with respect to ν. Up to the first order in ν, we have

$$\frac{\omega}{b} \approx 1 - \nu \cos\theta,$$

$$\sin\tilde{\theta} \approx \sin\theta + \frac{\nu}{2} \sin 2\theta,$$

$$\cos\tilde{\theta} \approx \cos\theta + \frac{\nu}{2} (\cos 2\theta - 1).$$

From this we conclude that the adiabatic approximation of the geometrical phase angle for general cyclic evolution (3.108) is identical with the Berry phase angle,

$$-\alpha_k^{\text{geom}} \approx -2\pi k(1 - \cos\theta) = \gamma_k^{\text{Berry}}. \tag{3.109}$$

These results justify the choice of (3.105) as a definition of the geometrical phase for the general cyclic evolution.

Next, we would like to show that the non-adiabatic geometric phase angle can be obtained from a connection one-from in the same way the Berry phase angle is obtained from the Mead–Berry connection one-form.

In analogy to (2.54) and (2.55), we define the following connection one-form

$$A^{\phi_k} := i \langle \phi_k | d | \phi_k \rangle, \tag{3.110}$$

and compute

3. Spinning Quantum System in an External Magnetic Field

$$\begin{aligned} A^{\phi_k} &= \langle k;\tilde{\theta},\varphi|d|k;\tilde{\theta},\varphi\rangle \\ &= \langle k;\tilde{\theta},\varphi|\frac{\partial}{\partial\tilde{\theta}}|k;\tilde{\theta},\varphi\rangle\, d\tilde{\theta} + \langle k;\tilde{\theta},\varphi|\frac{\partial}{\partial\varphi}|k;\tilde{\theta},\varphi\rangle\, d\varphi \quad (3.111) \\ &= \langle k;\tilde{\theta},\varphi|\frac{\partial}{\partial\tilde{\theta}}|k;\tilde{\theta},\varphi\rangle\frac{\partial\tilde{\theta}}{\partial\theta}\, d\theta + \langle k;\tilde{\theta},\varphi|\frac{\partial}{\partial\varphi}|k;\tilde{\theta},\varphi\rangle\, d\varphi. \end{aligned}$$

Here we have used in place of the single-valued eigenvectors of $h(R)$ of (2.55) the basis vectors given by (3.102), i.e., the eigenvectors of the operator $\tilde{h}(R)$. Comparing (3.111) with (3.29) and using (3.37) and (3.38), we immediately conclude

$$A^{\phi_k} = -k(1 - \cos\tilde{\theta})\, d\varphi = -k\left(1 - \frac{b}{\Omega}\cos\theta + \frac{\omega}{\Omega}\right) d\varphi. \quad (3.112)$$

The curvature two-form (field strength) that follows from this connection one-from is given by

$$F^{\phi_k} = dA = -k\frac{\left(1 - \frac{\omega}{b}\cos\theta\right)}{\left(\frac{\omega}{b}\right)^3}\sin\theta\, d\theta \wedge d\varphi.$$

The component

$$F^{\phi_k}_{\theta\varphi} = -k\frac{1 - \frac{\omega}{b}\cos\theta}{\left[1 - \frac{\omega}{b}\cos\theta + \left(\frac{\omega}{b}\right)^2\right]^{3/2}}\sin\theta$$

of F^{ϕ_k} can be used to define the field strength (three-)vector

$$\mathbf{F}^{\phi_k} = -\frac{k}{r^2}\frac{\left(1 - \frac{\omega}{b}\cos\theta\right)}{\left(\frac{\omega}{b}\right)^3}\hat{\mathbf{R}}(\theta,\varphi). \quad (3.113)$$

This is again a "monopole" type field similar to (3.60) but with a modified "monopole strength".

The connection one-form (3.112) has the same form as (3.38) for the Mead–Berry connection, except that here it depends on the angle $\tilde{\theta}$ which differs from the angle θ.

We can now use the analog of (2.65) and (3.46) for the adiabatic approximation and define a *geometric phase* angle also for the general cyclic evolution by

$$\gamma_k := \oint_C A^{\phi_k} = \oint i\langle\phi_k|d|\phi_k\rangle. \quad (3.114)$$

Then using (3.112), we obtain

$$\begin{aligned} \gamma_k &= -k\int_0^T (1 - \cos\tilde{\theta})\omega\, dt \\ &= -2\pi k(1 - \cos\tilde{\theta}) \\ &= -2\pi k\left(1 - \frac{b}{\Omega}\cos\theta + \frac{\omega}{\Omega}\right). \quad (3.115) \end{aligned}$$

3.5 Non-Adiabatic Geometric Phase

This agrees with the result (3.108) obtained from (3.105) and (3.106).

With these results we can rewrite the cyclic solution (3.104) of the Schrödinger equation in a form which completely resembles the adiabatic approximation, i.e., (3.63),

$$\psi(t) = e^{-i\alpha_k^{\mathrm{dyn}}(t)} e^{i\gamma_k(t)} |k; \tilde{\theta}, \varphi\rangle, \qquad (3.116)$$

where

$$\alpha_k^{\mathrm{dyn}}(t) := \int_0^t \langle \psi(t')|h(t')|\psi(t')\rangle dt' = \omega t k \left(\frac{\omega}{\Omega} + \cos\tilde{\theta}\right)$$

$$= \frac{t\alpha_k}{T},$$

$$\gamma_k(t) := \int_0^t i\langle \phi_k(t)|\frac{d}{dt}|\phi_k(t)\rangle dt$$

$$= -k\int_0^t (1 - \cos\tilde{\theta})\omega dt' = -\omega t k(1 - \cos\tilde{\theta})$$

$$= \frac{t\gamma_k}{T}.$$

The concepts introduced in this section for the cyclic evolution, (3.99), are the analogs of those of the adiabatic approximation. The single-valued basis vectors $|\phi_k\rangle$ are the analogs of the eigenvectors $|k; \theta, \varphi\rangle$ of the Hamiltonian. The connections A^{ϕ_k} of (3.112) are generalizations of the Mead–Berry connections (2.55). The geometric phase (3.114) is a generalization of the Berry phase (2.63).

The distinction between (3.116) and (3.63) is that in (3.116) we have an exact equality not an approximate one. This means that the pure state corresponding to $\psi(t)$ has indeed an exact cyclic evolution. If the frequency of the precession of the magnetic field ω is "small" enough, i.e., much smaller than the frequency b, then the curve (3.63) is "close enough" to (3.116), in order to provide an acceptable approximation for the geometric phase. Nevertheless an exact adiabatic cyclic evolution does not exist.

As we mentioned above, in addition to the cyclic evolutions of a pure state with period $\tau = \frac{2\pi}{\omega} = T$, there exist other cyclic evolutions with different periods. An example of this is the Class B solutions which correspond to the period $\tau = \frac{2\pi}{\omega}$ with Ω given by (3.73). A special case of class B cyclic evolutions is the cyclic evolutions corresponding to a time-independent Hamiltonian. These are obtained by setting $\omega = 0$, so that $\Omega = b$ and $\tau = \frac{2\pi}{b}$. This shows that indeed a time-independent Hamiltonian may also lead to non-trivial cyclic evolutions.

The geometrical phase (3.115) is not purely geometrical in the way that the Berry phase (3.109) is. Unlike the Berry phase that only depends on θ and is solely given by the path in the parameter space, the geometrical phase, for general cyclic evolution, also depends on the parameter $\nu = \omega/b$ of the Hamiltonian. To fully justify the name geometric phase also for the phase (3.115) of a non-adiabatic cyclic evolution, we must explore the mathematical structure of the state space $\mathcal{P}(\mathcal{H})$. We shall pursue this in the next four chapters.

Problems

3.1) Show that the rotation $\mathcal{R}(\theta, \varphi)$ defined by (3.13) transforms the vector $\mathbf{e}_3$ of (3.3) into the vector $\hat{\mathbf{R}}(\theta, \varphi)$ of (3.2). Use the standard representation for $\mathcal{R}_3(\varphi)$ and $\mathcal{R}_2(\theta)$, i.e.,

$$\mathcal{R}_3(\varphi) = \begin{pmatrix} \cos\varphi & -\sin\varphi & 0 \\ \sin\varphi & \cos\varphi & 0 \\ 0 & 0 & 1 \end{pmatrix}$$

$$\mathcal{R}_2(\theta) = \begin{pmatrix} \cos\theta & 0 & \sin\theta \\ 0 & 1 & 0 \\ -\sin\theta & 0 & \cos\theta \end{pmatrix}$$

(Note the change of sign for the angle θ.)

3.2) Find a rotation $\mathcal{R}'(\theta, \varphi)$ with the following properties:

a) It transforms $\mathbf{e}_3$ into the vector $\hat{\mathbf{R}}(\theta, \varphi)$.
b) The rotation $\mathcal{R}'(\theta = \pi, \varphi)$ is independent of φ.

3.3) Show that the state vector given by (3.26) is an eigenvector of J_3 with eigenvalue $(-k)$, and an eigenvector of $h(\hat{\mathbf{R}}(\theta = \pi, \varphi))$ which satisfies (3.11).

3.4) Use (3.69) and (3.71) to show the validity of (3.72) and (3.73). Hint: Note that $\mathbf{e}$ is a unit vector.

3.5) Explore the Class B cyclic solutions of (3.91), i.e., find the cyclic solutions with period $\tau = 2\pi/\Omega$. Obtain the dynamical and the geometrical phases. Show that there are cyclic states with non-vanishing geometrical phase even if the Hamiltonian is time-independent ($\omega = 0$).

4. Quantal Phases for General Cyclic Evolution

4.1 Introduction

In Chap. 3 we examined the phenomenon of the geometric phase for a class of non-adiabatic cyclic evolution for the specific example of the time-dependent Hamiltonian (3.1). In this chapter we shall uncover the general pattern that underlies the results obtained in Chap. 3.

In Sect. 4.2 we shall provide a discussion of the geometric phase for general cyclic evolution, and outline its relevance to the mathematics of fiber bundles.

In Sect. 4.3 we shall briefly point out the properties of cyclic states for the case of periodic Hamiltonians. In particular, we shall describe the geometric phase for the cyclic evolutions which have the same period as the Hamiltonian.

4.2 Aharonov–Anandan Phase

In general, we consider a time-dependent Hamiltonian $h(t) = h(R(t))$ whose time-dependence is given by a path $\mathbf{C}$ in a parameter space M

$$\mathbf{C} : R(0) \to R(t) \to R(\tau). \tag{4.1}$$

The time evolution of every pure physical state defines a curve in the space $\mathcal{P}(\mathcal{H})$ of all such states. In particular, a cyclic state $W(t) = |\psi(t)\rangle\langle\psi(t)|$ with period τ corresponds to a closed curve in $\mathcal{P}(\mathcal{H})$:

$$\mathcal{C} : W(0) \to W(t) \to W(\tau) = W(0). \tag{4.2}$$

In our example (3.1) (for Class A cyclic solutions) the cyclic states are labeled by integers or half-integers k. They correspond to the closed paths

$$\mathcal{C}_k : W_k(0) \to W_k(t) = |\phi_k(t)\rangle\langle\phi_k(t)| \to W_k(T) = W_k(0), \tag{4.3}$$

in the state space $\mathcal{P}(\mathcal{H})$, where $\phi_k(t)$ is given by (3.102).

Consider an arbitrary cyclic evolution with period τ. Then associated with the corresponding closed curve $\mathcal{C}$ in $\mathcal{P}(\mathcal{H})$ are three different curves in the Hilbert space $\mathcal{H}$:

1. The curve

$$C : |\psi(0)\rangle \to |\psi(t)\rangle \to |\psi(\tau)\rangle = e^{-i\alpha_\psi}|\psi(0)\rangle, \qquad (4.4)$$

where $|\psi(t)\rangle$ is the solution of the Schrödinger equation with the initial state vector $|\psi(0)\rangle$ being cyclic.

2. The closed curve

$$C^{\text{closed}} : |\phi(R(0))\rangle \to |\phi(R(t))\rangle \to |\phi(R(T))\rangle = |\phi(R(0))\rangle, \qquad (4.5)$$

where $|\phi(R(t))\rangle$ is the generalization of $|\phi_k(t)\rangle$ of (3.102), i.e., $|\phi(R)\rangle$ is a smooth single-valued function with values in $\mathcal{H}$ which has the property $W(t) = |\phi(R(t))\rangle\langle\phi(R(t))|$. $|\phi(R)\rangle$ is determined up to gauge transformations:

$$|\phi(R(t))\rangle \to |\phi'(R(t))\rangle = e^{i\zeta(R(t))}|\phi(R(t))\rangle,$$
$$\text{with} \quad e^{i\zeta(R(\tau))} = e^{i\zeta(0)}. \qquad (4.6)$$

In our example (3.1) the closed curve in $\mathcal{H}$ associated with the curve $\mathcal{C}_k$ of (4.3) is given by

$$C_k^{\text{closed}} : |\phi_k(0)\rangle \to |\phi_k(t)\rangle = |k; \tilde{\theta}, \omega t\rangle \to |\phi_k(T)\rangle = |\phi_k(0)\rangle, \qquad (4.7)$$

where the vectors $|\phi_k(t)\rangle$ are those of (3.102) and the gauge transformation (4.6) is the one in (3.101). The curve (4.4) in our example is given by the vectors (3.116).

3. The curve

$$\tilde{C} : |\tilde{\psi}(0)\rangle \to |\tilde{\psi}(t)\rangle := e^{i\int_0^t \langle\psi(t)|h(t')|\psi(t')\rangle dt'}|\psi(t)\rangle \to |\tilde{\psi}(\tau)\rangle, \qquad (4.8)$$

where $|\psi(t)\rangle$ is the solution of the Schrödinger equation.

These three curves C, C^{closed}, and $\tilde{C}$ have the property that under the projection of $\mathcal{H}$ onto $\mathcal{P}(\mathcal{H})$ (state vectors to states) they project onto the closed curve $\mathcal{C}$ of (4.2), i.e.,

$$|\psi(t)\rangle\langle\psi(t)| = |\phi(t)\rangle\langle\phi(t)| = |\tilde{\psi}(t)\rangle\langle\tilde{\psi}(t)| = W(t) \in \mathcal{P}(\mathcal{H}). \qquad (4.9)$$

For this reason we call these three curves *lifts* of $\mathcal{C}$. Figure 4.1 offers a schematic illustration of the situation. The curves C, C^{closed}, and $\tilde{C}$ will be called the *dynamical lift*, the *closed lift*, and the *Aharonov–Anandan (A-A) lift* of the closed curve $\mathcal{C}$, respectively. An important property of the A-A lift is that unlike the closed lift C^{closed} and the dynamical lift C, the A-A lift $\tilde{C}$ is uniquely determined by $\mathcal{C}$. The non-uniqueness of the closed lift is due to the fact that the single-valued state vectors $|\phi(R(t))\rangle$ are only defined up to gauge transformations (4.6). We shall discuss the non-uniqueness of the dynamical lift and the uniqueness of the A-A lift directly. The A-A lift is also

4.2 Aharonov–Anandan Phase

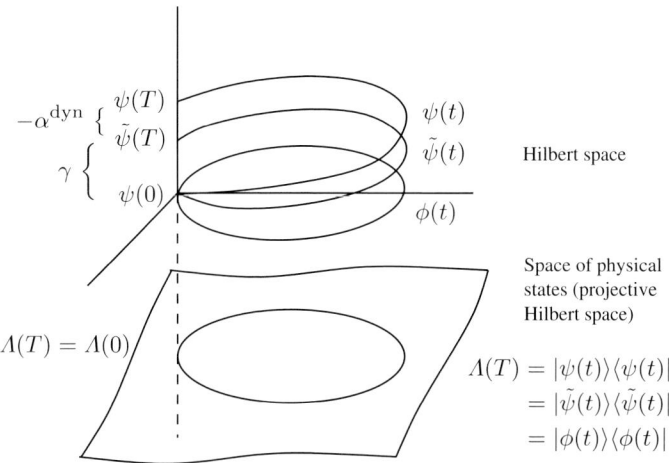

Fig. 4.1. Closed path in the space of physical states and its lifts.

called the *horizontal lift*. This terminology reflects the relevance of the A-A lift to the geometry of fiber bundles.

In our example (3.1), for the class A cyclic solutions with $\tau = T$, the A-A lift is given by

$$|\tilde{\psi}(t)\rangle := e^{i\int_0^t \langle \psi(t')|h(t')|\psi(t')\rangle dt'}|\psi(t)\rangle = e^{i\Omega k(1+\frac{\omega}{\Omega}\cos\tilde{\theta})t}|\psi(t)\rangle. \quad (4.10)$$

It is not a closed curve in $\mathcal{H}$, but has the property

$$\tilde{\mathcal{C}}_k : |\tilde{\psi}(0)\rangle \to |\tilde{\psi}(t)\rangle \to |\tilde{\psi}(T)\rangle = e^{i\gamma_k(T)}|\tilde{\psi}(0)\rangle, \quad (4.11)$$

where $\gamma_k(T) = -\alpha_k^{\text{geom}}$ is the geometrical phase (3.108) or (3.115).

The relation between the A-A lift, the dynamical lift and the closed lift in our example is expressed by

$$|\tilde{\psi}_k(t)\rangle = e^{i\alpha_k^{\text{dyn}}(t)}|\psi(t)\rangle = e^{i\gamma_k(t)}|k;\tilde{\theta},\varphi(t)\rangle, \quad (4.12)$$

where

$$\alpha_k^{\text{dyn}}(t) := \int_0^t \langle \psi(t')|h(t')|\psi(t')\rangle dt'.$$

It is a straightforward calculation (Problem 4.1) to show that the A-A lift,

$$|\tilde{\psi}(t)\rangle := e^{i\int_0^t \langle \psi(t')|h(t')|\psi(t')\rangle dt'}|\psi(t)\rangle, \quad (4.13)$$

satisfies the following equation and initial condition:

$$i\frac{d}{dt}|\tilde{\psi}(t)\rangle = [h(t) - \langle \psi(t)|h(t)|\psi(t)\rangle\mathbf{1}]\,|\tilde{\psi}(t)\rangle, \quad (4.14)$$
$$|\tilde{\psi}(0)\rangle = |\psi(0)\rangle,$$

provided that $\psi(t)$ is a solution of the Schrödinger equation (2.4).

Taking the inner product of the right-hand side of (4.14) with $|\tilde{\psi}(t)\rangle$ or $|\psi(t)\rangle$, we immediately obtain

$$\langle\tilde{\psi}(t)|\frac{d}{dt}|\tilde{\psi}(t)\rangle = \langle\psi(t)|\frac{d}{dt}|\tilde{\psi}(t)\rangle = 0\,. \tag{4.15}$$

This means that the tangent vector $\frac{d}{dt}|\tilde{\psi}(t)\rangle$ of $|\tilde{\psi}(t)\rangle$ is orthogonal (in the Hilbert space sense) to both $|\psi(t)\rangle$ and $|\tilde{\psi}(t)\rangle$.[1]

Although the dynamical lift C is uniquely determined by the Hamiltonian $h(t)$, it is not uniquely determined by the physical problem. For example, the simple substitution

$$h(t) \to h'(t) = h(t) - \kappa(t)\mathbf{1} \quad \text{with } \kappa(t) \in \mathbb{R} \tag{4.16}$$

leads to a new Hamiltonian $h'(t)$ which describes the same physics, i.e., it has the same (closed) curves of physical states $t \to W(t) = |\psi(t)\rangle\langle\psi(t)|$ in the projective Hilbert space $\mathcal{P}(\mathcal{H})$ as $h(t)$ does (Problem 4.2). But $h(t)$ and $h'(t)$ define different dynamical lifts $t \to \psi(t)$ and $t \to \psi'(t)$. Moreover, two Hamiltonians $h(t)$ and $h'(t)$ which have the same curves of physical states $t \to |\psi(t)\rangle\langle\psi(t)|$ and $t \to |\psi'(t)\rangle\langle\psi'(t)|$ differ by a multiple $\kappa(t)\mathbf{1}$ of the unit operator $\mathbf{1}$ (Problem 4.3). Thus the dynamical lift does not uniquely correspond to the physical problem. A lift that is uniquely associated with the closed curve $\mathcal{C} : t \to |\psi(t)\rangle\langle\psi(t)|$ in $\mathcal{P}(\mathcal{H})$, and therefore reflects the physical effects, is the A-A lift $\tilde{C} : t \to \tilde{\psi}(t)$. To see this, we define, in addition to $\tilde{\psi}(t)$ of (4.13), the state vector

$$|\tilde{\psi}'(t)\rangle := e^{i\int_0^t \langle\psi'(t')|h'(t')|\psi'(t')\rangle dt'}\,|\psi'(t)\rangle, \tag{4.17}$$

where $|\psi'(t)\rangle$ is a solution of the Schrödinger equation with $h'(t)$ as the Hamiltonian, namely

$$i\frac{d}{dt}|\psi'(t)\rangle = h'(t)\,|\psi'(t)\rangle. \tag{4.18}$$

Then one can show that $|\tilde{\psi}'(t)\rangle = |\tilde{\psi}(t)\rangle$ (Problem 4.4). This completes the proof of the uniqueness of the A-A lift.

We will now discuss the generalizations of (4.7), i.e., the closed lifts (4.5) of the curve $\mathcal{C}$. As seen from (3.114), $|\phi_k(t)\rangle$ can be used to calculate the geometric phase, if one does not have a solution $|\psi(t)\rangle$ of the Schrödinger equation. There are various ways to obtain these single-valued vectors $|\phi(t)\rangle \in \mathcal{H}$.[2] In general they are curves of *local sections*. A local section is a smooth function of an open patch O of $\mathcal{P}(\mathcal{H})$ into $\mathcal{H}$, i.e.,

---

[1] For a definition of the tangent vectors see Appendix A.
[2] The method used in Chap. 3 can be generalized in a straightforward way for Hamiltonians which are linear functions of the generators of a (finite-dimensional) Lie algebra. For more details see [176, 262], and references therein.

4.2 Aharonov–Anandan Phase

$$|\phi\rangle : O \to \mathcal{H}.$$

We will assume that our curve $\mathcal{C}$ lies in such an open patch O.[3] Then for any closed curve $\mathcal{C} : t \to W(t)$ one has the closed (single-valued) lift

$$t \to |\phi(t)\rangle = |\phi(W(t))\rangle, \qquad (4.19)$$

with

$$W(t) = |\phi(t)\rangle\langle\phi(t)|, \quad \text{and} \quad |\phi(\tau)\rangle = |\phi(0)\rangle.$$

This is however determined only up to a gauge transformation (4.6). The importance of the local sections $|\phi\rangle$ or alternatively the closed lifts $\mathcal{C}^{\text{closed}}$ is due to their utility in the calculation of the geometric phase.

Since $t \to |\psi(t)\rangle$, $t \to |\tilde{\psi}(t)\rangle$ and any $t \to |\phi(t)\rangle$ are lifts of the same closed curve $t \to W(t)$, there must exist a phase factor $w(t)$ such that

$$|\tilde{\psi}(t)\rangle = w(t)|\phi(t)\rangle. \qquad (4.20)$$

Next we calculate $w(t)$. Differentiating both sides of (4.20), we obtain

$$\frac{d}{dt}|\tilde{\psi}(t)\rangle = \frac{dw(t)}{dt}|\phi(t)\rangle + w(t)\frac{d}{dt}|\phi(t)\rangle.$$

Taking the scalar product of this expression with $|\tilde{\psi}(t)\rangle$, we have

$$w(t)\frac{dw(t)}{dt} + w^2(t)\langle\phi(t)|\left(\frac{d}{dt}|\phi(t)\rangle\right) = \langle\tilde{\psi}(t)|\left(\frac{d}{dt}|\tilde{\psi}(t)\rangle\right) = 0, \qquad (4.21)$$

where in the last equality we have used (4.15). Next we write (4.21) in the form

$$\frac{1}{w(t)}\frac{dw(t)}{dt} = -\langle\phi(t)|\frac{d}{dt}|\phi(t)\rangle,$$

which can be immediately integrated to yield

$$\frac{w(t)}{w(0)} = e^{-\int_0^t \langle\phi(t')|\frac{d}{dt'}|\phi(t')\rangle\, dt'}. \qquad (4.22)$$

We shall denote the phase factor on the right-hand side of (4.22) by $e^{i\gamma(t)}$ and write (4.20) as

$$|\tilde{\psi}(t)\rangle = w(0) e^{i\int_0^t i\langle\phi(t')|\frac{d}{dt'}|\phi(t')\rangle\, dt'} |\phi(t)\rangle =: w(0)\, e^{i\gamma(t)}\, |\phi(t)\rangle. \qquad (4.23)$$

---

[3] If this is not the case, i.e., if $\mathcal{C}$ lies in two or more open patches, then one must choose different local sections $|\phi\rangle$ on each patch. These will however be related by gauge transformations on the intersections of the patches. Thus the method we are to describe does indeed generalize to arbitrary curves. The only additional step is to take into account the necessary gauge transformations in going from one patch to the other.

If we choose the arbitrary constant phase $\omega(0)$ such that $|\tilde{\psi}(0)\rangle = |\psi(0)\rangle = |\phi(0)\rangle$, i.e., choose $\omega(0) = 1$, then we have

$$|\tilde{\psi}(t)\rangle = e^{i\gamma(t)} |\phi(t)\rangle. \tag{4.24}$$

This is the generalization of (4.12). Combining (4.13) and (4.24), we find

$$|\psi(t)\rangle = e^{-i \int_0^t \langle \psi(t')|h(t')|\psi(t')\rangle \, dt'} e^{i\gamma(t)} |\phi(t)\rangle. \tag{4.25}$$

Equation (4.25) is the general relation between the solution of the Schrödinger equation $|\psi(t)\rangle$ and the closed lift $|\phi(t)\rangle$. Equation (2.57) is the adiabatic approximation of this relation, and (3.104) is its special case for the spinning quantum system in a precessing external magnetic field. The adiabatic approximation uses in place of the closed lifts $|\phi(t)\rangle$ the single-valued eigenvectors of $h(R(t))$ which are more easily accessible than $|\phi(t)\rangle$ (by solving (2.6)). According to (4.23), for the closed path $\mathcal{C}$ in $\mathcal{P}(\mathcal{H})$ the phase angle $\gamma(\tau)$ is given by

$$\gamma(\tau) = \gamma(\mathcal{C}) = \oint_0^\tau i\langle\phi(t)|\frac{d}{dt}|\phi(t)\rangle dt = \oint i\langle\phi|d|\phi\rangle \quad \text{mod } 2\pi, \tag{4.26}$$

where $|\phi(t)\rangle$ corresponds to any of the closed lifts of $\mathcal{C}$.

The phase angle $\gamma(\tau)$ is independent of the choice of the time parameterization of $|\phi(t)\rangle$, i.e., the speed with which $|\phi(t)\rangle$ traverses its closed path. It is gauge invariant. It is independent of the choice of the Hamiltonian as long as these Hamiltonians describe the same closed path $\mathcal{C}$ in $\mathcal{P}(\mathcal{H})$. It depends on the closed curve $\mathcal{C}$. It is therefore considered to be a "geometric" property of $\mathcal{C}$, thus has the name *geometric phase*.

The one-form appearing in the integrand of (4.26), namely

$$\mathcal{A} := i\langle\phi|d|\phi\rangle, \tag{4.27}$$

is called the *Aharonov–Anandan connection one-form*. It is the analog of the Mead–Berry connection one-form (2.55). It satisfies the following transformation rule under the gauge transformation (4.6) (Problem 4.5).

$$\mathcal{A} \to \mathcal{A}' = \mathcal{A} - d\zeta. \tag{4.28}$$

The formula (4.27) for the A-A connection was obtained from the requirement that $\tilde{\psi}(t)$ is the A-A lift, i.e., the lift fulfilling (4.15) (which in turn followed from its definition (4.13)). This was the only possible definition of a lift which depended solely on the physics of the problem.

Equations (4.24) and (4.19) lead immediately to

$$|\tilde{\psi}(\tau)\rangle = e^{i\gamma(\mathcal{C})} |\tilde{\psi}(0)\rangle = e^{i\gamma(\mathcal{C})} |\psi(0)\rangle. \tag{4.29}$$

Then, using (4.13) one obtains for the cyclic evolution of a state vector

4.2 Aharonov–Anandan Phase

$$|\psi(\tau)\rangle = e^{-i\int_0^\tau \langle \psi(t')|h(t')|\psi(t')\rangle dt'} e^{i\gamma(\mathcal{C})} |\psi(0)\rangle. \tag{4.30}$$

This relation is the non-adiabatic generalization of (2.66).

Figure 4.1 shows a graphical representation of our description of the general cyclic evolution. It shows a *base space* representing the space $\mathcal{P}(\mathcal{H})$ of physical states and including a closed curve $\mathcal{C}$. This curve represents the cyclic evolution of a pure state $W(t)$. Above $\mathcal{C}$ are shown the lifts $C^{\text{closed}}, \tilde{C}$ and C of $\mathcal{C}$ which belong to the Hilbert space $\mathcal{H}$ (depicted in the figure by the three-dimensional space). Also shown is a *fiber* above the base point $W(0)$ (depicted by the positive z-axis with the z coordinate representing the phase angle modulo 2π or the element $e^{i\zeta(0)}$ of the (gauge) group $U(1)$). The fiber in this case (though it has been drawn as a straight line) is a copy of the unit circle S^1 or the group $U(1)$. We can attach a copy of this S^1 not only to the point $W(0)$ but to every one-dimensional projection operator $\Lambda \in \mathcal{P}(\mathcal{H})$. In this way we get a *bundle* of $U(1)$ fibers attached (which means loosely associated) to each point of $\mathcal{P}(\mathcal{H})$.

The mathematical structure that we have encountered here is an example of a *principal fiber bundle* (PFB). We shall refer to this PFB as the *Aharonov–Anandan principal bundle*. We shall develop the basic concepts of fiber bundle theory in the next chapter. Here we briefly mention its relevance to the phenomenon of the geometric phase.

Like any fiber bundle the A-A PFB consists of a base space, a set of fibers associated with the points of the base space and forming a larger space called the *total* or *bundle space*, and a *structure group* which acts on the fibers. As the total space consists of fibers over points of the base space, we can define a projection map from the total space to the base space. The fibers are then viewed as the inverse images[4] of points of the base space under this projection.

For the A-A PFB, the base space is the projective Hilbert space $\mathcal{P}(\mathcal{H})$ also denoted by $\mathbb{C}P^{N-1}$ where $N \leq \infty$ is the dimension of the Hilbert space $\mathcal{H}$. The total space $S(\mathcal{H})$ is the set of all normalized state vectors in $\mathcal{H}$, i.e.

$$S(\mathcal{H}) := \{\psi \in \mathcal{H} : \langle \psi|\psi\rangle = 1\} = S^{2N-1},$$

The projection map

$$\pi : S(\mathcal{H}) \to \mathcal{P}(\mathcal{H}) \quad (\text{or} \quad \pi : S^{2N-1} \to \mathbb{C}P^{N-1}), \tag{4.31}$$

is the obvious projection of state vectors onto states,

$$\pi(|\psi\rangle) = |\psi\rangle\langle\psi|. \tag{4.32}$$

The fibers are the normalized rays in the Hilbert space. They consist of all the normalized state vectors associated with a given pure state and hence

---

[4] Let $f : A \to B$ be a function and $B' \subset B$. Then the inverse image of B' under f is defined by $f^{-1}(B') := \{x \in A : f(x) \in B'\}$. In particular, the inverse image of a point $y \in B$ is the subset $\{x \in A : f(x) = y\}$ of A.

60   4. Quantal Phases for General Cyclic Evolution

differing by a phase factor. Therefore the fibers are copies of the unit circle S^1. Finally the structure group is the Abelian group $U(1)$ which has the manifold structure of S^1 as well. We shall denote the A-A PFB by

$$\eta : U(1) \to S(\mathcal{H}) \xrightarrow{\pi} \mathcal{P}(\mathcal{H}) \qquad (\text{or } U(1) \to S^{2N-1} \to \mathbb{C}P^{N-1}). \quad (4.33)$$

A PFB may be endowed with a geometric structure. The latter provides a notion of *parallel transportation* or alternatively the notion of a *horizontal lift*. A geometric structure on a PFB is also called a *connection*. For an Abelian PFB, i.e., a PFB with $U(1)$ as its structure group, a connection may be expressed by a differential one-form satisfying a set of (gauge) transformation rules. The A-A connection one-form $\mathcal{A}$ of (4.27) defines a particular connection on the A-A PFB η. The A-A phase also has a very well-known mathematical counterpart called *holonomy*.

It turns out that the A-A bundle η and the A-A connection $\mathcal{A}$ which were both introduced by physicists in their investigation of geometric phases, play a unique role in the mathematical theory of fiber bundles. In fact, both η and $\mathcal{A}$ have long been known by mathematicians and used by them to classify principal fiber bundles and their geometries. However, before revealing the special properties of η and $\mathcal{A}$, we shall present an elementary introduction to fiber bundles and gauge theories in the next chapter. These provide the basic mathematical and physical theories underlying the phenomenon of the geometric phase.

In the next section we shall briefly discuss the properties of cyclic states for the case of periodic Hamiltonians. Particularly interesting are the cyclic states whose period is the same as the period of the Hamiltonian.

4.3 Exact Cyclic Evolution for Periodic Hamiltonians

If the time-dependent Hamiltonian is periodic, i.e.,

$$h(t+T) = h(t), \quad (4.34)$$

for some time period T, then we can use the results of what is known as Floquet theory[5] in mathematics. In particular we next show that the evolution operator for any periodic Hamiltonian has precisely the same form as the evolution operator (3.87) for the spin system (3.1). More precisely we prove that *for any T-periodic Hamiltonian, satisfying (4.34), there exists a time-independent Hermitian operator $\tilde{h}$ and a T-periodic unitary operator $Z = Z(t)$ with $Z(0) = 1$, such that*

$$U^{\dagger}(t) = Z(t)\, e^{-it\tilde{h}}. \quad (4.35)$$

---

[5] The original reference to Floquet theory is [81]. For more recent reviews see for example [107, 112]. A discussion of the application of Floquet theory to the computation of the geometric phase can be found in [175] and references therein.

4.3 Exact Cyclic Evolution for Periodic Hamiltonians

To see this, consider the unitary operator $V(t) := U^\dagger(t+T)$. It is not difficult to check that V satisfies the following Schrödinger equation:

$$\frac{d}{dt}V(t) = -i\,h(t)\,V(t) \qquad (4.36)$$

$$V(0) = U^\dagger(T) =: V_0.$$

The operator $V'(t) := U^\dagger(t)\,V_0$ also satisfies (4.36). However we know from the uniqueness theorem for initial value linear differential equations that the solution of (4.36) is unique. Therefore we have $V(t) = V'(t)$, i.e.,

$$U^\dagger(t+T) = U^\dagger(t)V_0 = U^\dagger(t)\,U^\dagger(T). \qquad (4.37)$$

This relation is sufficient to construct a pair of operators $(Z(t), \tilde{h})$ which satisfies (4.35). To see this, we first write $t = nT + t_0$ for some integer n and $t_0 \in [0, T)$ and then apply (4.37) repeatedly. This yields

$$U^\dagger(t) = U^\dagger(t_0)\left[U^\dagger(T)\right]^n = U^\dagger(t_0)\,V_0^n. \qquad (4.38)$$

Next we assume that the unitary operator V_0 can be expressed as the exponential of a Hermitian operator[6], i.e.,

$$V_0 = e^{-i\tilde{h}'}. \qquad (4.39)$$

In view of (4.38), (4.39) and $n = \frac{t-t_0}{T}$, we have

$$U^\dagger(t) = U^\dagger(t_0)\,e^{\frac{-i(t-t_0)}{T}\tilde{h}'}$$

$$= U^\dagger(t_0)\,e^{\frac{it_0}{T}\tilde{h}'}\,e^{\frac{it}{T}\tilde{h}'}$$

$$= Z(t)\,e^{-it\tilde{h}}, \qquad (4.40)$$

where we have defined

$$Z(t) := U^\dagger(t_0)\,e^{\frac{it_0}{T}\tilde{h}'}, \quad \text{and} \quad \tilde{h} := \frac{\tilde{h}'}{T}.$$

Clearly, $\tilde{h}$ is Hermitian and $Z(t)$ is unitary. Furthermore, $Z(t)$ satisfies

---

[6] This can always be done if the Hilbert space is finite dimensional. For an infinite dimensional Hilbert space there are unitary operators which cannot be written as the exponent of ($-i$ times) some Hermitian operator. However, the set of unitary operators which are exponentials of Hermitian operators is a dense subset of the set of unitary operators with an appropriate choice of topology on the latter set. This means that one can always find a sequence of unitary operators $\{e^{-i\tilde{h}'_k} : k = 0, 1, 2, \cdots\}$ which converges to V_0 in this topology. Therefore, one can approximate V_0 with $e^{-i\tilde{h}'}$ where $\tilde{h}' := \tilde{h}'_k$ for some large value of k. Furthermore, one can improve this approximation by taking larger and larger values for k. For a discussion of this point see for example [42] and references therein.

$$Z(t+T) = Z(t), \quad \text{and} \quad Z(0) = U^\dagger(0) = 1.$$

For $t = T$, (4.40) reproduces the general result, namely

$$U^\dagger(T) = e^{-iT\tilde{h}}. \tag{4.41}$$

In particular, $\tilde{h}$ yields the exact cyclic states with period $\tau = T$ as its eigenstates.

We must note that the pair of operators $(Z(t), \tilde{h})$, so constructed, is not unique [175]. The non-uniqueness of $\tilde{h}$ is reminiscent of the non-uniqueness of the Hamiltonian $h = h(t)$ discussed in Sect. 4.2.

If the Hamiltonian is parameter dependent ($h = h(R)$) and its time dependence is realized by changing the parameters R in time, then the operator $\tilde{h}$ will be a function of the initial parameter, $R_0 = R(t = 0)$. Thus in general it also depends on a set of parameters belonging to the same parameter space M.

As in the case of the spin system of Chap. 3, the operator $\tilde{h}(R)$ generally differs from the Hamiltonian $h(R)$. In general, the dependence of $\tilde{h}(R)$ on the Hamiltonian is not explicitly known. The only established fact is that for an adiabatic evolution $\tilde{h}(R)$ can be approximated by $h(R)$.

In the remainder of this section we briefly discuss the relation between the operators $\tilde{h}$ and $Z(t)$ of (4.35) and the geometric phase of Aharonov and Anandan.

Consider the evolution of a cyclic state vector $|\psi(0)\rangle$. By definition $|\psi(0)\rangle$ is an eigenvector of the evolution operator:

$$U^\dagger(\tau)|\psi(0)\rangle = Z(\tau) e^{-i\tau\tilde{h}} |\psi(0)\rangle = e^{-i\alpha(\tau)} |\psi(0)\rangle, \tag{4.42}$$

where we have employed (4.35). On the other hand the evolution of $|\psi(0)\rangle$ is governed by the Schrödinger equation (2.4). In terms of the operators $Z(t)$ and $\tilde{h}$, these equations take the form

$$|\psi(t)\rangle = Z(t) e^{-it\tilde{h}} |\psi(0)\rangle, \tag{4.43}$$
$$h(t) = i\dot{Z}(t) Z^\dagger(t) + Z(t) \tilde{h} Z^\dagger(t), \tag{4.44}$$

where we have used the unitarity of $Z(t)$, and $\dot{Z}(t)$ stands for the time-derivative of $Z(t)$.

Next let us compute the dynamical phase angle. Substituting (4.43) and (4.44) in the definition of the dynamical phase angle (3.106)

$$\alpha^{\text{dyn}}(\tau) := \int_0^\tau \langle \psi(t')|h(t')|\psi(t')\rangle \, dt' \mod 2\pi, \tag{4.45}$$

we have (Problem 4.6):

$$\alpha^{\text{dyn}}(\tau) =$$
$$\int_0^\tau i\langle\psi(0)|e^{it'\tilde{h}} Z^\dagger(t') \dot{Z}(t') e^{-it'\tilde{h}} |\psi(0)\rangle \, dt' + \tau \langle\psi(0)|\tilde{h}|\psi(0)\rangle \mod 2\pi.$$
$$\tag{4.46}$$

4.3 Exact Cyclic Evolution for Periodic Hamiltonians

For a cyclic state with the same period as the Hamiltonian ($\tau = T$), the cyclic state vector $|\psi(0)\rangle$ is an eigenvector of the operator $\tilde{h}$. Hence the second term on the right-hand side of (4.46) is precisely the total phase angle

$$T\langle\psi(0)|\tilde{h}|\psi(0)\rangle = \alpha(T).$$

In view of this equation we can directly express the geometric phase in terms of the operator $Z(t)$ and its time derivative:

$$\gamma(T) = \gamma(\mathcal{C}) := \alpha^{\text{dyn}}(T) - \alpha(T) \mod 2\pi$$
$$= \int_0^T i\langle\psi(0)|Z^\dagger(t')\dot{Z}(t')|\psi(0)\rangle dt' \mod 2\pi. \quad (4.47)$$

Equation (4.47) is of some practical importance. Since both the Hamiltonian $h(t)$ and the operator $Z(t)$ are periodic with the same period T, one can in principle expand them in Fourier series. There are procedures to relate the Fourier components of $Z(t)$ to those of $h(t)$ which may be used to yield a series expansion of the geometric phase for arbitrary periodic systems. Such a procedure is outlined in [175] and references therein.

The operator $Z(t)$ is also of particular interest since it may be used to yield the single-valued vectors $|\phi(t)\rangle$ of (4.19), namely

$$|\phi(t)\rangle = Z(t)|\psi(0)\rangle. \quad (4.48)$$

In view of this identification, we can write (4.47) in the form

$$\gamma(T) = \int_0^T i\langle\psi(0)|Z^\dagger(t')\frac{d}{dt'}[Z(t')|\psi(0)\rangle] dt' \mod 2\pi$$
$$= \int_0^T i\langle\phi(t')|\frac{d}{dt'}|\phi(t')\rangle dt' \mod 2\pi,$$

which is identical with (4.26). The non-uniqueness of $Z(t)$, mentioned above, corresponds to the gauge freedom of $|\phi(t)\rangle$.

Furthermore, we can use (4.48) to express the evolution of a cyclic state vector according to

$$|\psi(t)\rangle = Z(t) e^{-it\tilde{h}} |\psi(0)\rangle$$
$$= e^{-i\frac{\alpha(T)}{T}t} Z(t) |\psi(0)\rangle$$
$$= e^{-i\frac{\alpha(T)}{T}t} |\phi(t)\rangle. \quad (4.49)$$

Here we have used the fact that $|\psi(0)\rangle$ is an eigenvector of $\tilde{h}$ with eigenvalue $\alpha(T)/T$.

The fact that the evolving state vector $|\psi(t)\rangle$ is expressed as the product of a phase factor and a periodic state vector[7] $|\phi(t)\rangle$ is a direct consequence

---
[7] Note that $|\phi(T+t)\rangle = |\phi(t)\rangle$ since $Z(t)$ is periodic and $|\phi(t)\rangle$ satisfies (4.48).

of the periodicity of the Hamiltonian. The state vectors of the form (4.49) are encountered in almost all physical systems with periodic features. The corresponding wave functions are known as the *Bloch wave functions*.

We conclude this chapter by noting that in the remainder of this book we shall only be concerned with the cyclic evolutions of periodic Hamiltonians which have the same period as the Hamiltonian.

Problems

4.1) Show that if the state vector $\psi(t)$ is the solution of the Schrödinger equation (2.4), then the state vector $\tilde{\psi}(t)$ defined by (4.13) satisfies (4.14).

4.2) Let ψ and ψ' be the solutions of the Schrödinger equations (2.4) and (4.18) corresponding to the Hamiltonians $h(t)$ and $h'(t)$, where $h'(t) = h(t) - \kappa(t)\mathbf{1}$, $\kappa(t)$ is a scalar function of time, and $\mathbf{1}$ is the identity operator. Show that both the Hamiltonians define the same curve $\mathcal{C}$ in $\mathcal{P}(\mathcal{H})$, i.e., show that for all $t \in [0, T]$, $|\psi'(t)\rangle\langle\psi'(t)| = |\psi(t)\rangle\langle\psi(t)|$.

4.3) Show that the dynamical lift of the closed curve $\mathcal{C}$ of (4.2) changes under the substitution $h(t) \to h'(t) = h(t) - \kappa(t)\mathbf{1}$, and calculate the phase difference between the dynamical lift $\psi(t)$ belonging to $h(t)$ and the dynamical lift $\psi'(t)$ belonging to $h'(t)$. Conversely, show that if two Hamiltonians $h(t)$ and $h'(t)$ define the same curve of physical states: $t \to |\psi(t)\rangle\langle\psi(t)|$, for every initial state $|\psi(0)\rangle\langle\psi(0)|$, then they must differ by a multiple of the identity operator $\mathbf{1}$.

4.4) Show that the horizontal (A-A) lift $\tilde{C}$ of $\mathcal{C}$ is not affected by the substitution $h(t) \to h'(t) := h(t) - \kappa(t)\mathbf{1}$.

4.5) Find the transformation rule for the connection one-form $\mathcal{A}^\phi$ of (4.27), under the gauge transformation of the local sections: $\phi(t) \to \phi'(t) = e^{i\zeta(t)}\phi(t)$.

4.6) Use (4.43) and (4.44) to prove (4.46).

5. Fiber Bundles and Gauge Theories

5.1 Introduction

In the preceding chapter we introduced the concept of the geometric phase and showed its natural emergence from the basic principles of quantum mechanics. We emphasized the geometric nature of the phase and briefly outlined a holonomy interpretation of the geometric phase using the Aharonov–Anandan principal fiber bundle. This involved several differential geometric concepts such as those of a *fiber bundle*, a *horizontal lift*, a *connection*, a *parallel transport* and a *holonomy*.

In this chapter, we shall introduce the essential concepts of the theory of fiber bundles. Specifically, we shall use the example of the geometric phase as our starting point to describe the basic elements and to develop the necessary results of fiber bundle theory. Our main objective will be to familiarize the reader with the most basic ideas, motivate the important definitions and quote the necessary mathematical results. We shall avoid discussing the technical details of the proofs since these may be found in textbooks on fiber bundles.[1]

The theory of fiber bundles is one of the most important mathematical achievements of the twentieth century. It has numerous applications in other branches of mathematics and theoretical physics. There is extensive literature on fiber bundles. Among these are several recent books and review articles intended for physicists, such as [64, 78, 114, 186, 190]. An elementary review of more basic differential geometric concepts is presented in Appendix A.

5.2 From Quantal Phases to Fiber Bundles

In our discussion of the geometric phase, the main assumption was the existence of the cyclic states. These were defined by the condition:

$$|\psi(\tau)\rangle\langle\psi(\tau)| = |\psi(0)\rangle\langle\psi(0)|.$$

We argued that the evolution of a cyclic state could be viewed as the motion of a point on a closed curve in the space of pure quantum states,

---

[1] Two of the standard texts on fiber bundles are [111] and [128].

$$\mathcal{P}(\mathcal{H}) = \{\Lambda \ : \ \Lambda = |\psi\rangle\langle\psi|, |\psi\rangle \in \mathcal{H}\}.$$

In a sense $\mathcal{P}(\mathcal{H})$ is recognized as physically more basic than the Hilbert space $\mathcal{H}$.[2] Obviously, if we work with the states, i.e., $\mathcal{P}(\mathcal{H})$, information about the phase will be automatically lost. Nonetheless, using the states has its own advantages which one should usually like to retain. This forces one to seek a compromise that would allow $\mathcal{P}(\mathcal{H})$ to remain as the basic quantum space and simultaneously keep track of the phases. In other words, one needs to supplement $\mathcal{P}(\mathcal{H})$ with a mechanism that stores the phase information.

Ordinary phase is a complex number with unit modulus, $e^{i\alpha}$, $\alpha \in \mathbb{R}$. Thus, for a given normalized state vector $|\psi\rangle$, the phase of $|\psi\rangle$ is an element of the unit circle

$$S^1 = \{z \in \mathbb{C} \ : \ |z| = 1\}$$

in the complex plane $\mathbb{C}$. Therefore, we can think of the state vector $|\psi\rangle$ as the state $\Lambda = |\psi\rangle\langle\psi|$ in $\mathcal{P}(\mathcal{H})$ and the phase $e^{i\alpha}$ in S^1. Let us now attach a copy of S^1 to each point Λ in $\mathcal{P}(\mathcal{H})$. Such a construction may be conveniently used for bookkeeping of the phases. We denote this construction by η. As we pointed out in Chap. 4, η is an example of a *fiber bundle*.

The projective Hilbert space $\mathcal{P}(\mathcal{H})$ is generally infinite dimensional. Thus the example presented above is by no means the simplest example of a fiber bundle. We would like to stress, however, that besides its remarkable ties with quantum mechanics, η plays an important role in the mathematical theory of fiber bundles and in this sense its study is of utmost importance both in physics and mathematics. We shall give the precise definition of η and explain its special properties in Chap. 6.

Let us now return to the example of a magnetic dipole in a precessing magnetic field. We shall first concentrate on the adiabatic evolution of the system as discussed in Chap. 2. The parameter space of this system is the unit sphere, S^2. A periodic change of the Hamiltonian (2.1) is given by a closed curve in S^2; a circular loop corresponds to the precessing magnetic field. We also introduced the single-valued state vectors $|n; R\rangle$ which were eigenstate vectors of the Hamiltonian $h(R)$. These were however defined up to a smooth but otherwise arbitrary phase, (2.16), over the open neighborhoods (patches) of S^2.

Once again, we are in a situation where there is an arbitrariness of phases. On the one hand, we would like to keep S^2 as the physical space of the parameters of the system, and on the other hand, we need to be able to fix the phases in a convenient manner and perform the necessary computations involving $|n; R\rangle$. The problem is quite similar to the case of $\mathcal{P}(\mathcal{H})$. It has a similar solution as well. Namely, we attach a copy of S^1, where the phases are defined, to each point R of S^2. Then, a choice of $|n; R\rangle$ corresponds to the point R in S^2 and a point on the copy of S^1 attached to R. The latter fixes

---

[2] It is precisely for this reason that it took half a century for physicists to pay the necessary attention to the phenomenon of geometric phase.

the phase of $|n;R\rangle$. This leads to another example of a fiber bundle which we call λ. The space S^2 and the copies of S^1 attached to its points are called the *base manifold* and *fibers*, respectively.

As a geometric or rather a topological object, λ has the structure of a three-dimensional manifold. However, with the limited local information that we have provided, we cannot identify λ with a known three-dimensional manifold. In fact, there are an infinite number of three-dimensional manifolds that fit our crude description of λ. A couple of the simplest examples of these are $S^2 \times S^1$ and the three-dimensional sphere S^3. For further details of smooth manifolds, we refer the reader to Appendix A.

At this point, we wish to clarify what we mean by similar and different manifolds. There are many ways in which we can distinguish between manifolds. For instance, we can distinguish between two copies of a single manifold, or we can put all of these copies in the equivalence classes of diffeomorphic manifolds and only distinguish between two manifolds that are not diffeomorphic.[3] In the latter case, an ellipse can be correctly identified with the unit circle S^1. We shall adopt the second notion of equivalence for manifolds. It is in this sense that we mean λ may be $S^2 \times S^1$, S^3 or something else. This definition of equality for manifolds is sensitive to their global properties. In general, every fiber bundle is a smooth manifold. The converse of this statement is of course false, for otherwise there would not be any use in calling manifolds fiber bundles. These issues, quite clearly, indicate the necessity of a precise definition of a fiber bundle. We shall devote the next section to fulfill this necessity.

5.3 An Elementary Introduction to Fiber Bundles

So far, we have seen how fiber bundles occur as natural mathematical constructions in the context of the geometric phase. As we can deduce from the two examples of fiber bundles that we considered, any fiber bundle E consists of a *total space* which is also denoted by the same symbol (E), a *base manifold* X, and a set of *fibers*. The fibers live over the points of the base manifold. We shall denote the fiber over a point $x \in X$ by F_x. All of the fibers look alike. More precisely, they are (diffeomorphic) copies of a smooth manifold F of dimension $\mathcal{N}$ which is called the *typical fiber*. The integer $\mathcal{N}$ is called the *fiber dimension* or the *rank* of the bundle.

The bundle E is in fact a collection of its fibers, i.e., any point $p \in E$ belongs to one of the fibers, say, F_x. To express this statement in a more precise language, one defines a projection map

$$\pi : E \to X, \tag{5.1}$$

---

[3] Two smooth manifolds are said to be diffeomorphic if they can be smoothly deformed to one another. See Appendix A for a precise definition.

by setting
$$\pi(p) := x \quad \text{if and only if} \quad p \in F_x.$$

Having defined π, we can refer to the fiber F_x over x as the inverse image of x under π, i.e., $F_x = \pi^{-1}(x)$. The projection map π has two important properties:

1. It is an onto function, so that there are fibers over all the points of X.
2. It is a smooth (continuous)[4] function.

The smoothness requirement on π can be viewed, among other things, as a necessary condition for having smooth curves in E projected to smooth curves in X. We have already encountered some examples of curves in the bundle space (Fig. 4.1) which project onto a given smooth curve C_X in the base space X. These are called the lifts of C_X. Briefly, a *lift* of a smooth curve $C_X : [0,T] \to X$ in X is a smooth curve $C_E : [0,T] \to E$ in E, such that $C_X = \pi(C_E) = \pi \circ C_E$.

There is, of course, more structure to a fiber bundle than what we extracted from our physical examples, η and λ. These structures or properties are also natural and help make fiber bundles more practical. Probably, the most essential property of a fiber bundle is that it is locally the Cartesian product of an open subset of the base manifold X and the typical fiber F. More precisely, for any point $x \in X$, there is an open neighborhood U_x of x, such that[5]

$$\pi^{-1}(U_x) \simeq U_x \times F, \tag{5.2}$$

where the symbol "$\simeq$" means "diffeomorphic". The motivation behind this requirement is indeed a simple and logical one. It means that we can encode the global or topological properties of a fiber bundle in a set of so-called *transition functions*. The idea is a direct generalization of the one used in the definition of a differentiable manifold.

Locally any m-dimensional manifold resembles (an open neighborhood[6] in) $\mathbb{R}^m$. Thus all the manifolds of the same finite dimension are locally identical. It is the global structure of a manifold which distinguishes it from other manifolds of the same dimension. The global structure of a manifold depends on a set of rules according to which the open neighborhoods of $\mathbb{R}^m$ are glued

---

[4] In what follows, we can usually relax the smoothness condition to continuity of the corresponding functions. The distinction between smoothness and continuity is of little interest in our discussions. Therefore, we shall always require the stronger condition of smoothness unless explicitly stated otherwise.

[5] Note that here the Cartesian product "$\times$" also induces a topology on the resulting product set $U_x \times F$. This is called the *product topology*. As we discuss in Appendix A a topology on a set of points determines the collection of its open subsets. The product topology chooses the open subsets of the product of two spaces to be the unions of products of their open subsets.

[6] As discussed in Appendix A, by an open neighborhood in $\mathbb{R}^m$, we mean a subset that is diffeomorphic to $\mathbb{R}^m$ itself.

5.3 An Elementary Introduction to Fiber Bundles

together to construct the manifold. These rules are stored in a set of functions which are also called the *transition* or *overlap functions*. To see how this is done, let us cover the manifold X with a set of open *patches*. These are subsets O_α of $X = \cup_\alpha O_\alpha$ each of which is diffeomorphic to (an open neighborhood in) $\mathbb{R}^m$. Thus for each O_α, there is a diffeomorphism $\varphi_\alpha : O_\alpha \to \mathbb{R}^m$ which identifies O_α with its image $\varphi_\alpha(O_\alpha)$ in $\mathbb{R}^m$:

$$O_\alpha \simeq \varphi_\alpha(O_\alpha) \subseteq \mathbb{R}^m.$$

Now let us consider two intersecting patches O_α and O_β. To study the global structure of the union $O_\alpha \cup O_\beta$, we must know how to glue, i.e., to identify, the images $\varphi_\alpha(O_\alpha \cap O_\beta)$ and $\varphi_\beta(O_\alpha \cap O_\beta)$. This is done by the overlap function

$$g_{\alpha\beta} := \varphi_\alpha \circ \varphi_\beta^{-1} : \varphi_\beta(O_\alpha \cap O_\beta) \to \varphi_\alpha(O_\alpha \cap O_\beta).$$

Using $g_{\alpha\beta}$, we can recover the union $O_\alpha \cup O_\beta$ as the two open subsets $\varphi_\alpha(O_\alpha)$ and $\varphi_\beta(O_\beta)$ of $\mathbb{R}^m$ with the points $x \in \varphi_\beta(O_\alpha \cap O_\beta)$ identified with (glued to) $g_{\alpha\beta}(x) \in \varphi_\alpha(O_\alpha \cap O_\beta)$. This procedure can be continued by adding more patches to the union $O_\alpha \cup O_\beta$ and eventually reconstructing the whole manifold. We have explored the manifold structure of several interesting examples in Appendix A. Among these is the manifold S^2 that can be described using two patches and a single overlap function (A.2).

We have a similar construction for fiber bundles. Once (5.2) is enforced, we can divide the base manifold X into a collection $\{U_\alpha\}$ of its open subsets, such that for each α the part of E over U_α is a copy of $U_\alpha \times F$. That is, there exist diffeomorphisms

$$\Phi_\alpha : \pi^{-1}(U_\alpha) \to U_\alpha \times F. \tag{5.3}$$

The pair (U_α, Φ_α) is called a *chart* or a *local trivialization* of the bundle E. If we restrict Φ_α to a point $x \in U_\alpha$, we will obtain a diffeomorphism from the fiber F_x onto the typical fiber,

$$\Phi_\alpha(x) := \Phi_\alpha|_{\{x\}} : F_x \longrightarrow \{x\} \times F \equiv F \tag{5.4}$$

or briefly

$$\Phi_\alpha(x) : F_x \longrightarrow F.$$

The global structure of E is given by the *transition functions*

$$G_{\alpha\beta}(x) := \Phi_\alpha(x) \circ \Phi_\beta^{-1}(x) : F \longrightarrow F, \quad \forall x \in U_\alpha \cap U_\beta. \tag{5.5}$$

These determine the necessary rules for gluing the regions $\pi^{-1}(U_\alpha)$ of E over the intersections of U_α, i.e., identifying $\Phi_\alpha(\pi^{-1}(U_\alpha \cap U_\beta))$ with $\Phi_\beta(\pi^{-1}(U_\alpha \cap U_\beta))$.

A fiber bundle E is completely specified by a smooth manifold X as the base manifold, a smooth manifold F as the typical fiber, and a set of functions (diffeomorphisms of F) $G_{\alpha\beta}(x)$ as the transition functions. The latter functions must satisfy two simple conditions that are directly deduced from (5.5); they are:

1. The inverses are given by

$$G_{\alpha\beta}^{-1}(x) = G_{\beta\alpha}(x), \quad \forall x \in U_\alpha \cap U_\beta, \tag{5.6}$$

where by inverse we mean the inverse function.

2. The following cyclic identity is satisfied

$$G_{\alpha\beta}(x) \circ G_{\beta\gamma}(x) \circ G_{\gamma\alpha}(x) = \mathbf{1}, \quad \forall x \in U_\alpha \cap U_\beta \cap U_\gamma, \tag{5.7}$$

where the symbols "$\circ$" and "$\mathbf{1}$" denote the operation of composition of functions and the identity function (on F), respectively.

One of the consequences of these properties is that the transition functions $G_{\alpha\beta}(x)$ form a group under composition. In fact, they belong to a Lie group of transformations (diffeomorphisms) of F. This group is called the *structure group* G of the bundle E. We will adopt the point of view that in general G does not consist only of the transition functions and that besides its left action on F, as the action of the transition functions, it may also act on the total space E from the right.[7]

Having introduced the basic ingredients of fiber bundles, we address the question of the equivalence of fiber bundles. Like in the study of other mathematical structures, in order to distinguish fiber bundles from one another, we need a precise definition of an equivalence relation for fiber bundles. Since a fiber bundle is also a manifold, the desired notion of equivalence for fiber bundles must be a special case of the notion of equivalence for manifolds. This suggests that we seek a diffeomorphism that also preserves the bundle structure. A function with this property is called a *bundle map* or a *bundle morphism*. By definition, a bundle morphism $\mathcal{F}: E_1 \to E_2$ is a smooth function between two fiber bundles $E_1 : (E_1, X_1, \pi_1, G_1)$ and $E_2 : (E_2, X_2, \pi_2, G_2)$, such that $\mathcal{F}$ maps the fibers of E_1 into the fibers of E_2. This allows us to induce a smooth function $f : X_1 \to X_2$ from $\mathcal{F}$, namely we use the fact that for each $x_1 \in X_1$, $\mathcal{F}(\pi_1^{-1}(x_1))$ is a fiber of E_2, thus under π_2 it is mapped onto a single point $x_2 \in E_2$. The induced function $f : X_1 \to X_2$ is defined to map x_1 to x_2, i.e., $f(x_1) := x_2$. Clearly, we have $f \circ \pi_1 = \pi_2 \circ \mathcal{F}$. This property may be demonstrated by a commutative diagram:

$$\begin{array}{ccc} E_1 & \xrightarrow{\mathcal{F}} & E_2 \\ \pi_1 \downarrow & \bigcirc & \downarrow \pi_2 \\ X_1 & \xrightarrow{f} & X_2. \end{array}$$

The commutativity of this diagram, emphasized by the symbol "$\bigcirc$", means that we can follow the arrows in any of the two alternative directions. If a bundle morphism is also a diffeomorphism, then it is called a *bundle isomorphism*. The concept of bundle isomorphism is analogous to the concepts of group and vector space isomorphisms. It defines an equivalence relation

---

[7] As we will see, this is the case for principal fiber bundles.

on the set (category) of all fiber bundles. The isomorphic bundles are then viewed as *equivalent bundles*.

Similarly to the case of manifolds, the set of all bundle isomorphisms that map a fiber bundle (E, X, π, G) onto itself, forms a group under the operation of composition of functions. This is by construction a subgroup of the diffeomorphism group of the total space E. There is a particular class of bundle isomorphisms which are of utmost importance in fiber bundle theory. These are bundle isomorphisms which leave the base manifold X unaltered, i.e., the induced map f of all such bundle isomorphisms is the identity map on X (for all $x \in X$, $f(x) = x$). The set of these "special" bundle isomorphisms forms a subgroup of the group of all bundle isomorphisms of E. This subgroup is called *the group of gauge transformations* of E. We shall discuss the importance of this group in the context of gauge theories.

The simplest and the most familiar of all fiber bundles are the so-called *trivial bundles*. A trivial bundle is basically the product manifold[8] of two smooth manifolds,

$$E \simeq X \times F. \tag{5.8}$$

The trivial bundles are already "trivialized". This means that we can choose to have a single chart with $U_\alpha = X$.[9] Consequently, there is no need to introduce transition functions. An additional structure that distinguishes a trivial bundle from a product manifold is that a trivial bundle E has a structure group G that may act on its fibers, $F_x := \{x\} \times F \simeq F$. We shall see in the context of principal bundles that this additional structure allows a trivial principal bundle to support a geometric structure. Thus in general a trivial bundle may have a quite non-trivial geometry.

In physics, one most often tries to describe physical quantities as functions or fields of space-time. For example, we have scalar fields such as the temperature field of a gas, vector fields such as the velocity field of a fluid, and tensor fields such as the electromagnetic field. Let us examine a real scalar field f. By definition f is a function from space-time, e.g., Minkowski space M^4, into the real numbers $\mathbb{R}$

$$f : M^4 \longrightarrow \mathbb{R}.$$

The field f can also be described in terms of its graph. This is the function $\mathcal{F} : M^4 \to M^4 \times \mathbb{R}$, defined by

$$\mathcal{F}(x) := (x, f(x)), \quad \forall x \in M^4.$$

The target space $M^4 \times \mathbb{R}$ is a typical example of a trivial bundle. Its typical fiber is the real line $\mathbb{R}$. Similarly, the graph of, say, a complex vector field $\varphi = (\varphi^1(x), \cdots, \varphi^N(x))$ is a function from M^4 into the trivial bundle $M^4 \times \mathbb{C}^N$. These two examples of trivial fiber bundles have the interesting property that

---

[8] See Appendix A.2 for definitions.
[9] Such a trivialization is called a global trivialization.

their fibers are vector spaces. They are called *trivial vector bundles*. Most often the fields employed in ordinary field theories are associated with trivial vector bundles. Most generally, the (matter) fields correspond to smooth functions s from a subset U of the base manifold X of a general fiber bundle (E, X, π, G) into its total space E, such that

$$\pi(s(x)) = x, \quad \forall x \in U \subseteq X. \tag{5.9}$$

These functions are called *sections* of the fiber bundle. If the domain U of such a function is not the entire base manifold, the section is called a *local section*, otherwise it is called a *global section*.

Next we would like to review the subject of (general – not necessarily trivial) *vector bundles*. We explained that a fiber bundle was a collection of four objects:

$$(E, X, \pi, G),$$

with fibers given by $F_x := \pi^{-1}(x)$ for all $x \in X$. A vector bundle (E, X, π, G) is a fiber bundle whose fibers V_x and typical fiber V are complex (real) vector spaces and whose structure group G is the general linear group $GL(\mathcal{N}, \mathbb{C})$ (respectively $GL(\mathcal{N}, \mathbb{R})$). The fibers of a vector bundle are mapped to its typical fiber $V = \mathbb{C}^{\mathcal{N}}$ (respectively $\mathbb{R}^{\mathcal{N}}$) by vector space isomorphisms.

A simple example of a non-trivial vector bundle is the *Möbius strip*. In fact, the fibers of the Möbius strip are finite intervals, but we shall consider a Möbius strip of infinite width. The bundle structure of the Möbius strip is discussed in [64, 190] and [186] in detail. It is not difficult to see that the Möbius strip is not a trivial bundle. The base manifold is the circle S^1 and the fibers are copies of the real line $\mathbb{R}$. The (unique) trivial bundle with the same base manifold and typical fiber is the cylinder $S^1 \times \mathbb{R}$. As two-dimensional manifolds, $S^1 \times \mathbb{R}$ is orientable, whereas the Möbius strip is not orientable. Hence the Möbius strip has a different topological structure and as a vector bundle it is not trivial.[10] The Möbius strip can be constructed using two charts which are glued together via transition functions belonging to the group $\mathbb{Z}_2 = O(1)$. The structure group of our Möbius strip is the group $GL(1, \mathbb{R})$. This is the Abelian group formed by non-zero real numbers under multiplication. Note that $\mathbb{Z}_2 = O(1)$ is a subgroup of $GL(1, \mathbb{R})$.

If the fibers of a vector bundle are one-dimensional, i.e., the rank $\mathcal{N} = 1$, then the vector bundle is called a *line bundle*. For example the Möbius strip (with infinite width) is a real line bundle over S^1. Complex line bundles have numerous applications in mathematics and theoretical physics. Among their applications in physics are areas such as Abelian gauge theories [64, 78, 186], geometric quantization [273], and geometric phases. The simplest non-trivial complex line bundles have $X = S^2$ as their base manifold. These are directly

---

[10] A simple test of orientability at least for two-dimensional manifolds is that one cannot paint the whole manifold (surface) with a brush that always touches it. Accordingly, a cylinder is orientable whereas the Möbius strip is not.

related to the quantum mechanical system of (2.1), and are discussed in Sect. 6.5.

Closely related to vector bundles are another class of fiber bundles called *principal fiber bundles*. A principal fiber bundle (PFB) is a fiber bundle whose typical fiber and the structure group are identical, i.e., they are diffeomorphic as smooth manifolds. Let us consider a PFB

$$(P, X, \pi, G), \quad \text{with} \quad \pi^{-1}(x) = F_x \simeq G, \quad \forall x \in X,$$

and a complete set of local trivializations $\{(U_\alpha, \Phi_\alpha)\}$. The transition functions $G_{\alpha\beta}(x) \in G$, (5.5), act on the typical fiber $F = G$ by group multiplication from the left,

$$g_1 \xrightarrow{G_{\alpha\beta}(x)} g_2 := G_{\alpha\beta}(x) \bullet g_1, \quad g_1, g_2 \in F = G.$$

Furthermore, G has a right action $R : G \times P \to P$,

$$R(g, p_1) =: R_g(p_1), \quad \forall g \in G, \forall p_1 \in P,$$

on the total space P that is defined by

$$p_1 \xrightarrow{g} p_2 = R_g(p_1) := \Phi_\alpha^{-1}(x) \left[\Phi_\alpha(x)[p_1] \bullet g \right], \tag{5.10}$$

where $x := \pi(p_1)$ and $\bullet$ denotes the group multiplication in G. Equation (5.10) can be described in the following manner. First, we map the point $p_1 \in F_x$ to $F = G$ via $\Phi_\alpha(x)$. The resulting object, $\Phi_\alpha(x)[p_1]$, is a group element. We multiply the latter by the chosen element $g \in G$ from the right and finally map the product back to F_x via $\Phi_\alpha^{-1}(x)$. In this way the right action of G moves the points of P within individual fibers. There is a rather confusing but widely used notation for the right action of G on P:

$$p_2 := R_g(p_1) \stackrel{\text{notation}}{=\!=\!=} p_1 \cdot g.$$

An important observation is that the definition of the right action (5.10) is independent of the choice of the local chart. This means that if we choose a different chart (U_β, Φ_β) with $x \in U_\beta$ in (5.10), we will arrive at the same result (Problem 5.1).

The canonical right action of the group G on itself which is employed in the definition of the right action of G on P has certain desirable properties. Among these is the fact that this action is both *transitive* and *free*.[11] This leads immediately to the fact that G acts on the fibers F_x transitively and freely. Briefly this means that every pair of points p and p' of a fiber F_x can be connected via a group element g, i.e., there is some $g \in G$ such that $p' = p \cdot g$ (transitivity), and that there is no group element besides the identity element that leaves all the points of a fiber unchanged (freedom!).

---

[11] For a detailed discussion of this point see Appendix A.3.

The structure group of a PFB is also called the *gauge (symmetry) group* in gauge theories. The subject of PFBs is directly related to the gauge theories of physics. In fact, the early work on PFBs was simultaneously conducted by theoretical physicists and differential geometers in the 1950s and 1960s.

The simplest types of PFBs are the *trivial principal fiber bundles*. A typical example of a trivial PFB is the two-dimensional torus $T^2 = S^1 \times S^1$. The base manifold and the typical fiber are both S^1. The structure group is $U(1)$ which has the manifold structure of S^1 as well.

The bundles η and λ that we introduced in connection with the quantum mechanical phases are examples of $U(1)$ PFBs. The fact that the group $U(1)$ is an Abelian Lie group simplifies the study of $U(1)$ PFBs appreciably.

We mentioned that PFBs are closely related to vector bundles. To see this, let us consider a vector bundle $E : (E, X, \pi, G)$, with G being the matrix group $GL(\mathcal{N}, \mathbb{C})$ or $GL(\mathcal{N}, \mathbb{R})$. We can associate a PFB $\mathcal{E} : (\mathcal{E}, X, \pi', G)$ to E by replacing the fibers E_x of E by copies of the group G and using the transition functions of E to glue these fibers, i.e., identifying the transition functions of $\mathcal{E}$ with those of E. This is always possible since the transition functions of E belong to G. The PFB $\mathcal{E}$ is called the *associated principal bundle* to the vector bundle E. An intuitive description of this construction is to identify the points of the fibers $\mathcal{E}_x$ of $\mathcal{E}$ with the frames on the fibers E_x of E. Note that a frame on E_x is a collection of $\mathcal{N}$ linearly independent vectors. These can be viewed as the rows or columns of an $\mathcal{N} \times \mathcal{N}$ matrix which in turn corresponds to an element of G. If we view the action of G on E_x as linear transformations of the vectors (points of E_x), then the action of G on $\mathcal{E}_x$ corresponds to linear transformations of the frames. Thus, the relation between a vector bundle and its associated PFB resembles the relation between the active and the passive transformations of linear algebra.

Conversely, we can associate vector bundles to a given PFB. This is however not a unique process. We must first choose a representation of the structure group G. This is because in general G is an abstract Lie group; it does not have to be a general linear group. A choice of a representation allows us to identify G with a subgroup of a general linear group. In particular, the transition functions are represented by matrices that can act on vectors through the simple multiplication of a matrix by a vector from the left. The fibers of the associated vector bundle are copies of the representation space. They are glued together using the representation of the transition functions of the original PFB. Thus to each PFB there corresponds a class of associated vector bundles that are labeled by the representations of the group G.

In physics we most often encounter the unitary and orthogonal groups. It turns out that in fact these groups play a central role in the study of vector bundles and their associated PFBs. The mathematical result leading to this assertion is known as the theory of *bundle reduction* [64]. According to this theory, the transition functions of any complex vector bundle of rank $\mathcal{N}$ may be chosen from the unitary group $U(\mathcal{N})$ rather than the group $GL(\mathcal{N}, \mathbb{C})$.

Similarly, the transition functions of a real vector bundle of rank $\mathcal{N}$ may be chosen from the orthogonal group $O(\mathcal{N})$. Thus, in general, we can use the groups $U(\mathcal{N})$ and $O(\mathcal{N})$ as the structure groups of the complex and real vector bundles, respectively. In this case, we will call such vector bundles, $U(\mathcal{N})$-*vector bundles*, respectively, $O(\mathcal{N})$-*vector bundles*. The corresponding associated PFBs will naturally have the groups $U(\mathcal{N})$ and $O(\mathcal{N})$ as their structure groups. They are called $U(\mathcal{N})$-*principal bundles* and $O(\mathcal{N})$-*principal bundles* for short.

So far, we have defined the notion of a fiber bundle and studied the main categories of fiber bundles. We showed that the topological properties of a fiber bundle are determined by a set of gluing or transition functions. For trivial fiber bundles, the topology is uniquely defined by the Cartesian product. However, in general, one needs some qualitative measure of the topology. To get a feeling about the difference between the topology and the geometry of a fiber bundle (which also is a manifold), we consider the two simple examples of the finite cylinder $S^1 \times [0,1]$, and the torus $T^2 = S^1 \times S^1$. Figures 5.1 and 5.2 illustrate two copies of the cylinder and the torus, respectively. Although both copies of each space describe the same global or topological structure, their geometrical properties are different. This can be verified by noting that the objects in each figure can be deformed in a continuous (smooth) fashion into one another, whereas the objects in each figure cannot be smoothly deformed into those of the other. A smooth deformation or more precisely a diffeomorphism does not change the global properties of a manifold, but it does move its points. Thus, in general, the distance between the points of a manifold and the tangent vectors change under a diffeomorphism.

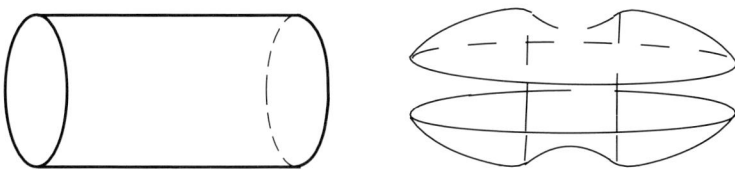

Fig. 5.1. Two geometrically inequivalent copies of the cylinder $S^1 \times [0,1]$.

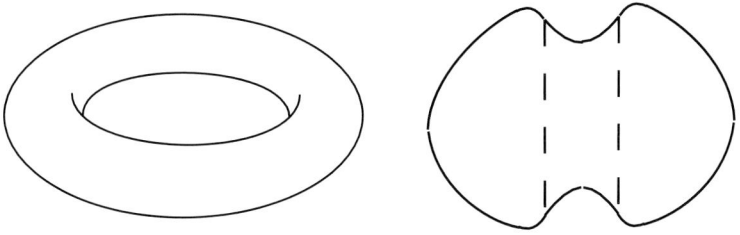

Fig. 5.2. Two geometrically inequivalent copies of the torus.

This implies that the geometric structure of a manifold is not invariant under general diffeomorphisms.

In the previous paragraph, we have assumed that the manifolds under consideration possess geometric structures. For the cylinders and the tori depicted in Figs. 5.1 and 5.2, the geometric structure is fixed to be the one induced by the Euclidean space $\mathbb{R}^3$ in which they are embedded. However, we can define and study the geometry of a manifold independently of the geometry of an embedding Euclidean space. In any case, a geometry on a manifold is in general given by a symmetric tensor field

$$g = g_{ij}(x)\, dx^i \otimes dx^j,$$

which is called the *metric tensor*. The metric tensor can be used to determine the length of the curves and the distances between the points of the manifold. Furthermore, it specifies the angle between two vectors, and defines the concept of parallel transport of vectors along the curves of the manifold. The latter can also be done using yet another quantity called a *connection* on a manifold. This is expressed by a three-indexed field $\Gamma = (\Gamma_{ij}^k(x))$. Γ_{ij}^k are called the *Christoffel symbols* of the manifold. They do not however transform like the components of a tensor field. The geometry of smooth manifolds is discussed in many textbooks [64, 186, 190]. We shall only offer a brief discussion of parallel transport on a manifold in the following section.

The geometry of an m-dimensional smooth manifold X is in fact identical with the geometry of its tangent bundle or alternatively its frame bundle. The *tangent bundle* TX of X is a real vector bundle of rank m over X, whose fibers are the tangent spaces to the manifold.[12] The *frame bundle* FX is the associated PFB to the tangent bundle TX. Its fibers consist of the frames over the corresponding fibers of TX, i.e., the tangent spaces. One of the most important applications of the geometry of tangent bundles (respectively frame bundles) is in the general theory of relativity where the fundamental physical field is the metric tensor of a space-time manifold.

5.4 Geometry of Principal Bundles and the Concept of Holonomy

By the word *geometry* one usually means a structure which allows one to compare two points, i.e., to measure their distance, or two directions, i.e., to measure the angle between them, or both. Just having the ability to compare directions suffices to define many interesting geometric concepts. For a smooth manifold, the directions are quantified by tangent vectors. A priori, one cannot compare two tangent vectors unless they belong to the same

---

[12] For a more detailed discussion of tangent vectors and tangent spaces of a manifold, see Appendix A.2.

5.4 Geometry of Principal Bundles and the Concept of Holonomy

tangent space. Let us consider two distinct points x_1 and x_2 of a smooth manifold X, and let $v_1 \in TX_{x_1}$ and $v_2 \in TX_{x_2}$ be two tangent vectors at x_1 and x_2, respectively. To compare v_1 with v_2, we proceed by finding a way to associate with v_2 a vector which belongs to TX_{x_1}. Then we can compare v_1 with this vector directly. The associated vector is also called a *parallel vector* to v_2. In a Euclidean space, we can use the notion of Euclidean distance to define parallel vectors to a given vector. This turns out not to apply for general manifolds since, to begin with, the notion of distance is not universally determined as it is in the case of a Euclidean space. Nevertheless, we can use our knowledge about Euclidean geometry to define a notion of parallelism for arbitrary manifolds. In fact, we will conclude that there does not exist a unique notion of parallelism for a given smooth manifold. The notion of parallelism is uniquely defined if an additional structure is given to the manifold. This additional structure is called a *connection*. We shall first describe an intuitive procedure to construct the parallel vector to v_2 at x_1.

In order to determine the parallel vector to v_2 at x_1, we first choose a smooth curve C_X which joins x_1 and x_2 and try to solve the problem locally; that is we first replace the role of x_1 by a point x which is "nearby" x_2 and belongs to C_X. If we can find a way to determine uniquely the parallel vector to v_2 that is tangent to X at x, then we can repeat the same procedure until we arrive at x_1. Here, by the word "nearby" we mean that $x := C_X(t=T-dt)$, where $x_1 = C_X(t=0)$, $x_2 = C_X(t=T)$, and $t \in [0, T]$ is the parameter of the curve $C_X : [0, T] \to X$. Next let us imagine that the manifold is embedded in a Euclidean space[13] $\mathbb{R}^d$ and view C_X as sitting inside $\mathbb{R}^d$. To find the parallel vector at x we treat the vector v_2 as an element of $\mathbb{R}^d$ (note that $T\mathbb{R}^d = \mathbb{R}^d$) and find the parallel vector to v_2 at the point $x \in \mathbb{R}^d$. The resultant vector is in general not tangent to the manifold, so we choose the component of it which is tangent to X and discard the other (normal) components. The parallel vector $v_2(x)$ that is defined in this way belongs to TX_x. It differs from the parallel vector to v_2 in the sense of $\mathbb{R}^d$ by a factor which is proportional to dt. It also depends on the point $x \in C_X$. Next we choose a point x' which is "nearby" x and which belongs to the portion of C_X that lies between x and x_1. We use the same method to define the parallel vector $v_2(x')$ to the vector $v_2(x)$. This procedure can be carried out successively until we finally define the desirable parallel vector to v_2 at x_1. The vector v_2 is said to be *parallel transported* to x_1 along the curve C_X.

As we can clearly see, there is a major difference between the parallel transportation in $\mathbb{R}^d$ and the one we described for a general manifold X. Our prescription depended on the particular curve C_X in X. Furthermore, in order to follow this prescription we must know the precise embedding of X in $\mathbb{R}^d$. If we deform a portion of X which includes C_X and deform back after we parallel transported v_2, the result will in general be different. Hence

---

[13] This is always possible, due to the Whitney embedding theorem. See Appendix A.2 for a brief review.

78   5. Fiber Bundles and Gauge Theories

in this general case, the deformed and undeformed copies of X have different *geometric structures*.

A geometry on a manifold can be studied in terms of the parallel transport of tangent vectors. These belong to the tangent bundle TX. Alternately we can speak of the parallel transport of a frame from x_2 to x_1 along a curve C_X. This does not introduce any additional difficulties, for a frame is nothing but a collection of tangent vectors. As we discussed in the previous section frames of TX constitute the frame bundle FX of X. This is the associated PFB to the tangent bundle. Thus the geometry of TX can be understood in terms of the geometry of FX and vice versa. The same is valid for arbitrary vector bundles. We can always construct the associated PFB to a given vector bundle and translate the geometric content of the vector bundle into that of the associated PFB. This signifies the importance of the subject of this section: *the geometry of principal fiber bundles*.

The notion of *parallelism* that we developed using an embedding of the manifold X in a Euclidean space $\mathbb{R}^d$, has the same general properties as parallelism in Euclidean geometry. More precisely it defines an equivalence relation between parallel vectors. It is not difficult to see that the above procedure ensures that:

1) Every vector is parallel to itself [reflexivity].
2) If a vector v_1 is parallel to another vector v_2 (v_1 is parallel transported to v_2 via the curve $C_X : [0,T] \to X$), then v_2 is parallel to v_1 (v_2 is parallel transported to v_1 via the curve $C_X^{-1} : [0,T] \to X$ with $C_X^{-1}(t) = C_X(T-t)$ for all $t \in [0,T]$) [symmetry].
3) For three vectors v_1, v_2, and v_3, if v_1 is parallel to v_2 and v_2 is parallel to v_3, then v_1 is parallel to v_3 [transitivity].

These properties, which may be shown to hold for Euclidean geometry or the geometry we outlined using an embedding of a manifold in a Euclidean space, serve as the postulates of an abstract geometric structure on a principal fiber bundle.

In our discussion of the Aharonov–Anandan phase, we briefly mentioned that a geometric structure on a PFB is determined by a quantity also called a *connection*. In the following paragraphs, we shall try first to motivate and then present the definition(s) of a connection on a PFB. Again, there are many excellent references on the subject [64, 114, 186].

We shall first try to understand what one means by *parallelism* for PFBs. Let us consider a PFB $P : (P, X, \pi, G)$ and let C_X be a smooth curve in X with parameter $t \in [0,T]$. We will view the concept of *parallel transportation* on P as a means to find a curve C_P in the total space P that follows the curve C_X as the parameter t traverses the interval $[0,T]$, such that

1. C_P projects onto C_X under the projection $\pi : P \to X$.

5.4 Geometry of Principal Bundles and the Concept of Holonomy

2. The tangent vectors $w_t \in TP$ to C_P project onto the tangent vectors $v_t \in TX$ to C_X under the push-forward (differential) map[14] $\pi_* : TP \to TX$.

These conditions are, in fact, not restrictive enough as to specify C_P uniquely. The structure that enables one to determine the curve

$$C_P : [0, T] \ni t \longrightarrow p(t) \in P,$$

uniquely for a given curve

$$C_X : [0, T] \ni t \longrightarrow x(t) \in P$$

and an initial condition $p(0) \in F_{x(0)}$, is called a *geometry* or a *connection* on the PFB P. The curve C_P is called the *horizontal lift* of C_X associated with this connection. The tangent vectors w_t to C_P are called the *horizontal vectors*.

For different choices of the curve C_X in X, a connection yields different horizontal lifts C_P and hence different collections of horizontal vectors. Let us focus our attention on a point $p \in P$, and consider the horizontal lifts of all the curves which pass through the point $x := \pi(p) \in X$. These will determine a set of horizontal vectors that belong to the tangent space T_pP of P at p. These horizontal vectors form a vector subspace of T_pP that is called the *horizontal subspace* and denoted by H_pP. On the other hand, we can consider another type of curves in P that pass through the point p but lie entirely on the fiber P_x. The tangent vectors to these curves (at p) project onto the zero vector in T_xX. These vectors are called the *vertical vectors*. They also form a vector subspace V_pP of T_pP, which is naturally called the *vertical subspace*. More importantly, as a vector space, T_pP is just the direct sum of its horizontal and vertical subspaces

$$T_pP = H_pP \oplus_p V_pP, \qquad \forall p \in P. \tag{5.11}$$

We would like to add that by definition the vertical subspaces are uniquely defined by the projection map $\pi : P \to X$, whereas the horizontal subspaces are only determined if the bundle P is endowed with a connection. The direct sum $\oplus_p$ carries an index p to indicate that the decomposition depicted in (5.11) is valid pointwise.

So far, we have discussed the utility of a connection in defining the horizontal lifts and horizontal subspaces. We will next try to give more precise definitions of a connection. There are three equivalent definitions of a connection on a PFB. We have already outlined the first definition which identifies a connection as a rule which determines the horizontal lifts. This point of view is the one originally used in the definition of the horizontal lift in the context of the A-A phase in Chap. 4. There we also argued that the horizontal lift and therefore the A-A phase could be expressed in terms of a connection

---

[14] A definition of a push-forward map is given in Appendix A.2.

one-form – the A-A connection one-form. In the following we present the other two definitions of a connection on a PFB. The purpose of these alternative definitions is to emphasize the fact that in general a connection may be expressed in terms of a connection one-form.

In the remainder of this section we shall assume that the structure group G is a matrix group. The general case is obtained quite similarly. Nevertheless, it is the case of matrix groups that is of interest in ordinary gauge theories and particularly in the holonomy interpretations of the geometric phase.

Definition 5.4.1. *A **connection** on a principal fiber bundle (P, X, π, G) is a collection of vector subspaces $H_p P \subset T_p P$, such that for all $p \in P$:*

1. *The linear map $\pi_*|_{H_p P} : H_p P \to T_x X$, with $x := \pi(p)$ is an isomorphism of vector spaces.*
2. *The subspaces $H_p P$ depend smoothly on p.*
3. *The right action of G on P transforms the horizontal subspaces according to*
$$H_{p \cdot g} P = R_{g*}(H_p P).$$

In this definition the subscript $*$ denotes the push-forward (differential) of the corresponding maps. For example, $\pi_* : T_p P \to T_x X$ is the push-forward map associated with $\pi : P \to X$, and $R_{g*} : T_p P \to T_{p \cdot g} P$ is the push-forward map associated with the right action map $R_g : P \to P$ defined by (5.10).

In fact, this definition is of limited practical use. A more practical approach would be to find a mathematical object that yields these horizontal subspaces algebraically. This is done by a differential one-form ω on P with values in the Lie algebra[15] $\mathcal{G}$ of the structure group G. This means that if we evaluate ω on a point $p \in P$, the result, ω_p, is a covariant vector whose components belong to the Lie algebra $\mathcal{G}$ of G. The idea is to obtain the horizontal subspaces $H_p P$ as the space of solutions of the equation

$$\omega_p(w_p) = 0. \qquad (5.12)$$

That is, the tangent vectors $w_p \in T_p P$ that satisfy (5.12) are defined to be the horizontal vectors. Thus given such a one-form we ought to be able to obtain the horizontal vectors and hence the horizontal subspaces. However, we must first ensure that ω is well-defined. Even more importantly we wish to understand the reason for choosing ω to be a Lie algebra-valued one-form.

To make sure that ω is well-defined, we must determine its values on the vertical vectors. This is the main reason for choosing ω to be a $\mathcal{G}$-valued one-form. In our discussion of the vertical vectors, we introduced them as the tangent vectors to the curves in P that lay on a single fiber. Therefore, a vertical subspace $V_p P$ is the tangent space to the fiber F_x, at $p \in F_x$. On the

---

[15] Recall that a differential one-form is a linear function which maps the tangent vectors to real numbers. Similarly, a Lie algebra-valued one-form is a linear function which maps the tangent vectors to the elements of a Lie algebra.

other hand, the fibers F_x are copies of the structure group G. Consequently, there is a correspondence between $V_p P$ and the tangent spaces $T_g G$ of G. For every $g \in G$, $T_g G$ is (canonically) isomorphic to the tangent space to G at the identity element $e \in G$, but this space is easily identified with the Lie algebra of G.[16]

The relation between the vertical subspaces and the structure Lie algebra can be easily understood by considering a typical curve Γ in the fiber F_x and trying to find the corresponding curve in G. Let us choose an arbitrary vertical vector v_p at p. Then, there is a smooth curve $\Gamma : [0, T] \to F_x$ which starts off at p, i.e., $\Gamma(0) = p$, and is tangent to v_p at p, namely

$$v_p = \left.\frac{d\Gamma(t)}{dt}\right|_{t=0}. \tag{5.13}$$

Since for all $t \in [0, T]$, $\Gamma(t) \in F_x$, the transitivity of the right action of G on F_x implies the existence of $g(t) \in G$, such that

$$\Gamma(t) = p \cdot g(t), \qquad \forall t \in [0, T]. \tag{5.14}$$

The group elements $g(t)$ trace a smooth curve $\tilde{\Gamma} : [0, T] \to G$ in G, with $\tilde{\Gamma}(0) = e$. Substituting (5.14) in (5.13), and noting that p does not depend on the parameter t, we can immediately associate to the vertical vector v_p the element

$$\mathcal{X} := \left.\frac{dg(t)}{dt}\right|_{t=0} \tag{5.15}$$

of the Lie algebra $T_e G = \mathcal{G}$. It is obvious that our construction works both ways. That is, given $\mathcal{X}$ and a point $p \in P$, we can obtain the corresponding vertical vector v_p.

Having established the relationship between the vertical subspaces and the structure Lie algebra, we can present the third definition of a connection on a PFB. This definition involves a set of conditions on the Lie algebra-valued one-form ω that makes it unique and reasonably transformed under the right action of G on P.

Definition 5.4.2. *A **connection** on a principal fiber bundle (P, X, π, G) is a one-form ω on P with values in the Lie algebra $\mathcal{G}$ of G, such that for all $p \in P$*

1. *The vertical vectors $v_p \in V_p P$ satisfy the equation $\omega_p(v_p) = \mathcal{X}$, where $\mathcal{X} \in \mathcal{G}$ is the element of $\mathcal{G}$ defined by (5.15).*
2. *ω_p depends smoothly on p.*
3. *Under the right action of G on P, ω transforms according to*[17]

---

[16] We have presented a simple justification of this identification in Appendix A.3.
[17] Note that here G is a matrix group and the elements $g \in G$ and $\mathcal{X} \in \mathcal{G}$ are finite-dimensional matrices.

$$\omega_{p.g}\left(R_{g*}(v_p)\right) = g \cdot (\omega_p(v_p)) \cdot g^{-1} = g \cdot \mathcal{X} \cdot g^{-1}, \tag{5.16}$$

where $v_p \in V_pP$ and $g \in G$ are arbitrary elements.

Given a Lie algebra-valued one-form that satisfies these conditions, we identify the horizontal subspaces by

$$H_pP := \{w_p \in T_pP \; : \; \omega_p(w_p) = 0\}. \tag{5.17}$$

In Def. 5.4.2, the first condition is enforced to make ω well-defined on V_pP. The smoothness condition (in both Def. 5.4.1 and Def. 5.4.2) is to ensure that the horizontal lifts are smooth curves in P. Finally, the last requirement may be viewed as a consistency condition the details of which will be explained momentarily.

A fact of practical importance about the $\mathcal{G}$-valued one-form ω is that *locally* it can be determined by a $\mathcal{G}$-valued one-form on X [64, 186]. In other words, for any $x_0 \in X$, there is an open chart (U_α, Φ_α) including x_0, a local section $s : U_\alpha \to P$, and a one-form A_α^s on X with values in $\mathcal{G}$, such that on U_α

$$s^*(\omega) = A_\alpha^s, \tag{5.18}$$

where the superscript $*$ denotes the pullback operation for the differential forms.[18] The Lie algebra-valued one-form A_α^s is called the *local connection one-form*.

As the notation suggests, A_α^s depends on the choice of the local section s and the chart (U_α, Φ_α). In fact, it is possible to choose a *canonical local section* s_α associated with each chart. The local connection one-form A_α associated with the canonical local section depends only on the chart.

The canonical local sections are defined by

$$\Phi_\alpha\left(s_\alpha(x)\right) := e \in G, \qquad \forall x \in U_\alpha. \tag{5.19}$$

In fact, it is not difficult to see the generality of the use of the canonical local sections. Any local section $s : U_\alpha \to P$ can be obtained from the canonical one by an appropriate right action of G on P; that is for any $x \in U_\alpha$, there is $g(x) \in G$ such that

$$s(x) = s_\alpha(x) \cdot g(x).$$

Hence, if we know the transformation properties of A_α^s under the right action of $g(x) \in G$ on the fibers F_x, we can express A_α^s in terms of A_α. We can deduce the transformation rule from the defining conditions of Def. 5.4.2 and (5.18). In general, if two local sections s_1 and s_2 are related by

$$s_1(x) \;\to\; s_2(x) = s_1(x) \cdot g(x), \qquad \forall x \in U_\alpha, \; g(x) \in G, \tag{5.20}$$

then the corresponding local connection one-forms satisfy [186]

---

[18] Check Appendix A.2 for a definition of the pullback operation for differential forms.

5.4 Geometry of Principal Bundles and the Concept of Holonomy

$$A_\alpha^{s_2}(x) = g^{-1}(x) \cdot A_\alpha^{s_1}(x) \cdot g(x) + g^{-1}(x) \cdot dg(x). \tag{5.21}$$

In particular, we can select s_1 to be the canonical local section and express A_α^s in terms of A_α.

So far, we have defined the local connection one-form on a single chart. As suggested by the qualification "local", we cannot extend the domain of the definition of a local connection one-form outside its local chart. Instead, we can try to find the transformation rule that it satisfies upon going from one chart to another. This amounts to the investigation of the behavior of local connection one-forms under the left action of the transition functions on the fibers. We shall again be content to quote the result.

Let us choose two intersecting charts (U_α, Φ_α) and (U_β, Φ_β). Then for all $x \in U_\alpha \cap U_\beta$ we have [64, 186]

$$A_\beta^s(x) = G_{\alpha\beta}^{-1}(x) \cdot A_\alpha^s(x) \cdot G_{\alpha\beta}(x) + G_{\alpha\beta}^{-1}(x) \cdot dG_{\alpha\beta}(x). \tag{5.22}$$

In a sense, the advantage of expressing a connection on a PFB by $\mathcal{G}$-valued one-forms on the base manifold is balanced by the unfortunate fact that we need several of these objects which are tied together according to (5.22). In contrast, the $\mathcal{G}$-valued one-form ω is globally defined, and therefore it is more useful in abstract studies of the geometry of PFBs.

Next we would like to demonstrate the utility of the local connection one-forms in defining the horizontal lifts in practice.

Let $C_X : [0, T] \to X$ be a curve in the base manifold X. To obtain the horizontal lift $C_P : [0, T] \to P$ of C_X with $C_P(0) = p_0$, we proceed as follows:

1. Choose a complete set of local trivializations $\{(U_\alpha, \Phi_\alpha)\}$.
2. Divide C_X into its segments C_X^α which belong to the U_α's.
3. For each U_α, which includes a segment of C_X, choose a local section $s : U_\alpha \to P$. Then the horizontal lift of C_X is the horizontal lifts of its segments. This reduces the problem to the case where C_X lies in a single chart U_α.
4. Define the horizontal lift of $C_X \subset U_\alpha$ by

$$C_P(t) := s\left(C_X(t)\right) \cdot g_s(t), \tag{5.23}$$

where $g_s(t) \in G$ is the solution of

$$\frac{dg_s(t)}{dt} = -A_\alpha^s(v_t) \cdot g_s(t), \tag{5.24}$$

$$g_s(0) = e, \quad \text{with} \quad v_t := \frac{dC_X(t)}{dt}.$$

In (5.24), the right-hand side is the product of two matrices. The first of these is the element of the Lie algebra $\mathcal{G}$ which is obtained by evaluating the local connection one-form A_α^s on the tangent vector v_t to the curve C_X at $C_X(t)$. The resemblance of (5.24) to the Schrödinger equation (2.4) is quite remarkable. We can use this resemblance to give the solution of (5.24), namely

84   5. Fiber Bundles and Gauge Theories

$$g_s(t) = \mathcal{P} \exp\left[-\int_{C_X(0)}^{C_X(t)} A_\alpha^s(v_t)\right], \tag{5.25}$$

where the symbol $\mathcal{P}$ means that the quantity on the right-hand side is *path-ordered*. To clarify what we mean by a path-ordered product, let us write the local connection one-form in its components,

$$A_\alpha^s =: -i A_j(x)\, dx^j. \tag{5.26}$$

Here we have introduced a factor of $-i$ to unify our notation with those of the previous chapters, and we have suppressed the dependence of A_i on the local section s and the chart α for convenience. Furthermore, note that the coordinates x^i correspond to the points $x \in U_\alpha$, and that A_i belong to the Lie algebra $\mathcal{G}$ (up to a multiplicative factor of i).

Physicists' definition of a "connection", also called a "gauge potential", corresponds to the Lie algebra-valued one-form[19]

$$A := A_i(x)\, dx^i = i A_\alpha^s. \tag{5.27}$$

Under the right action of the group element $g(x) \in G$, A transforms according to

$$A(x) \xrightarrow{g} A'(x) = g^{-1}(x) \cdot A(x) \cdot g(x) + i g^{-1} \cdot dg(x), \tag{5.28}$$

where we have used (5.21) and (5.27). If G is an Abelian group, then transformation (5.28) reduces to

$$A(x) \xrightarrow{g} A'(x) = A(x) + d(i \ln g(x)), \tag{5.29}$$

with "ln" denoting the natural logarithm.

In view of (5.26) and the definition of the tangent vector v_t, namely

$$v_t := \frac{dC_X(t)}{dt} = \frac{dx^i(t)}{dt} \frac{\partial}{\partial x^i}, \qquad C_X(t) =: (x^i(t)),$$

Equation (5.25) can be written in the form

$$g_s(t) = \mathcal{T} \exp\left[i \int_0^t A_j(C_X(t')) \frac{dx^i(t')}{dt'} dt'\right] \tag{5.30}$$

$$= \mathcal{P} \exp\left[i \int_{C_X(0)}^{C_X(t)} A_i(x) dx^i\right], \qquad x \in C_X$$

$$= \mathcal{P} \exp\left[i \int_{C_X(0)}^{C_X(t)} A(x)\right], \tag{5.31}$$

---

[19] There is also an alternative convention which differs from ours by a minus sign in the definition of A.

5.4 Geometry of Principal Bundles and the Concept of Holonomy

where $\mathcal{T}$ denotes the time-ordering operator that we described in Chap. 2. Note that the exponents in (5.25), (5.30), and (5.31) belong to the Lie algebra $\mathcal{G}$ and "exp" denotes the *exponential map* which maps the algebra elements to the group elements.[20]

In (5.23) we have used the local section $s : U_\alpha \to P$ to define the horizontal lift of the curve C_X. Next we repeat the same construction using two different local sections $s_1 : U_\alpha \to P$ and $s_2 : U_\beta \to P$, with $C_X \subset U_\alpha \cap U_\beta$. Then according to (5.20) and (5.23), we have

$$\begin{aligned} s_1(C_X(t)).g_1(t) &= C_P(t) \\ &= s_2(C_X(t)).g_2(t) \\ &= s_1(C_X(t)).g(C_X(t)).g_2(t) , \end{aligned} \quad (5.32)$$

where g_1 and g_2 are associated with the local sections s_1 and s_2, respectively. In other words, they are defined as the solutions of (5.24) corresponding to the local connection one-forms $A := is_1^*(\omega)$ and $A' := is_2^*(\omega)$, namely

$$\frac{dg_1(t)}{dt} = iA(v_t).g_1(t), \quad \text{with} \quad g_1(0) = e, \quad (5.33)$$

$$\frac{dg_2(t)}{dt} = iA'(v_t).g_2(t), \quad \text{with} \quad g_2(0) = e. \quad (5.34)$$

Equations (5.32) indicate that $g_2(t) = g^{-1}(C_X(t)).g_1(t)$. Substituting this relation in (5.34) and using (5.33) to simplify the result, we have

$$A'(v_t) = g^{-1}(C_X(t)).A(v_t).g(C_X(t)) + ig^{-1}(C_X(t)).\left[\frac{dg(C_X(t))}{dt}\right].$$

This equation may be further simplified, if we recall the definition of the vector v_t, namely that v_t is the tangent vector to C_X at $C_X(t)$,

$$v_t = \frac{dx^i(t)}{dt}\frac{\partial}{\partial x^i}.$$

The result is

$$\frac{dx^i(t)}{dt}A'_i = \frac{dx^i(t)}{dt}\left(g^{-1}.A_i.g + ig^{-1}.\frac{\partial g}{\partial x^i}\right),$$

where A_i, A'_i, and g are computed at $x = C_X(t)$. Since C_X is an arbitrary curve, we have

$$A'_i = g^{-1} \cdot A_i \cdot g + ig^{-1} \cdot \partial_i g,$$

which is precisely (5.28) written in terms of the components of the local connection one-forms A and A'. This shows that the transformation property of

---

[20] For a brief discussion of the exponential map, see Appendix A.3. For the matrix groups "exp" reduces to the ordinary exponential of matrices.

the local connection one-form (5.28) – alternatively (5.21) which is equivalent to the defining condition (5.16) for the global connection one-form ω – is a logical consequence of the arbitrariness of the choice of a local section s in the definition of the local connection one-form (5.18) and the horizontal lift (5.23). In other words, we have shown that transformations (5.20) and (5.28) are naturally intertwined.

For a $U(1)$ principal bundle, the local connection one-form takes its values in the Lie algebra $u(1) = i\mathbb{R}$. Thus its components are numbers and we can ignore the ordering rules in the definition of the horizontal lift. Therefore the right-hand sides of (5.25), (5.30), and (5.31) are the exponentials of ordinary integrals.

Another simplification can be made if the basic curve C_X is a closed curve, i.e, $C_X(T) = C_X(0) =: x_0$. In this case, the end point $C_E(T)$ also belongs to the fiber F_{x_0} and in view of (5.23), we have

$$C_P(T) = C_P(0) \cdot g_s(T), \tag{5.35}$$

where we have used the single-valuedness of the local section s, namely $s(C_X(T)) = s(C_X(0)) =: C_P(0)$. The group element $g_s(T)$ is still given by any of (5.25), (5.30) or (5.31). However, the integrals in these equations are now evaluated over the closed curve C_X,

$$g_s(T) = \mathcal{P}\exp\left[i\oint_{C_X} A\right]. \tag{5.36}$$

If we choose different closed curves C_X in X, we will obtain different group elements $g_s(T)$. These group elements depend on the connection on the PFB, and the initial point $p_0 := C_P(0)$ of the horizontal lift. The collection of all such elements which correspond to a point $p_0 \in P$ form a subgroup of G. This subgroup is called the *holonomy group* of the connection ω (respectively A) associated with the point p_0. The element $g_s(T)$ of (5.36) is called the *holonomy element* associated with the point $p_0 \in P$, the connection ω (respectively A), and the curve C_X. We should like to note that (5.36) is valid even if C_X does not lie completely in one chart. In this case, it must be understood as the product of the path-ordered exponentials corresponding to the segments C_X^α of C_X that lie in different charts. We shall not proceed with the discussion of this general situation, as it is completely understood in view of the preceding argument.

We conclude this section by noting that the defining equation (5.24) for horizontal lifts in general and holonomy elements in particular, may also be expressed in a more compact form:

$$\frac{\mathcal{D}}{\mathcal{D}t} g_s(t) = 0,$$

where the operator $\frac{\mathcal{D}}{\mathcal{D}t}$ is defined by

$$\frac{\mathcal{D}}{\mathcal{D}t} := \frac{dx^i(t)}{dt}\mathcal{D}_i,$$
$$\mathcal{D}_i := \partial_i - iA_i. \qquad (5.37)$$

The operators $\mathcal{D}_i$ and $\frac{\mathcal{D}}{\mathcal{D}t}$ are called the *covariant derivative* and *covariant time derivative* operators, respectively.

5.5 Gauge Theories

Earlier in this chapter, we pointed out that the theory of PFBs is directly related to the gauge theories of physics. The historical development of gauge theories, however, has been quite independent of the development of the theory of fiber bundles. This is the main reason why physicists usually avoid any reference to fiber bundles in discussing gauge theories. Nevertheless, exploring the relevance of fiber bundles to gauge theories has proven very helpful in understanding some of the more subtle aspects of these theories. For this reason, we shall develop gauge theories without employing the mathematics of fiber bundles. We shall then point out how the same constructions are explained and generalized using fiber bundles in the next section.

Gauge theories have many important applications in theoretical physics. In particular, they are probably the most successful theories of fundamental interactions. In general, a gauge theory is a field theory involving two types of fields. These are *gauge fields* (also called gauge potentials) and *matter fields*. In the context of particle physics, the gauge fields represent, after quantization, the particles which mediate the fundamental forces. The best-known example of such particles is the *photon* which mediates electromagnetic interactions. There are also other particles which interact through the gauge fields. These are described by the matter fields. The *electron* field is a typical example of a matter field. Another area of application of gauge theories is in the context of geometric phases. We shall explore an example of this in Chap. 8, where we discuss a *gauge theory of molecular physics*.

We start from the gauge theory of electromagnetism. We consider a non-relativistic (spinless) charged particle subject to an electromagnetic field in three-dimensional Euclidean space $\mathbb{R}^3$. In the position (coordinate) representation, the particle is described by a scalar wave function $\psi = \psi(\mathbf{x}, t)$ where $\mathbf{x} \in \mathbb{R}^3$. The dynamics of the wave function is determined by the Schrödinger equation,

$$i\frac{\partial \psi(\mathbf{x};t)}{\partial t} = H^{(\text{el})}\psi(\mathbf{x};t). \qquad (5.38)$$

Here $H^{(\text{el})}$ is the Hamiltonian for the particle of electric charge e, in an electromagnetic field

$$\mathbf{E}^{(\text{el})} = -\nabla A_0^{(\text{el})} - \frac{1}{c}\frac{\partial \mathbf{A}^{(\text{el})}}{\partial t}, \qquad \mathbf{B}^{(\text{el})} = \nabla \wedge \mathbf{A}^{(\text{el})}. \qquad (5.39)$$

It is given by

$$H^{(\text{el})} = \frac{1}{2m}\left(\mathbf{p} - \frac{e\mathbf{A}^{(\text{el})}}{c}\right)^2 + eA_0^{(\text{el})} \tag{5.40}$$

where $A_0^{(\text{el})}$ and $\mathbf{A}^{(\text{el})}$ are the electromagnetic scalar and vector potentials, respectively.

The Hamiltonian contains, in place of the usual canonical momentum $\mathbf{p}$, the *covariant* (also called gauge-covariant, kinematical or mechanical) momentum

$$\boldsymbol{\pi} := \mathbf{p} - \frac{e}{c}\mathbf{A}^{(\text{el})} \tag{5.41}$$

which fulfills the commutation relation (Problem 5.2):

$$[\pi_a, \pi_b] = i\frac{e}{c}\epsilon_{abc}(B^{(\text{el})})^c, \qquad a,b,c = 1,2,3. \tag{5.42}$$

The Schrödinger equation (5.38) may also be written in the form:

$$\left(i\frac{\partial}{\partial t} - eA_0^{(\text{el})}\right)\psi = \frac{1}{2m}\left(\frac{1}{i}\nabla - \frac{e}{c}\mathbf{A}^{(\text{el})}\right)^2\psi. \tag{5.43}$$

As is well known, the electric field $\mathbf{E}^{(\text{el})}$ and the magnetic field $\mathbf{B}^{(\text{el})}$ are unaltered if the potentials are transformed in the following way

$$\begin{aligned}\mathbf{A}^{(\text{el})} &\to \mathbf{A}'^{(\text{el})} = \mathbf{A}^{(\text{el})} + \nabla\alpha(\mathbf{x},t),\\ A_0^{(\text{el})} &\to A_0'^{(\text{el})} = A_0^{(\text{el})} - \frac{1}{c}\frac{\partial\alpha}{\partial t}\end{aligned} \tag{5.44}$$

where $\alpha(\mathbf{x},t)$ is a well-behaved (smooth and single-valued) function of $\mathbf{x}$ and t. The scalar and the vector potential may be collectively viewed as a four-vector $(A^\mu) := (A_0^{(\text{el})}, \mathbf{A}^{(\text{el})})$ in the Minkowski space-time M^4 of special relativity. Alternatively they may be used to define a covariant vector field (a one-form),

$$A^{(\text{el})} = A_\mu^{(\text{el})}(x)dx^\mu, \tag{5.45}$$

on M^4, where $\mu = 0, 1, 2, 3$ and $x^0 := ct$. This is done by lowering the space-time index μ of A^μ by the Minkowski metric

$$\eta = (\eta_{\mu\nu}) = \text{diag}(1, -1, -1, -1),$$

i.e., by setting $A_a^{(\text{el})} = -A^a$ for $a = 1, 2, 3$. In terms of the one-form $A^{(\text{el})}$, transformations (5.44) take the form:

$$A^{(\text{el})} \to A'^{(\text{el})} = A^{(\text{el})} - d\alpha, \tag{5.46}$$

where d denotes the exterior derivative,

$$d\alpha := \partial_\mu \alpha \, dx^\mu = \frac{\partial\alpha}{\partial t}dt + \nabla\alpha.d\mathbf{x}. \tag{5.47}$$

In order that the Schrödinger equation (5.43) also remains unaltered under these transformations, the wave function $\psi(\mathbf{x}, t)$ has to undergo the (phase) transformation (Problem 5.3):

$$\psi(\mathbf{x}, t) \rightarrow \psi'(\mathbf{x}, t) = e^{i\frac{e}{c}\alpha(\mathbf{x},t)} \psi(\mathbf{x}, t). \tag{5.48}$$

This may easily be shown by checking that the Schrödinger equation (5.43) has the same form in terms of either of $A_0^{'(\mathrm{el})}$, $\mathbf{A}^{'(\mathrm{el})}$, ψ' or $A_0^{(\mathrm{el})}$, $\mathbf{A}^{(\mathrm{el})}$, ψ.

The combination of the two transformations (5.44), alternatively (5.46), and (5.48) is called a *gauge transformation* and the four-vector $(A^\mu) = (A_0^{(\mathrm{el})}, \mathbf{A}^{(\mathrm{el})})$ is called the *gauge potential*. In particular, with the interpretation of the above quantities as the electromagnetic scalar and vector potential and of ψ as the wave function of a charged particle, transformations (5.46) and (5.48) are called the *electromagnetic gauge transformations*.

If one defines in analogy to the covariant momentum (5.41) the gauge-covariant derivatives $\mathbf{D}$ and D_t by

$$\begin{aligned}\frac{1}{i}\mathbf{D} &= \frac{1}{i}\mathbf{D}(\mathbf{A}^{(\mathrm{el})}) := \frac{1}{i}\nabla - \frac{e}{c}\mathbf{A}^{(\mathrm{el})}, \\ i\, D_t &= i\, D_t(A_0^{(\mathrm{el})}) := i\frac{\partial}{\partial t} - e\, A_0^{(\mathrm{el})},\end{aligned} \tag{5.49}$$

then one can write the Schrödinger equation (5.43) as

$$i\, D_t \psi = \frac{-1}{2m}\mathbf{D}^2 \psi \tag{5.50}$$

Equations (5.44) can also be written in terms of the covariant derivatives,

$$\begin{aligned}\frac{1}{i}\mathbf{D}(\mathbf{A}^{'(\mathrm{el})})\, e^{i\frac{e}{c}\alpha}\psi(\mathbf{x},t) &= e^{i\frac{e}{c}\alpha}\frac{1}{i}\mathbf{D}(\mathbf{A}^{(\mathrm{el})})\psi(\mathbf{x},t), \\ i\, D_t(A_0^{'(\mathrm{el})})\, e^{i\frac{e}{c}\alpha}\psi(\mathbf{x},t) &= e^{i\frac{e}{c}\alpha} i\, D_t(A_0^{(\mathrm{el})})\psi(\mathbf{x},t),\end{aligned} \tag{5.51}$$

where $\mathbf{D}(\mathbf{A}^{'(\mathrm{el})})$ and $D_t(A_0^{'(\mathrm{el})})$ are the covariant derivatives defined as in (5.49) but in terms of the transformed vector and scalar potentials $\mathbf{A}^{'(\mathrm{el})}$ ($A_0^{'(\mathrm{el})}$).

One may also define the covariant four-derivative $D(A^{(\mathrm{el})})$ with the components

$$D_\mu(A^{(\mathrm{el})}) := \partial_\mu + \frac{ie}{c} A_\mu^{(\mathrm{el})}. \tag{5.52}$$

Here the zeroth component is given by

$$D_0(A^{(\mathrm{el})}) := \frac{1}{c} D_t = \frac{1}{c}\frac{\partial}{\partial t} + \frac{ie}{c} A_0^{(\mathrm{el})}.$$

In terms of the covariant four-derivative D, (5.51) take the simple form:

$$D_\mu(A^{'(\mathrm{el})}) e^{i\frac{e}{c}\alpha}\psi(\mathbf{x},t) = e^{i\frac{e}{c}\alpha} D_\mu(A^{(\mathrm{el})})\psi(\mathbf{x},t). \tag{5.53}$$

The significance of this transformation property is that it ensures the covariant derivatives of the wave function, $D_\mu \psi$, transform in exactly the same way as the wave function ψ transforms, namely,

$$D_\mu(A'^{(\mathrm{el})})\psi' = e^{i\frac{e}{c}\alpha} D_\mu(A^{(\mathrm{el})})\psi.$$

Since $\psi = \psi(\mathbf{x}, t)$ is an arbitrary wave function, (5.53) is equivalent to

$$D_\mu(A'^{(\mathrm{el})}) = e^{i\frac{e}{c}\alpha} D_\mu(A^{(\mathrm{el})}) e^{-i\frac{e}{c}\alpha}. \tag{5.54}$$

This equation indicates that the operators D_μ transform *covariantly* under the gauge transformation, hence the name *covariant derivative*.

Covariant derivatives are also directly linked to the electromagnetic field strength $(\mathbf{E}^{(\mathrm{el})}, \mathbf{B}^{(\mathrm{el})})$. A simple calculation shows that

$$[D_\mu, D_\nu] = \frac{ie}{c}(\partial_\mu A_\nu^{(\mathrm{el})} - \partial_\nu A_\mu^{(\mathrm{el})}) =: \frac{ie}{c} F_{\mu\nu}^{(\mathrm{el})}, \tag{5.55}$$

where $F_{\mu\nu}^{(\mathrm{el})}$ are the (space-time) components of Maxwell's electromagnetic field strength tensor. These are related to the electric and magnetic fields according to

$$F_{0a}^{(\mathrm{el})} = (E^{(\mathrm{el})})^a; \quad F_{ab}^{(\mathrm{el})} = -\epsilon_{abc}(B^{(\mathrm{el})})^c.$$

In order to generalize the above gauge theory we introduce the following notation

$$\mathcal{U}(\alpha(\mathbf{x}, t)) := e^{i\frac{e}{c}\alpha(\mathbf{x}, t)} \tag{5.56}$$

Then the transformation (5.48) is written as

$$\psi(\mathbf{x}, t) \to \psi'(\mathbf{x}, t) = \mathcal{U}(\alpha(\mathbf{x}, t)) \psi(\mathbf{x}, t). \tag{5.57}$$

Now requiring the covariant derivatives $D_\mu(A^{(\mathrm{el})})\psi$ to satisfy the same transformation rule as ψ, one finds

$$D_\mu(A'^{(\mathrm{el})})\psi' = \mathcal{U}(\alpha)\left(D_\mu(A^{(\mathrm{el})})\psi\right), \tag{5.58}$$

and

$$D_\mu(A'^{(\mathrm{el})}) = \mathcal{U}(\alpha) D_\mu(A^{(\mathrm{el})}) \mathcal{U}^{-1}(\alpha). \tag{5.59}$$

This in turn is reflected in the transformation rule for the one-form,

$$A^{(\mathrm{el})} \to A'^{(\mathrm{el})} = \mathcal{U} A^{(\mathrm{el})} \mathcal{U}^{-1} - \frac{ie}{c} \mathcal{U} d\mathcal{U}^{-1} \tag{5.60}$$

$$= A^{(\mathrm{el})} - d\alpha,$$

where d is the exterior derivative defined by (5.47). In particular for the vector potential, one has

$$\mathbf{A}'^{(g)} = u\mathbf{A}^{(g)}u^{-1} + i\frac{c}{e}u\nabla u^{-1}$$
$$= \mathbf{A}^{(g)} + \nabla\alpha \tag{5.61}$$

The transformations (5.48), for each particular space-time point $(t, \mathbf{x})$, form the group $U(1)$. Since $U(1)$ is an Abelian Lie group, the electromagnetic gauge theory is called an *Abelian gauge theory*. The group $U(1)$ is called the *gauge group* or the *symmetry group* of the gauge theory. A gauge transformation (5.48) is associated with a function

$$u : M^4 \to U(1)$$

from Minkowski space-time M^4 (or a region in it) onto the symmetry group $U(1)$,

$$u(t, \mathbf{x}) = e^{i\frac{e}{c}\alpha(t,\mathbf{x})}.$$

The set of all gauge transformations (all such functions) also form a Lie group. This is called the *group of gauge transformations*. For the gauge theory of electromagnetism, this is basically an infinite direct product of copies of $U(1)$ corresponding to space-time points. We should like to emphasize the difference between the group of gauge transformations and the symmetry group. The former is in general infinite dimensional whereas the latter is usually finite dimensional.[21]

The Abelian gauge theory of electromagnetism has been generalized to non-Abelian or Yang–Mills gauge theories. In place of the symmetry group $U(1)$ one takes an arbitrary Lie group G (usually a compact and often a simple group). We shall encounter examples of gauge theories with the unitary symmetry groups $G = U(N)$ in Chap. 8.

In general the ingredients of an abstract gauge theory are

1. A parameter space X which is in general a smooth manifold. Usually X is a subset of three-dimensional Euclidean space $\mathbb{R}^3$ or Minkowski space M^4. But it can also be any other manifold whose elements $x \in X$ we call parameters. The coordinates of x are denoted by x^i, where $i = 1, 2, \cdots, \dim(X)$.
2. The *gauge* or *symmetry group* G which locally defines all the gauge transformations. These transformations form an infinite-dimensional group called the *group of gauge transformations*. Locally, i.e., on open patches U_α of the parameter manifold X, the elements of the group of gauge transformations correspond to functions $g = g(x)$ from U_α onto the symmetry group G. Note however that the group of gauge transformations is globally defined and in general can be a quite complicated infinite-dimensional group.[22] Here we suffice to indicate that locally it is an infi-

---

[21] In some texts the group of gauge transformations is called the *gauge group*. We shall use the name *symmetry group* to avoid any misunderstanding.

[22] We have given the global definition of a group of gauge transformations in the language of abstract fiber bundles in Sect. 5.3. More details may be found in [163].

nite direct product of copies of G each of which is labeled by a parameter $x \in U_\alpha$, i.e., locally it is the product group $\times_{x \in U_\alpha} G_x$.

3. The *matter fields* which are represented by a collection $\psi = (\psi^n)$ of smooth functions $\psi^n : U_\alpha \to \mathbb{C}$ and which transform according to an $\mathcal{N}$-dimensional unitary representation[23]

$$\mathcal{U} : G \ni g \to \mathcal{U}(g) \in U(\mathcal{N})$$

of the symmetry group G,

$$\psi(x) \to \psi'(x) = \mathcal{U}(g(x))\psi(x). \tag{5.62}$$

In terms of the components we have

$$\psi'^n(x) = \sum_{\ell=1}^{\mathcal{N}} \mathcal{U}^{n\ell}(g(x))\psi^\ell(x), \quad n = 1, \cdots \mathcal{N}. \tag{5.63}$$

4. The *gauge field* or *potential* whose components $A_i^{(\text{g})}(x) = (A_i^{nm}(x))$ that are $\mathcal{N} \times \mathcal{N}$ Hermitian matrices, with $\mathcal{N}$ being the dimension of the representation space, are postulated to transform under the transformation (5.62) according to

$$A_i^{'(\text{g})}(x) = \mathcal{U}(g(x)) \cdot A_i^{(\text{g})}(x) \cdot \mathcal{U}^{-1}(g(x))$$
$$+ \frac{i}{\text{g}} \left[\partial_i \mathcal{U}(g(x)) \cdot \mathcal{U}^{-1}(g(x)) \right]. \tag{5.64}$$

Furthermore under coordinate transformations of the parameter space X, $A_i^{(\text{g})}$ transform as components of a covariant vector. Hence they define a matrix-valued one-form

$$A^{(\text{g})} = A_i^{(\text{g})}(x)\, dx^i. \tag{5.65}$$

In terms of $A^{(\text{g})}$, the transformation rule (5.64) takes the form:

$$A^{'(\text{g})}(x)$$
$$= \mathcal{U}(g(x)) \cdot A^{(\text{g})}(x) \cdot \mathcal{U}^{-1}(g(x)) + \frac{i}{\text{g}} \left[d\mathcal{U}(g(x)) \cdot \mathcal{U}^{-1}(g(x)) \right]. \tag{5.66}$$

In this postulated gauge transformation property there occurs a constant g which has a dimension, $\dim \text{g} = (\dim x \cdot \dim A^{(\text{g})})^{-1}$. It is needed to adjust the dimension of $\frac{\partial}{\partial x^i}$ and the dimension of $A_i^{(\text{g})}$ which in general will be different from that of $\frac{\partial}{\partial x^i}$, if $A_i^{(\text{g})}$ has already been given a physical interpretation as, e.g., in electromagnetic gauge theory. This constant g is called the *coupling*

---

[23] See Appendix A.3 for a definition of a representation of a group.

constant of the gauge theory and it is characteristic of the gauge theory under consideration.

With the gauge potential one defines the *gauge-covariant derivatives*:

$$D_i(A^{(g)}) := \partial_i + i\,g\,A_i^{(g)}(x). \tag{5.67}$$

The transformation property of the gauge-covariant derivative follows from the postulated transformation property of the $A_i^{(g)}(x)$. It can be shown (Problem 5.4) immediately using (5.64) and (5.62), that

$$D_i(A^{'(g)}) := \partial_i + ig A_i^{'(g)} = \mathcal{U}(g(x)) \cdot D_i(A^{(g)}) \cdot \mathcal{U}^{-1}(g(x)) \tag{5.68}$$

and that

$$D_i(A^{(g)})\psi \to D_i(A^{'(g)})\psi' = \mathcal{U}(g(x))\, D_i(A^{(g)})\psi. \tag{5.69}$$

This means that the covariant derivatives $D_i\psi$ of ψ transform in the same way as ψ does. The gauge potentials with their postulated transformation property (5.64) have, indeed, been introduced with the intention of obtaining a derivative which has the transformation property (5.69) (the ordinary derivatives $\partial_i\psi$ do not possess this property). For the special case of an Abelian gauge theory for which $A_i^{(g)}$ and $\mathcal{U}$ are 1×1 matrices we obtain by inserting (5.56) into (5.66)

$$A^{'(g)}(x) = A^{(g)}(x) - \frac{1}{g}\frac{e}{c}\,d\alpha(x). \tag{5.70}$$

This is identical with (5.46) for $g = \frac{e}{c}$. Electromagnetic gauge theory is therefore a special case of a general gauge theory with gauge group $U(1)$ and coupling constant $\frac{e}{c}$.

In analogy to the electromagnetic field strength ($F_{ij}^{(el)}$) of (5.55), one can define the non-Abelian gauge field strength tensor

$$F_{ij}^{(g)} := \frac{1}{ig}[D_i, D_j] = \partial_i A_j^{(g)} - \partial_j A_i^{(g)} + ig\left[A_i^{(g)}, A_j^{(g)}\right]. \tag{5.71}$$

The brackets in (5.71) mean the commutator of the $\mathcal{N} \times \mathcal{N}$ matrices with components

$$\left[A_i^{(g)}, A_j^{(g)}\right]^{nm} := \sum_{l=1}^{\mathcal{N}}(A_i^{n\ell}A_j^{\ell m} - A_j^{n\ell}A_i^{\ell m}). \tag{5.72}$$

Using the transformation property (5.64) of the gauge potentials one can show (Problem 5.4) that the transformed field strength tensor (i.e. the field strength tensor defined using the transformed gauge potentials (5.64)),

$$F_{ij}^{'(g)}(x) := \frac{1}{ig}\left[D_i(A^{'(g)}), D_j(A^{'(g)})\right] \tag{5.73}$$

is obtained from the original $F_{ij}^{(g)}$ of (5.71) by

$$F_{ij}^{'(g)}(x) = \mathcal{U}(g(x)) \cdot F_{ij}^{(g)}(x) \cdot \mathcal{U}^{-1}(g(x)). \tag{5.74}$$

In terms of the components, one has

$$F_{ij}^{'nm} = \sum_{\ell,k=1}^{\mathcal{N}} \mathcal{U}^{n\ell}(g) F_{ij}^{\ell k} \mathcal{U}^{km}(g).$$

This means that the field strength tensor ($F_{ij}^{(g)}$) of a non-Abelian gauge theory is gauge covariant, not gauge invariant. Therefore unlike in an Abelian gauge theory, the components F_{ij}^{nm} of the field strength of a non-Abelian gauge theory are not physical quantities. They may however be used to construct gauge-invariant (physically measurable) quantities. Examples of these are the pure Yang–Mills Lagrangian density $\mathrm{tr}(F_{\mu\nu}^{(g)} F^{(g)\mu\nu})$ and the charge density $\mathrm{tr}(\epsilon^{\mu\nu\sigma\tau} F_{\mu\nu}^{(g)} F_{\sigma\tau}^{(g)})$ of instanton physics, where "tr" denotes the trace of the corresponding matrices [186].

There are other gauge-invariant quantities associated with a gauge theory. These are called the *Wilson loop integrals* [30, 58]. They are defined for closed loops C of the parameter manifold X as the traces of the path-ordered exponentials,

$$\mathrm{tr}\left[\mathcal{P} e^{ig \oint_C A_i^{(g)} dx^i}\right]. \tag{5.75}$$

We have indeed encountered an example of the electromagnetic Wilson loop integrals in the context of the Dirac phases and the Aharonov–Bohm effect, in Sect. 2.4.

According to the definition (5.71) of $F_{ij}^{(g)}$, it may happen that $F_{ij}^{(g)} = 0$, even though $A_i^{(g)} \neq 0$. A gauge theory for which $F_{ij}^{(g)}$ is identically zero is called *trivial*. One can show (Problem 5.5.) that if

$$A_i^{nm}(x) = -i \sum_\ell \left(\partial_i \mathcal{U}^{\dagger n\ell}(x)\right) \mathcal{U}^{\ell m}(x) = i \sum_\ell \mathcal{U}^{\dagger n\ell}(x) \left(\partial_i \mathcal{U}^{\ell m}(x)\right) \tag{5.76}$$

for some unitary matrix $\mathcal{U}(x)$, then the gauge theory is trivial. A gauge potential of the form (5.76) is called a *pure gauge*. There are cases where a gauge theory is trivial but the gauge potential is not a pure gauge. Such theories are based on topologically non-trivial parameter spaces. They usually provide interesting information about the topological structure of the parameter space. A simple example of such theories is the system studied in Sect. 2.4, in the context of the Aharonov–Bohm effect. More interesting examples are found in the context of topological field theories [36].

If the parameter space M of a gauge theory is a subset of a Euclidean space (with metric: $\mathrm{diag}(1,\cdots,1)$), then we can identify $A_i^{(g)}$ with $(A^{(g)})^i$ and view them as components of a (matrix-valued) vector $\mathbf{A}^{(g)}$. Similarly, we can define a gauge-covariant derivative vector $\mathbf{D}(\mathbf{A}^{(g)})$,

$$\mathbf{D}(\mathbf{A}^{(g)}) = \nabla + i\,\mathrm{g}\,\mathbf{A}^{(g)}(x). \tag{5.77}$$

In terms of $\mathbf{A}^{(g)}$ the transformation (5.66) takes the form:

$$\mathbf{A}'^{(g)} = u \cdot \mathbf{A}^{(g)} \cdot u^{-1} + \frac{i}{g} (\nabla u) \cdot u^{-1}. \tag{5.78}$$

Similarly we can introduce (contravariant) vector analog of the field strength tensor $F^{(g)}$. We shall denote this by $\mathbf{F}^{(g)}$. This notation will be used later in Chap. 8 in our discussion of the gauge theory of molecular physics.

5.6 Mathematical Foundations of Gauge Theories and Geometry of Vector Bundles

In the preceding section we offered an essentially local description of gauge theories. By a local description we mean that the ingredients of the gauge theories that we listed involved local fields that were defined on open patches of the parameter space. The gauge theories may however have interesting global properties. These may be most conveniently understood using the machinery of fiber bundles. This is one of the motivations for studying the relationship between gauge theories and fiber bundles. Another important motivation is to try to understand the true meaning of *gauge symmetry*. So far we have introduced this symmetry as a generalization of an appealing symmetry of classical electrodynamics which seemed to carry through to the quantum domain. As we explain in the following, fiber bundles provide a most natural interpretation of gauge symmetry.

In general, an abstract gauge theory is associated with a principal fiber bundle $(\mathcal{E}, X, \pi, G)$ and its associated vector bundles $(E, X, \check{\pi}, \check{G})$. These are defined by the unitary representations of the structure group G. The base space X and the structure group G provide the parameter space and the symmetry group of the gauge theory. The matter fields and the gauge potential are identified with the global sections and a local connection one-form of an associated vector bundle. Finally, the group of gauge transformations is the group consisting of all bundle isomorphisms $\mathcal{F} : \mathcal{E} \to \mathcal{E}$ of the PFB $\mathcal{E}$ which project to the identity map on X, i.e., $\pi(\mathcal{F}(\mathcal{E}_x)) = x$ (respectively $\check{\pi}(\check{\mathcal{F}}(E_x)) = x$), for all $x \in X$.

In order to explain the mathematical structure of gauge theories, we first treat the subject of the *geometry of vector bundles*.

A geometric structure on a vector bundle may be defined in terms of its associated PFB. There is also a closely related alternative approach in which one defines the notion of the parallel transportation of the vectors (points of the vector bundle) directly.

In this alternative approach, a connection on a vector bundle E is identified with a first-order differential operator

$$\check{D} : C^\infty(E) \longrightarrow C^\infty(TX^*) \otimes C^\infty(E), \tag{5.79}$$

acting on the (vector) space of global (smooth or C^∞) sections $C^\infty(E)$ of E. Here $\otimes$ denotes the tensor product of vector spaces, and $C^\infty(TX^*)$ is the space of smooth sections of the cotangent bundle of the base space X, i.e., the space of one-forms on X.

In the following we shall concentrate on the case of $U(\mathcal{N})$ bundles, $(E, X, \check{\pi}, U(\mathcal{N}))$. In general the structure group of a vector bundle need not be $U(\mathcal{N})$. The unitary (and orthogonal) groups however play a central role in physics, particularly in the subject of the geometric phase. Our analysis may readily be generalized to other subgroups of the general linear group.

Let us consider a global section

$$\Psi : X \longrightarrow E, \tag{5.80}$$

and a local trivialization (U_α, Φ_α) of E. For each $x \in U_\alpha$, Φ_α induces a (vector space) isomorphism $\Phi_\alpha(x) : E_x \to \mathbb{C}^\mathcal{N}$, between the fiber E_x over x and the typical fiber $\mathbb{C}^\mathcal{N}$. The inverse $\Phi_\alpha^{-1}(x)$ may be used to define a complete orthonormal basis $\{s_n(x)\}$ on E_x. This is done by choosing a complete orthonormal (unitary) basis $\{e_n\}$ for $\mathbb{C}^\mathcal{N}$ and using $\Phi_\alpha^{-1}(x)$ to map it to E_x. Similarly, we can define the dual basis $\{\bar{s}^n(x)\}$ to $\{s_n(x)\}$. We shall view s_n and $\bar{s}^n$ as column and row vectors belonging to E_x and its dual E_x^*, respectively. By construction, we have

$$\bar{s}^n(x)\,[s_m(x)] = \bar{s}^n(x) \cdot s_m(x) = \delta_m^n, \tag{5.81}$$

$$\sum_{n=1}^{\mathcal{N}} s_n(x) \cdot \bar{s}^n(x) = \mathbf{1}_{\mathcal{N} \times \mathcal{N}}, \tag{5.82}$$

where δ_m^n is the Kronecker delta function and $\mathbf{1}_{\mathcal{N} \times \mathcal{N}}$ is the $\mathcal{N} \times \mathcal{N}$ unit matrix.

The basis vectors s_n are indeed local sections,

$$s_n : U_\alpha \longrightarrow \check{\pi}^{-1}(U_\alpha) \subseteq E, \tag{5.83}$$

of E. They may be defined independently of the functions Φ_α. In general a complete orthonormal local basis is a collection of $\mathcal{N}$ local sections defined over an open neighborhood of X and satisfying the orthonormality (5.81) and completeness (5.82) conditions. The local basis sections defined using a given basis $\{e_n\}$ of $\mathbb{C}^\mathcal{N}$ is called a canonical basis. This is analogous to the canonical local sections (5.19) of a PFB. In fact, the frame consisting of the canonical local basis sections s_n of E, viewed as a matrix whose columns are defined by s_n, corresponds to the canonical local section (5.19) of the associated PFB $(\mathcal{E}, X, \pi, U(\mathcal{N}))$ to the vector bundle $(E, X, \check{\pi}, U(\mathcal{N}))$.

In some cases, it might be possible to extend the domain of definition of a (smooth) local basis $\{s_n\}$ to the whole base manifold X, i.e., choose s_n to be global sections. This is however not always the case. In general, one can show (Problem 5.6) that if the vector bundle E (and therefore $\mathcal{E}$) is non-trivial, then there is no global basis.

5.6 Gauge Theories and Vector Bundles

Local bases are useful because they provide local (coordinate) representations of the globally defined objects. For example for a global section Ψ, (5.80), of E, we have

$$\Psi(x) = \psi^n(x)\, s_n(x), \qquad \forall x \in U_\alpha, \tag{5.84}$$

where $\psi^n : U_\alpha \to \mathbb{C}$ are scalar functions. They are called the local components or fiber-coordinates of Ψ associated with the local basis $\{s_n\}$. We can also view ψ^n, with $n = 1, 2, \cdots, \mathcal{N}$, as components of a local vector-valued function $\psi : U_\alpha \to \mathbb{C}^\mathcal{N}$, which we demonstrate by a column vector

$$\psi = \begin{pmatrix} \psi^1 \\ \psi^2 \\ \vdots \\ \psi^\mathcal{N} \end{pmatrix}. \tag{5.85}$$

The local components ψ^n of a global section Ψ depend on the particular basis used for the local description of Ψ. The global sections themselves, of course, are independent of the choice of a basis. This is the main reason for their utility as mathematical realizations of physical matter fields in gauge theories.

Next consider an alternative local basis,

$$s'_n : U_\beta \longrightarrow \check{\pi}^{-1}(U_\beta) \subseteq E, \tag{5.86}$$

defined over some open neighborhood U_β with $U_\alpha \cap U_\beta \neq \emptyset$. In terms of s'_n, the global section Ψ (5.80) is represented by

$$\Psi(x) = \psi'^n(x)\, s'_n(x), \qquad \forall x \in U_\beta. \tag{5.87}$$

On the other hand, since both $\{s_n(x)\}$ and $\{s'_n(x)\}$ are unitary bases, for every $x \in U_\alpha \cap U_\beta$, there exists a unitary matrix $\mathcal{U}(x) \in U(\mathcal{N})$ such that $s'_n(x) =: \mathcal{U}(x)\, s_n(x)$. This equation is reminiscent of the fact that the structure group $U(\mathcal{N})$ of E acts on the fibers E_x from the left by matrix multiplication. It may be viewed as a transformation of bases in E_x,

$$s_n(x) \xrightarrow{\mathcal{U}(x)} s'_n(x) = \mathcal{U}(x)\, s_n(x). \tag{5.88}$$

Under this transformation the components transform according to (Problem 5.7)

$$\psi^n(x) \xrightarrow{\mathcal{U}(x)} \psi'^n(x) = \sum_{m=1}^{\mathcal{N}} (\mathcal{U}^{-1}(x))^{nm}\, \psi^m(x). \tag{5.89}$$

Alternatively, we have

$$\psi(x) \xrightarrow{\mathcal{U}(x)} \psi'(x) = \mathcal{U}^{-1}(x)\, \psi(x). \tag{5.90}$$

In (5.89), $(\mathcal{U}^{-1}(x))^{nm}$ are the matrix elements

$$(\mathcal{U}^{-1}(x))^{nm} := \bar{s}^n(x) \cdot \mathcal{U}^{-1}(x) \cdot s_m(x),$$

of the inverse $\mathcal{U}^{-1}(x)$ of $\mathcal{U}(x)$.

Having reviewed the utility of local bases in representing global sections Ψ of a vector bundle by local $\mathbb{C}^{\mathcal{N}}$-valued functions ψ, we proceed with a discussion of a connection on a vector bundle. We mentioned that a connection on a vector bundle may be given by a first-order differential operator $\check{D}$, (5.79). In a local basis $\{s_n\}$, $\check{D}$ is expressed by an ordinary first-order differential operator acting on the local representation ψ of Ψ. Namely for all $x \in U_\alpha$, we have

$$\check{D}\Psi(x) = (\check{D}_i \psi(x))^n \, dx^i \otimes s_n(x), \tag{5.91}$$

where $\check{D}_i$ are the matrix-valued differential operators

$$\check{D}_i := \partial_i - i\check{A}_i(x), \tag{5.92}$$

which act on the vector-valued function $\psi = \psi(x)$ of (5.85). These operators are also called *covariant derivative operators*. They are analogs of the covariant derivatives (5.37), that we encountered in our study of PFBs. In fact, they are used for the same purpose, i.e., for defining parallel transportation.

In the expression for the covariant derivatives (5.92), there appear matrix-valued functions $\check{A}_i$ which are assumed to be Hermitian. Under a coordinate transformation of the base manifold X, the global quantities such as Ψ and $\check{D}\Psi$ remain invariant. This implies that the entries $\check{A}_i^{nm}$ of $\check{A}_i$ transform like the components of a (covariant vector or a) one-form on X. Hence we can define the one-forms $\check{A}^{nm} := \check{A}_i^{nm} dx^i$ and the matrix-valued one-form

$$\check{A} = \check{A}_i(x)\, dx^i. \tag{5.93}$$

By definition $-i\check{A}$ is a one-form whose components are anti-Hermitian matrices. These belong to the Lie algebra $u(\mathcal{N})$ of the unitary group $U(\mathcal{N})$. Thus $-i\check{A}$ is a Lie algebra-valued one-form. It is the analog of the local connection one-form (5.18) on a PFB.

Before discussing the utility of the connection operator $\check{D}$ in defining a notion of distant parallelism on a vector bundle, we wish to examine the properties of the local representation of $\check{D}$, i.e., the covariant derivatives $\check{D}_i$ or alternatively the matrix-valued one-form $\check{A}$, under a transformation of local bases (5.88). Requiring $\check{D}\Psi$ to be independent of the choice of a local basis, and using (5.90), we can easily show that under the transformation (5.88), $\check{D}_i$ and $\check{A}$ must transform according to

$$\check{D}_i \xrightarrow{\mathcal{U}(x)} \check{D}_i' = \mathcal{U}^{-1}(x) \cdot \check{D}_i \cdot \mathcal{U}(x), \tag{5.94}$$

$$\check{A} \xrightarrow{\mathcal{U}(x)} \check{A}' = \mathcal{U}^{-1}(x) \cdot \check{A} \cdot \mathcal{U}(x) + i\mathcal{U}^{-1}(x) \cdot d\mathcal{U}(x), \tag{5.95}$$

where d denotes the exterior derivative,

$$d\mathcal{U}(x) := \partial_i \mathcal{U}(x)\, dx^i.$$

We shall shortly see that the resemblance of the transformation (5.95) to (5.28) is by no means accidental.

The parallel transport of a vector $\Psi_0 \in E_{x_0}$, along a curve $C_X : [0,T] \to X$, with $x_0 := C_X(t=0)$, is defined to be the end point of the horizontal lift $C_E : [0,T] \to E$ of C_X which, in turn, is defined as the solution of the differential equation

$$\frac{\check{D}}{Dt}\Psi(x(t)) = 0, \quad \text{with} \quad \Psi(t=0) = \Psi_0, \tag{5.96}$$

where $\Psi(x(t)) =: C_E(t)$, $x(t) = (x^1(t), \cdots, x^m(t))$ are coordinates of $C_X(t)$, and $\frac{\check{D}}{Dt}$ is locally expressed by

$$\frac{\check{D}}{Dt}\psi(x(t)) := \frac{dx^i(t)}{dt}\check{D}_i\psi(x(t)), \quad \forall C_X(t) \in U_\alpha. \tag{5.97}$$

The transformation rules (5.94) and (5.95) ensure that the horizontal lift C_E is independent of the choice of a local basis. This can be easily checked by writing (5.97) in a new basis and using (5.90) and (5.94).

In order to make the analogy with the geometry of PFBs transparent, we need first note that for the vector bundle E, we perform a transformation of local bases by multiplying the local basis sections s_n by a unitary matrix-valued function $\mathcal{U} : U_\alpha \to U(n)$ from the left. The same is done for a PFB by the action of a group-valued function $g : U_\alpha \to G$ from the right. To clarify the relation between g and $\mathcal{U}$, we recall the definition of the association of vector bundles with PFBs.

Consider an associated vector bundle $E : (E, X, \check{\pi}, \check{G})$ to a given PFB $\mathcal{E} : (\mathcal{E}, X, \pi, G)$. By definition E is determined by a representation (ρ, V) of G; the typical fiber of E is the representation space V, and the structure group $\check{G}$ and the transition functions $\check{G}_{\alpha\beta}(x)$ of E are the representation of those of $\mathcal{E}$, namely, $\check{G} := \rho(G)$ and $\check{G}_{\alpha\beta}(x) := \rho(G_{\alpha\beta}(x)) \in GL(V)$. Here $GL(V)$ denotes the group of all vector space isomorphisms mapping V onto itself. $GL(V)$ is also called the *automorphism group* of V.

The representation (ρ, V) also induces a representation of the Lie algebra $\mathcal{G}$ of G. This is done by the push-forward (differential) map, $\rho_* : TG_e = \mathcal{G} \to \mathcal{G}\ell(V) = TGL_1(V)$, where $\mathcal{G}\ell(V)$ is the Lie algebra of $GL(V)$; TG_e and $TGL_1(V)$ are the tangent spaces at the identity element of G and $GL(V)$ respectively. The push-forward map ρ_* is induced by the representation: $\rho : G \to GL(V)$. [24]

If E is a $U(\mathcal{N})$ vector bundle, then $V = \mathbb{C}^\mathcal{N}$ and $\rho : G \to U(\mathcal{N})$, so that

---
[24] For a brief review of the necessary definitions, see Appendix A.2.

$$\rho(g(x)) = \mathcal{U}(x) \in U(\mathcal{N}).$$

Then comparing (5.28) and (5.95), we can identify the matrix-valued one-form $\check{A}$ of (5.93) with the representation of the local connection one-form A (5.27) on the PFB $\mathcal{E}$, i.e.,

$$\check{A} = \check{A}_i(x)dx^i = \rho_*(A_i(x))dx^i =: \rho_*(A). \tag{5.98}$$

If the group G is itself the unitary group $U(\mathcal{N})$ and $(\rho, \mathbb{C}^\mathcal{N})$ is the defining representation of $U(\mathcal{N})$, or if E is a given vector bundle and $\mathcal{E}$ is defined as its associated PFB, then $\check{A}$ and A are identical.

In abstract Abelian or non-Abelian gauge theories one starts from a given PFB $(\mathcal{E}, M, \pi, G)$ which is endowed with a geometric structure and a set of irreducible unitary representations $(\mathcal{U}_\lambda, \mathbb{C}^{\mathcal{N}_\lambda})$ of the structure group G. These representations are then used to define a set of associated vector bundles E_λ whose global sections are identified with different species of matter fields Ψ_λ present in the theory.

For simplicity we concentrate on a particular species of matter fields and drop the label λ without loss of generality.

Usually a matter field $\Psi \in C^\infty(E)$ is expressed in terms of the (local) vector-valued function ψ, (5.85). Then a transformation of a local basis (5.88) appears as a passive transformation in E_x. The corresponding active transformation is given by (5.90). The latter is precisely the first of the pair of transformations which we called a gauge transformation (5.62).

The logical consistency of the geometric theory of vector bundles demands a local basis transformation (5.88) to be accompanied by the transformation (5.95) of the matrix-valued one-forms $\check{A}$, (5.93). The one-form $\check{A}$ is related to the local connection one-form A, (5.27), of $\mathcal{E}$ according to (5.98). The physical field corresponding to $\check{A}$ is the one-form $A^{(g)}$ of gauge fields or gauge potentials, (5.65). More precisely, we have

$$\check{A} = -\mathrm{g}\, A^{(g)}. \tag{5.99}$$

In view of this relation, the transformation (5.64) which together with (5.62) form a gauge transformation, is identified with the transformation (5.95) of $\check{A}$.

Hence the gauge transformations (5.62) and (5.64) are logical consequences of the requirement that the physical (matter) fields must be independent of the choice of a local basis. In other words, a (local) gauge transformation is identical with a (local) coordinate transformation in the fibers of E. Thus, *gauge symmetry* of a field theory means the natural requirement that the matter fields (and their dynamics) are invariant under fiber-coordinate transformations.

In general, the dynamics of matter fields is given by a set of differential equations. These involve ordinary partial derivatives of the components of the matter fields supplemented with the necessary gauge fields, so that the equations respect the gauge symmetry. The appropriate combination of the ordinary derivatives and the gauge fields which appear in the dynamical field equations, are the gauge-covariant derivatives D_i of (5.67). In view of (5.99), they are identical to the covariant derivatives $\check{D}_i$ of (5.92).

A typical example of the utility of covariant derivatives in dynamical field equations is in the non-relativistic quantum mechanics of charged particles. It is well known that non-relativistic quantum mechanics (in the position representation) may be viewed as a classical field theory [76]. The matter fields correspond to the wave functions and the gauge fields are given for example by the electromagnetic gauge field $A^{(\text{el})}$ (5.45). The appearance of the corresponding covariant derivatives (5.52) in the Schrödinger equation (5.50) is a special case of the application of covariant derivatives in classical field theories.

In our description of gauge theories in Sect. 5.5, we also introduced the gauge field strength $F_{ij}^{(\text{g})}$. These also have a very well-known mathematical counterpart. They correspond to the components F_{ij} of the *local curvature two-form* F of the connection one-form A on $\mathcal{E}$. The local connection two-form F is a Lie-algebra-valued two-form given by

$$F = \frac{1}{2} F_{ij}(x)\, dx^i \wedge dx^j$$
$$F_{ij}(x) := \partial_i A_j - \partial_j A_i - i[A_i, A_j]. \tag{5.100}$$

It can be directly expressed in terms of the local connection one-form A according to

$$F = dA - \frac{i}{2}[A \stackrel{\wedge}{,} A], \tag{5.101}$$

where $[A \stackrel{\wedge}{,} A] := [A_i, A_j]\, dx^i \wedge dx^j$.

Under the right action of a group element $g(x) \in G$, F transforms according to

$$F(x) \xrightarrow{g} F'(x) = g(x)^{-1} \cdot F(x) \cdot g(x). \tag{5.102}$$

Equation (5.102) can be derived using (5.28) and (5.100).

It is clear from (5.101) and (5.27) that the local curvature two-form F also depends on the local basis section s that defines A as a pullback one-form of a global connection one-form ω. Therefore F is only locally meaningful. In fact, we could have defined F as a pullback two-form of a *global curvature two-form* in the total space $\mathcal{E}$. Again the global curvature two-form has the advantage of being globally defined whereas the local curvature two-form depends on the choice of a local section. It can be extended to the whole base manifold X only in the sense that one can switch from one local section to the other in the intersection of the neighboring charts. Under a transformation of charts, i.e., under the action of the transition functions, $G_{\alpha\beta}(x)$, from the left, the local curvature two-form transforms according to

$$F(x) \to F'(x) = G_{\alpha\beta}(x)^{-1} \cdot F(x) \cdot G_{\alpha\beta}(x).$$

102   5. Fiber Bundles and Gauge Theories

Here, the unprimed and primed quantities correspond to the charts (U_α, Φ_α) and (U_β, Φ_β), respectively.

A local curvature two-form may also be defined on the associated vector bundle E. This is done in full analogy with the case of the local connection one-form (5.98), namely we define

$$\check{F} := \rho_*(F), \tag{5.103}$$

where ρ_* is the push-forward map corresponding to the representation ρ of G. The local curvature two-forms F and $\check{F}$ naturally appear in the study of the geometric properties of PFBs and vector bundles. Two important situations where F and $\check{F}$ emerge naturally are in an extension of the global connection operator $\check{D}$ to act on the tensor product of sections of a vector bundle and arbitrary differential forms (Problem 5.8) and in the non-Abelian generalization of Stokes' theorem (Problem 5.9). Another application of the local curvature two-forms is in gauge theories, where they play the role of the gauge field strength tensor $F_{ij}^{(g)}$ of (5.71). More precisely, we have

$$\check{F}_{ij} = -g\, F_{ij}^{(g)}. \tag{5.104}$$

The only ingredient of gauge theories which does not have a mathematical counterpart is the coupling constant g which can be conveniently absorbed in the definition of the one-form $\check{A}$, (5.99).

We conclude this chapter by noting that every Lie group G has a natural representation on its Lie algebra $\mathcal{G}$. This representation is called the *adjoint representation*,[25] $\text{Ad}: G \to GL(\mathcal{G})$. The associated vector bundle defined by the adjoint representation is called the *adjoint vector bundle*. This vector bundle is of utmost importance in gauge theories of particle physics. In particular the Higgs fields are global sections of the adjoint bundle.[26]

Problems

5.1) Let (P, X, π, G) be a PFB. Show that the definition of the right action of G on P given by (5.10) is independent of the local trivialization. Hint: Choose $x \in U_\alpha \cap U_\beta$ and use the transition function $G_{\alpha\beta}$ to transform (5.10) into a form in which the α's are replaced by β's. Note that the group multiplication from the right commutes with the group multiplication from the left even if the group is non-Abelian.

5.2) Show that the covariant momentum of electromagnetism

$$\pi = \mathbf{p} - \frac{e}{c}\mathbf{A}^{(\text{el})}(\mathbf{x})$$

---
[25] A definition of the adjoint representation is given in Appendix A.2.
[26] For more details of this subject, refer to [163].

fulfills the commutation relation

$$[\pi_a, \pi_b] = i\frac{e}{c}\epsilon_{abc}(B^{(el)})^c$$

where the canonical momentum p_a and position x_a fulfill the usual Heisenberg commutation relations and where $\mathbf{B}^{(el)} = \nabla \wedge \mathbf{A}^{(el)}(\mathbf{x})$.

5.3) a) Show that the Schrödinger equation

$$\left(i\frac{\partial}{\partial t} - eA_0^{(el)}\right)\psi = \frac{1}{2m}\left(\frac{1}{i}\nabla - \frac{e}{c}\mathbf{A}^{(el)}\right)^2\psi$$

remains unaltered when $A_0^{(el)}$, $\mathbf{A}^{(el)}$ and ψ undergo the following gauge transformations

$$\mathbf{A}^{(el)} \to \mathbf{A}'^{(el)} := \mathbf{A}^{(el)} + \nabla\alpha(\mathbf{x},t)$$

$$A_0^{(el)} \to A_0'^{(el)} := A_0^{(el)} - \frac{1}{c}\frac{\partial\alpha}{\partial t}$$

$$\psi \to \psi' := e^{if}\psi,$$

where $f = f(\mathbf{x},t)$ is related to $\alpha = \alpha(\mathbf{x},t)$. Calculate f.

b) Obtain the transformation property of the gauge-covariant derivatives

$$D_\mu(A^{(el)}) := \partial_\mu + \frac{ie}{c}A_\mu^{(el)},$$

under the above gauge transformations, i.e., find the operator $\mathcal{U}(\alpha)$ in the transformation

$$D_\mu(A'^{(el)}) = \mathcal{U}(\alpha)\, D_\mu(A^{(el)})\, \mathcal{U}^{-1}(\alpha).$$

5.4) Obtain transformation property of the gauge-covariant derivatives $D_i(A^{(g)})$ from the transformation property of the gauge potential $A^{(g)}$ with respect to a non-Abelian gauge transformation. Use this to derive the transformation property of the field strength tensor ($F_{ij}^{(g)}$).

5.5) Show that if the gauge field is a pure gauge, i.e., $A_i^{(g)}(x) = i\mathcal{U}^\dagger(x)\partial_i\mathcal{U}(x)$ for some $\mathcal{U}(x) \in U(\mathcal{N})$, then the gauge theory is trivial, i.e., $F_{ij}^{nm} = 0$.

5.6) Let E be a vector bundle of rank $\mathcal{N}$, and $(\mathcal{E}, X, \pi, G)$ be its associated PFB. Show that
   a) the existence of a global section of $\mathcal{E}$ is a necessary and sufficient condition for $\mathcal{E}$ to be a trivial PFB;
   b) the existence of $\mathcal{N}$ linearly independent global sections of E is a necessary and sufficient condition for E to be a trivial vector bundle.

5.7) Use (5.84), (5.88), and (5.82) to show the validity of (5.89). Hint: Denote the sections Ψ and s_a by Dirac kets, e.g.,

$$|\Psi\rangle = \psi_a |s_a\rangle.$$

5.8) Let $(E, X, \check{\pi}, \check{G})$ be a vector bundle, with $C^\infty(E)$ and $\Omega(X)$ denoting the spaces of global sections of E and arbitrary differential forms on X, respectively. A generalization of the global connection operator $\check{D}$ to an operator

$$\check{D} : \Omega(X) \otimes C^\infty(E) \to \Omega(X) \otimes C^\infty(E)$$

is given by requiring $\check{D}$ to be linear and satisfy

$$\check{D}(\sigma^{(p)} \otimes \Psi) := (d\sigma^{(p)}) \otimes \Psi + (-1)^p \sigma^{(p)} \otimes (\check{D}\Psi),$$

for every differential p-form $\sigma^{(p)}$ of X and $\Psi \in C^\infty(E)$. Here $\check{D}\Psi$ is defined by (5.91). Show that $\check{D}^2$ is a zeroth-order differential operator. In particular, show that there exist linear transformations $\mathcal{V}_{ij} : C^\infty(E) \to C^\infty(E)$ satisfying

$$\check{D}^2 \Psi = (dx^i \wedge dx^j) \otimes (\mathcal{V}_{ij} \Psi).$$

Obtain an explicit local expression for $\mathcal{V} := \mathcal{V}_{ij}(x) dx^i \wedge dx^j$ in terms of the matrix-valued one-form $\check{A}$. Express $\mathcal{V}$ in terms of the local curvature two-form $\check{F}$.

5.9) Consider parallel transporting a point p_0 of a PFB $(\mathcal{E}, X, \pi, G)$ along an infinitesimally small closed curve $C_X : [0, T] \to X$. Suppose that a particular coordinate system is chosen on X such that the curve C_X is given according to

$$C_X(t) := \begin{cases} (x_0^1, \cdots, x_0^i + t\delta x^i, \cdots, x_0^m) \\ \quad \text{for } 0 \leq t \leq \frac{T}{4}; \\ (x_0^1, \cdots, x_0^i + \frac{T}{4}\delta x^i, \cdots, x_0^j + (t - \frac{T}{4})\delta x^j, \cdots, x_0^m) \\ \quad \text{for } \frac{T}{4} \leq t \leq \frac{T}{2}; \\ (x_0^1, \cdots, x_0^i + (\frac{3T}{4} - t)\delta x^i, \cdots, x_0^j + \frac{T}{4}\delta x^j, \cdots, x_0^m) \\ \quad \text{for } \frac{T}{2} \leq t \leq \frac{3T}{4}; \\ (x_0^1, \cdots, x_0^i, \cdots, x_0^j + (T - t)\delta x^j, \cdots, x_0^m) \\ \quad \text{for } \frac{3T}{4} \leq t \leq T; \end{cases}$$

where $x_o := \pi(p_0) = C_X(0)$, i and j are fixed coordinate labels, and δx^i and δx^j are infinitesimal real numbers. Compute the corresponding holonomy element

$$\mathcal{P} \exp\left(i \oint_{C_X} A\right),$$

i.e., explicitly evaluate the contribution of each side of C_X expanding the corresponding exponential and retaining only the terms of up to order $(\delta x)^2$, in the result. Finally recombine the total contribution as a single exponential and express the result in terms of the local curvature two-form F. (Optional: Try to generalize this analysis to closed loops which are not infinitesimal. This leads to a non-Abelian generalization of Stokes' theorem. See [17, 50].)

6. Mathematical Structure of the Geometric Phase I: The Abelian Phase

6.1 Introduction

In the preceding chapter, we have developed the parts of the theory of fiber bundles which are relevant to our study of geometric phases and briefly described gauge theories. We introduced abstract gauge theories as generalizations of the Abelian gauge theory of electromagnetism. There is also another Abelian gauge theory which we encountered in Chap. 4. We call the latter the *Abelian gauge theory of quantum mechanics*. The parameter space of this gauge theory is the projective Hilbert space $\mathcal{P}(\mathcal{H})$ associated with a Hilbert space $\mathcal{H}$, the matter fields are the pure state vectors which belong to $\mathcal{H}$, the gauge or symmetry group is the group $U(1)$ of the phases of the state vectors, and the gauge potential is the Aharonov–Anandan (A-A) connection.

The defining PFB associated with this gauge theory is the A-A bundle η whose structure is determined by the Hilbert space $\mathcal{H}$. The A-A connection defines a natural geometric structure on η. The associated vector bundle to η that yields the state vectors as its global sections is the one defined by the standard representation of $U(1)$. Thus it is a complex line bundle over $\mathcal{P}(\mathcal{H})$.

In this chapter, we shall present a detailed description of the mathematical structure of the Abelian gauge theory of quantum mechanics. In particular we offer different holonomy interpretations of the Abelian geometric phase and reveal their relationship.

6.2 Holonomy Interpretations of the Geometric Phase

In Chap. 4, we outlined a holonomy interpretation of the geometric phase. This interpretation used the $U(1)$ PFB η (4.33) and identified the phase with the holonomy of a particular connection which we called the Aharonov–Anandan (A-A) connection. We shall devote this section to a more systematic discussion of the holonomy interpretations of the geometric phase. We shall start our analysis by first describing an alternative interpretation of the adiabatic phase.

The first holonomy interpretation of Berry's adiabatic phase was suggested by Simon [230]. The idea was to construct a complex line bundle L^n:

$$(L^n, M, \check{\pi}, U(1)) \tag{6.1}$$

over the parameter space M of an adiabatically evolving system. We shall refer to this interpretation of Berry's phase as the Berry–Simon (B-S) interpretation.

Let us consider an arbitrary quantum mechanical system whose evolution is governed by the parameter-dependent Hamiltonian, $h = h(R)$. Suppose that the time dependence of the Hamiltonian justifies the adiabatic approximation. Under this condition, the eigenstates $|n;R(0)\rangle\langle n;R(0)|$ of the initial Hamiltonian $h(R(0))$ undergo (approximate) cyclic evolutions. Here we assume that the eigenvalues $E_n(R)$ of $h(R)$ are non-degenerate and that both $|n,R\rangle$ and $E_n(R)$ depend smoothly on $R \in M$, where M is the parameter space. Moreover, we assume that the evolution of each state is such that there are no level crossings. We can express this statement by imposing the condition: If $E_n(R(0)) \neq E_{n'}(R(0))$, then for all $R \in M$, $E_n(R) \neq E_{n'}(R)$. This condition is implicitly enforced by appropriately choosing the parameter space M.

For example, for the system (3.1) the parameter space is chosen to be the unit sphere $S^2 \subset \mathbb{R}^3$. Actually, the (non-compact) manifold $\mathbb{R}^3 - \{0\}$ or any of its submanifolds could be an acceptable parameter space. The origin $\mathbf{R} = 0 \in \mathbb{R}^3$ must be excluded because all the energy levels collapse to zero there (in particular, they become degenerate). The parameter space $\mathbb{R}^3 - \{0\}$ allows, for example, the more general case of a spin in a magnetic field which has both time-dependent orientation and magnitude, with the condition that the magnitude never be zero. Nonetheless, the choice made in Chap. 2, i.e., $M = S^2$, is not quite accidental. It turns out that this choice includes all the physically interesting phenomena. A classification of the "appropriate" parameter spaces for all finite-dimensional quantum systems (finite-dimensional Hilbert spaces) has been given in [177]. This classification uses some group theoretical techniques which are beyond the scope of the present text.

For each value of n, i.e., each eigenstate of the initial Hamiltonian, the fibers of the corresponding B-S line bundle L^n are defined by

$$L_R^n := \{\psi \in \mathcal{H} : |\psi\rangle = c|n;R\rangle, c \in \mathbb{C}\} \subset \mathcal{H}, \qquad R \in M. \tag{6.2}$$

Note that unlike $|n;R\rangle$ that are only defined up to a multiplicative factor, the fibers L_R^n are uniquely determined by the eigenvalue equation (2.6). They are simply the rays in the Hilbert space associated with the eigenstates $|n;R\rangle\langle n;R|$. To specify the bundle structure of L^n, we need in addition to fibers and the base manifold, the transition functions. To choose the transition functions, we re-examine the eigenvalue equation (2.6). As we remarked in Sect. 2.1, the single-valued, normalized eigenvectors $|n;R\rangle$ are only defined locally. Namely, they are defined on open neighborhoods U_α of M. To emphasize the local nature of $|n;R\rangle$, we attach the label α to these eigenvectors, i.e., label them by $|n;R,\alpha\rangle$. Now let us consider another open neighborhood U_β intersecting U_α, and the associated single-valued eigenvectors $|n;R,\beta\rangle$. On the intersection, i.e., for all $R \in U_\alpha \cap U_\beta$, both $|n; R, \alpha\rangle$ and $|n'; R, \beta\rangle$ satisfy the same equation (2.6). Hence they may only differ by a phase factor,

$$|n; R, \beta\rangle = e^{i\zeta_{\beta\alpha}(R)} |n; R, \alpha\rangle, \qquad \zeta_{\beta\alpha}(R) \in \mathbb{R}. \tag{6.3}$$

Now let us consider a covering $\{U_\alpha\}$ of M such that for each U_α, we have a complete set of smooth and single-valued eigenvectors $|n; R, \alpha\rangle$. The phase factors $e^{i\zeta_{\beta\alpha}}$ relate the local eigenvectors associated with different U_α's. Since the fibers are defined by the eigenrays, they are also related by the same phase factors. Hence, we define

$$G_{\beta\alpha}(R) := e^{\zeta_{\beta\alpha}(R)} \in U(1) \tag{6.4}$$

to be the transition functions of L^n.

The associated $U(1)$ PFB to L^n is obtained by considering only the normalized state vectors in the Hilbert space and using the same transition functions. We shall denote this bundle by λ^n:

$$(\lambda^n, M, \pi, U(1)). \tag{6.5}$$

We have introduced an example of this PFB in Sect. 5.2, where the parameter space was the two-dimensional sphere S^2 and the bundle structure was determined by the Hamiltonian (3.1). In general, the fibers of λ^n are defined by

$$\lambda_R^n := \left\{ \psi \in \mathcal{H} \ : \ |\psi\rangle = e^{i\alpha} |n; R\rangle, \ \alpha \in [0, 2\pi) \right\}. \tag{6.6}$$

In other words, λ^n is obtained by replacing the fibers L_R^n of L^n by their subsets λ_R^n.

A periodic change of the Hamiltonian is characterized by a closed curve

$$\mathbf{C} : [0, T] \ni t \longrightarrow R(t) \in M, \qquad R(0) = R(T). \tag{6.7}$$

in the base manifold. As in Chap. 2, we consider the evolution of the initial state vector $|\psi(0)\rangle = |n; R(0)\rangle$ which is an eigenstate vector of the initial Hamiltonian $h(R(0))$. By construction, $|\psi(0)\rangle$ belongs to the fiber $\lambda_{R(0)}^n$ over $R(0) \in M$. Under the adiabatic approximation, after a complete cycle, $\psi(0)$ returns to the same ray in the Hilbert space but in the process it acquires a phase factor. We demonstrated this in (2.66). The geometric part of the total phase (2.53) could be expressed as a "line" integral of a one-form A^n (2.55) over the closed loop $\mathbf{C}$.

Comparing (2.53) and (5.36) and noting the fact that λ^n is a $U(1)$ (Abelian) PFB, we can immediately identify Berry's phase $e^{i\gamma_n(\mathbf{C})}$ of (2.66) with the holonomy of the connection A^n over λ^n. In fact, we have shown in Sect. 2.3 that under a gauge transformation (2.58), A^n transforms as a local connection one-form. The single-valued energy eigenvectors $|n; R\rangle$ which are used to define the bundles λ^n and specially determine the B-S connection one-form A^n are indeed identical with the local sections s of (5.18). The gauge (phase) transformation (2.16) is an example of (5.20).

It is indeed quite remarkable that the Hamiltonian operator of a quantum mechanical system defines not only a family of topological fiber bundle structures on the space of parameters, but it also determines a geometric structure on each of these bundles. In the following sections we shall see the amazing generality of this construction.

The topological structure of λ^n can be studied by calculating the transition functions (6.4). This can be quite difficult depending on the complexity of the Hamiltonian. There is another much more practical way of determining the topologies of these bundles. Namely, one seeks a set of quantities which are invariant under all the smooth deformations (bundle isomorphisms) of the bundle, and tries to classify fiber bundles according to the values of these quantities. Naturally, under such a deformation the topological structure of the bundle is invariant. Thus the corresponding invariant quantities are called *topological invariants* of the bundle. For a $U(1)$ PFB, there is a topological invariant which determines its topology uniquely. This is a two-form on the base manifold and is called the first *Chern class*. We shall demonstrate this process by working out a specific example, namely the system of (3.1). We shall also discuss the physical significance of the corresponding topological information.

The second holonomy interpretation of the geometric phase, and in particular Berry's adiabatic phase, is the Aharonov–Anandan (A-A) interpretation [7, 9]. In the A-A approach, one considers a complex line bundle E:

$$(E, \mathbb{C}P^\infty, \Pi, U(1)), \tag{6.8}$$

over the projective Hilbert space $\mathcal{P}(\mathcal{H}) = \mathbb{C}P^\infty$. If the Hilbert space is finite dimensional ($\mathcal{H} = \mathbb{C}^N$), then $\mathbb{C}P^\infty$ is replaced by $\mathbb{C}P^{N-1}$.

We have provided a detailed discussion of the complex projective space $\mathbb{C}P^N$ in Appendix A.2, and we shall not repeat its description here. However, before describing the line bundle E, we would like to comment on the infinite-dimensional case, $\mathcal{H} = \mathbb{C}P^\infty$.

Recalling the definition of $\mathbb{C}P^N$, a point $x \in \mathbb{C}P^{N-1}$ is a complex line in $\mathbb{C}^N$ which passes through the origin. Therefore, we have the following sequence of inclusions

$$\mathbb{C}P^1 \subset \mathbb{C}P^2 \subset \cdots \subset \mathbb{C}P^{N-1} \subset \mathbb{C}P^N \subset \cdots . \tag{6.9}$$

We will use this relation to extend the definition of $\mathbb{C}P^N$ to the case $N = \infty$. In view of (6.9), we have the set theoretical equality

$$\bigcup_{i=1}^N \mathbb{C}P^i = \mathbb{C}P^N, \tag{6.10}$$

which can be immediately generalized to define

$$\mathbb{C}P^\infty := \bigcup_{i=1}^\infty \mathbb{C}P^i. \tag{6.11}$$

6.2 Holonomy Interpretations of the Geometric Phase

We can see this as a limiting case of (6.10) with $N \to \infty$. This limit is called a *direct* or an *inductive limit*. Equation (6.11) defines $\mathbb{C}P^\infty$ as a point set without a topological or manifold structure. We choose a natural topology on $\mathbb{C}P^\infty$ by defining the open subsets to be the unions of open subsets of $\mathbb{C}P^i \subset \mathbb{C}P^\infty$, for all $i = 1, 2, \cdots$.

An alternative choice for a topology on $\mathbb{C}P^\infty$ is the one induced from the Hilbert space. The situation again resembles the finite-dimensional case. We can view $\mathbb{C}P^\infty$ as the set of the equivalence classes of rays in the Hilbert space. In this picture $\mathbb{C}P^\infty = \mathcal{H}/\sim$, where the equivalence relation "$\sim$" is defined on $\mathcal{H}$ by

$$\psi \sim \psi' \quad \text{if and only if} \quad \psi' = z \cdot \psi$$

for some non-zero complex number z. Hence, $\mathbb{C}P^\infty$ is identified with the quotient space and there is an onto projection $\rho : \mathcal{H} \to \mathbb{C}P^\infty$. The open subsets of $\mathbb{C}P^\infty$ are then defined to be the subsets whose inverse image under ρ are open subsets of $\mathcal{H}$. This is an example of a *quotient topology*. The two topological descriptions of $\mathbb{C}P^\infty$ are in fact not equivalent. However the line bundles (respectively $U(1)$ PFB's) based on $\mathbb{C}P^\infty$ with any of these two topological structures turn out to be in one-to-one correspondence.[1] Similarly, one can use the Hilbert space structure of $\mathcal{H}$ to endow the latter topological description of $\mathbb{C}P^\infty$ with an infinite-dimensional (Hilbert) manifold structure.[2]

Having discussed subtleties of the structure(s) of the projective Hilbert space $\mathcal{P}(\mathcal{H}) = \mathbb{C}P^\infty$, we return to the description of the A-A line bundle E. Each point $\Lambda \in \mathcal{P}(\mathcal{H})$ is a pure quantum state. The fiber E_Λ over this state is defined to be the corresponding ray in the Hilbert space $\mathcal{H}$. Since each ray is a copy of $\mathbb{C}$, the fibers are one dimensional, and E is a complex line bundle. Unlike λ^n of the B-S approach, the topological structure of E is uniquely defined by the structure of $\mathbb{C}P^\infty$. We shall refer to E as the A-A line bundle for the moment.

The A-A PFB η of (4.33) is the associated $U(1)$ principal bundle to E. The fibers η_Λ of η are obtained as the subsets of those of E which consist of normalized state vectors, namely

$$\eta_\Lambda := \{\psi \in \mathcal{H} \,:\, |\psi\rangle\langle\psi| = \Lambda \,,\, \langle\psi|\psi\rangle = 1\}. \tag{6.12}$$

It is not difficult to see that each fiber η_Λ is a copy of the unit circle S^1 or the group $U(1)$. In particular the total space of η consists of all the unit vectors in $\mathcal{H}$. As a manifold it is identical with the unit sphere $S^\infty \subset \mathcal{H}$. The fiber bundle

---

[1] In precise mathematical language, one says that the two topological spaces are of the same homotopy type. Two topological spaces X_1 and X_2 are said to be of the same homotopy type if there exist continuous functions $f : X_1 \to X_2$ and $g : X_2 \to X_1$, such that $f \circ g$ and $g \circ f$ are homotopic (continuously deformable) to the identity functions $Id_2 : X_2 \to X_2$ and $Id_1 : X_1 \to X_1$, respectively. A precise definition of the homotopy relation for continuous functions between two topological spaces is given following (6.18), below.

[2] A description of the Hilbert manifold structure of $\mathbb{C}P^\infty$ may be found in [6].

112   6. Mathematical Structure of the Geometric Phase I: The Abelian Phase

$$\eta : (S^\infty, \mathbb{C}P^\infty, \pi, U(1)) \tag{6.13}$$

is a direct generalization of its finite-dimensional counterpart:

$$\left(S^{2N-1}, \mathbb{C}P^{N-1}, \pi, U(1)\right). \tag{6.14}$$

It can be described quite conveniently by viewing $\mathbb{C}P^\infty$ as the set of all equivalence classes of the equivalence relation

$$\psi' \sim \psi \quad \text{iff} \quad \psi' = e^{i\alpha}\psi,$$

where $\psi, \psi' \in S^\infty$ and $\alpha \in [0, 2\pi)$. It is clear from this definition that each equivalence class is a copy of $U(1)$. In the fiber bundle picture, the equivalence classes play the role of the fibers. We have presented a detailed discussion of the finite-dimensional case in Appendix A.2.

The evolution of an exact cyclic state corresponds to a closed loop

$$\mathcal{C} : [0, T] \ni t \longrightarrow \Lambda(t) \in \mathcal{P}(\mathcal{H}), \quad \Lambda(0) = \Lambda(T), \tag{6.15}$$

in $\mathcal{P}(\mathcal{H}) = \mathbb{C}P^\infty$. The geometric phase is then identified with the holonomy of this loop that is determined by the local connection one-form $\mathcal{A}$ of (4.27). Recall that $\mathcal{A}$ depends on the choice of the vectors $|\phi\rangle$. The latter is also an example of the local section s of (5.18). It satisfies the expected gauge (phase) transformation rule (4.6) which is a special case of (5.20). Similarly, we can check that the gauge transformation rule (4.28) for $\mathcal{A}$ is identical with the gauge transformation rule (5.28) for a local connection one-form on a $U(1)$ PFB. The local sections $|\phi\rangle$ may also be used to define the transition functions of E or η explicitly. This is done in the same way we derived (6.4) using $|n; R\rangle$.

This concludes our description of the A-A interpretation of the geometric phase. We should like to emphasize that the A-A principal bundle η has a "universal" character as both its topological and geometrical structures are independent of the details of the Hamiltonian operator. The role of the Hamiltonian, in this interpretation, is to determine the curve $\mathcal{C}$ of the pure quantum states. Another major advantage of this approach is that it does not assume the adiabaticity of the evolution of the system.

In the case that a quantum system undergoes an adiabatic change, the geometric phase reduces to Berry's phase, and either of the A-A or B-S interpretations may be applied. In contrast to the A-A fiber bundles, the fiber bundles of the B-S approach are defined over the space of the physical parameters of the system. Thus they are all finite-dimensional spaces. In particular, the curve $\mathbf{C}$ (6.7) can be directly observed in the laboratory, whereas the loop $\mathcal{C}$ (6.15) lives in the projective Hilbert space $\mathcal{P}(\mathcal{H})$ and is merely a mathematical entity.

In the context of the adiabatic approximation, one can in principle try to understand the mathematical relationship between the two constructions.

It is also interesting to find a B-S type interpretation for the general nonadiabatic geometric phase [180, 181]. We shall next offer a description of the mathematical relationship between the B-S and A-A interpretations of the adiabatic phase.

6.3 Classification of $U(1)$ Principal Bundles and the Relation Between the Berry–Simon and Aharonov–Anandan Interpretations of the Adiabatic Phase

The purpose of seeking mathematical interpretations of the geometric phase is to elucidate the geometric nature of the phase by relating it to known geometric constructions, such as holonomy elements of fiber bundles. For the special case of Berry's adiabatic phase both the B-S and A-A interpretations provide equally admissible geometric descriptions. The fiber bundles used in these two approaches are however quite different. The B-S approach involves the $U(1)$ PFBs $\lambda^n \to M$ (6.5) over the space of physical parameters, whereas the A-A approach employs the $U(1)$ PFB $\eta \to \mathcal{P}(\mathcal{H})$ (6.13) over the projective Hilbert space. As both approaches describe a single physical phenomenon, it seems a relevant question to ask whether there is a deeper relationship between the corresponding mathematical constructions.

Indeed, the bundle η, (6.13), of the A-A interpretation turns out to play a central role in the mathematical theory of fiber bundles. It is named quite appropriately the *universal $U(1)$ principal bundle* [63, 78, 186] by mathematicians. Here the adjective "universal" reflects the utility of η in the classification of all $U(1)$ PFBs. Furthermore, there is a quite special connection on η which is also used in the classification of all possible geometric structures (connections) on a given $U(1)$ PFB. This connection is called the *universal connection* [188]. It is quite remarkable that the universal connection on η turns out to be precisely the A-A connection of (4.27).

In order to describe the mathematical theory underlying the relation between the B-S and A-A interpretations, we shall first recall the definitions of a few related concepts. We begin by describing the concept of a *pullback bundle*.

Consider two smooth manifolds X and X', a smooth function $f : X \to X'$, and a fiber bundle E': (E', X', π', G) over X'. Then there is a unique procedure to use f and E' to induce a fiber bundle structure on X. The induced bundle is called the *pullback bundle* and denoted by f^*E': (f^*E', X, π, G). The fibers of f^*E' are defined according to the following assignment: For each $x \in X$, the fiber $f^*E'_x$ over x is defined to be identical with the fiber $E'_{f(x)}$ of E' over $f(x) \in X'$. This however does not specify the global (topological) bundle structure of f^*E'. To furnish this collection of fibers over X with a bundle structure we must determine the transition functions (5.5) of

f^*E'. Let us choose a complete set of local trivializations $\{(U'_\alpha, \Phi'_\alpha)\}$ of E', and denote the corresponding transition functions by $G'_{\alpha\beta}$. Since $f : X \to X'$ is a smooth (continuous) function, the inverse images of open subsets of X' under f are open subsets[3] of X. Thus, the inverse images $U_\alpha := f^{-1}(U'_\alpha)$ form an open covering of X. They are used to define a complete set of local trivializations of f^*E'. The corresponding transition functions $G_{\alpha\beta}$ of f^*E' are then naturally defined by

$$G_{\alpha\beta}(x) := G'_{\alpha\beta}(f(x)), \quad \forall x \in U_\alpha \cap U_\beta.$$

The construction of the pullback bundle f^*E' can be illustrated by the following diagram,

$$\begin{array}{ccc} f^*E' & \xleftarrow{f^*} & E' \\ \downarrow & & \downarrow \\ X & \xrightarrow{f} & X', \end{array} \quad (6.16)$$

where f^* denotes the pullback operation. We can also define an induced function $f_* : f^*E' \to E'$ that identifies the points of the fibers of f^*E' with those of E'. f_* is the symbol used to represent the procedure according to which we defined fibers of f^*E'. The advantage of defining this apparently trivial function is to summarize the definition of f^*E' as a commutative diagram, namely

$$\begin{array}{ccc} f^*E' & \xrightarrow{f_*} & E' \\ \pi \downarrow & \circlearrowleft & \downarrow \pi' \\ X & \xrightarrow{f} & X'. \end{array} \quad (6.17)$$

The commutativity of this diagram is algebraically expressed by

$$f \circ \pi = \pi' \circ f_*. \quad (6.18)$$

Therefore, the function f_* is by construction a bundle morphism.

Another important concept of interest in the classification of PFBs is the concept of *homotopy*. Homotopy is an equivalence relation on the set of all continuous functions between two topological spaces. In simple words, two such maps are called *homotopic* if they can be "deformed" into one another in a "continuous manner." The precise definition of homotopy is as follows. Let $f, g : X_1 \to X_2$ be two continuous functions between two topological spaces X_1 and X_2. We can view X_1 and X_2 as manifolds. Then f and g are said to be homotopic: $f \sim g$, if there is a continuous function

$$\mathcal{F} : X_1 \times [a, b] \longrightarrow X_2, \quad [a, b] \subset \mathbb{R},$$

such that

$$f = \mathcal{F}|_{X_1 \times \{a\}}, \quad \text{and} \quad g = \mathcal{F}|_{X_1 \times \{b\}}.$$

---
[3] See Appendix A for a clarification of this point.

This means that $\mathcal{F}$ yields f and g as its restrictions to $X_1 \times \{a\}$ and $X_2 \times \{b\}$, respectively. The function $\mathcal{F}$ is called a *homotopy* between f and g. It is easy to verify that the homotopy relation $\sim$ is an equivalence relation (Problem 6.1). Thus we can classify all the continuous functions between two topological spaces X_1 and X_2 by placing them in the equivalence classes of this equivalence relation. These are naturally called the *homotopy classes*. The set of all homotopy classes of maps between X_1 and X_2 is denoted by $[X_1, X_2]$.

The importance of the concept of homotopy is that it is only sensitive to the topological structures of the corresponding spaces. More precisely, the set $[X_1, X_2]$ is a topological invariant of both X_1 and X_2. In practice, one usually fixes one of these spaces and investigates the dependence of the set $[X_1, X_2]$ on the other. In particular, it is possible to choose one of these two spaces such that the set $[X_1, X_2]$ possesses a group structure [189]. The most well-known choices for X_1 which give $[X_1, X_2]$ a group structure are the spheres S^d, $d = 1, 2, \cdots$. $[S^d, X_2]$ is called the d-th *homotopy group* [190] of the space X_2 and is denoted by $\pi_d(X_2)$. The groups $\pi_d(X_2)$ are topological invariants of X_2. They play an important role in modern mathematics and theoretical physics. A typical example of the application of homotopy groups in physics is in the subject of magnetic monopoles and multimonopoles [87].

A crucial disadvantage of the homotopy groups is that they are very difficult and sometimes almost impossible to compute. There are similar and closely related topological invariants which are computed much more easily. These are called the *homology* and *cohomology* groups [64, 78, 186, 190]. There exist many different realizations of the homology and cohomology groups. We shall briefly outline the most well-known cohomology groups, the *de Rham cohomology* groups $H^p_{dR}(X)$.

Let us consider a smooth manifold X and a differential p-form ω^p on X. As we discussed in Sect. 2.4, if the exterior derivative of ω^p vanishes, $d\omega^p = 0$, then ω^p is called a *closed p-form*. A special class of closed forms are the so-called *exact forms*. A p-form ω^p is called an exact p-form if there is a $(p-1)$-form τ^{p-1} such that $\omega^p = d\tau^{p-1}$. Clearly since $d^2 = 0$, any exact form is closed.[4] The converse is in general not true. In fact, the failure of some of the closed forms to be exact is related to the global (topological) properties of the manifold X. Thus, the extent to which closed forms differ from being exact can be used as a measure of the topological properties of a manifold. This is precisely what the de Rham cohomology represents.

Let us define an equivalence relation $\approx$ on the set of all closed differential p-forms by

$$\omega_1^p \approx \omega_2^p \quad \text{iff} \quad (\omega_1^p - \omega_2^p) \text{ is exact.}$$

If we denote the set of all closed p-forms by $\mathcal{Z}^p(X)$ and the set of all exact p-forms by $\mathcal{B}^p(X)$, then the set of all the equivalence classes of the relation

---

[4] See Appendix A.2 for a discussion of differential forms and exterior differentiation.

≈ is the quotient set $\mathcal{Z}^p(X)/\mathcal{B}^p(X)$. This set is called the p-th de Rham cohomology $H_{dR}^p(X)$ of X. In fact, the sets $\mathcal{Z}^p(X)$, $\mathcal{B}^p(X)$, and $H_{dR}^p(X)$ have a vector space structure under ordinary addition of forms and multiplication by real numbers. In particular, $H_{dR}^p(X)$ has the structure of an Abelian group under addition.

In general a cohomology group $H^p(X,\mathbb{Z})$ of a finite-dimensional manifold has only the structure of a finitely generated Abelian group [190]. This means that it has the form

$$\mathbb{Z} \oplus \cdots \oplus \mathbb{Z} \oplus \mathbb{Z}_{r_1} \oplus \cdots \oplus \mathbb{Z}_{r_k},$$

where $\mathbb{Z}$ and $\mathbb{Z}_{r_i}$ denote the Abelian groups of integers and integers modulo r_i respectively,[5] and there are only a finite number of terms in the direct sum of these groups. $H^p(X,\mathbb{Z})$ reduces to $H_{dR}^p(X)$, if we drop the part of the above direct sum that involves finite groups and change $\mathbb{Z}$'s to $\mathbb{R}$'s in the rest. This concludes our discussion of cohomology groups. The interested reader may refer to [186, 190] or textbooks on algebraic topology for more detailed discussions of the cohomology groups.

Before we start our presentation of the classification theorem for $U(1)$ PFBs, we wish to list some introductory texts on this subject. These are the books [63, 114, 186] and the article [78]. A more advanced and classic text is *The Topology of Fiber Bundles* [232]. Furthermore, we wish to suggest that the reader recall the definitions of the pullback of a fiber bundle and the pullback of a differential form. For a review of the latter, see Appendix A.2.

The central idea of the classification theory of PFBs is the existence of the so-called *universal bundles*. These are special PFBs which yield all other PFBs, with the same structure group, as pullback bundles. The construction for the $U(1)$ bundles is as follows.

Let X be a smooth manifold and P be an arbitrary $U(1)$ PFB over X. Then there is a $U(1)$ PFB, $\eta(U(1))$, over a manifold $B(U(1))$ and a smooth function $f : X \to B(U(1))$, such that P is isomorphic to the pullback bundle $f^*\eta(U(1))$; $P \cong f^*\eta(U(1))$. This is expressed as a commutative diagram:

$$\begin{array}{ccc} f^*\eta(U(1)) \cong P & \xrightarrow{f_*} & \eta(U(1)) \\ \downarrow & \circ & \downarrow \\ X & \xrightarrow{f} & B(U(1)). \end{array} \quad (6.19)$$

Furthermore, since the topological fiber bundle structure of $f^*\eta(U(1))$ depends only on the homotopy class of the map f, $U(1)$ PFBs over X are classified by the elements of $[X, B(U(1))]$ [186]. In other words, there is a one-to-one correspondence between the set of all $U(1)$ PFBs and the set $[X, B(U(1))]$ of the homotopy classes of maps from X to $B(U(1))$.

As we mentioned earlier, it turns out that the A-A bundle η (6.13) is a universal bundle. Thus we may take

---

[5] See Appendix A for the definitions.

6.3 Relation Between the A-A and B-S Interpretations

$$\eta(U(1)) \equiv \eta \quad \text{and} \quad B(U(1)) \equiv \mathbb{C}P^\infty.$$

An important property of $\mathbb{C}P^\infty$ is that all its homotopy groups $\pi_d(\mathbb{C}P^\infty)$ are trivial except $\pi_2(\mathbb{C}P^\infty)$ which is given by[6]

$$\pi_2(\mathbb{C}P^\infty) = H^2(\mathbb{C}P^\infty, \mathbb{Z}) = \mathbb{Z}. \tag{6.20}$$

This result can be used to show that, for example, all $U(1)$ PFBs on S^2 are classified by integers. This is a direct consequence of (6.20) and the definition: $\pi_2(\mathbb{C}P^\infty) := [S^2, \mathbb{C}P^\infty]$

The classification theorem also extends to the classification of the geometric structures, i.e., connections, on PFBs [188]. This is done by constructing a *universal connection* $\mathcal{A}(U(1))$ on $\eta(U(1)) = \eta$ and expressing all possible connections on a PFB as pullback connection one-forms. In other words, for any $U(1)$ PFB P with a local connection one-form A we can choose the function $f : X \to \mathbb{C}P^\infty$ (in the appropriate homotopy class) such that not only $P = f^*\eta$, but also

$$A = f^*\mathcal{A}(U(1)). \tag{6.21}$$

Comparing the expression for the universal connection $\mathcal{A}(U(1))$ of the mathematical literature [188] with that of the A-A connection one-form (4.27), one immediately notices their coincidence.

In view of these most astonishing identifications, the problem of the investigation of the relation between the B-S and A-A interpretations of the geometric phase reduces to seeking an appropriate function $f_n : M \to \mathbb{C}P^\infty$ that induces the B-S PFBs λ^n (6.5) as the pullback bundles from the universal A-A bundle η (6.13). The natural choice for f_n which allows for both

$$\lambda^n = f_n^*\eta, \tag{6.22}$$
$$A^n = f_n^*\mathcal{A}, \tag{6.23}$$

is the function

$$f_n(R) := |n; R\rangle\langle n; R|, \quad \forall R \in M. \tag{6.24}$$

In (6.23) and (6.24), A^n denotes the Berry connection one-form (2.55) and the state $|n; R\rangle\langle n; R|$ is the energy eigenstate defined by (2.6).

It is remarkable that f_n maps the closed loop $\mathbf{C}$ (6.7) in the parameter space onto the closed loop $\mathcal{C}$ (6.15) in the projective Hilbert space $\mathcal{P}(\mathcal{H}) = \mathbb{C}P^\infty$, i.e.,

$$\mathcal{C} \stackrel{\text{adiabatic}}{=\!=\!=} f_n \circ \mathbf{C}. \tag{6.25}$$

Therefore, f_n carries all the information necessary for the computation of the adiabatic phase.

The relation between the B-S and A-A interpretations of Berry's adiabatic phase is briefly demonstrated by the following set of identities,

---

[6] The spaces that have only a single non-trivial homotopy group are called the Eilenberg–McLane spaces.

$$\gamma_n(\mathbf{C}) := \oint_{\mathbf{C}} A^n = \oint_{\mathbf{C}} f_n^* \mathcal{A} = \oint_{f_n \circ \mathbf{C}} \mathcal{A} = \oint_{\mathcal{C}} \mathcal{A} =: \gamma(\mathcal{C}). \qquad (6.26)$$

Here we have used (2.65), (6.23), (6.25), and (4.26) in the first, second, fourth, and fifth equalities, respectively. The third equality in (6.26) is a consequence of a property of the pullback operation for differential forms (Problem 6.2).

This completes our investigation of the relation between the two apparently different interpretations of the adiabatic geometric phase. Although the use of the classification theorem has not yet led to any physically interesting results, our observations suggest the existence of unprecedented fundamental relationships between quantum mechanics and the differential geometry of fiber bundles. Indeed, we shall see in the next section that the mathematical tools developed in this section are quite helpful in providing us with a better understanding of the non-adiabatic geometric phase.

6.4 Holonomy Interpretation of the Non-Adiabatic Phase Using a Bundle over the Parameter Space

In general every time-dependent Hamiltonian involves an *evolution parameter* that characterizes the adiabaticity of the time-dependence of the system. We have already seen an example of an evolution parameter in our analysis of a spin system (3.1). There, it was the frequency ω of the precession of the magnetic field that determined whether the system underwent an adiabatic time evolution. In fact, the precession frequency ω alone was not sufficient to fulfill this task and an intrinsic frequency scale, namely the frequency b, was also needed. Specifically, the ratio $\nu = \omega/b$ played the role of an evolution parameter. The condition $\nu \ll 1$ characterized the domain in which the adiabatic approximation was reliable. We shall adopt the same notation to denote the evolution parameter of a general quantum mechanical system. In general, there may be several intrinsic frequencies which may lead to several evolution parameters. For a given environmental process and a given initial state, however, only one of these evolution parameters determines the adiabaticity of the evolution.[7] We shall label this evolution parameter by ν.

In our discussion of the A-A interpretation of the geometric phase, we emphasized that this interpretation was valid for both adiabatic and non-adiabatic systems. This is easily seen by recalling the definition of the A-A bundle η (4.33). By construction, the $U(1)$ PFB η and its geometric structure are independent of the dynamics. In particular they do not depend on the evolution parameter.

Unlike the A-A interpretation, the B-S interpretation is based on the adiabatic approximation. The fibers of the B-S bundle λ^n (6.5) consist of the eigenstate vectors of the Hamiltonian $h(R)$. We shall see that a B-S type

---

[7] A precise definition of the relevant evolution parameter may be found in [179].

6.4 Holonomy Interpretation of the Non-Adiabatic Phase

construction for the non-adiabatic case, if it exists, depends on the evolution parameter ν. Moreover, this construction must reduce to its adiabatic counterpart, i.e., the B-S bundle λ^n and connection A^n, in the adiabatic domain. We shall denote the non-adiabatic analog of λ^n by $\tilde{\lambda}^n_\nu$. If we symbolize the domain of validity of the adiabatic approximation by $\nu \to 0$, then $\tilde{\lambda}^n_{\nu \to 0} \to \lambda^n$. This clearly shows that *we are seeking a $U(1)$ PFB $\tilde{\lambda}^n_\nu$ over the parameter manifold M, and an appropriate connection one-form $\tilde{A}^n_\nu$ on $\tilde{\lambda}^n_\nu$ such that the holonomy element associated with a closed loop $\mathbf{C}$ (6.7) in M yields the non-adiabatic phase.*

To obtain the desired non-adiabatic B-S bundle and connection, we appeal to the classification theorem of the previous section. According to this theorem, the bundle $\tilde{\lambda}^n_\nu$, as any other $U(1)$ PFB, can be obtained as a pullback bundle from the universal A-A bundle η. Therefore, we must seek a smooth function $\tilde{f}_{n,\nu} : M \to \mathbb{C}P^\infty$ such that

$$\tilde{\lambda}^n_\nu = \tilde{f}^*_{n,\nu} \eta, \tag{6.27}$$

$$\tilde{A}^n_\nu = \tilde{f}^*_{n,\nu} \mathcal{A}. \tag{6.28}$$

The classification theorem does not however offer a prescription for determining the form of $\tilde{f}_{n,\nu}$. In general, $\tilde{f}_{n,\nu}$ carries all the information about the geometric phase. Thus, it depends on the initial state and the evolution parameter ν. An important property of $\tilde{f}_{n,\nu}$ is that it maps the closed curve $\mathbf{C}$ of (6.7) onto the closed curve $\mathcal{C}$ of (6.15). This is necessary to ensure that the holonomy elements of the non-adiabatic B-S approach coincide with the A-A phase. To see this, it is sufficient to realize that the above condition means

$$\mathcal{C} = \tilde{f}_{n,\nu} \circ \mathbf{C}, \tag{6.29}$$

and replace the role of f_n in (6.26) by $\tilde{f}_{n,\nu}$. Equation (6.29) is the exact form of the approximate equality (6.25).

The converse of the requirement that "all" closed curves[8] in M are mapped onto the closed curves in $\mathcal{P}(\mathcal{H})$ is not generally true. In other words, there may be open curves in M that are also mapped onto closed curves in $\mathcal{P}(\mathcal{H})$. These correspond to the possibility of a non-periodic quantum system (Hamiltonian) possessing cyclic states. We shall not discuss such systems here.

In order to find an expression for $\tilde{f}_{n,\nu}$, we use the analogy with the adiabatic case and our knowledge of the dynamics of the spin system (3.1). In the adiabatic case the function f_n (6.24) was defined to map the point $R \in M$ to the eigenstate $|n; R\rangle\langle n; R| \in \mathcal{P}(\mathcal{H})$ of $h(R)$. This suggests that we can implicitly define $\tilde{f}_{n,\nu}$ to yield the eigenstates of a Hermitian operator $\tilde{h}(R)$ with the property that the eigenstates of the initial value of $\tilde{h}$ undergo exact

---

[8] Here by "all" closed curves, we mean all the closed curves of interest. For example in Chap. 3, the closed curves of interest were all the circular paths in S^2 which belonged to the planes perpendicular to the x^3-direction.

cyclic evolutions, i.e., the initial state $W(0) = |\psi(0)\rangle\langle\psi(0)| \in \mathcal{P}(\mathcal{H})$, (4.2), is an eigenstate of $\tilde{h}(R(0))$. Conversely we demand that all the initial cyclic states of interest are obtained in this way[9], i.e., as eigenstates of a Hermitian operator, $\tilde{h}(R(0))$.

In general $\tilde{h}$ will also depend on the particular time-dependence of the Hamiltonian, i.e., the closed curve traversed by the parameters of the Hamiltonian in the parameter space M. For the particular example of a spin system in a precessing magnetic field, we saw in Chap. 3 that the operator $\tilde{h}$ for all the circular loops in the parameter space S^2 corresponding to the precession of the magnetic field about a fixed axis, depended only on the initial parameters $(R(0) = (\theta, 0))$. In the following we shall consider such curves in the parameter space. In other words, we shall consider only those periodic changes of the environment that can be specified with the initial values of the parameters $R(0)$ of the Hamiltonian and the evolution parameter ν. For such changes the operator $\tilde{h}$ is a function of the same quantities. Hence for each value of the evolution parameter ν it is a single-valued function of the parameters $R \in M$. In this case we can immediately define

$$\tilde{f}_{n,\nu}(R) := |\tilde{n}; R\rangle\langle\tilde{n}; R| := \tilde{\Lambda}_n(R), \quad \forall R \in M \tag{6.30}$$

where the numbers $\tilde{n}$ label the spectrum of $\tilde{h}(R)$, and we have assumed that the eigenvalue corresponding to $\tilde{n}$ is non-degenerate. Once again we emphasize that the operator $\tilde{h}(R)$ and its eigenstates $|\tilde{n}; R\rangle\langle\tilde{n}; R|$ depend also on the evolution parameter ν.[10]

By definition, the eigenstates $\tilde{\Lambda}_n(R(0)) \in \mathcal{P}(\mathcal{H})$ of $\tilde{h}(R(0))$ undergo exact cyclic evolutions. This means that $\tilde{\Lambda}_n(R(t))$ trace a closed curve $\mathcal{C}$ in the projective Hilbert space. The geometric phase can be originally obtained as the holonomy of the universal A-A connection associated with the closed curve $\mathcal{C} = \tilde{f}_{n,\nu} \circ \mathbf{C}$. Hence, in order to compute the phase, one must either solve the Schrödinger equation and obtain $\mathcal{C}$ directly, or find the expression for the map $\tilde{f}_{n,\nu}$ and use (6.28). For the system (3.1), both of these have been done. In this case, the operator $\tilde{h}(R(0))$ is given by (3.93):

$$\tilde{h}(\theta, 0) = b\,\mathbf{e} \cdot \mathbf{J}, \tag{6.31}$$

where $R(0) = (\theta = \text{const.}, \varphi = 0) =: R_0$. It is convenient to relabel the vector $\mathbf{e}$ of (3.72), by $\tilde{R}(0)$ or briefly $\tilde{R}_0$. Then we can define

$$\tilde{h}_0 := \tilde{h}(R_0) := b\,\hat{\tilde{\mathbf{R}}}_0 \cdot \mathbf{J}. \tag{6.32}$$

---

[9] We have already shown in Sect. 4.3 that for a periodic Hamiltonian such an operator can be constructed.

[10] The operator $\tilde{h}(R(t))$ has the property that it has a set of eigenvectors that are solutions of the Schrödinger equation for the Hamiltonian $h(R(t))$. Such an operator is called a *dynamical invariant* [155]. There is an alternative formulation of geometric phases in terms of dynamical invariants. For a discussion of this formulation and a list of references see [183].

6.4 Holonomy Interpretation of the Non-Adiabatic Phase

Comparing (3.92) and (6.32), we arrive at

$$\tilde{h}_0(R_0) = h(\tilde{R}_0), \tag{6.33}$$

where $h(\tilde{R}_0)$ is the operator obtained by evaluating the Hamiltonian $h(R)$ at $R = \tilde{R}_0 = \mathbf{e}$. It is also helpful to recall that according to (3.72), (3.73), and (3.75), we have $\mathbf{e} = (\tilde{\theta}, \tilde{\varphi} = 0)$. This justifies our new notation of labeling $\mathbf{e}$ by $\tilde{R}_0$.

Since in general our results must be independent of the choice of the origin of the spherical coordinates, we could label the initial (and the final) point of the curve $\mathbf{C}$ (6.7) by $R = (\theta, \varphi)$ rather than $R_0 = (\theta, 0)$. In this case, we would obtain

$$\tilde{h}(R) = h(\tilde{R}). \tag{6.34}$$

We can immediately use (6.34) to write down the formula for the function $\tilde{f}_{\nu,\nu}$. The result is

$$\tilde{f}_{n,\nu}(R) := \tilde{\Lambda}_n(R) = \Lambda_n(\tilde{R}) = f_n(\tilde{R}). \tag{6.35}$$

Here use has been made of (6.30), (6.34), and $f_n : M \to \mathcal{P}(\mathcal{H})$ is the function defined in (6.24):

$$f_n(R) := \Lambda_n(R) = |n; R\rangle\langle n; R|. \tag{6.36}$$

Therefore, (6.35) can be written in the form

$$\tilde{f}_{n,\nu}(R) := f_n(\tilde{R}) = |n; \tilde{R}\rangle\langle n; \tilde{R}|. \tag{6.37}$$

In Sect. 3.4 we also introduced a function $F_\nu : S^2 \to S^2$ in (3.81). This function simply plays the role of the tilde, namely

$$F_\nu(R) := \tilde{R}. \tag{6.38}$$

In terms of F_ν, (6.37) is written in the form

$$\tilde{f}_{n,\nu} = f_n \circ F_\nu. \tag{6.39}$$

An advantage of introducing the function F_ν is that it leads to a direct construction of the bundle $\tilde{\lambda}_\nu^n$ and the connection $\tilde{A}_\nu^n$. Using a simple (functorial) property of the pullback operation:

$$(f_n \circ F_\nu)^* = F_\nu^* \circ f_n^*, \tag{6.40}$$

we find (Problem 6.3)

$$\tilde{\lambda}_\nu^n = \tilde{f}_{n,\nu}^* \eta = F_\nu^*(f_n^* \eta) = F_\nu^* \lambda^n, \tag{6.41}$$
$$\tilde{A}_\nu^n = \tilde{f}_{n,\nu}^* \mathcal{A} = F_\nu^*(f_n^* \mathcal{A}) = F_\nu^* A_n, \tag{6.42}$$

where we have used (6.22), (6.23), (6.39), and (6.40).[11]

Equations (6.41) and (6.42) indicate that we can obtain the non-adiabatic B-S bundle $\tilde{\lambda}^n_\nu$ and connection one-form $\tilde{A}^n_\nu$ directly as the pullback bundle and pullback form of their adiabatic analogs. The function $F_\nu : S^2 \to S^2$, in this respect, plays an important role. On the one hand, now we can compare two finite-dimensional PFBs λ^n and $\tilde{\lambda}^n_\nu$ over the parameter space X and try to understand the meaning of *adiabaticity* directly. On the other hand, we can use (6.42) to obtain the expression for the non-adiabatic connection one-form without making use of the original definition of the geometric phase as the difference of the total and dynamical phases. We can actually compare the result obtained in this way with our earlier derivation of (3.112). Direct calculation shows that indeed (6.42) does reproduce the results of Sect. 3.5. We have (Problem 6.4)

$$\tilde{A}^n_\nu = F^*_\nu(A_n) = -k\left[1 - \frac{b}{\Omega}(\cos\tilde{\theta} - \omega/b)\right]d\varphi, \qquad (6.43)$$

where we have only used (3.42) and the definition of the pullback of a differential one-form. Equation (6.43) is in complete agreement with (3.112). This marks the practical importance of (6.42) in the computation of the geometric phase.

The construction of $\tilde{\lambda}^n_\nu$ (6.27) for the system (3.1) depends directly on (6.34). This is the sole reason why we could write down the function $\tilde{f}_{n,\nu}$ (6.37) so easily. In general, (6.34) fails to hold and there is no explicit formula for $\tilde{f}_{n,\nu}$. However, there is a class of quantum systems for which (6.34) is satisfied. In this case, our analysis of the spin system (3.1) directly generalizes.

In view of (6.38), we can rewrite (6.34) in the form

$$\tilde{h} = h \circ F_\nu. \qquad (6.44)$$

Let us concentrate on the systems whose cyclic states are the eigenstates of the operator $\tilde{h}$ of (6.44), where h is the Hamiltonian and $F_\nu : M \to M$ is a smooth function. For these systems, the construction of the non-adiabatic B-S bundle and connection is realized according to (6.41) and (6.42).

There are two important conditions on the function F_ν. These are

1. $F_\nu : M \to M$ is a smooth and single-valued function.
2. In the adiabatic domain of the evolution parameter $(\nu \to 0)$, F_ν can be approximated by the identity map, $\text{Id} : M \to M$ on M.

We ought to emphasize that in practice, there is no exact procedure for obtaining the function F_ν, nor is there any analytic criterion for distinguishing the quantum systems which satisfy (6.44) from those which do not. Yet, there

---

[11] Note that the identity (6.40) holds for both the pullback of bundles and differential forms. In the fancy but sometimes useful language of category theory, the operation * is said to be a *covariant functor*. This means that it satisfies (6.40).

are examples which fit the scheme described here. A class of quantum systems that satisfy this condition is the systems whose Hamiltonian belongs to an irreducible representation of a compact semisimple Lie algebra and whose time evolution is governed by the action of a one-parameter subgroup of the Lie group. In fact, the system (3.1) is a typical example of such a system, where the Lie algebra is $su(2)$. The study of these systems leads to many interesting relationships between the foundations of the geometric phase and modern representation theory of compact semisimple Lie groups [177].

This concludes our treatment of the non-adiabatic B-S interpretation of the geometric phase. The method suggested here applies to a limited number of quantum systems. For these systems, however, it offers a direct approach for comparing the adiabatic and non-adiabatic evolutions. Moreover, it is important to note that unlike the original adiabatic B-S interpretation, the construction of the non-adiabatic B-S bundles does not correspond to an approximate description of the geometric phase.

In the next section, we shall discuss the topological aspects of the geometric phase. In particular, we shall examine the topological properties of the bundles associated with the spin system (3.1).

6.5 Spinning Quantum System and Topological Aspects of the Geometric Phase

The A-A interpretation of the geometric phase involves the universal bundle η (6.13). The topological and geometrical properties of η are independent of the details of the Hamiltonian, e.g., the evolution parameter, and the initial cyclic state under evolution. The latter information is encoded in the definition of the closed curve $\mathcal{C}$ (6.15). The situation is the opposite in the B-S approach. As we discussed in the previous section the exact (non-adiabatic) B-S bundle $\tilde{\lambda}_\nu^n$ and connection $\tilde{A}_\nu^n$ depend on the evolution parameter ν explicitly. The dependence of the (non-adiabatic) B-S connection one-form on ν is manifestly seen in (6.43). However, the topological structure of the corresponding $U(1)$ bundles and its dependence on ν remain to be investigated.

In fact, even in the adiabatic case the B-S bundles λ^n may and actually do have different topological structures. In the present section, we shall explore the topological properties of these bundles and try to find the physical implications of the topological content of the geometric phase.

In general, the study of the topological structures of arbitrary PFBs is a difficult problem. The topology of a PFB, or as a matter of fact any fiber bundle with given base manifold, typical fiber, and structure group, is determined by its transition functions. However, if we perturb the transition functions the topological structure of the bundle will not change. Thus, the topological structure of a bundle does not correspond to a unique set of transition functions. It can be shown that it is the homotopy classes of the transition functions that influence the topology of a fiber bundle. This is the main reason for the importance of the homotopy classes of the maps f of (6.19) in the classification theorem for the $U(1)$ PFBs.

Let us consider the case of $U(1)$ PFBs.[12] If the PFB $(P, X, \pi', U(1))$ in question is induced from the universal bundle $(\eta, \mathbb{C}P^\infty, \pi, U(1))$ via a known inducing function $f : X \to \mathbb{C}P^\infty$, then the topology of this bundle is uniquely determined by the homotopy class of f.

In general, there is no algorithm for determining the set of homotopy classes of functions between two arbitrary topological spaces (manifolds). However, due to a special topological property of the space $\mathbb{C}P^\infty$, namely

$$\pi_d(\mathbb{C}P^\infty) = \begin{cases} \{0\} & \text{for } d \neq 2 \\ \mathbb{Z} & \text{for } d = 2, \end{cases} \tag{6.45}$$

the following useful identity holds,

$$[X, \mathbb{C}P^\infty] = H^2(X, \mathbb{Z}). \tag{6.46}$$

Here $[X, \mathbb{C}P^\infty]$ denotes the set of homotopy classes of maps from X into $\mathbb{C}P^\infty$, and the equality means that there is a one-to-one correspondence between the two sets. Equation (6.46) suggests that globally every $U(1)$ PFB is associated with an element of the cohomology group $H^2(X, \mathbb{Z})$. This element is the first *Chern class* C_1 of the bundle. The first Chern class can be most easily described in terms of the de Rham cohomology. In fact C_1 can be viewed as an element of $H^2_{\text{dR}}(X)$, and can thus be represented by a closed two-form on X. Most importantly, it can be analytically expressed in terms of the curvature two-form of a connection on the $U(1)$ bundle P. The formula for the first Chern class is

$$C_1 = \frac{F}{2\pi}, \tag{6.47}$$

where F denotes the curvature two-form defined by a connection on P. We must note that the gauge invariance of F plays an important role in its appearance in the formula for the first Chern class.

One may naively argue that according to the definition of the curvature two-form

$$F := dA, \tag{6.48}$$

both F and C_1 are exact two-forms, and consequently C_1 represents the identity element in $H^2_{\text{dR}}(X)$. This is a fallacy, because (6.48) is only valid locally.[13] This argument would only be true if the bundle P had a single

---

[12] The more general case of $U(\mathcal{N})$ PFBs will be discussed in the next chapter. In fact, it turns out that the topological properties of a general PFB, with finite-dimensional fibers, can be studied by first reducing it to a PFB with a compact structure group. On the other hand, every compact Lie group can be viewed as a subgroup of some $U(\mathcal{N})$ group. Thus the transition functions can in general be chosen from the groups $U(\mathcal{N})$.

[13] Note that A is only defined on an open neighborhood of X.

chart, in which case P would be the trivial bundle $X \times S^1$ and C_1 would correctly be identified with the identity element of $H^2_{\mathrm{dR}}(X)$.

In practice, one integrates C_1 on the two-dimensional submanifolds of X and obtains a set of numbers which by construction are independent of the geometrical structure of the bundle P. In particular, one can find certain embedded copies of S^2 in X which reflect its cohomology properties,[14] and are used to yield integer numbers upon integrating C_1. These numbers are called the *first Chern numbers*.

For example, for $X = S^2$, there is just one such copy of S^2. It is clearly the whole manifold S^2. Therefore the topological structure of any $U(1)$ bundle over S^2 is determined by a single *Chern number*

$$c_1(P) := \int_{S^2} C_1(P) = \frac{1}{2\pi} \int_{S^2} F \in \mathbb{Z}. \tag{6.49}$$

By now we have developed all the machinery that we need to explore the topology of the B-S PFBs for the spinning quantum system (3.1). We can either pursue investigating the homotopy classes of the maps f_k (6.24) and $\tilde{f}_{k,\nu}$ (6.30), or directly compute the first Chern number (6.49) for each of these $U(1)$ bundles. The former approach is rather lengthy and we shall not present it here.[15] The latter method is by far the most straightforward. One needs to use the expression for the adiabatic (3.45) and non-adiabatic (3.113) B-S curvature two-forms and (6.49) to compute the corresponding first Chern numbers. The integrations can be performed exactly. They lead to

$$c_1(\lambda^k) = -2k,$$
$$c_1(\tilde{\lambda}^k_\nu) = \begin{cases} -2k & \text{if } \nu < 1 \\ 0 & \text{if } \nu > 1, \end{cases} \tag{6.50}$$

where k is introduced in (3.11) and (3.93).

In view of the fact that Chern numbers are integers, expression (6.50) implies the label k to take integer or half-integer values. We have directly verified this statement in Sect. 3.3. For the case where the energy eigenstates $|k; R\rangle$ are also eigenstates of the total angular momentum operator $\mathbf{J}^2$, this is a manifestation of the quantization of angular momentum. Here, however, it arises independently as a result of the application of the classification theorem for $U(1)$ PFBs to B-S bundles. Equation (6.50) is an example of a *topological quantization* of angular momentum. This is analogous to the topological quantization of the monopole charge in the context of magnetic monopoles[16] [87].

---

[14] These are known as the two-cells of the manifold in its cellular decomposition [190].
[15] The interested reader can find a detailed discussion of this approach in [184].
[16] In fact, it turns out that the subjects of geometric phase, Abelian and non-Abelian monopoles, classification of $U(1)$ bundles, and the representation theory of compact Lie groups are quite interrelated. A by-product of these interrelations is a series of topological quantization schemes. These are discussed in [177].

Another interesting observation is that for the value $\nu = 1$ of the evolution parameter, the function F_ν of (3.81) which is used to define $\tilde\lambda_\nu^n$ (6.41) is ill-defined at the north pole ($\theta = 0$). Thus it fails to be single-valued and our construction of the non-adiabatic B-S bundles does not apply. This is a typical example of a case for which one cannot find a function F_ν, and therefore $\tilde f_{n,\nu}$, to yield a B-S type interpretation for the non-adiabatic phase. The value $\nu = 1$ is in a sense a *critical value* for the evolution parameter at which the non-adiabatic B-S bundles undergo a *topological phase transition*.

For $\nu > 1$, all the B-S bundles become topologically trivial. Of course, this does not mean that the geometric phases vanish. As is the case in electromagnetism, a trivial bundle can support non-trivial geometric structures. The geometric phase is essentially a geometric phenomenon. It depends on the values of the connection one-form and curvature two-form at each point, and not on the first Chern numbers that are basically the integrals of the curvature two-form.

Nevertheless, there are rather important implications of the topological content of the geometric phase. The first of these is the topological quantization that we have just mentioned. Another interesting result is that if $c_1 \neq 0$, that is the bundle is non-trivial, then there are holonomy elements that are different from the identity element of the structure group. This means that it is impossible to avoid the geometric phases in the study of these systems. The converse is obviously not true.

We conclude this chapter by noting that the results presented here are easily generalized to the case of non-Abelian geometrical phases. We have devoted the next chapter to a discussion of the non-Abelian phases and the classification of $U(N)$ PFBs. There are also similar results for the cases where the Hilbert space is a real vector space (rather than a complex one); in this case the groups $U(N)$ are replaced by $O(N)$ and one arrives at the classification theorem for $O(N)$ PFBs [44].

Problems

6.1) Show that the homotopy relation is an equivalence relation.

6.2) Let X_1 and X_2 be two smooth manifolds, $f: X_1 \to X_2$ be a smooth function, $C_1: [0,T] \to X_1$ be a curve in X_1, and A_2 be a one-form on X_2. Using the definition of the pullback operation for differential forms show that

$$\int_{C_1} f^*(A_2) = \int_{f \circ C_1} A_2,$$

where $f^*(A_2)$ is the pullback one-form induced by the function f.

6.3) Let X_1, X_2, and X_3 be smooth manifolds, (E_3, X_3, π_3, G) be a fiber bundle on X_3, A_3 be a one-form on X_3, and $f: X_1 \to X_2$ and $g: X_2 \to$

X_3 be smooth functions. Using the definition of the pullback operation for differential forms and the pullback bundle, prove the validity of

$$(g \circ f)^*(A_3) = (f^* \circ g^*)(A_3),$$
$$(g \circ f)^*(E_3) = (f^* \circ g^*)(E_3).$$

6.4) Use the definition of the pullback operation for differential forms and (3.42) and (3.81) to prove (6.43), i.e., compute the pullback one-form $F_\nu^*(A)$.

7. Mathematical Structure of the Geometric Phase II: The Non-Abelian Phase

7.1 Introduction

In our discussion of the adiabatic and non-adiabatic phases we had made certain assumptions about the non-conservative quantum systems under investigation. We had specifically required the Hamiltonian $h = h(R)$ to depend smoothly on the parameters $R \in M$ and that its eigenvalues $E_n(R)$ be non-degenerate. This had direct implications on the description of the adiabatic phase. A clear manifestation of the latter requirement is (2.76), where the non-degeneracy of the energy eigenvalue $E_n(R)$ renders the right-hand side finite. In the general case of an exact cyclic evolution the role of the Hamiltonian in generating the approximate cyclic states as its eigenstates is played by another parameter-dependent Hermitian operator, $\tilde{h} = \tilde{h}(R)$. In this case, the non-degeneracy requirement applies to $\tilde{h}$. In this chapter, we shall lift this requirement and consider the initial states which are associated with an N-fold degenerate eigenvalue of a Hermitian operator $\tilde{h}(R(0))$. We shall first employ the adiabatic approximation, $\tilde{h}(R(0)) \stackrel{\text{adiabatic}}{=} h(R(0))$.

7.2 The Non-Abelian Adiabatic Phase

Consider an eigenvalue $E_n(R)$ of the Hamiltonian $h(R)$ which is N-fold degenerate. Suppose that N does not depend on the parameter R. Namely, as $R = R(t)$ varies in time, N remains constant. In particular, assume that during the evolution of the quantum system there are no level crossings. This means that if $E_n(R_0) \neq E_m(R_0)$ for some $R_0 \in M$, then we require $E_n(R) \neq E_m(R)$ for all $R \in M$. Alternatively, we demand the degeneracy subspaces $\mathcal{H}_n(R)$ and $\mathcal{H}_m(R)$ corresponding to two distinct energy eigenvalues $E_n(R)$ and $E_m(R)$ not to intersect.

Under these conditions, one can easily generalize the results of Sect. 2.2 on the adiabatic approximation. In this case the adiabatic evolution of the quantum system means that an energy eigenstate vector $|n, a; R(0)\rangle \in \mathcal{H}_n(R(0))$ evolves in time in such a way that at every moment of time $t \in [0, T]$, it remains an eigenstate vector of the Hamiltonian $h(R(t))$; that is, it is an element of $\mathcal{H}_n(R(t))$.[1] In particular, if the Hamiltonian is periodic, i.e., $h(R(T)) = h(R(0))$, then after each cycle the initial eigenvector returns to the same degeneracy eigenspace, $\mathcal{H}_n(R(0))$. This can be conveniently stated in terms of the corresponding quantum states. The energy eigenstates (eigenprojectors) $\Lambda_n(R)$ are defined as the solutions of the eigenvalue equation

$$h(R)\Lambda_n(R) = \Lambda_n(R)h(R) = E_n(R)\Lambda_n(R). \tag{7.1}$$

Alternatively, we can introduce the orthonormal eigenvectors $|n, a; R\rangle$ satisfying

$$h(R)|n, a; R\rangle = E_n(R)|n, a; R\rangle, \quad \forall a = 1, 2, \cdots, \mathcal{N}, \tag{7.2}$$
$$\mathcal{H}_n(R) := \text{Span}\{|n, a; R\rangle : a = 1, 2, \cdots, \mathcal{N}\}, \tag{7.3}$$

and write

$$\Lambda_n(R) = \sum_{a=1}^{\mathcal{N}} |n, a; R\rangle\langle n, a; R|. \tag{7.4}$$

The vectors $|n, a; R\rangle$ are defined only locally ($R \in U \subset M$). They satisfy

$$\langle n, a; R|m, b; R\rangle = \delta_{nm}\delta_{ab}. \tag{7.5}$$

Thus for each value of n, the eigenvectors $|n, a; R\rangle$ provide an orthonormal basis for the degeneracy subspace $\mathcal{H}_n(R)$. Although the choice of these basic eigenvectors is quite arbitrary, every two choices $|n, a; R\rangle$, $|n, a; R\rangle'$ are related by a unitary transformation:

$$|n, a; R\rangle' = \sum_{b=1}^{\mathcal{N}} |n, b; R\rangle \, \mathcal{U}^{ba}(R), \tag{7.6}$$

where $\mathcal{U}^{ab}(R)$ are the matrix elements of an $\mathcal{N} \times \mathcal{N}$ unitary matrix $\mathcal{U}(R)$. It is a straightforward calculation to show that the new basic eigenvectors $|n, a; R\rangle'$ also form an orthonormal basis for $\mathcal{H}_n(R)$ and both $|n, a; R\rangle$ and $|n, a; R\rangle'$ define the same eigenprojector,

$$\sum_{a=1}^{\mathcal{N}} |n, a; R\rangle'\langle n, a; R|' = \sum_{a=1}^{\mathcal{N}} |n, a; R\rangle\langle n, a; R| = \Lambda_n(R). \tag{7.7}$$

Equation (7.6) is a direct generalization of (2.16).

Given a set of basic eigenvectors $|n, a; R\rangle$, we can form the frame

$$s^{(n)}(R) := (|n, 1; R\rangle, |n, 2; R\rangle, \cdots, |n, \mathcal{N}; R\rangle). \tag{7.8}$$

---

[1] The condition of the validity of the adiabatic approximation in the presence of degeneracies is again given by (2.45), provided that the labels n and m in (2.45) label distinct eigenvalues. Further details can be found in [170] and [227].

7.2 The Non-Abelian Adiabatic Phase

As is the case for the individual basic eigenvectors $|n, a; R\rangle$, the frames $s^{(n)}(R)$ are smoothly defined only over an open subset of M. The orthonormality condition (7.5) implies that in fact $s^{(n)}(R)$ is a unitary frame for the eigenspace $\mathcal{H}_n(R)$; it can be expressed by a unitary matrix. To see this it is sufficient to view the basic vectors $|n, a; R\rangle$ as $\mathcal{N}$ ordinary column vectors (Problem 6.1). Under the transformation (7.6), the frame $s^{(n)}(R)$ transforms into another unitary frame:

$$s^{(n)}(R) \longrightarrow s'^{(n)}(R) = s^{(n)}(R)\mathcal{U}(R), \tag{7.9}$$

where the right-hand side is the product of two unitary matrices. This is a typical example of a gauge transformation (5.20).

Once more we are considering a periodic quantum system whose time-dependence is described by a closed curve $\mathbf{C}$ (6.7) in the parameter space M, i.e., $h(t) = h(R(t))$ with $R(t) \in \mathbf{C}$. According to the adiabatic approximation, an initial state vector $|\psi(0)\rangle = |n, a; R(0)\rangle$ (for fixed n and a) undergoes an approximate cyclic evolution. This means that it returns to the same degeneracy subspace $\mathcal{H}_n(R(0))$. Thus, in general the resultant state vector $|\psi(t = T)\rangle$, which is the solution of the Schrödinger equation (2.4), is related to the initial state vector through the action of an $\mathcal{N} \times \mathcal{N}$ unitary matrix,

$$|\psi(T)\rangle = |\psi(0)\rangle \, \mathcal{U}_\psi. \tag{7.10}$$

This is simply because both the initial and final state vectors are normalized state vectors belonging to the same $\mathcal{N}$-dimensional Hilbert subspace $\mathcal{H}_n(R(0))$.

The unitary matrix $\mathcal{U}_\psi \in U(\mathcal{N})$ is the generalization of the total phase $e^{-i\alpha_\psi} \in U(1)$ of (2.47). It is determined by the solution of the Schrödinger equation and consists of a dynamical and a geometrical part.

According to the adiabatic theorem, if the initial state vector $|\psi(0)\rangle$ belongs to an energy eigenspace $\mathcal{H}_n(R(0))$, then at any later time $t > 0$, $|\psi(t)\rangle$ will belong to $\mathcal{H}_n(R(t))$. Let us choose an arbitrary initial state vector

$$|\psi(0)\rangle = \sum_{a=1}^{\mathcal{N}} c_a^n(0) |n, a; R(0)\rangle. \tag{7.11}$$

Then we can use the ansatz

$$|\psi(t)\rangle = \sum_{a=1}^{\mathcal{N}} c_a^n(t) |n, a; R(t)\rangle \tag{7.12}$$

to solve the Schrödinger equation (2.4). The situation is quite analogous to the non-degenerate case discussed in Chap. 2. Substituting (7.12) in (2.4), one arrives at the following differential equation for the coefficients c_a^n,

$$\frac{dc_b^n(t)}{dt} + \sum_{a=1}^{\mathcal{N}} \left[iE_n(R(t))\delta_{ab} + \langle n, b; R(t)| \frac{d}{dt} |n, a; R(t)\rangle \right] c_a^n(t) = 0. \tag{7.13}$$

The solution of this equation is given by

$$c_b^n(t) = \sum_{a=1}^{\mathcal{N}} \left[\mathcal{T} \exp \int_0^t (-iE_n(R(\tau))\mathbf{1} d\tau + iA_{\mathcal{N}}^n(R(\tau))) \right]^{ba} c_a^n(0), \quad (7.14)$$

where $\mathcal{T}$ is the time-ordering operator (2.30), $\mathbf{1}$ is the $\mathcal{N} \times \mathcal{N}$ unit matrix, and $A_{\mathcal{N}}^n$ is the non-Abelian analog of the Berry connection one-form (2.53). It is defined in terms of its matrix elements,

$$[A_{\mathcal{N}}^n]^{ba}(R(\tau)) := i\langle n, b; R(\tau)| \frac{d}{d\tau} |n, a; R(\tau)\rangle d\tau$$
$$= i\langle n, b; R|d|n, a; R\rangle. \quad (7.15)$$

It is quite straightforward (Problem 6.2) to show that $A_{\mathcal{N}}^n$ is an $\mathcal{N} \times \mathcal{N}$ Hermitian matrix, and that under gauge transformations (7.6) and (7.9), it transforms according to

$$A_{\mathcal{N}}^n \to A_{\mathcal{N}}'^n = \mathcal{U}^{-1}(R) \cdot A_{\mathcal{N}}^n \cdot \mathcal{U}(R) + i\mathcal{U}^{-1}(R) \cdot d\mathcal{U}(R). \quad (7.16)$$

This equation is identical with the gauge transformation rule satisfied by a local connection one-form, i.e., (5.28).

Equation (7.14) can be further simplified by noting that the first term in the integrand commutes with all the matrices. Thus it can be pulled out of the time-ordered exponential. In the remaining term we can suppress the explicit time dependence and change the time integration into an integration over the parameters $R \in \mathbf{C}$. In this way we can also replace the time-ordering operator $\mathcal{T}$ by the path-ordering operator $\mathcal{P}$. The final result is

$$c_b^n(t) = e^{-i\int_0^t E_n(R(\tau))d\tau} \sum_a \left[\mathcal{P} \exp \left(i \int_{R(0)}^{R(t)} A_{\mathcal{N}}^n(R) \right) \right]^{ba} c_a^n(0).$$

In view of the last equality and (7.12), we have

$$|\psi(t)\rangle = \sum_{a,b=1}^{\mathcal{N}} |n, b; R(0)\rangle e^{-i\int_0^t E_n(R(\tau))d\tau}$$
$$\times \left[\mathcal{P} \exp \left(i \int_{R(0)}^{R(t)} A_{\mathcal{N}}^n(R) \right) \right]^{ba} c_a^n(0). \quad (7.17)$$

After a complete cycle, $t = T$, we obtain

$$|\psi(T)\rangle = \sum_{a,b=1}^{\mathcal{N}} |n, b; R(0)\rangle e^{-i\int_0^T E_n(R(t))dt} \left[\mathcal{P} \exp \left(i \oint_{\mathbf{C}} A_{\mathcal{N}}^n \right) \right]^{ba} c_a^n(0). \quad (7.18)$$

7.2 The Non-Abelian Adiabatic Phase

If we choose the initial state vector $|\psi(0)\rangle$ to be one of the basis vectors, then all $c_a^n(0)$ vanish except one which takes the value 1: $|\psi(0)\rangle = |n, a; R(0)\rangle =: |\psi_a(0)\rangle$. In this case the evolving state vector is given by

$$|\psi_a(t)\rangle = \sum_{b=1}^{\mathcal{N}} |n, b; R(0)\rangle e^{-i\int_0^t E_n(R(\tau))d\tau} \left[\mathcal{P}\exp\left(i\int_{R(0)}^{R(t)} A_{\mathcal{N}}^n(R)\right)\right]^{ba}. \tag{7.19}$$

Evaluating $|\psi_a(t)\rangle$ at $t = T$, we find

$$|\psi_a(T)\rangle = \sum_{b=1}^{\mathcal{N}} |n, b; R(0)\rangle e^{-i\int_0^T E_n(R(t))dt} \left[\mathcal{P}\exp\left(i\oint_C A_{\mathcal{N}}^n\right)\right]^{ba} \tag{7.20}$$

Furthermore, we define the $\mathcal{N} \times \mathcal{N}$ unitary matrices $\mathcal{U}^{ab}(R)$ that transform the basic eigenvectors into one another,

$$|n, a; R\rangle = |n, b; R\rangle \mathcal{U}^{ba}(R). \tag{7.21}$$

Using these matrices, we can write (7.18) and (7.20) in the form (7.10). As this does not serve any practical purpose, we shall not pursue it further. Instead, we shall use (7.20) to yield the expression for the evolution of the initial basic frame $\psi^n(t=0) := s^{(n)}(R(0))$ introduced in (7.8). The result is given by

$$\psi^n(t) := (|\psi_1(t)\rangle, \cdots, |\psi_{\mathcal{N}}(t)\rangle) \tag{7.22}$$

$$= \psi^n(0) \cdot \left[e^{-i\int_0^t E_n(R(\tau))d\tau} \mathcal{P}\exp\left(i\int_{R(0)}^{R(t)} A_{\mathcal{N}}^n(R)\right)\right], \tag{7.23}$$

$$\psi^n(T) = (|\psi_1(T)\rangle, \cdots, |\psi_{\mathcal{N}}(T)\rangle)$$

$$= \psi^n(0) \cdot \left[e^{-i\int_0^T E_n(R(t))dt} \mathcal{P}\exp\left(i\oint_C A_{\mathcal{N}}^n\right)\right], \tag{7.24}$$

where the dot means the multiplication of $\mathcal{N} \times \mathcal{N}$ matrices (Problem 6.3).

The first exponential in (7.24) is the familiar *dynamical phase* factor. The path-ordered exponential

$$\mathcal{P}\exp\left(\oint_C A_{\mathcal{N}}^n\right) \tag{7.25}$$

is by construction an $\mathcal{N} \times \mathcal{N}$ unitary matrix. It is the non-Abelian generalization of Berry's adiabatic phase [272].

We can, more generally, study the evolution of the degenerate states $\Lambda_n(R)$ (7.7). There is a one-to-one correspondence between these states and the energy eigenspaces $\mathcal{H}_n(R)$. In general, an $\mathcal{N}$-fold degenerate state W can be identified with the corresponding $\mathcal{N}$-dimensional degeneracy Hilbert subspace. Thus, these states can be viewed as the points of the Grassmann manifold:

$$Gr_\mathcal{N} = \mathcal{H}/\mathbb{C}^\mathcal{N}. \tag{7.26}$$

$Gr_\mathcal{N}$ is defined to be the set of all $\mathcal{N}$-dimensional Hilbert subspaces of $\mathcal{H}$. In fact, if the Hilbert space is finite dimensional, say $\mathcal{H} = \mathbb{C}^N$, then the space of all $\mathcal{N}$-fold degenerate states is the Grassmann manifold [186]

$$Gr_\mathcal{N}(N) = \mathbb{C}^N/\mathbb{C}^\mathcal{N}. \tag{7.27}$$

The infinite-dimensional Grassmann manifold $Gr_\mathcal{N}$ is the inductive limit of $Gr_\mathcal{N}(N)$ where $N \to \infty$. The special case of non-degenerate states is obtained by taking $\mathcal{N} = 1$. In this case, one recovers the following well-known identities:

$$\mathbb{C}P^{N-1} = Gr_1(N) \quad \text{and} \quad \mathbb{C}P^\infty = Gr_1.$$

We shall see another realization of Grassmann manifolds shortly.

Each state W can be represented by an $\mathcal{N}$-frame in the corresponding degeneracy subspace. The correspondence between the $\mathcal{N}$-frames and the $\mathcal{N}$-fold degenerate states are many (in fact infinite) to one. This is in complete analogy with the relation between the normalized state vectors and pure states. Since every unitary $\mathcal{N}$-frame of an $\mathcal{N}$-dimensional Hilbert subspace is obtained from a fixed one by the action of the group $U(\mathcal{N})$, the space of all the unitary $\mathcal{N}$-frames can be easily identified with a principal $U(\mathcal{N})$-bundle over $Gr_\mathcal{N}$.

Let us consider for a moment the case of the finite-dimensional Hilbert space $\mathcal{H} = \mathbb{C}^N$. The space of unitary $\mathcal{N}$-frames on $\mathbb{C}^N$ can be parameterized by the quotient space $U(N)/U(N - \mathcal{N})$. To see this, note first that every unitary $\mathcal{N}$-frame in $\mathbb{C}^N$ can also be viewed as an N-frame. This is simply a consequence of the obvious inclusion of $U(\mathcal{N}) \subseteq U(N)$, for $\mathcal{N} \leq N$. Secondly, it is not difficult to observe that choosing a unitary frame in $\mathbb{C}^N$ which corresponds to a given $\mathcal{N}$-frame, is equivalent to fixing $\mathcal{N}$ basic vectors in $\mathbb{C}^N$. The remaining $N - \mathcal{N}$ basic vectors can be arbitrarily chosen. Each choice of these $N - \mathcal{N}$ basic vectors corresponds to an element of $U(N - \mathcal{N})$. This suggests the definition of an equivalence relation on all unitary N-frames, i.e., $U(N)$, by identifying two N-frames which correspond to a given $\mathcal{N}$-frame. Clearly, each equivalence class is a copy of $U(N - \mathcal{N})$. Then the set of all unitary $\mathcal{N}$-frames is identified with the set of all the equivalence classes of this equivalence relation. This justifies the use of the quotient set $U(N)/U(N-\mathcal{N})$ to represent the space of all unitary $\mathcal{N}$-frames in $\mathbb{C}^N$.

This space serves as the total space of the above-mentioned PFB. The projection map of this PFB is naturally defined by recognizing the identity

$$Gr_\mathcal{N}(N) = U(N)/U(N - \mathcal{N}) \times U(\mathcal{N}). \tag{7.28}$$

This is best understood by noting that all the unitary $\mathcal{N}$-frames of a given $\mathcal{N}$-dimensional Hilbert subspace represent a single $\mathcal{N}$-fold degenerate state.

The case of an infinite-dimensional Hilbert space is treated similarly by taking the direct limit of $N \to \infty$.

The PFB over $Gr_\mathcal{N}$ is the generalization of the A-A PFB (6.13). We shall denote it by

$$\eta_\mathcal{N} : (\eta_\mathcal{N}, Gr_\mathcal{N}, \pi, U(\mathcal{N})). \tag{7.29}$$

$\eta_\mathcal{N}$ can also be obtained as the associated $U(\mathcal{N})$ PFB to a complex vector bundle

$$E_\mathcal{N} : (E_\mathcal{N}, Gr_\mathcal{N}, \check{\pi}, U(\mathcal{N})). \tag{7.30}$$

The fibers $E_{\mathcal{N}_W}$ of $E_\mathcal{N}$ are precisely the degeneracy subspaces corresponding to the states $W \in Gr_\mathcal{N}$.

In terms of the degenerate states, the adiabatic theorem states that the Liouville–von Neumann equation (2.5) with $h(t) = h(R(t))$ and $W(0) = \frac{1}{\mathcal{N}}\Lambda_n(R(0))$ is solved by

$$W(t) \stackrel{\text{adiabatic}}{=\!=\!=} \frac{1}{\mathcal{N}}\Lambda_n(R(t)) \in Gr_\mathcal{N}. \tag{7.31}$$

Since $R(T) = R(0)$, the states $W(t)$ trace a closed curve in the space of all $\mathcal{N}$-fold degenerate states $Gr_\mathcal{N}$. We shall denote this closed curve by

$$\mathcal{C}_\mathcal{N} : [0,T] \ni t \longrightarrow W(t) \in Gr_\mathcal{N}. \tag{7.32}$$

In other words, given a closed curve **C** (6.7) in the parameter manifold M and an initial $\mathcal{N}$-fold degenerate energy eigenstate $\Lambda_n(R(0))$, we obtain an (approximately) closed curve $\mathcal{C}_\mathcal{N}$ in $Gr_\mathcal{N}$.

As $R(t)$ traverses the loop **C**, the unitary frame $\psi^n(t)$ (7.23) defines a curve $C_\mathcal{N}$ in the space of all unitary $\mathcal{N}$-frames, i.e., in the total space $\eta_\mathcal{N}$ of the A-A PFB (7.29). This curve is the analog of the curve C of (4.4). It projects onto the closed curve $\mathcal{C}_\mathcal{N}$ under the projection π and is therefore a lift of this curve.

Furthermore, let us consider the frame

$$\tilde{\psi}^n(t) := e^{i\int_0^t E_n(R(\tau))d\tau}\psi^n(t)$$
$$= \psi^n(0) \cdot \mathcal{P}\exp\left(i\int_{R(0)}^{R(t)} A_\mathcal{N}^n(R)\right) \tag{7.33}$$

which defines the curve

$$\tilde{C}_\mathcal{N} : \tilde{\psi}^n(0) \to \tilde{\psi}^n(t) \to \tilde{\psi}^n(T) \tag{7.34}$$

in $\eta_\mathcal{N}$. For all $t \in [0,T]$, the frame $\tilde{\psi}^n(t)$ projects onto the state $W(t)$. Thus the curve $\tilde{C}_\mathcal{N}$ is a lift of the curve $\mathcal{C}_\mathcal{N}$ as well. Another interesting lift of the curve $\mathcal{C}_\mathcal{N}$ is given by the basic frames $s^{(n)}$ of (7.8),

$$C_\mathcal{N}^{\text{closed}} : s^{(n)}(R(0)) \to s^{(n)}(R(t)) \to s^{(n)}(R(T)) = s^{(n)}(R(0)). \tag{7.35}$$

Besides the fact that we are considering adiabatically evolving systems, the lifts $C_\mathcal{N}$, $\tilde{C}_\mathcal{N}$ and $C_\mathcal{N}^{\text{closed}}$ are the non-Abelian counterparts of the dynamical (4.4), Aharonov–Anandan (4.8), and closed (4.7) lifts of Chap. 4, respectively. We shall next drop the adiabaticity condition and find the exact (non-adiabatic) non-Abelian counterparts of these lifts.

7.3 The Non-Abelian Geometric Phase

A general cyclic evolution of an $\mathcal{N}$-fold degenerate state corresponds to a closed curve

$$\mathcal{C}_\mathcal{N} : [0, T] \ni t \longrightarrow W(t) \in Gr_\mathcal{N}, \tag{7.36}$$

where $W(t)$ satisfies the Liouville–von Neumann equation (2.5) with an initial condition $W(0)$.

In general, for a quantum system undergoing an exact cyclic evolution, the initial cyclic $\mathcal{N}$-fold degenerate state $W(0)$ cannot be an eigenprojector of the initial Hamiltonian. This follows directly from the fact that if we choose $W(t) = \frac{1}{\mathcal{N}}\Lambda_n(R(t))$, then

$$i\frac{d}{dt}W(t) = \frac{1}{\mathcal{N}}[h(R(t)), \Lambda_n(R(t))] = 0. \tag{7.37}$$

Hence the initial state is indeed a stationary state,

$$W(t) = W(0).$$

This means that the choice of $W(0) = \frac{1}{\mathcal{N}}\Lambda_n(R(0))$ for the initial condition leads to a curve (7.32) in $Gr_\mathcal{N}$ that is not closed. However if the time-dependence of the Hamiltonian is adiabatic then (7.31) holds approximately, and one has an approximately closed curve in $Gr_\mathcal{N}$.

We shall be interested in the cyclic states $W(0)$ that are not stationary. As in the Abelian case, these can be obtained as the eigenprojectors of another Hermitian operator $\tilde{h}(R(0))$ which in general does not commute with the Hamiltonian $h(R(0))$. We shall first focus our attention on the closed curve $\mathcal{C}_\mathcal{N}$ (7.36). The state $W(t) \in \mathcal{C}_\mathcal{N}$ can be represented by an $\mathcal{N}$-frame

$$s(W(t)) := (|\phi_1(W(t))\rangle, \cdots, |\phi_\mathcal{N}(W(t))\rangle), \tag{7.38}$$

where $|\phi_a(W)\rangle$ are $\mathcal{N}$ orthonormal pure basic vectors with

$$W = \frac{1}{\mathcal{N}}\sum_{a=1}^{\mathcal{N}} |\phi_a(W)\rangle\langle\phi_a(W)|. \tag{7.39}$$

In general the basic vectors $|\phi_a(W)\rangle$ and hence the basic frame $s(W) = (|\phi_1(W)\rangle, \cdots, |\phi_\mathcal{N}(W)\rangle)$ can be smoothly defined only over open subsets U of $Gr_\mathcal{N}$. In fact, $s : U \to \eta_\mathcal{N}$ is a local section (5.9) of the A-A bundle $\eta_\mathcal{N}$ (7.29).

The situation is quite similar to the adiabatic case. The frames s and the basic vectors $|\phi_a\rangle$ play the role of the frames $s^{(n)}$ of (7.8) and the basic eigenvectors $|n, a; R\rangle$ of (7.2). They are also determined up to $U(\mathcal{N})$ gauge transformations. Namely, we can choose another set of orthonormal basic vectors

7.3 The Non-Abelian Geometric Phase

$$|\phi_a(W)\rangle' := \sum_{b=1}^{\mathcal{N}} |\phi_b(W)\rangle \mathcal{U}^{ba}(W), \tag{7.40}$$

where again $\mathcal{U}^{ab}(W)$ are the matrix elements of an $\mathcal{N} \times \mathcal{N}$ unitary matrix $\mathcal{U}(W)$. The frame $s'(W)$ associated with $|\phi_a(W)\rangle'$ is related to $s(W)$ according to

$$s'(W) = s(W) \cdot \mathcal{U}(W). \tag{7.41}$$

The only difference with the adiabatic case is that here the basic vectors $|\phi_a(W)\rangle$ and the frames $s(W)$ depend on the states W. As functions their domain of definition is the open patches of $Gr_\mathcal{N}$.

Following the approach of the preceding section, let us first study the dynamics of an initial pure state vector $|\psi(0)\rangle$ which belongs to the degeneracy subspace corresponding to the initial state $W(0)$. We shall denote the degeneracy subspaces by $\mathcal{H}(W)$, for all $W \in Gr_\mathcal{N}$. By construction, since $W = W(t)$ is the exact solution of the Liouville–von Neumann equation (2.5), $|\psi(0)\rangle$ evolves in time in such a way that for all $t \in [0, T]$, $|\psi(t)\rangle \in \mathcal{H}(W(t))$.

Moreover, since

$$\mathcal{H}(W) := \text{Span}\{|\phi_a(W)\rangle : a = 1, \cdots, \mathcal{N}\}, \tag{7.42}$$

the exact solution of the Schrödinger equation (2.4) is given by

$$|\psi(t)\rangle = \sum_{a=1}^{\mathcal{N}} c_a(t)|\phi_a(W(t))\rangle. \tag{7.43}$$

Substituting the last equation in (2.4), we obtain a differential equation for the coefficients $c_a(t)$,

$$\frac{dc_b(t)}{dt} + \sum_{a=1}^{\mathcal{N}} [i\langle \phi_b(W(t))|h(R(t))|\phi_a(W(t))\rangle$$
$$+ \langle \phi_b(W(t))|\frac{d}{dt}|\phi_a(W(t))\rangle]c_a(t) = 0. \tag{7.44}$$

This equation is the non-adiabatic analog of (7.13). Solving (7.44) and substituting the result in (7.43), we obtain

$$|\psi(t)\rangle = \sum_{a,b=1}^{\mathcal{N}} |\phi_b(W(0))\rangle \left[\mathcal{T} \exp \int_0^t (\ i\mathcal{E}_\mathcal{N}(\tau)d\tau \ | \ i\mathcal{A}_\mathcal{N}(W(\tau)))\right]^{ba} c_a(0). \tag{7.45}$$

Here $\mathcal{E}_\mathcal{N}$ and $\mathcal{A}_\mathcal{N}$ are $\mathcal{N} \times \mathcal{N}$ Hermitian matrices defined by the following matrix elements,

$$[\mathcal{E}_\mathcal{N}(\tau)]^{ab} := \langle \phi_a(W(\tau))|h(R(\tau))|\phi_b(W(\tau))\rangle, \tag{7.46}$$

$$[\mathcal{A}_\mathcal{N}(W(\tau))]^{ab} := i\langle \phi_a(W(\tau))|\frac{d}{d\tau}|\phi_b(W(\tau))\rangle \, d\tau. \tag{7.47}$$

Using (7.45), we can determine the evolution of the initial frame $\psi(0) := s(0) = (|\phi_1(W(0))\rangle, \cdots, |\phi_\mathcal{N}(W(0))\rangle)$. The result is

$$\psi(t) = \psi(0) \cdot \left[\mathcal{T} \exp \int_0^t (-i\mathcal{E}_\mathcal{N}(\tau)d\tau + i\mathcal{A}_\mathcal{N}(W(\tau)))\right]. \quad (7.48)$$

At the end of a complete cycle, we have

$$\psi(T) = \psi(0) \cdot \left[\mathcal{T} \exp \int_0^T (-i\mathcal{E}_\mathcal{N}(t)dt + i\mathcal{A}_\mathcal{N}(W(t)))\right]. \quad (7.49)$$

An important difference between (7.49) and its adiabatic counterpart (7.23) is that here the matrices $\mathcal{E}_\mathcal{N}(\tau)$ and $\mathcal{A}_\mathcal{N}(W(\tau))$ do not generally commute. Thus it is not possible to write the total $U(\mathcal{N})$ phase as the product of the generalization of the dynamical phase,

$$U^{\text{dynamical}} := \mathcal{T} \exp\left(-i \int_0^T \mathcal{E}_\mathcal{N}(t)dt\right), \quad (7.50)$$

and the geometrical phase,

$$U^{\text{geometric}} := \mathcal{T} \exp\left(i \int_0^T \mathcal{A}_\mathcal{N}(W(t))\right). \quad (7.51)$$

Nevertheless, the latter quantity can be written as a path-ordered integral,

$$U^{\text{geometric}} = \mathcal{P} \exp\left(i \oint_{C_\mathcal{N}} \mathcal{A}_\mathcal{N}\right), \quad (7.52)$$

and identified with a holonomy element of the A-A principal bundle $\eta_\mathcal{N}$ (7.29).

In (7.52), we have viewed $\mathcal{A}_\mathcal{N}$ as a Hermitian matrix (Problem 6.2) whose entries are the one-forms

$$[\mathcal{A}_\mathcal{N}]^{ab} := i\langle \phi_a(W) | d | \phi_b(W) \rangle \quad (7.53)$$

defined locally on $Gr_\mathcal{N}$. Hence, $-i\mathcal{A}_\mathcal{N}$ is a (local) one-form on $Gr_\mathcal{N}$ with values in the Lie algebra of $U(\mathcal{N})$.[2] Furthermore, as $|\phi_a(W)\rangle$ undergo a gauge transformation (7.40), $\mathcal{A}_\mathcal{N}$ transforms according to (Problem 6.2)

$$\mathcal{A}_\mathcal{N} \to \mathcal{A}'_\mathcal{N} = \mathcal{U}^{-1}(W) \cdot \mathcal{A}_\mathcal{N} \cdot \mathcal{U}(W) + i\mathcal{U}^{-1}(W) \cdot d\mathcal{U}(W), \quad (7.54)$$

where $\mathcal{U}(W)$ (7.41) belongs to the group $U(\mathcal{N})$. Once again, the transformation rule satisfied by $\mathcal{A}_\mathcal{N}$ is identical with the one satisfied by a local connection one-form (5.28) on the A-A PFB $\eta_\mathcal{N}$ (7.29). This in turn implies

---

[2] Recall that the Lie algebra of $U(\mathcal{N})$ consists of all anti-Hermitian $\mathcal{N} \times \mathcal{N}$ matrices.

that indeed the non-Abelian phase defined by (7.52) is the holonomy element associated with the closed curve $\mathcal{C}_\mathcal{N}$ (7.36) and defined by the connection $\mathcal{A}_\mathcal{N}$. $\mathcal{A}_\mathcal{N}$ is the non-Abelian analog of the A-A connection one-form $\mathcal{A}$ (4.27) of Chap. 4.

We must emphasize that the dynamical phase (7.50) and the geometrical phase (7.51) are defined independently. As we mentioned above, their product does not yield the total $U(\mathcal{N})$ phase (7.49). These phases are defined as direct generalizations of the dynamical and geometrical phases of the adiabatic and Abelian cases. Moreover, the geometric phase, as defined by (7.51), is a geometric quantity which depends only on the closed curve $\mathcal{C}_\mathcal{N}$ and the geometric structure of the bundle $\eta_\mathcal{N}$.

An important distinction between the Abelian and non-Abelian phases is that unlike the Abelian geometric phase, the non-Abelian geometric phase is not a gauge-invariant quantity. It is however gauge covariant (Problem 6.4). This means that one cannot measure all the matrix elements of $U^{\text{geometric}}$. Two examples of related quantities which are gauge invariant and thus experimentally measurable are the eigenvalues of $U^{\text{geometric}}$ and its trace,

$$\text{tr}\left[\mathcal{P}\exp\left(i\oint_{\mathcal{C}_\mathcal{N}} \mathcal{A}_\mathcal{N}\right)\right].$$

The latter is an example of a Wilson loop integral (5.75). It is associated with a gauge theory which might be called a *non-Abelian gauge theory of quantum mechanics*.

7.4 Holonomy Interpretations of the Non-Abelian Phase

The holonomy interpretations of the non-Abelian geometric phase are direct generalizations of those of the Abelian geometric phase. The only difference is that the structure groups of the fiber bundles used in the non-Abelian case are the non-Abelian unitary groups $U(\mathcal{N})$ ($\mathcal{N} \neq 1$). Thus some of the results which hold exclusively for $U(1)$ bundles do not apply.

In the preceding sections, we introduced the non-Abelian A-A PFB $\eta_\mathcal{N}$ (7.29) and showed that the A-A connection one-form $\mathcal{A}_\mathcal{N}$ (7.53) provides $\eta_\mathcal{N}$ with a geometric structure. Furthermore, we pointed out that the geometric phase (7.52) is identical with the holonomy element (5.36) of the closed curve $\mathcal{C}_\mathcal{N}$ (7.36) in the state space. We shall naturally call this holonomy interpretation of the non-Abelian geometric phase, the *Aharonov–Anandan interpretation*. For the case $\mathcal{N} = 1$, we immediately recover the A-A interpretation of the Abelian phase.

In view of this interpretation, we can define the exact (non-adiabatic) non-Abelian analogs of the *dynamical, Aharonov–Anandan,* and *closed* lifts of the curve $\mathcal{C}_\mathcal{N} \subset Gr_\mathcal{N}$. These are the curves in the total space $\eta_\mathcal{N}$ of the A-A PFB which project onto $\mathcal{C}_\mathcal{N}$ in $Gr_\mathcal{N}$:

dynamical lift ≡
$$C_\mathcal{N} : \psi(0) \longrightarrow \psi(t) \longrightarrow \psi(T),$$
A-A lift ≡
$$\tilde{C}_\mathcal{N} : \tilde{\psi}(0) \longrightarrow \tilde{\psi}(t) \longrightarrow \tilde{\psi}(T),$$
closed lift ≡
$$C_\mathcal{N}^{\text{closed}} : s(W(0)) \to s(W(t)) \to s(W(T)) = s(W(0)).$$

Here $\psi(t)$ and $s(W(t))$ are defined by (7.48) and (7.38), $W(t) = \mathcal{C}_\mathcal{N}(t) \in \mathcal{C}_\mathcal{N}$, and

$$\tilde{\psi}(t) := \psi(0) \cdot \mathcal{P} \exp\left(i \int_{W(0)}^{W(t)} \mathcal{A}_\mathcal{N}(W) \right). \tag{7.55}$$

If we endow the A-A PFB $\eta_\mathcal{N}$ with the geometric structure given by the A-A connection one-form $\mathcal{A}_\mathcal{N}$, we can identify the A-A lift $\tilde{C}_\mathcal{N}$ with the horizontal lift of the basic curve $\mathcal{C}_\mathcal{N}$. This is a clear indication of the geometric nature of the A-A lift and consequently the geometric phase.

This concludes our description of the A-A interpretation of the non-Abelian geometric phase. If the quantum system of interest undergoes an adiabatic evolution, we can employ the adiabatic approximation successfully. In this case there is an alternative interpretation of the (adiabatic) geometric phase. This is the non-Abelian generalization of the *Berry–Simon interpretation* of the adiabatic phase.

The B-S interpretation of the non-Abelian phase involves a $U(\mathcal{N})$ principal bundle over the parameter space M. We shall first introduce the B-S vector bundles

$$L_\mathcal{N}^n : (L_\mathcal{N}^n, M, \check{\pi}, U(\mathcal{N})). \tag{7.56}$$

The fibers $L_{\mathcal{N}_R}^n$ of $L_\mathcal{N}^n$ are defined to be the degeneracy eigenspaces $\mathcal{H}_\mathcal{N}(R)$ (7.3),

$$L_{\mathcal{N}_R}^n := \mathcal{H}_\mathcal{N}(R), \qquad \forall R \in M. \tag{7.57}$$

Thus the A-A vector bundle $L_\mathcal{N}^n$ is a complex vector bundle of rank (fiber dimension) $\mathcal{N}$ over the parameter space M.

The non-Abelian B-S PFB

$$\lambda_\mathcal{N}^n : (\lambda_\mathcal{N}^n, M, \pi, U(\mathcal{N})) \tag{7.58}$$

is the associated $U(\mathcal{N})$ PFB to $L_\mathcal{N}^n$. The fibers $\lambda_{\mathcal{N}_R}^n$ of $\lambda_\mathcal{N}^n$ consist of the unitary frames over $\mathcal{H}_\mathcal{N}(R)$.

The non-Abelian B-S connection one-form $A_\mathcal{N}^n$ (7.15) satisfies a gauge transformation rule (7.16) which is identical with the gauge transformation rule (5.28) for a local connection one-form on $\lambda_\mathcal{N}^n$ (Problem 6.2). Hence, $A_\mathcal{N}^n$ provides $\lambda_\mathcal{N}^n$ with a geometric structure. The holonomy elements (5.36) defined by this connection are identical with the non-Abelian adiabatic phases (7.25) derived in Sect. 7.2.

Once again, we can raise the question whether there is a deeper relationship between the mathematical structures used in the A-A and the B-S interpretations of the non-Abelian phase. The answer to this question is given by the classification theorem for the $U(\mathcal{N})$ PFBs. The situation is almost exactly identical with the Abelian case. We shall briefly outline this relationship in the next section.

7.5 Classification of $U(\mathcal{N})$ Principal Bundles and the Relation Between the Berry–Simon and Aharonov–Anandan Interpretations of Non-Abelian Phase

The main purpose of the present section is to extend the results of Sects. 6.4 and 6.5 to the case of the non-Abelian phase. Specifically we shall investigate the mathematical relationship between the Berry–Simon (B-S) and the Aharonov–Anandan (A-A) interpretations of the non-Abelian adiabatic phase. The idea is to recognize the A-A PFB (7.29) as the universal classifying $U(\mathcal{N})$ bundle, and employ the classification theorem for $U(\mathcal{N})$ PFBs.

In general for any Lie group G, there is a *universal classifying space* $B(G)$ and a *universal classifying principal bundle*

$$\eta(G) : (\eta(G), B(G), \pi, G)$$

over $B(G)$, such that any G-PFB is obtained as a pullback bundle from $\eta(G)$. This means that for any G-PFB $P : (P, X, \pi', G)$ there is a continuous (smooth) function $f : X \to B(G)$, such that $P = f^*\eta(G)$. This is demonstrated by the following commutative diagram,

$$\begin{array}{ccc} f^*\eta(G) \cong P & \xrightarrow{f_*} & \eta(G) \\ \downarrow & \circlearrowleft & \downarrow \\ X & \xrightarrow{f} & B(G), \end{array} \qquad (7.59)$$

where f^* is the pullback operation and f_* is the bundle morphism induced by f. Diagram (7.59) is the general form of (6.19). In general the topological structure of $f^*\eta(G)$ depends only on the homotopy class $[f] \in [X, B(G)]$ of the function $f : X \to B(G)$. Thus the problem of classifying all G-PFBs over a manifold X is equivalent to determining $[X, B(G)]$.

The groups of interest in the context of the non-Abelian phase are the unitary groups $U(\mathcal{N})$. It turns out that for $G = U(\mathcal{N})$, one can take

$$B(U(\mathcal{N})) = Gr_{\mathcal{N}}, \qquad (7.60)$$

$$\eta(U(\mathcal{N})) = \eta_{\mathcal{N}}. \qquad (7.61)$$

In other words, the A-A bundle $\eta_{\mathcal{N}}$ over $Gr_{\mathcal{N}}$ is precisely the universal $U(\mathcal{N})$-bundle [63, 114, 232]. Moreover, there is a *universal connection* on $\eta(U(\mathcal{N})) =$

$\eta_\mathcal{N}$ that yields all the connection one-forms on the bundle P as pullback connection one-forms [188]. Remarkably it further follows that the universal connection one-form on $\eta_\mathcal{N}$ is identical with the A-A connection one-form $\mathcal{A}_\mathcal{N}$ (7.53). In view of these surprising identifications, it is quite easy to explore the relation between the B-S and A-A interpretations. This relation is established via a smooth function

$$f_n : M \longrightarrow Gr_\mathcal{N}, \tag{7.62}$$

and illustrated by the following commutative diagram,

$$\begin{array}{ccc} f_n^* \eta_\mathcal{N} \cong \lambda_\mathcal{N}^n & \xrightarrow{f_{n*}} & \eta_\mathcal{N} \\ \downarrow & \circ & \downarrow \\ M & \xrightarrow{f_n} & Gr_\mathcal{N}. \end{array} \tag{7.63}$$

The function f_n is defined by

$$f_n(R) := \frac{1}{\mathcal{N}} \Lambda_n(R). \tag{7.64}$$

In the non-degenerate case $\mathcal{N} = 1$, and (7.64) reduces to (6.24).

The main motivation for the definition of f_n is that, for an adiabatically evolving system, it maps the closed curve $\mathbf{C}$ (6.7) of parameters into the closed curve $\mathcal{C}_\mathcal{N}$ (7.32) of states, namely

$$f_n \circ \mathbf{C} \stackrel{\text{adiabatic}}{=} \mathcal{C}_\mathcal{N}. \tag{7.65}$$

The B-S and A-A PFBs and their connections are linked by f_n according to

$$\lambda_\mathcal{N}^n = f_n^* \eta_\mathcal{N}, \tag{7.66}$$
$$A_\mathcal{N}^n = f_n^* \mathcal{A}_\mathcal{N}. \tag{7.67}$$

Equations (7.65), (7.66), and (7.67) are the non-Abelian analogs of (6.25), (6.22), and (6.23) respectively.

In terms of this relationship, the (approximate) equivalence of Berry's adiabatic phase (7.25) and A-A's (geometric) phase (7.52) is expressed by the following series of identities,

$$\begin{aligned} \text{adiabatic phase} &:= \mathcal{P} \exp\left(i \oint_\mathbf{C} A_\mathcal{N}^n\right) \\ &= \mathcal{P} \exp\left(i \oint_\mathbf{C} f_n^* \mathcal{A}_\mathcal{N}\right) \\ &= \mathcal{P} \exp\left(i \oint_{f_n \circ \mathbf{C}} \mathcal{A}_\mathcal{N}\right) \\ &= \mathcal{P} \exp\left(i \oint_{\mathcal{C}_\mathcal{N}} \mathcal{A}_\mathcal{N}\right) =: \text{A-A phase.} \end{aligned} \tag{7.68}$$

7.5 B-S and A-A Interpretations of the Non-Abelian Phase

One of the applications of this relationship is in the construction of a B-S type interpretation for the general (non-adiabatic) non-Abelian phase. In fact, the analysis of Sect. 6.4 can be easily extended to the non-Abelian case.

In general, the initial exact cyclic state $W(0)$ is an eigenstate of a Hermitian operator $\tilde{h}(R(0))$. Subsequently, for all $t \in [0, T]$ the solution

$$W(t) = \frac{1}{\mathcal{N}} \sum_{a=1}^{\mathcal{N}} |\phi_a(W(t))\rangle\langle\phi_a(W(t))| \tag{7.69}$$

of the Liouville–von Neumann equation (2.5) remains an eigenstate of $\tilde{h}(R(t))$. In other words, the single-valued basis vectors $|\phi_a(W(t))\rangle$ are the eigenvectors of $\tilde{h}(R(t))$. They correspond to an $\mathcal{N}$-fold degenerate eigenvalue $\tilde{E}_{\tilde{n}}(R(t))$ of $\tilde{h}(R(t))$.

Introducing the following convenient notation:

$$|\tilde{n}, a; R(t)\rangle := |\phi_a(W(t))\rangle \tag{7.70}$$

and using the analogy with the Abelian case, we define the function $\tilde{f}_{n,\nu} : M \to Gr_{\mathcal{N}}$ by

$$\tilde{f}_{n,\nu}(R) := \frac{1}{\mathcal{N}} \sum_{a=1}^{\mathcal{N}} |\tilde{n}, a; R\rangle\langle\tilde{n}, a; R|, \tag{7.71}$$

where ν denotes the evolution parameter of the system. $\tilde{f}_{n,\nu}$ is the generalization of (6.37). It can be used to construct the non-adiabatic generalization of the non-Abelian B-S PFB and connection one-form. Denoting these by $\tilde{\lambda}_{\mathcal{N},\nu}^n$ and $\tilde{A}_{\mathcal{N},\nu}^n$ respectively, we have

$$\tilde{\lambda}_{\mathcal{N},\nu}^n := \tilde{f}_{n,\nu}^* \eta_{\mathcal{N}}, \tag{7.72}$$

$$\tilde{A}_{\mathcal{N},\nu}^n := \tilde{f}_{n,\nu}^* \mathcal{A}_{\mathcal{N}}. \tag{7.73}$$

By definition (7.71) ensures the validity of the following exact identity,

$$\mathcal{C}_{\mathcal{N}} = \tilde{f}_{n,\nu} \circ \mathbf{C}. \tag{7.74}$$

This relation can be directly used to show that the holonomy element of the closed curve $\mathbf{C}$ in $\tilde{\lambda}_{\mathcal{N},\nu}^n$ is identical with the geometric phase (7.52). This is done by repeating the calculations performed in (7.68) after taking

$$\text{adiabatic phase} \longrightarrow \text{non-adiabatic phase,}$$
$$A_{\mathcal{N}}^n \longrightarrow \tilde{A}_{\mathcal{N},\nu}^n,$$
$$f_n \longrightarrow \tilde{f}_{n,\nu}.$$

As was the case in the discussion of the Abelian phase, the explicit dependence of the operator $\tilde{h}(R(t))$ on the Hamiltonian $h(R(t))$ is not generally known. However, there is a special class of systems for which this dependence is given by

$$\tilde{h}(R) = h(F_\nu(R)), \tag{7.75}$$

where $F_\nu : M \to M$ is a smooth function. Clearly, in the adiabatic domain of the evolution parameter ν, F_ν is approximated by the identity function on M. If such a function exists then we can immediately write down the inducing function $\tilde{f}_{n,\nu}$. The result is similar to the Abelian case:

$$\tilde{f}_{n,\nu} = f_n \circ F_\nu. \tag{7.76}$$

This can be most easily justified by introducing the notation

$$\tilde{R} := F_\nu(R)$$

and noting that for such systems

$$\tilde{h}(R) = h(\tilde{R}), \quad \text{and} \quad |\tilde{n}, a, R\rangle = |n, a; \tilde{R}\rangle. \tag{7.77}$$

Following the analysis of Sect. 6.4, we have

$$\tilde{\lambda}^n_{\mathcal{N},\nu} = \tilde{f}^*_{n,\nu}\eta_{\mathcal{N}} = F^*_\nu(f^*_n\eta_{\mathcal{N}}) = F^*_\nu \lambda^n_{\mathcal{N}}, \tag{7.78}$$

$$\tilde{A}^n_{\mathcal{N},\nu} = \tilde{f}^*_{n,\nu}\eta_{\mathcal{N}} = F^*_\nu(f^*_n\eta_{\mathcal{N}}) = F^*_\nu A^n_{\mathcal{N}}. \tag{7.79}$$

Equations (7.78) and (7.79) are the non-Abelian generalizations of (6.41) and (6.42). Equation (7.78) can be summarized by the following commutative diagram

$$\begin{array}{ccc} \tilde{\lambda}^n_{\mathcal{N},\nu} & \xrightarrow{F_{\nu *}} & \lambda^n_{\mathcal{N}} \\ \downarrow & \circ & \downarrow \\ M & \xrightarrow{F_\nu} & M. \end{array}$$

The last equality in (7.79) relates the adiabatic and non-adiabatic B-S connection one-forms. In view of (7.77), it can be explicitly expressed by

$$[\tilde{A}^n_{\mathcal{N},\nu}]^{ab} = i\langle \tilde{n}, a; R|d|\tilde{n}, b; R\rangle = i\langle n, a; \tilde{R}|d|n, b; \tilde{R}\rangle.$$

We would like to conclude this chapter by indicating that the study of the topological structure of the non-Abelian B-S bundles $\tilde{\lambda}^n_{\mathcal{N},\nu}$ is in general a difficult problem. This is because unlike $\mathbb{C}P^\infty$, $Gr_{\mathcal{N}}$ have many non-trivial homotopy groups. Thus the set $[M, Gr_{\mathcal{N}}]$ of homotopy classes of maps from M into $Gr_{\mathcal{N}}$ is not as easily determined as in the Abelian case. In particular, there is no analog of the first Chern class that would classify all such bundles. Nevertheless, one can investigate the known topological invariants of these bundles. In this regard, we wish to refer the interested reader to the article by Avron et al. [21] where a number of specific examples have been studied in some detail. A through discussion of the non-Abelian adiabatic geometric phase for general spin 1 systems may be found in [178].

Another difference between the Abelian and non-Abelian cases is that although in both cases one can define the local curvature two-form F (5.100), it is not possible to express the non-Abelian geometric phase (holonomy elements) solely in terms of F. This is in spite of the fact that there is a well-known non-Abelian generalization of Stokes' theorem. The utility of this generalization is however rather limited. The interested reader may consult [17]. The original idea of the non-Abelian Stokes' theorem is due to Halpern [96]. Further references are provided in [50].

Problems

7.1) Let $\mathcal{H}_n(R)$ be the degeneracy eigenspace defined by (7.3) and $\{|a\rangle\}$, $a = 1, 2, \cdots, \mathcal{N}$ be an arbitrary basis of $\mathcal{H}_n(R)$.
   a) Express $|a\rangle$ as column vectors and write the eigenstate vectors $|n, a; R\rangle$ as a linear combination of $|a\rangle$.
   b) Use the components of $|n, a; R\rangle$ in this basis to express the frame $s^{(n)}(R)$ of (7.9) as an $\mathcal{N} \times \mathcal{N}$ matrix and show that the corresponding matrix is unitary.
   c) Use the result of part (c) to show that (7.6) implies (7.9).

7.2) Show that both $A_\mathcal{N}^n(R)$ (7.15) and $\mathcal{A}_\mathcal{N}$ (7.53) are Hermitian matrices and derive their gauge transformation rules, i.e., prove (7.16) and (7.54).

7.3) Use (7.19), (7.20), and (7.22) to show the validity of (7.23) and (7.24).

7.4) Show that both the adiabatic (7.25) and non-adiabatic (7.52) non-Abelian geometric phases are gauge covariant quantities.

7.5) Using the analogy with the Abelian case, specify the transition functions of the non-Abelian B-S bundles $L_\mathcal{N}^n$ and $\lambda_\mathcal{N}^n$.

8. A Quantum Physical System in a Quantum Environment – The Gauge Theory of Molecular Physics

8.1 Introduction

In Chap. 3 we considered a complicated physical system which consisted of a quantum physical system (spinning particle with magnetic moment) and a classical physical system (external, time-dependent magnetic field); the latter we called the environment. Now we want to consider a complicated physical system which is entirely quantum mechanical. To understand it we divide it into simpler parts which are again quantum systems. A natural division for a large class of quantum systems is the separation into a fast moving part and a slow moving part. The slowly moving part usually consists of the collective motion of the quantum system as a whole. The paradigm of this procedure is the Born–Oppenheimer method in molecular physics. The swiftly moving electron or electrons corresponds to the quantum system of Chap. 3 and the slowly moving nuclei, which determine the collective motion of the molecule as a whole, take the place of the quantum mechanical environment.

The conventional Born–Oppenheimer approximation [48] used an assumption that did not allow for any of the effects connected with the geometric phase. In this sense the conventional Born–Oppenheimer approximation should be called the "crude adiabatic approximation" [160]. The true adiabatic approximation that takes into account the geometric phase naturally introduces the gauge structure in the Born–Oppenheimer approximation. The importance of using the true adiabatic approximation was first recognized in *vibronic* (vibrational + electronic) problems in Jahn–Teller systems, where the geometric phase was noticed as the sign change of an electronic wave function [105, 158, 159] after a circular transport around a conical intersection of potential energy surfaces (PESs). Potential energy surfaces are determined by the eigenvalues of the electronic Hamiltonian which in turn depend on nuclear coordinates. Conical intersections of the PESs are singularities of the Born–Oppenheimer approximation where derivatives of the electronic wave functions with respect to nuclear coordinates are not well defined. The amplitude of a wave function is zero at those points due to the vibronic centrifugal (or Born–Huang) term. Later it was recognized that the sign-change of the electronic wave function could be correctly taken into account by incorporating a gauge potential that created a delta function like a fictitious magnetic flux at the location of the conical intersection. The sign-change caused by this fictitious magnetic flux was originally called the *molecular Aharonov–Bohm effect* [165–167].

The geometric phase is a global effect in the sense that it changes the interference pattern of the wave function. In general, neglecting geometric phases leads to serious problems in determining the correct interference patterns. The existing experimental results, however, indicate that errors caused by neglecting the geometric phase are quite small. This is mostly because of the fact that the de Broglie wavelength associated with the nuclear motion is very short (see Chap. 11). The detection of more enhanced geometric phase effects in molecular systems would require more sensitive experiments.

Geometric phases also play a significant role in the study of other manybody systems where there are singularities that are not present in the Hamiltonian but arise from the dynamics (or correlation) of the subsystems. For these systems the geometric phase effects (more precisely the topological phase effects of the Aharonov–Bohm type) also take the form of fictitious magnetic fluxes located at the singularities. A particular realization of this mechanism is in the fractional quantum Hall systems where zeros of the wave function and the fictitious magnetic fluxes situated at these points arise from the correlation of the electrons in a strong magnetic field [151].

Molecular systems have the advantage of being quite realistic and having well-known interactions and constituents. The fact that these systems have been subject to extensive theoretical and experimental studies makes them especially useful for the study of geometric phases. In this chapter we provide a general discussion of the improved Born–Oppenheimer approximation that accounts for the geometric phase effects in molecular systems.

8.2 The Hamiltonian of Molecular Systems

In this section we give a brief introduction to the structure of molecules. In particular, we discuss the standard form of the Hamiltonian that is the starting point of the Born–Oppenheimer method.

In the classical constituent picture, the molecule consists of the nuclei and electrons that are in constant Coulomb interaction with one another. Therefore, the total potential energy of the system is the sum of the Coulomb potentials associated with pairs of particles.

Suppose that the molecule consists of N_e electrons and N_n nuclei[1] of charge $Z_n e$ where e stands for the magnitude of the electron charge, and denote the electronic coordinates and momenta by $\mathbf{r}_i, \mathbf{p}_i (i = 1, 2, \cdots, N_\mathrm{e})$ and the nuclear coordinates and momenta by $\mathbf{R}_j, \mathbf{P}_j (j = 1, 2, \cdots, N_\mathrm{n})$, respectively. Then the classical total energy for the system has the form

$$E = T_\mathrm{nuclei} + T_\mathrm{electrons} + U_\mathrm{nuclei} + U_\mathrm{nuclei-electrons} + U_\mathrm{electrons}, \qquad (8.1)$$

---

[1] Here the subscript "n" in N_n stands for *nuclei*. It should not be confused with n used to label an electronic state.

where the first two terms are, respectively, the kinetic energies of the nuclei and the electrons,

$$T_{\text{nuclei}} = \sum_{i=1}^{N_n} \frac{1}{2m_i} \mathbf{P}_i^2,$$

$$T_{\text{electrons}} = \frac{1}{2m_e} \sum_{i=1}^{N_e} \mathbf{p}_i^2, \tag{8.2}$$

and the last three terms are, respectively, the Coulomb potential energies representing the interaction of the nuclei, the nuclei and the electrons, and the electrons,

$$U_{\text{nuclei}} = \sum_{j=i+1}^{N_n} \sum_{i=1}^{N_n} \frac{k(Z_j e)(Z_i e)}{|\mathbf{R}_j - \mathbf{R}_i|}, \tag{8.3a}$$

$$U_{\text{nuclei-electron}} = \sum_{i=1}^{N_e} \sum_{j=1}^{N_n} \frac{k(Z_j e)(-e)}{|\mathbf{r}_i - \mathbf{R}_j|}, \tag{8.3b}$$

$$U_{\text{electron}} = \sum_{j=i+1}^{N_e} \sum_{i=1}^{N_e} \frac{k(-e)(-e)}{|\mathbf{r}_i - \mathbf{r}_j|}. \tag{8.3c}$$

We obtain a quantum mechanical description of the molecule by quantizing the classical system, i.e., replacing the classical observables by the corresponding linear operators. For example, the quantum Hamiltonian operator H has the same form as the classical total energy (8.1), provided that we identify the positions and momenta of constituent particles with the corresponding quantum operators.

The energy levels E of the molecule are determined by eigenvalue equation for the Hamiltonian,

$$H|\psi^E\rangle = E|\psi^E\rangle, \tag{8.4}$$

which in the position representation takes the form of the $N = N_n + N_e$ body time-independent Schrödinger equation:

$$\langle \mathbf{R}_1 \cdots \mathbf{R}_{N_n}, \mathbf{r}_1 \cdots \mathbf{r}_{N_e} | H | \psi^E \rangle = E \langle \mathbf{R}_1, \cdots \mathbf{r}_{N_e} | \psi^E \rangle. \tag{8.4'}$$

This equation is obviously too complicated to admit an exact solution.

In order to devise a suitable approximation scheme for treating the N-body quantum system, i.e., solving (8.4), we consider a simplified picture where we do not distinguish between the constituent particles but we take account of the types of motion that occur in the molecule. As a first step, we treat the motion of the electrons and nuclei separately. This is because the light and swiftly moving electrons whirl around the much heavier and slower nuclei. Therefore, we obtain a good approximation by confining our attention to the motion of the electrons in the presence of N_n almost stationary nuclei. In this approximation, the total energy of the electrons has the form:

$$E_{\text{electrons}} \cong T_{\text{electrons}} + U_{\text{nuclei-electrons}} + U_{\text{electrons}}. \quad (8.5)$$

As seen from (8.3b), the potential energy term $U_{\text{nuclei-electrons}}$ and consequently the total energy $E_{\text{electrons}}$ of the electrons depend on the slowly changing positions $\mathbf{R}_j$ of the nuclei.

Because the electrons are much lighter than the nuclei, they instantaneously follow the motion of the nuclei. Therefore, the slowly changing nuclear variables are essentially identical with the variables describing the molecule as a whole. The dynamical variables corresponding to the slow motion of the molecule as a whole are called the collective variables. In order to distinguish between different types of collective motions, it is convenient to introduce three kinds of collective variables.[2] These are

1. The center of mass coordinates of the molecule:

$$\mathbf{R}_G = \frac{\sum_{i=1}^{N_n} m_i \mathbf{R}_i}{\sum_{i=1}^{N_n} m_i}; \quad (8.6)$$

2. The angular velocity describing the overall rotation of the molecule: $\boldsymbol{\omega} = (\omega_x, \omega_y, \omega_z)$. This is the angular velocity of the molecule-fixed moving axis system $Oxyz$ having its origin at $\mathbf{R}_G$, with respect to the non-rotating coordinate system moving with the center of mass.
3. The internal coordinates describing the vibration of the molecule around an equilibrium configuration: $\mathbf{Q} = (Q_1, \cdots, Q_{3N_n-6})$. These coordinates are often chosen to satisfy

$$\sum_{i=1}^{N_n} m_i \dot{\mathbf{u}}_i^2 = \sum_{k=1}^{3N_n-6} \dot{Q}_k^2, \quad (8.7)$$

where $\mathbf{u}_i = (u_{ix}, u_{iy}, u_{iz}) = \mathbf{R}_i - \mathbf{R}_G$ are the positions of the atoms in the $Oxyz$ frame, and a dot denotes the total time derivative. Note that we have excluded six degrees of freedom from the sum in (8.7). Three of these correspond to $\mathbf{R}_G$ which is fixed by the condition

$$\sum_{i}^{N_n} m_i \mathbf{u}_i = 0.$$

The other three degrees of freedom that are excluded from the sum in (8.7) are associated with the choice of the molecule-fixed frame $Oxyz$ (alternatively the choice of $\boldsymbol{\omega}$). This is determined by the condition

---

[2] Here we ignore the contribution of the electrons to the collective motions, because their masses are much smaller than those of the nuclei.

8.2 The Hamiltonian of Molecular Systems

$$\sum_i^{N_\mathrm{n}} m_i \mathbf{u}_i^\mathrm{eq} \times \mathbf{u}_i = 0,$$

where $\mathbf{u}_i^\mathrm{eq}$ are equilibrium (or some typical constant) values of $\mathbf{u}_i$.

Next, we consider the molecule from the point of view of the nuclei: The (classical) electrons move in complicated orbits creating a smeared-out charge distribution. This charge distribution gives rise to a potential energy $E_n^e(\mathbf{Q})$ for the orbiting electrons that is obtained by averaging the value of $E_\mathrm{electrons}$ of (8.5) for all possible electron orbits. The subscript n in E_n^e indicates that in the quantum description of the system E_n^e depends on the quantum number n of the electronic state. The total potential energy $U_n(\mathbf{Q})$ experienced by the nuclei in the molecule-fixed frame $Oxyz$ is the sum of the potential energy $E_n^e(\mathbf{Q})$ due to the electrons and the repulsive Coulomb interaction potential energy $U_\mathrm{nuclei}(\mathbf{Q})$ of the nuclei, (8.3a). That is

$$U_n(\mathbf{Q}) = E_n^e(\mathbf{Q}) + U_\mathrm{nuclei}(\mathbf{Q}). \tag{8.8}$$

If the potential $E_n^e(\mathbf{Q})$ created by the whirling electrons was absent, then N_n positively charged nuclei would repel each other.[3]

The potential energy $U_n(\mathbf{Q})$ can take different forms depending on the net effect of the electron's motion, alternatively the particular electronic state (this is indicated by the label n), and the type of the molecule. In general, the collective motions (of the nuclei) may lead to a transition of the electronic state, i.e., a change of the label n. We shall always assume that such a transition does not occur. That is the electronic state is labeled by a fixed label n. This assumption is equivalent to the adiabatic approximation. In many cases, especially in the electronic ground states, the potential energy curve has a minimum at an equilibrium nuclear separation (Fig. 8.1). As a consequence, the nuclei remain bound to each other as long as their energy does not exceed the dissociation energy D_e. If one expands $U_n(\mathbf{Q})$ about the equilibrium atomic configuration $\mathbf{Q}_e$ (see Sect. 8.6), one obtains as the leading term the parabolic potential curve depicted by the dashed line in Fig. 8.1. This curve provides a reliable qualitative approximation near the potential minimum – also known as the harmonic region. Because of the particular shape of the potential energy, the relative nuclear coordinates perform harmonic oscillations.

The total energy of the molecule has the form

$$E = T_\mathrm{nuclei} + U_n(\mathbf{Q}) = T_\mathrm{nuclei} + E_n^e(\mathbf{Q}) + U_\mathrm{nuclei}(\mathbf{Q}), \tag{8.9}$$

where T_nuclei denotes the nuclear contribution to the kinetic energy,

---

[3] Note that here we neglect the nuclear interactions.

152   8. The Gauge Theory of Molecular Physics

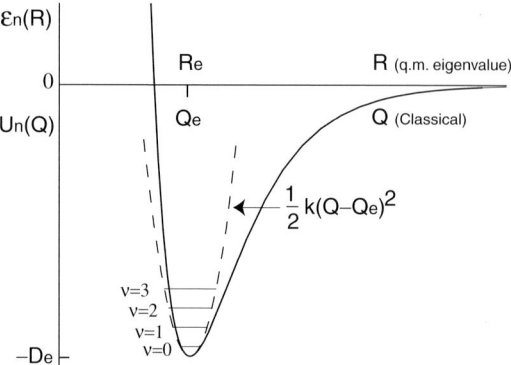

Fig. 8.1. Typical potential energy curve for bound states of a diatomic molecule: The classical potential energy of the nuclei is plotted as a function of the internuclear distance. This function is the same as the induced scalar potential $\varepsilon_n(R)(R = Q)$, in the time-independent Schrödinger equation (8.58) for the collective (nuclear) motion. Because of its approximate parabolic shape near the minimum, the collective energy levels are the vibrational excitations $\nu = 0, 1, 2, \ldots$. As discussed in Sects. III.5 and V.4 of [43], each of these vibrational levels split into doubly degenerate rotational bands (not shown here).

$$\begin{aligned}
2T_{\text{nuclei}} &= \sum_{i=1}^{N_n} m_i \dot{\mathbf{R}}_i^2 = \sum_{i=1}^{N_n} m_i (\dot{\mathbf{R}}_G + \dot{\mathbf{u}}_i + \boldsymbol{\omega} \times \mathbf{u}_i)^2 \\
&= \dot{\mathbf{R}}_G^2 \sum_{i=1}^{N_n} m_i + \sum_{\alpha,\beta=x,y,z} I_{\alpha\beta}\omega_\alpha\omega_\beta + \sum_{k=1}^{3N_n-6} \dot{Q}_k^2 \\
&\quad + 2 \sum_{\alpha=x,y,z} \sum_{k=1}^{3N_n-6} \omega_\alpha \xi_k^\alpha \dot{Q}_k,
\end{aligned} \quad (8.10)$$

$I_{\alpha\beta}$ are the elements of the inertia tensor,

$$I_{\alpha\beta} := \sum_{i=1}^{N_n} \sum_{\gamma=x,y,z} m_i (u_{i\gamma}^2 \delta_{\alpha\beta} - u_{i\alpha} u_{i\beta}), \quad (8.11)$$

and ξ_k^α are defined by

$$\sum_k \xi_k^\alpha \dot{Q}_k := \sum_i m_i (\mathbf{u}_i \times \dot{\mathbf{u}}_i)_\alpha. \quad (8.12)$$

The above picture in which one treats the electronic motion and the motion of the nuclei separately is the essence of the Born–Oppenheimer approximation. The quantum analogs of (8.5) and (8.9) yield the Hamiltonian operator and the energy spectrum associated with the motion of the electrons and the collective motion of the molecule, respectively.

8.2 The Hamiltonian of Molecular Systems

As seen in (8.10), there are three types of collective motion corresponding to three sets of collective coordinates:

1. translation of the center of mass of the molecule which is essentially the center of mass motion of the nuclei (the first term on the right-hand side of (8.10));
2. rotation of the molecule as a whole about its center of mass (the second term on the right-hand side of (8.10));
3. vibration of the molecule about its equilibrium configuration (the third term on the right-hand side of (8.10)).

The translation of the center of mass leads to a continuum of energy eigenvalues. This is similar to the translational motion of a free particle. In the following, we shall use the center mass reference frame of the molecule where this type of motion does not occur. We shall therefore ignore the corresponding translational energy.

The pure rotational motion[4] of a molecule occurs provided that the nuclear configuration is considered to be fixed at $\mathbf{Q} = \mathbf{Q}_e$. This condition is often fulfilled when the molecule is in the vibrational ground state. It is also fulfilled to a good approximation for the excited vibrational levels. However, the breakdown of the separation of the rotational and vibrational motions occurs frequently due to centrifugal stretching and the Coriolis coupling (the last term on the right-hand side of (8.10)).

Next, we consider quantizing (8.9) to obtain the electronic Hamiltonian operator. That is we promote the classical observables of position and momentum of the constituent particles to the quantum mechanical operators. We shall use a ˆ to emphasize that the corresponding quantity is an operator.[5]

So far we have employed the internal coordinates $\mathbf{Q}$ to describe the nuclear motion. It turns out that the quantum mechanical kinetic energy operator has a very complicated form in terms of $\mathbf{Q}$ (see Problem 8.1). Therefore, we shall next introduce and use the so-called Jacobi coordinate system in terms of which the expression for the kinetic energy operator simplifies.[6] In fact, we shall not offer a general discussion of the Jacobi coordinates, but consider three illustrative examples (Fig. 8.2), namely the Jacobi coordinate systems for the diatomic, triatomic, and tetra-atomic molecules.

The Jacobi coordinate system for a diatom is defined by the vector connecting the two atoms,[7]

---

[4] The energy spectrum of a molecule which is associated with its rotational motion is called the *rotational spectrum* [43, §III.4].
[5] We leave the derivation of the quantum mechanical analog of (8.10) as an exercise.
[6] It turns out that the Mead–Berry vector potential also has a simpler form in Jacobi coordinates. Furthermore, the Jacobi coordinate system is a good coordinate system in dealing with situations where large-amplitude vibrations cause significant vibrational–rotational coupling.
[7] The Jacobi coordinates are the components of this vector.

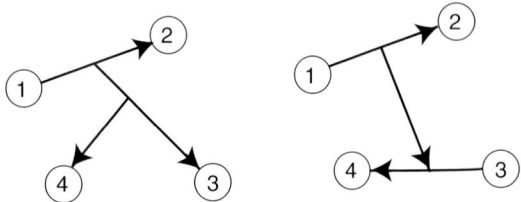

Fig. 8.2. Two types of Jacobi coordinates for a tetra-atom.

$$\mathbf{R}_{2,1} := \mathbf{R}_2 - \mathbf{R}_1. \tag{8.13}$$

This may be viewed as the position vector for a point particle whose mass coincides with the reduced mass:

$$\mu_{2,1} := \frac{m_1 m_2}{m_1 + m_2}, \tag{8.14}$$

where m_i are the masses of the constituent atoms. In these coordinates, the kinetic energy operator takes the form

$$\hat{T}_{\text{nuclei}} = -\frac{1}{2\mu_{2,1}} \nabla^2_{\mathbf{R}_{2,1}}. \tag{8.15}$$

In this equation $\nabla^2_{\mathbf{R}_{2,1}}$ stands for the Laplacian associated with the Jacobi coordinates $R_{2,1_j}$.

The Jacobi coordinate system for a triatom is defined by the two vectors: $\mathbf{R}_{2,1}$ and

$$\mathbf{R}_{3,12} := \mathbf{R}_3 - \mathbf{R}_{12}, \tag{8.16}$$

where 1, 2 and 3 are used to label the three atoms, and $\mathbf{R}_{12}$ is the position of the center of mass of the subsystem composed of the two atoms labeled by 1 and 2, i.e.,

$$\mathbf{R}_{12} := \frac{m_1 \mathbf{R}_1 + m_2 \mathbf{R}_2}{m_1 + m_2}. \tag{8.17}$$

Now, viewing $\mathbf{R}_{3,12}$ as the position of a point particle of mass

$$\mu_{3,12} := \frac{m_3 m_{12}}{m_3 + m_{12}}, \tag{8.18}$$

with $m_{12} := m_1 + m_2$, we can express the kinetic energy operator for the system in the form

$$\hat{T}_{\text{nuclei}} = -\frac{1}{2\mu_{2,1}} \nabla^2_{\mathbf{R}_{2,1}} - \frac{1}{2\mu_{3,12}} \nabla^2_{\mathbf{R}_{3,12}}. \tag{8.19}$$

Again $\nabla^2_{\mathbf{R}_{3,12}}$ stands for the Laplacian associated with the components of $\mathbf{R}_{3,12}$.

The situation for a tetra-atom is rather different, as there are two types of Jacobi coordinate systems for a tetra-atom. These are shown in Fig. 8.2.

The first Jacobi coordinate system (depicted on the left in Fig. 8.2) is an extension of the triatomic system. It is defined by the three vectors: $\mathbf{R}_{2,1}$, $\mathbf{R}_{3,12}$ and

$$\mathbf{R}_{4,123} := \mathbf{R}_4 - \mathbf{R}_{123}. \tag{8.20}$$

Similarly to the case of the diatom and the triatom, we identify $\mathbf{R}_{4,123}$ with the position of a point particle of mass

$$\mu_{4,123} := \frac{m_4 m_{123}}{m_4 + m_{123}}, \tag{8.21}$$

where $\mathbf{R}_{123} := (m_1 \mathbf{R}_1 + m_2 \mathbf{R}_2 + m_3 \mathbf{R}_3)/m_{123}$ and $m_{123} := m_1 + m_2 + m_3$. Then, the kinetic energy for the system takes the form

$$\hat{T}_{\text{nuclei}} = -\frac{1}{2\mu_{2,1}} \nabla^2_{\mathbf{R}_{2,1}} - \frac{1}{2\mu_{3,12}} \nabla^2_{\mathbf{R}_{3,12}} - \frac{1}{2\mu_{4,123}} \nabla^2_{\mathbf{R}_{4,123}}. \tag{8.22}$$

The other Jacobi coordinate system (depicted on the right in Fig. 8.2) is determined by the three vectors: $\mathbf{R}_{2,1}$, $\mathbf{R}_{4,3} := \mathbf{R}_4 - \mathbf{R}_3$, and

$$\mathbf{R}_{34,12} := \mathbf{R}_{34} - \mathbf{R}_{12}. \tag{8.23}$$

These vectors may be viewed as position vectors of three point particles of mass $\mu_{2,1} := m_2 m_1/(m_2 + m_1)$, $\mu_{4,3} := m_3 m_4/(m_3 + m_4)$, and $\mu_{34,12} := m_{12} m_{34}/(m_{12} + m_{34})$, respectively. The kinetic energy of the system is then given by

$$\hat{T}_{\text{nuclei}} = -\frac{1}{2\mu_{2,1}} \nabla^2_{\mathbf{R}_{2,1}} - \frac{1}{2\mu_{4,3}} \nabla^2_{\mathbf{R}_{4,3}} - \frac{1}{2\mu_{34,12}} \nabla^2_{\mathbf{R}_{34,12}}. \tag{8.24}$$

Generalizing the cases of the diatomic, triatomic, and tetra-atomic molecule, we define a Jacobi coordinate system for a general molecule in terms of the position vectors $\mathbf{R}_{\alpha,\beta}$ of the center of mass of pairs (α, β) of different subsystems of atoms in the molecule. If we denote by $\mathbf{R}_\alpha$ (respectively $\mathbf{R}_\beta$) and m_α (respectively m_β) the center of mass and the reduced mass of subsystem α (respectively β), then

$$\mathbf{R}_{\alpha,\beta} := \frac{m_\alpha \mathbf{R}_\alpha + m_\beta \mathbf{R}_\beta}{m_\alpha + m_\beta}. \tag{8.25}$$

The kinetic energy operator of the molecule then takes the form

$$\hat{T}_{\text{nuclei}} = -\sum_{(\alpha,\beta)} \frac{1}{2\mu_{\alpha,\beta}} \nabla^2_{\mathbf{R}_{\alpha,\beta}}, \tag{8.26}$$

where $\mu_{\alpha,\beta}$ is the reduced mass for the pair of subsystems (α, β), namely

$$\mu_{\alpha,\beta} = \frac{m_\alpha m_\beta}{m_\alpha + m_\beta}. \tag{8.27}$$

The expression (8.26) for the kinetic energy operator may be slightly simplified by scaling the Jacobi coordinates, i.e., the vectors $\mathbf{R}_{\alpha,\beta}$, so that the associated reduced masses coincide. This yields

$$\hat{T}_\text{nuclei} = \frac{\hat{\mathbf{P}}^2}{2\mu}, \tag{8.28}$$

where $\hat{\mathbf{P}}$ is the nuclear momentum operator conjugate to the position vector $\hat{\mathbf{R}}$ of the scaled Jacobi coordinates. In the center of mass frame, $\hat{\mathbf{R}}$ represents the $3N_\text{n} - 3$ degrees of freedom of the molecule.

Having obtained the expression (8.28) for the kinetic energy, we can express the quantum mechanical Hamiltonian in the form

$$H = \frac{\hat{\mathbf{P}}^2}{2\mu} + h(\hat{\mathbf{p}}, \hat{\mathbf{r}}, \hat{\mathbf{R}}), \tag{8.29}$$

where h is the electronic Hamiltonian,

$$h(\hat{\mathbf{p}}, \hat{\mathbf{r}}, \hat{\mathbf{R}}) = \sum_{i=1}^{N_\text{e}} \frac{\hat{\mathbf{p}}_i^2}{2m_\text{e}} + V(\hat{\mathbf{R}}, \hat{\mathbf{r}}), \tag{8.30}$$

$\hat{\mathbf{p}}_i$ and $\hat{\mathbf{r}}_i$ respectively denote the momentum and position operator associated with the i-th electron, $\hat{\mathbf{p}} := (\hat{\mathbf{p}}_1, \cdots \hat{\mathbf{p}}_{N_\text{e}})$, $\hat{\mathbf{r}} := (\hat{\mathbf{r}}_1, \cdots \hat{\mathbf{r}}_{N_\text{e}})$, and $V(\hat{\mathbf{R}}, \hat{\mathbf{r}})$ stands for the potential energy operator that describes all the electron–electron, electron–nucleus, and nucleus–nucleus interactions. The potential $V(\hat{\mathbf{R}}, \hat{\mathbf{r}})$ depends on both $\hat{\mathbf{R}}$ and $\hat{\mathbf{r}}$.[8] The electronic Hamiltonian h may be replaced by $U(\hat{\mathbf{R}})$ which is the operator obtained by quantizing $U_n(\mathbf{Q})$.

The observables $\hat{\mathbf{p}}$, $\hat{\mathbf{r}}$ describe the fast (electronic) degrees of freedom and are called the *fast variables*. The observables $\hat{\mathbf{P}}$, $\hat{\mathbf{R}}$ describe the slow degrees of freedom of the nuclei or the molecule as a whole and are called the *slow (collective) variables*. In the quantum mechanical description of the molecule, both the fast and slow variables are operators, and $h(\hat{\mathbf{p}}, \hat{\mathbf{r}}, \hat{\mathbf{R}})$ describes a quantum system in a quantum environment.

In an intermediate quasi-classical theory, one can identify the slow variables with slowly changing classical parameters. Then (8.30) is a parameter-dependent Hamiltonian (with parameters $\mathbf{R}$) describing a quantum system (electron) in a slowly changing classical environment (the molecule as a whole). If the motion of the (classical) nuclei is slow enough, then the quantum number describing the electronic state does not change in time, and we

---

[8] $V(\hat{\mathbf{R}}, \hat{\mathbf{r}})$ may also depend on other relevant operators such as spin. These will however be suppressed.

have an adiabatic time evolution. It is to this intermediate theory that the results of Chaps. 2 and 3 apply; the eigenvectors and eigenvalues of (8.30) depend on $\mathbf{R}$ according to (2.6); the evolution is cyclic, if the relative position of the nuclei $\mathbf{R}(t)$ returns after a period T to its original position; and the Mead–Berry connection (2.54) and Berry phase (2.65) are obtained in terms of the eigenvectors $|n;\mathbf{R}\rangle$ of the fast electronic Hamiltonian $h(\hat{\mathbf{R}})$ of (8.30).

In quantum theory the observables are operators acting in the space of physical state vectors. The physical system is a combination of two subsystems: the subsystem of the fast or electronic motions and the subsystem of the slow or nuclear motions. Denoting the space of state vectors for the fast motion by $\mathcal{H}^{\text{fast}}$ and the space of state vectors for the slow motion by $\mathcal{H}^{\text{slow}}$, the space of physical state vectors for the molecule is the direct (tensor) product space[9]

$$\mathcal{H} = \mathcal{H}^{\text{slow}} \otimes \mathcal{H}^{\text{fast}}. \tag{8.31}$$

$\mathcal{H}^{\text{slow}}$ and $\mathcal{H}^{\text{fast}}$ are in general both infinite dimensional but often one obtains a good approximation if one restricts oneself to a finite-dimensional subspace of $\mathcal{H}^{\text{fast}}$. We will denote dim $\mathcal{H}^{\text{fast}}$ by $\mathcal{N}$ which can be one, finite, or infinite. The Hamiltonians (8.29), (8.30), the observables $\hat{\mathbf{p}}$, $\hat{\mathbf{r}}$, $\hat{\mathbf{P}}$, $\hat{\mathbf{R}}$, and the other relevant observables, such as the electron's spin, which we have not explicitly mentioned above, are operators acting in $\mathcal{H}$.

8.3 The Born–Oppenheimer Method

Now that the problem has been set up we can proceed to the solution [41, 45], which is to solve the eigenvalue problem

$$H\psi^E = E\psi^E, \tag{8.32}$$

where H is the Hamiltonian (8.29). This can, of course, only be done approximately. In the Born–Oppenheimer approximation [48] one first solves the eigenvalue problem for the fast Hamiltonian (8.30) and obtains the eigenvalues $\varepsilon_n(\mathbf{R})$ (the nth potential energy surface, PES) as functions of the eigenvalues $\mathbf{R}$ of the operator $\hat{\mathbf{R}}$ (slow position, i.e., relative position of the nuclei).[10]

In order to obtain these $\varepsilon_n(\mathbf{R})$[11] – which we will not calculate explicitly – we introduce two different complete systems of basis vectors of the space $\mathcal{H}$ in (8.31). The first basis system $|\mathbf{R},\mathbf{r}\rangle$ are generalized eigenvectors of the operators $\hat{\mathbf{R}}$ and $\hat{\mathbf{r}}$, i.e.,

---

[9] See [43, §III.5] for a definition of the direct product.
[10] In the conventional Born–Oppenheimer approach $\mathbf{R}$ itself is the classical parameter; here we treat immediately the fully quantum mechanical version.
[11] Actually, h and ε_n depend on $\hat{\mathbf{Q}}$ and $\mathbf{Q}$ rather than $\hat{\mathbf{R}}$ and $\mathbf{R}$. There are some redundancies of coordinates in $\hat{\mathbf{R}}$ and $\mathbf{R}$ to specify atomic configurations but we use them for the convenience of expression.

$$\hat{\mathbf{R}}|\mathbf{R},\mathbf{r}\rangle = \mathbf{R}|\mathbf{R},\mathbf{r}\rangle, \tag{8.33a}$$

$$\hat{\mathbf{r}}|\mathbf{R},\mathbf{r}\rangle = \mathbf{r}|\mathbf{R},\mathbf{r}\rangle. \tag{8.33b}$$

These eigenvectors are the standard (δ-function normalized) position eigenvectors,

$$\langle \mathbf{R}',\mathbf{r}'|\mathbf{R},\mathbf{r}\rangle = \delta^{3N_n-3}(\mathbf{R}'-\mathbf{R})\delta^{3N_e}(\mathbf{r}'-\mathbf{r}). \tag{8.34}$$

The second basis system $|\mathbf{R},n\rangle$ consists of the generalized eigenvectors of the operators $\hat{\mathbf{R}}$ and $h(\hat{\mathbf{R}})$, i.e., they satisfy

$$\hat{\mathbf{R}}|\mathbf{R},n\rangle = \mathbf{R}|\mathbf{R},n\rangle, \tag{8.35}$$

$$h(\hat{\mathbf{R}})|\mathbf{R},n\rangle = \varepsilon_n(\mathbf{R})|\mathbf{R},n\rangle, \tag{8.36}$$

and are normalized according to

$$\langle \mathbf{R}',n'|\mathbf{R},n\rangle = \delta^{3N_n-3}(\mathbf{R}'-\mathbf{R})\delta_{n'n}. \tag{8.37}$$

Notice that $h(\hat{\mathbf{R}})$ is an operator which acts in both $\mathcal{H}^{\text{fast}}$ and $\mathcal{H}^{\text{slow}}$. The existence of these simultaneous eigenvectors follows from the fact that $[\hat{\mathbf{R}}, h(\hat{\mathbf{R}})] = 0$. The label n stands for the set of electronic quantum numbers, all or part of which label the electronic energy value ε_n. In general other quantum numbers are needed in addition to $\mathbf{R}$ and n to label the basis vectors completely, but those quantum numbers are suppressed here. The eigenvalue $\varepsilon_n(\mathbf{R})$ is obtained by solving (8.36) for each value of $\mathbf{R}$.

The basis vectors $|\mathbf{R},n\rangle$ can be expanded in terms of the basis vectors $|\mathbf{R}',\mathbf{r}'\rangle$,

$$|\mathbf{R},n\rangle = \int d^{3N_n-3}R' \int d^{3N_e}r'|\mathbf{R}',\mathbf{r}'\rangle\langle \mathbf{R}',\mathbf{r}'|\mathbf{R},n\rangle. \tag{8.38}$$

The transition matrix elements between these two sets of basis vectors are given by

$$\langle \mathbf{R}',\mathbf{r}'|\mathbf{R},n\rangle = \delta^{3N_n-3}(\mathbf{R}'-\mathbf{R})\langle \mathbf{r}'|n(\mathbf{R})\rangle. \tag{8.39}$$

This equation is a statement and a definition. It states that the $\mathbf{R}$-dependence is given by the δ-function, and it defines the reduced transition matrix: $\langle \mathbf{r}|n(\mathbf{R})\rangle := \langle \mathbf{r}|n\rangle_{\mathbf{R}}$ which for every $\mathbf{r}$ and n is a function of $\mathbf{R}$. For every value of $\mathbf{R}$, this reduced transition matrix determines a basis,

$$\{|n(\mathbf{R})\rangle \; : \; n = \text{all electronic quantum numbers}\},$$

in $\mathcal{H}^{\text{fast}}$. The basis vectors (8.36) of $\mathcal{H}$ can be represented in terms of the basis vectors $|n(\mathbf{R})\rangle$ as

$$|\mathbf{R},n\rangle = |\mathbf{R}\rangle\tilde{\otimes}|n(\mathbf{R})\rangle, \tag{8.40a}$$

where according to (8.37),

$$\langle n(\mathbf{R})|m(\mathbf{R})\rangle = \delta_{nm}, \tag{8.40b}$$

and $|\mathbf{R}\rangle$ is the basis system of (generalized, δ-function normalized) eigenvectors of $\hat{\mathbf{R}}$ for the space $\mathcal{H}^{\text{slow}}$. The symbol $\tilde{\otimes}$ in (8.40a) indicates that the right hand side of this equation is not the usual direct (tensor) product [43, §III.5] of the vectors $|\mathbf{R}\rangle$ and $|n(\mathbf{R})\rangle$. This is because according to (8.36) $\mathbf{R}$ and n are not independent quantum numbers.

We can now define $h(\mathbf{R})$ as the operator obtained from $h(\hat{\mathbf{R}})$ by replacing the operator $\hat{\mathbf{R}}$ by the number $\mathbf{R}$. Note that $h(\mathbf{R})$ acts in $\mathcal{H}^{\text{fast}}$ whereas $h(\hat{\mathbf{R}})$ acts in $\mathcal{H}^{\text{slow}} \otimes \mathcal{H}^{\text{fast}}$. From (8.37) and (8.40a), it follows that

$$h(\hat{\mathbf{R}})|\mathbf{R}\rangle\tilde{\otimes}|n(\mathbf{R})\rangle = |\mathbf{R}\rangle\tilde{\otimes}h(\mathbf{R})|n(\mathbf{R})\rangle,$$
$$= \varepsilon_n(\mathbf{R})|\mathbf{R}\rangle\tilde{\otimes}|n(\mathbf{R})\rangle. \tag{8.41}$$

Consequently, the operator $h(\mathbf{R})$ which depends on the parameter $\mathbf{R}$ has the eigenvector $|n(\mathbf{R})\rangle$ with eigenvalue $\varepsilon_n(\mathbf{R})$. Both of these also depend on the parameter[12] $\mathbf{R}$,

$$h(\mathbf{R})|n(\mathbf{R})\rangle = \varepsilon_n(\mathbf{R})|n(\mathbf{R})\rangle. \tag{8.42}$$

The symbol $n(\mathbf{R})$ in $|n(\mathbf{R})\rangle$ does not mean that the value of the quantum number changes as a function of $\mathbf{R}$, but that the physical meaning of the quantum number changes, i.e., the fast observable with eigenvalue n is different for different values of $\mathbf{R}$. We have already encountered a similar situation in Chap. 3, in particular in (3.11), where the eigenvalue k remains the same but the observable $\mathbf{e_R} \cdot \mathbf{J}$ changes with the parameter $\mathbf{e_R}$ ($\mathbf{e_R}$ is $\hat{\mathbf{R}}$ in (3.11)).

Unlike the $\mathbf{R}$ and $\mathbf{r}$ of (8.33a), the $\mathbf{R}$ and n of (8.35) are not independent quantum numbers. As a consequence, the matrix elements $\langle \mathbf{R}, n|\mathcal{O}|\psi\rangle$ of operators $\mathcal{O}$ do not have the familiar form. For instance, the basis vectors $|\mathbf{R}, \mathbf{r}\rangle$ satisfy

$$\langle \mathbf{R}, \mathbf{r}|\hat{\mathbf{P}}|\psi\rangle = \frac{1}{i} \nabla_{\mathbf{R}} \langle \mathbf{R}, \mathbf{r}|\psi\rangle \tag{8.43}$$

for any state vector $|\psi\rangle$, whereas a similar expression for the basis vectors $|\mathbf{R}, n\rangle$ does not exist. That is, in general, the relation

$$\langle \mathbf{R}, n|\hat{\mathbf{P}}|\psi\rangle = \frac{1}{i} \nabla_{\mathbf{R}} \langle \mathbf{R}, n|\psi\rangle \tag{8.44}$$

does not hold. Instead one derives for the matrix elements of $\hat{\mathbf{P}} = \hat{\mathbf{P}} \otimes 1^{\text{fast}}$:

---

[12] The value of the parameter $\mathbf{R}$ in both operators and vectors must be the same. $h(\mathbf{R})$ applied to $|n(\mathbf{R}')\rangle$, with $\mathbf{R}' \neq \mathbf{R}$, is a complicated expression obtained by expanding $|n(\mathbf{R}')\rangle$ with respect to the basis $\{|n(\mathbf{R})\rangle\}$ in $\mathcal{H}^{\text{fast}}$.

160   8. The Gauge Theory of Molecular Physics

$$\langle \mathbf{R}, n | \hat{\mathbf{P}} | \psi^E \rangle$$
$$= \langle \mathbf{R}, n | \hat{\mathbf{P}} \sum_{n'} \int d^{3N_n - 3} R' | \mathbf{R}', n' \rangle \langle \mathbf{R}' n' | \psi^E \rangle$$
$$= \left(\langle \mathbf{R} | \tilde{\otimes} \langle n(\mathbf{R}) | \right) \left(\hat{\mathbf{P}} \otimes 1^{\text{fast}} \right) \sum_{n'} \int d^{3N_n - 3} R' \left(| \mathbf{R}' \rangle \tilde{\otimes} | n'(\mathbf{R}') \rangle \right) \langle \mathbf{R}', n' | \psi^E \rangle.$$

Here we have used (8.40a) and the fact that $|\mathbf{R}, n\rangle$ form a complete system of basis vectors in $\mathcal{H} = \mathcal{H}^{\text{slow}} \otimes \mathcal{H}^{\text{fast}}$,

$$\sum_{n=1}^{\mathcal{N}} \int d^{3N_n - 3} R | \mathbf{R}, n \rangle \langle \mathbf{R}, n | = 1.$$

The operator $\hat{\mathbf{P}}$ acts only in $\mathcal{H}^{\text{slow}}$, and we obtain

$$\langle \mathbf{R}, n | \hat{\mathbf{P}} | \psi^E \rangle = \langle n(\mathbf{R}) | \sum_{n'} \int d^{3N_n - 3} R' | n'(\mathbf{R}') \rangle \langle \mathbf{R} | \hat{\mathbf{P}} | \mathbf{R}' \rangle \langle \mathbf{R}', n' | \psi^E \rangle.$$

The position representation of $\hat{\mathbf{P}}$ in $\mathcal{H}^{\text{slow}}$ is given by $\frac{1}{i} \nabla_{\mathbf{R}}$. Therefore[13]

$$\langle \mathbf{R}, n | \hat{\mathbf{P}} | \psi^E \rangle$$
$$= \langle n(\mathbf{R}) | \int d^{3N_n - 3} R' \sum_m | n'(\mathbf{R}') \rangle \left[\frac{-1}{i} \nabla_{\mathbf{R}'} \delta^{3N_n - 3}(\mathbf{R}' - \mathbf{R}) \right] \langle \mathbf{R}', n' | \psi^E \rangle$$
$$= \frac{1}{i} \sum_{n'} \langle n(\mathbf{R}) | \nabla_{\mathbf{R}} | n'(\mathbf{R}) \rangle \langle \mathbf{R}, n' | \psi^E \rangle. \tag{8.45}$$

Note that the differentiation $\nabla_{\mathbf{R}}$ in (8.45) applies to both $|n'(\mathbf{R})\rangle$ and $\langle \mathbf{R}, n' | \psi^E \rangle$. Hence,

$$\langle \mathbf{R}, n | \hat{\mathbf{P}} | \psi^E \rangle$$
$$= \sum_{n'} \left(\langle n(\mathbf{R}) | n'(\mathbf{R}) \rangle \frac{1}{i} \nabla_{\mathbf{R}} \langle \mathbf{R}, n' | \psi^E \rangle - i \left(\langle n(\mathbf{R}) | \nabla_{\mathbf{R}} | n'(\mathbf{R}) \rangle \right) \langle \mathbf{R}, n' | \psi^E \rangle \right).$$

Now using the orthogonality (8.40b), we write this as

---

[13] Precisely what we calculate is

$$\int dx' \langle x | \hat{P}_x | x' \rangle \varphi(x') = \int dx' \left(\frac{1}{i} \frac{d}{dx} \delta(x - x') \right) \varphi(x')$$
$$= \int dx' \frac{-1}{i} \delta'(x' - x) \varphi(x') = \frac{-1}{i} (-1) \varphi'(x),$$

where $\varphi(x)$ is a well-behaved function. In this derivation we have used results of [86, §I.2].

$$\langle \mathbf{R}, n | \hat{\mathbf{P}} | \psi^E \rangle = \sum_{n'} \left(\frac{1}{i} \nabla_{\mathbf{R}} \delta^{nn'} - \mathbf{A}^{nn'}(\mathbf{R}) \right) \langle \mathbf{R}, n' | \psi^E \rangle,$$

$$= \sum_{n'} \frac{1}{i} \mathbf{D}^{nn'} \langle \mathbf{R}, n' | \psi^E \rangle, \quad (8.46)$$

where we have introduced

$$\mathbf{A}^{nn'}(\mathbf{R}) := i \langle n(\mathbf{R}) | \nabla_{\mathbf{R}} | n'(\mathbf{R}) \rangle, \quad (8.47)$$

$$\frac{1}{i} \mathbf{D}^{nn'} := \frac{1}{i} \nabla_{\mathbf{R}} \delta^{nn'} - \mathbf{A}^{nn'}(\mathbf{R}). \quad (8.48)$$

This calculation shows that if n is the eigenvalue of an $\mathbf{R}$-dependent observable, the matrix elements of the operator $\hat{\mathbf{P}}$ for an arbitrary $\psi \in \mathcal{H}$ are *not* given by the derivative (8.44) but by the more complicated expression (8.46).

$\mathbf{A}^{nn'}$ and $\mathbf{D}^{nn'}$ appearing on the right-hand side of (8.46) respectively define the $\mathcal{N} \times \mathcal{N}$ matrix $\mathbf{A}$ and the $\mathcal{N} \times \mathcal{N}$ matrix-valued differential operator $\mathbf{D}$, where $\mathcal{N} := \dim \mathcal{H}^{\text{fast}}$ (with $\mathcal{N} = \infty$ in the exact theory). $\mathbf{A}$ is an example of a (non-Abelian for $\mathcal{N} \neq 1$) gauge potential, whereas $\mathbf{D}$ is the corresponding gauge-covariant derivative (5.77). The nomenclature has its origin in the gauge theory that underlies the new Born–Oppenheimer description (which does not use the conventional Born–Oppenheimer approximation). This will be discussed in detail in Sect. 8.4.

The gauge potential (8.47), for the case $\mathcal{N} = 1$, was originally introduced in [165] and [167]. It has the same form as the Mead–Berry vector potential[14] that we defined in (2.54) and used in (3.30) for a quantum system in a classical environment. In the quantum theory of the molecule, the fast quantum system is the electron(s) and the place of the slowly changing classical environmental parameters is taken by the eigenvalues of the observable $\mathbf{R}$ which determines the relative position of the nuclei. In Sect. 8.2 the adiabaticity assumption restricted us to one fixed value of the quantum number n, and we considered only the diagonal matrix elements. Here, we see that this is not necessary and the gauge potential is an $\mathcal{N} \times \mathcal{N}$ matrix where the value of $\mathcal{N}$ is determined by the required approximation. The Mead–Berry connection (8.47) describes many observable effects which could not be explained in the conventional Born–Oppenheimer approximation which consists of omitting the Mead–Berry potential in (8.46) and the appropriate places in the equations below.

Following the same approach as the one we pursued in the derivation of (8.46), we can compute the matrix elements of $\hat{\mathbf{P}}^2 = \hat{\mathbf{P}}^2 \otimes 1^{\text{fast}}$. The result is

---

[14] For $\mathcal{N} \neq 1$, the gauge potential is the non-Abelian analog (7.15) of the Mead–Berry vector potential (connection).

$$\langle \mathbf{R}, n | \hat{\mathbf{P}}^2 | \psi^E \rangle$$
$$= -\sum_{n'} \langle n(\mathbf{R}) | \nabla_\mathbf{R}{}^2 | n'(\mathbf{R}) \rangle \langle \mathbf{R}, n' | \psi^E \rangle$$
$$= \sum_{n',\ell} \left(\frac{1}{i} \delta^{n\ell} \nabla_\mathbf{R} - \mathbf{A}^{n\ell} \right) \cdot \left(\frac{1}{i} \delta^{\ell n'} \nabla_\mathbf{R} - \mathbf{A}^{\ell n'} \right) \langle \mathbf{R} n' | \psi^E \rangle. \quad (8.49)$$

Comparing (8.44) and (8.46), we see the difference between the usual or naively expected representations of the momentum operator and its correct representation. The difference arises from the $\mathbf{R}$-dependence of the eigenvectors $|n(\mathbf{R})\rangle$. If these basis vectors are independent of $\mathbf{R}$, then the $\mathbf{A}^{nn'}(\mathbf{R})$ vanish and the momentum operator is represented by the ordinary derivative. If the dependence of these eigenvectors on the slow variable $\mathbf{R}$ is not neglected, then the momentum operator is represented by the covariant derivative. The standard assumption of the ordinary Born–Oppenheimer approximation is that $|n(\mathbf{R})\rangle$ is a slowly changing function of $\mathbf{R}$ so that its derivative can be neglected.

We will now write the two eigenvalue equations (8.32) and (8.42) in the form of differential equations.

We can expand the eigenvector $|\psi^E\rangle$ of the molecular Hamiltonian in (8.32) with respect to the basis system $\{|\mathbf{R}, n\rangle\}$ of (8.35),

$$|\psi^E\rangle = \sum_n \int d^{3N_n - 3} R' |\mathbf{R}', n\rangle \langle \mathbf{R}', n | \psi^E \rangle. \quad (8.50)$$

The sum extends over all values of the electronic quantum numbers n which are in general infinite but often a small number $\mathcal{N}$ of terms in (8.50) provides a good approximation.

The wave function of the energy eigenvector $|\psi^E\rangle$ is its scalar product with the basis vector $|\mathbf{R}, \mathbf{r}\rangle$ of (8.33a),

$$\psi^E(\mathbf{R}, \mathbf{r}) := \langle \mathbf{R}, \mathbf{r} | \psi^E \rangle. \quad (8.51)$$

Next, we define
$$\psi_n^E(\mathbf{R}) := \langle \mathbf{R}, n | \psi^E \rangle, \quad (8.52)$$

and
$$\phi_n(\mathbf{R}, \mathbf{r}) := \langle \mathbf{r} | n(\mathbf{R}) \rangle. \quad (8.53)$$

If we take the scalar product of (8.50) with the basis vector $|\mathbf{R}, \mathbf{r}\rangle$ and use (8.39), we obtain
$$\psi^E(\mathbf{R}, \mathbf{r}) = \sum_n \phi_n(\mathbf{R}, \mathbf{r}) \psi_n^E(\mathbf{R}). \quad (8.54)$$

The functions $\phi_n(\mathbf{R}, \mathbf{r})$ are called the electronic eigenfunctions. The functions $\psi_n^E(\mathbf{R})$ are called the nuclear wave functions, though they are not wave functions in the usual sense in the slow space $\mathcal{H}^{\text{slow}}$. (Usually by wave function

8.3 The Born–Oppenheimer Method

in the position representation in the space $\mathcal{H}^{\text{slow}}$ one means the component $\langle\mathbf{R}|\psi\rangle$ of the vector $\psi \in \mathcal{H}^{\text{slow}}$ with respect to the basis system $|\mathbf{R}\rangle$, whereas the $\psi_n^E(\mathbf{R})$ are defined by (8.52).)

If we take the scalar product of (8.42) with the position eigenket $|\mathbf{r}\rangle$ of $\mathcal{H}^{\text{fast}}$ and use (8.30), we find

$$\langle\mathbf{r}|h(\mathbf{R})|n(\mathbf{R})\rangle = \langle\mathbf{r}|\left[\frac{\hat{\mathbf{p}}^2}{2m_e} + V(\mathbf{R},\hat{\mathbf{r}})\right]|n(\mathbf{R})\rangle$$

$$= \left[-\frac{\nabla_r^2}{2m_e} + V(\mathbf{R},\mathbf{r})\right]\langle\mathbf{r}|n(\mathbf{R})\rangle$$

$$= \varepsilon_n(\mathbf{R})\langle\mathbf{r}|n(\mathbf{R})\rangle. \tag{8.55}$$

Using (8.53), we can write this equation in the form

$$\left[-\frac{\nabla_r^2}{2m_e} + V(\mathbf{R},\mathbf{r})\right]\phi_n(\mathbf{R},\mathbf{r}) = \varepsilon_n(\mathbf{R})\phi_n(\mathbf{R},\mathbf{r}). \tag{8.56}$$

This is the time-independent Schrödinger equation for the electronic wave function which depends parametrically on the eigenvalue $\mathbf{R}$ of the slow observable $\hat{\mathbf{R}}$.

If we take the scalar product of both sides of (8.32) with the basis vectors $|\mathbf{R}, n\rangle$ of (8.35) and (8.36), then in view of (8.29) and (8.36) we obtain

$$\langle\mathbf{R}, n|\frac{\hat{\mathbf{P}}^2}{2\mu}|\psi^E\rangle + \varepsilon_n(\mathbf{R})\psi_n^E(\mathbf{R}) = E\psi_n^E(\mathbf{R}). \tag{8.57}$$

The first matrix element in the above equation is given by (8.49). In view of this equation, we have

$$\sum_m\left[\sum_l\frac{1}{2\mu}\left(\delta^{nl}\frac{\nabla_\mathbf{R}}{i} - \mathbf{A}^{nl}\right)\cdot\left(\delta^{lm}\frac{\nabla_\mathbf{R}}{i} - \mathbf{A}^{lm}\right) + \varepsilon_n(\mathbf{R})\delta^{nm}\right]\psi_m^E(\mathbf{R})$$

$$= E\psi_n^E(\mathbf{R}). \tag{8.58}$$

This is a system of an infinite number of coupled differential equations.

If we use in place of (8.46) the relation (8.44), which only holds for the case that the basis vectors $|n(\mathbf{R})\rangle$ are independent of $\mathbf{R}$, then we derive in place of (8.58) the following equation for every value n of the electronic quantum number.

$$\left[\frac{-1}{2\mu}\nabla_\mathbf{R}^2 + \varepsilon_n(\mathbf{R})\right]\psi_n^E(\mathbf{R}) = E\psi_n^E(\mathbf{R}). \tag{8.59}$$

Equation (8.59) is the conventional Born–Oppenheimer approximation for the Schrödinger equation of the slow motion.

In the following we shall first discuss the solution of the conventional Born–Oppenheimer problem which uses (8.56) and (8.59). The first step is to solve (8.56) for all values of the parameter **R**. This is done using various approximation methods and empirical assumptions [110, 240]. Once the eigenvalues $\varepsilon_n(\mathbf{R})$ are obtained as functions of **R**, they are treated as potential energy functions (potential energy surface PESs) in the Schrödinger equation (8.59). For this reason, one calls $\varepsilon_n(\mathbf{R})$ the induced scalar potential. It is the potential that is induced by the electrons in the n-th electronic state in the dynamical equation for the slow or collective motion.

Comparing (8.9) with (8.30) we see that $\varepsilon_n(\mathbf{R})$ corresponds to the classical potential function $U_n(\mathbf{Q})$. For a stable molecule, $\varepsilon_n(\mathbf{R})$ has a minimum at the position $\mathbf{Q} = \mathbf{Q}_e$, called the equilibrium position, and behaves like a harmonic oscillator in the neighborhood of $\mathbf{Q}_e$. See Fig. 8.1.

As we have already mentioned, in practice one uses various approximation methods to calculate $\varepsilon_n(\mathbf{R})$ by solving (8.56). Here we shall assume that this has been done and that $\varepsilon_n(\mathbf{R})$ is given to us for the values of the electronic quantum number n that we are concerned with. We shall then use this potential function $\varepsilon_n(\mathbf{R})$ for the solution of the Schrödinger equation (8.59) or (8.58) of the collective motion of the molecule. We shall present more details in the example of the next section. Here we shall describe the general picture.

If we treat (8.59) as the Schrödinger equation whose solution yields the energy spectrum of the collective motion,[15] then all the energy values E correspond to a single electronic state with a definite value of the electronic quantum number $n = A$ or $n = B$, etc., and the energy values E of the molecule are therefore labeled by this electronic quantum number $E_{n=A}$ or $E_{n=B}$, etc. In addition to the electronic quantum number n there is the quantum number N which distinguishes different solutions of (8.59) for a fixed value of n. This means that for example the electronic energy level $E_{n=A}$ splits into sublevels that are associated with different solutions of (8.59) for $n = A$. The standard picture of molecular spectra is discussed in [43, §III.5]: Each electronic state $n = A, B, \cdots$ splits into vibrational excitations and then into rotational bands. See Fig. 8.3.

Since all eigenvalues in (8.59) correspond to one electronic quantum number, say $n = A$, there is only one electronic state contributing to the eigenvector ψ^E in (8.50).[16] Thus,

$$\psi^E = \psi^{E_A} = \int d^{3N_n - 3} R' |\mathbf{R}', A\rangle \langle \mathbf{R}', A | \psi^{E_A} \rangle, \qquad (8.60)$$

and all other expansion coefficients in (8.50) are zero

$$\langle \mathbf{R}, n | \psi^E \rangle = 0 \quad \text{for} \quad n \neq A. \qquad (8.61)$$

In contrast to the conventional Born–Oppenheimer approximation (8.59), the exact[17] Schrödinger equation (8.58) is an infinite-dimensional matrix differential equation. In addition to the induced scalar potentials $\varepsilon_n(\mathbf{R})$, this

---

[15] This is the situation in the conventional Born–Oppenheimer approximation.
[16] Note that the $\varepsilon_n(\mathbf{R})$ are determined by solving (8.56).
[17] It is exact if the $\varepsilon_n(\mathbf{R})$ are exactly known.

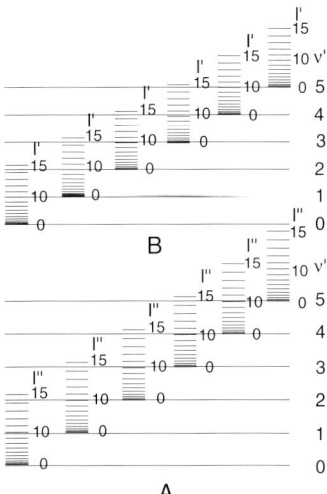

Fig. 8.3. Schematic picture of energy levels of a molecule. A and B denote two different electronic states. The vibrational quantum number is denoted by ν' or ν'', and the rotational quantum number is l' or l''.

equation also contains induced vector potentials $\mathbf{A}^{nm}(\mathbf{R})$ which couple all electronic quantum numbers to one given energy value E. That is in general the sum in (8.50) has an infinite number of non-zero terms.

The system of coupled differential equations (8.58) is of no practical value unless the infinite sums can be replaced by finite sums with a small number $\mathcal{N}$ of terms. This means that (8.50) and (8.54) can be approximated by a sum of $\mathcal{N}$ terms.[18] The simplest case is of course the case (8.60) with $\mathcal{N}=1$. More common is the case that two or three electronic states contribute. Of special interest is the case where two electronic quantum numbers n_1 and n_2 with non-zero contribution to (8.50) and (8.54) correspond to the same potential curve $\varepsilon_{n_1}(\mathbf{R}) = \varepsilon_{n_2}(\mathbf{R})$. In these approximations, which are often quire reliable, (8.58) takes the form of the $\mathcal{N}$-dimensional matrix equation:

$$\sum_m^{\mathcal{N}} \left[\sum_l^{\mathcal{N}} \frac{-1}{2\mu} \mathbf{D}^{nl} \cdot \mathbf{D}^{lm} + \tilde{\varepsilon}^{nm}(\mathbf{R}) \right] \psi_m^E(\mathbf{R}) = E\psi_n^E(\mathbf{R}), \qquad (8.62)$$

where

$$\tilde{\varepsilon}^{nm}(\mathbf{R}) := \varepsilon_n(\mathbf{R})\delta^{nm} + \frac{1}{2\mu} \sum_{l'=\mathcal{N}+1}^{\infty} \mathbf{A}^{nl'} \cdot \mathbf{A}^{l'm}, \qquad (8.63)$$

and $\mathbf{D}^{n\ell}$ and $\mathbf{A}^{n\ell}$ are respectively the covariant derivatives (8.48) and the Mead–Berry vector potential (8.47) with $n, \ell = 1, 2, \cdots \mathcal{N}$.

---

[18] This is called the Born–Huang approximation [47].

The Schrödinger equation (8.62) defines a Hamiltonian H_{eff} that governs the motion of the slow system. In the position representation of the slow system, H_{eff} is given by

$$H_{\text{eff}}^{nm}(\mathbf{R}) := \langle n(\mathbf{R})|H|m(\mathbf{R})\rangle = \frac{-1}{2\mu}\sum_l \mathbf{D}^{nl}\cdot\mathbf{D}^{lm} + \tilde{\varepsilon}^{nm}(\mathbf{R}). \qquad (8.64)$$

It is an $\mathcal{N}\times\mathcal{N}$ matrix called the effective Hamiltonian. The rows and columns of H_{eff} are labeled by the fast quantum numbers, and its matrix elements are differential operators acting on the functions $\psi_n^E(\mathbf{R}) = \langle \mathbf{R}, n|\psi^E\rangle$. (Note that the $\psi_n^E(\mathbf{R})$ are not wave functions in the slow space $\mathcal{H}^{\text{slow}}$ by which one means the component $\langle \mathbf{R}|\psi\rangle$ with respect to the basis system $|\mathbf{R}\rangle$.)

Equations (8.58) and (8.62) have been written in a form reminiscent of a gauge theory with $\mathbf{A}$ defined by (8.47) being the gauge potential and $\mathbf{D}$ defined by (8.48) being the gauge-covariant derivative. In matrix notation, (8.62) becomes

$$\left[\frac{-1}{2\mu}\mathbf{D}\cdot\mathbf{D} + \tilde{\varepsilon}(\mathbf{R})\right]\psi^E(\mathbf{R}) = E\psi^E(\mathbf{R}), \qquad (8.65)$$

where $\mathbf{D}$ and $\tilde{\varepsilon}(\mathbf{R})$ are matrices whose entries are respectively given by (8.48) and (8.63)), and $\psi^E(\mathbf{R})$ is a column ($\nu \times 1$) matrix whose rows are labeled by n.

8.4 The Gauge Theory of Molecular Physics

In (8.47) and (8.48) we have introduced quantities which appear in every gauge theory. Using these quantities we have written the the Schrödinger equations (8.62) and (8.65) for the collective motion of the molecule in a gauge-covariant way. The reason for this is that the Born–Oppenheimer description leads in a natural way to a gauge theory (provided that one does *not* use the conventional, drastic Born–Oppenheimer approximation). That the appearance of the Berry phase is a feature of an underlying gauge theory was already indicated in Chaps. 2, 5, and 6. In Chap. 2, above (2.69), we mentioned the analogy between the electromagnetic quantities of Maxwell's theory and the Mead–Berry vector potential (connection) and gauge field strength (curvature). With the Born–Oppenheimer wave function for the collective motion we have now also introduced the matter field of a gauge theory. For $\mathcal{N} > 1$, this is a non-Abelian gauge theory.

In order to devise *the gauge theory of molecular physics* we must choose for the ingredients of the abstract gauge theory defined in Sect. 5.5, the concrete physical quantities of the molecule which we discussed in Sect. 8.3, and show that they fulfill the defining relations (5.62), (5.78), etc., of the abstract quantities.

8.4 The Gauge Theory of Molecular Physics

We will distinguish three cases. The first case corresponds to the exact description where the sum in the expansions (8.50) and (8.54) extends over a complete system of fast basis vectors. As $\mathcal{H}^{\text{fast}}$ is in general infinite-dimensional, the sums in (8.50), (8.54), and (8.58) are also infinite. In this case the gauge covariant field strength F turns out to be zero.[19] The second and third cases, which are of practical importance, correspond to the Born–Huang approximation where the sum in (8.50) and (8.54) extends over an incomplete system $n = 1, 2, \ldots, \mathcal{N}$ with $\mathcal{N}$ a small number. It turns out that for these cases the field strength F is in general non-zero. We differentiate between the second and third cases according to the degeneracy properties of the energy levels.

In the second case, which is relevant to the Λ-doubling problem in diatoms, we will consider $\mathcal{N}$-fold degenerate levels

$$\varepsilon_1(\mathbf{R}) = \varepsilon_2(\mathbf{R}) = \cdots \varepsilon_n(\mathbf{R}) = \cdots \varepsilon_\mathcal{N}(\mathbf{R}) =: \varepsilon(\mathbf{R}). \tag{8.66}$$

This case is quite analogous to the systems we studied in Chap. 7. It might be viewed as a realization of the non-Abelian gauge theory of quantum mechanics that we introduced in Chap. 7.

The third case, which will be discussed in detail in the next chapter, is similar to the second case except that the degeneracy (crossing of potential energy surfaces) occurs only at certain points $\mathbf{R}_0$ of the nuclear coordinate space

$$\varepsilon_i(\mathbf{R}_0) = \varepsilon_j(\mathbf{R}_0). \tag{8.67}$$

The crossing is usually not isolated but a part of a line or a higher dimensional geometric object. The crossing point usually forms a conical intersection, and the molecular Aharonov–Bohm effect, i.e., the sign-change of the electronic wave function when it is transported around the crossing point, occurs

In each case the ingredients of gauge theory are given by the same physical quantities and satisfy similar equations. The main difference is in the dimension $\mathcal{N}$ of the matrices (alternatively the range of the summation in the matrix equations.) In the first case $\mathcal{N} = \dim \mathcal{H}^{\text{fast}} (= \infty)$, whereas in the second and third cases $\mathcal{N}$ is a small number.

In the gauge theory (5.62) of molecular physics, the matter field is the so-called nuclear wave function $\psi_n^E(\mathbf{R})$, and the gauge transformation is defined by the unitary matrix $\mathcal{U}(\mathbf{R}) \in U(\mathcal{N})$ which depends on the parameters of the slow motion $\mathbf{R}$,

$$\psi_n^E(\mathbf{R}) \rightarrow \psi_{n'}^{\prime E}(\mathbf{R}) = \sum_l \mathcal{U}^{n'l}(\mathbf{R}) \psi_l^E(\mathbf{R}). \tag{8.68}$$

The gauge potentials are the $\mathcal{N} \times \mathcal{N}$ matrix-valued Mead–Berry vector potentials (8.47). They are defined in terms of the basis vectors for the fast motion

---

[19] But this does not mean we only have trivial results. If the relevant parameter space is multiply-connected, Aharonov–Bohm type effects may occur.

but occur in the equations for the slow motion, e.g., in (8.62). The covariant derivatives of the molecular gauge theory are defined in the usual way (8.48) in terms of the Mead–Berry vector potentials.

We begin our discussion of the gauge theory of molecular physics by showing that the $\mathbf{A}$ defined by (8.47) is really a gauge potential, i.e., under the gauge transformation (8.68) it transforms according to (5.78). This follows from the postulated principle that the wave function of the total system $\psi^E(\mathbf{R},\mathbf{r}) = \langle \mathbf{R},\mathbf{r}|\psi^E\rangle$ must not be affected by the gauge transformation

$$\psi^E(\mathbf{R},\mathbf{r}) = \sum_n \phi_n(\mathbf{R},\mathbf{r})\psi_n^E(\mathbf{R}) = \sum_{n'} \phi'_{n'}(\mathbf{R},\mathbf{r})\psi'^E_{n'}(\mathbf{R}). \tag{8.69}$$

In view of (8.68), this condition is fulfilled provided that the electronic wave functions $\phi_n(\mathbf{R},\mathbf{r})$ transform according to

$$\phi_n(\mathbf{R},\mathbf{r}) \to \phi'_{n'}(\mathbf{R},\mathbf{r}) = \sum_l \phi_l(\mathbf{R},\mathbf{r})\mathcal{U}^{\dagger ln'}(\mathbf{R}). \tag{8.70}$$

This means that the basis vectors $|n(\mathbf{R})\rangle$ of $\mathcal{H}^{\text{fast}}$ transform according to

$$|n(\mathbf{R})\rangle \to |n'(\mathbf{R})\rangle = \sum_l |l(\mathbf{R})\rangle \mathcal{U}^{\dagger ln'}(\mathbf{R}). \tag{8.71}$$

If we insert (8.71) into the definition (8.47) of $\mathbf{A}^{nm}(\mathbf{R})$, we obtain

$$\begin{aligned}\mathbf{A}^{nm} \to \mathbf{A}'^{n'm'}(\mathbf{R}) &= \mathcal{U}^{n'n}(\mathbf{R})i\langle n(\mathbf{R})|\left(\nabla_\mathbf{R}|m(\mathbf{R})\rangle \mathcal{U}^{\dagger mm'}(\mathbf{R})\right)\\ &= \mathcal{U}^{n'n}(\mathbf{R})i\langle n(\mathbf{R})|\nabla_\mathbf{R}|m(\mathbf{R})\rangle \mathcal{U}^{\dagger mm'}(\mathbf{R})\\ &\quad + \mathcal{U}^{n'n}(\mathbf{R})\langle n(\mathbf{R})|m(\mathbf{R})\rangle i\left(\nabla_\mathbf{R}\mathcal{U}^{\dagger mm'}(\mathbf{R})\right),\end{aligned} \tag{8.72}$$

where we have omitted the summation symbol and summed over the repeated indices $n, m = 1, \cdots, \mathcal{N}$ which should be understood hereafter. Now, using (8.47) and (8.40b), we have

$$\mathbf{A}'^{n'm'}(\mathbf{R}) = \mathcal{U}^{n'n}\mathbf{A}^{nm}(\mathbf{R})\,\mathcal{U}^{\dagger mm'} + i\mathcal{U}^{n'n}\left(\nabla_\mathbf{R}\mathcal{U}^{\dagger nm'}(\mathbf{R})\right). \tag{8.73}$$

This is the desired transformation property (5.78) with the coupling constant set to $g = -1$. From this it follows, as in any gauge theory, that $\mathbf{D}$ of (8.48) is a covariant derivative. That is it transforms according to

$$\mathbf{D}\psi^E \to \mathbf{D}'\psi'^E = \mathcal{U}\,\mathbf{D}\psi^E, \tag{8.74}$$

$$\frac{1}{i}\mathbf{D} \to \frac{1}{i}\mathbf{D}' = \mathcal{U}\,\frac{1}{i}\mathbf{D}\mathcal{U}^\dagger. \tag{8.75}$$

The corresponding gauge-covariant field strength (the curvature two-form) $\check{F}$ is defined in the standard way, i.e., according to (5.71). Its components are given by:

$$F_{ij}^{nm} = \partial_i A_j^{nm} - \partial_j A_i^{nm} - i\left[A_i, A_j\right]^{nm} = i\left[D_i, D_j\right]^{nm}, \qquad (8.76)$$

where $n, m = 1, 2, \cdots, \mathcal{N}$ with $\mathcal{N} = \dim \mathcal{H}^{\text{fast}}$ (in the first exact case) or over a small number of values (in the second case).

We next wish to consider the effect of the $U(\mathcal{N})$ gauge transformations on the Schrödinger equation for the collective motion (8.62). Using (8.68) and (8.75) in (8.62), we obtain (after multiplying by $\mathcal{U}$ from the left)

$$\left(\frac{-1}{2\mu} \mathbf{D}' \cdot \mathbf{D}' + \mathcal{U}\tilde{\varepsilon}\,\mathcal{U}^\dagger\right)\psi'^E = E\psi'^E. \qquad (8.77)$$

In this equation $\mathbf{D}'$ is the gauge transformed $\mathbf{D}$, i.e., the covariant derivative calculated from (8.47) and (8.48) using the transformed basis vectors $|n'(\mathbf{R})\rangle$; these have been proved to be related to the original $\mathbf{D}$ (calculated using the original $|n(\mathbf{R})\rangle$) according to (8.75). The ψ'^E are the gauge transformed ψ^E of (8.68). In order that (8.77) has the same form in terms of the transformed quantities as (8.62) in terms of the original quantities, the term $\mathcal{U}\tilde{\varepsilon}\,\mathcal{U}^\dagger$ must be $\tilde{\varepsilon}'$, the gauge transformed $\tilde{\varepsilon}$, i.e., the matrix (8.63) calculated in terms of the transformed basis vectors $|n'(\mathbf{R})\rangle$.

The matrix $\tilde{\varepsilon}$ of (8.63) consists of two terms. The second term is not present in the exact case (for which usually $U(\mathcal{N}) = U(\infty)$). We will discuss this case first. The matrix $\tilde{\varepsilon}(\mathbf{R})$ is then diagonal with the matrix elements

$$\tilde{\varepsilon}^{nm} = \varepsilon_n(\mathbf{R})\delta^{nm}. \qquad (8.78)$$

The summation in the above sums extends over a complete system of quantum numbers n for which the following completeness relation holds:

$$\sum_{n=1}^{\infty} |n(\mathbf{R})\rangle\langle n(\mathbf{R})| = 1^{\text{fast}}. \qquad (8.79)$$

Using this completeness relation and inserting (8.47) into (8.76), one can show by a straightforward calculation (Problem 8.3) that

$$F_{ij}^{nm} = 0, \qquad (8.80)$$

where all the derivatives $\nabla_{\mathbf{R}}|n(\mathbf{R})\rangle$ are assumed to be well defined. This means that in this case the gauge theory is trivial if the relevant manifold is simply connected. However, crossings of PESs (a degeneracy of ε_n at some points $\mathbf{R}$) may cause the effective manifold for the nuclear motion to be multiply connected. This is because such crossings usually make $\nabla_{\mathbf{R}}|n(\mathbf{R})\rangle$ ill defined at the crossing point of $\varepsilon_n(\mathbf{R})$. Therefore even though $F_{ij}^{nm} = 0$, a pure gauge vector potential may give rise to non-trivial Aharonov–Bohm type effects.

We shall now move on to the second case: the Born–Huang approximation by $\mathcal{N}$-fold degenerate levels in the electronic energy eigenfunctions. In the spirit of the Born–Oppenheimer description we want to use as basis vectors $|n(\mathbf{R})\rangle$ of $\mathcal{H}^{\text{fast}}$ only the eigenvectors of the fast Hamiltonian $h(\mathbf{R})$. We shall therefore allow in (8.71) only gauge transformations which transform one eigenvector of $h(\mathbf{R})$ into another eigenvector of $h(\mathbf{R})$. The $U(\mathcal{N})$ transformation in (8.71) must therefore transform within the same energy eigenspace $\mathcal{H}_\varepsilon \subset \mathcal{H}^{\text{fast}}$ of $h(\mathbf{R})$. This means that we have the subcase (8.66). (The other subcase in which $U(\mathcal{N})$ is not a symmetry group of $h(\mathbf{R})$ leads to non-diagonal $\tilde{\varepsilon}$ and non-trivial $\mathbf{A}$.)

If all the $\varepsilon_n(\mathbf{R})$ are non-degenerate then the largest gauge group is $U(1)$. Usually for diatomic molecules the $\varepsilon_n(\mathbf{R})$ are doubly degenerate (Λ-doublets, corresponding to electrons orbiting or spinning clockwise or counterclockwise around the internuclear axis). In this case the gauge group is $U(2)$. In general, the gauge group is $U(\mathcal{N})$, and the sum in (8.68), (8.69), (8.70), (8.71), etc., extends only over the $\mathcal{N}$ different electronic quantum numbers n with the same eigenvalue (8.66). In particular the system of basis vectors that occur in (8.76) is incomplete. It satisfies

$$\sum_{l=1}^{\mathcal{N}} |l(\mathbf{R})\rangle\langle l(\mathbf{R})| = \Lambda_{\mathcal{H}_\varepsilon}(\mathbf{R}), \tag{8.81}$$

where $\Lambda_{\mathcal{H}_{\varepsilon_n}}$ is the projection operator onto the $\mathcal{N}$-dimensional subspace $\mathcal{H}_\varepsilon$. From (8.81) it can no longer be proven that (8.80) holds. Therefore, the $\mathcal{N}$-dimensional gauge potential $\mathbf{A}^{nm}$ defined by (8.47) with eigenvectors $|n(\mathbf{R})\rangle$ of the fast Hamiltonian $h(\mathbf{R})$ cannot in general be made to vanish by a $U(\mathcal{N})$ gauge transformation.

The matrix $\tilde{\varepsilon}$ in (8.62) is now more complicated and contains the second term of (8.63) where the summation over l' extends over all those electronic quantum numbers which do not belong to the electronic energy $\varepsilon(\mathbf{R})$, i.e., $l' = \mathcal{N}+1, \mathcal{N}+2, \cdots, \infty$.

Next, we write (8.63) in the form

$$\tilde{\varepsilon}^{nm}(\mathbf{R}) = \varepsilon(R)\langle n(\mathbf{R})|m(\mathbf{R})\rangle$$
$$+ \frac{i^2}{2\mu} \sum_{l=\mathcal{N}+1}^{\infty} \langle n(\mathbf{R})|\nabla|l(\mathbf{R})\rangle\langle l(\mathbf{R})|\nabla|m(\mathbf{R})\rangle, \tag{8.82}$$

$$= \varepsilon(\mathbf{R})\delta^{nm} - \frac{1}{2\mu}\tilde{\varepsilon}^{nm}, \quad n,m = 1,2,\cdots\mathcal{N}. \tag{8.83}$$

As the vectors $|l(\mathbf{R})\rangle$ in (8.82) of the subspace orthogonal to $\mathcal{H}_\varepsilon$ are not affected by our $U(\mathcal{N})$ gauge transformation, we can take any basis system for them and use the projection operator

$$\tilde{\Lambda} = 1 - \Lambda_{\mathcal{H}_\varepsilon} = \sum_{l=\mathcal{N}+1}^{\infty} |l(\mathbf{R})\rangle\langle l(\mathbf{R})|. \tag{8.84}$$

Then using the identity $\langle n(\mathbf{R})|\nabla_R|l(\mathbf{R})\rangle = -(\nabla_R\langle n(\mathbf{R})|)|l(\mathbf{R})\rangle$, we have

8.4 The Gauge Theory of Molecular Physics

$$\tilde{\varepsilon}^{nm} = -\sum_{l=\mathcal{N}+1}^{\infty} \left(\nabla_{\mathbf{R}} \langle n(\mathbf{R})|\right) l(\mathbf{R}) \rangle \langle l(\mathbf{R})| \nabla_{\mathbf{R}} | m(\mathbf{R})\rangle$$
$$= -\left(\nabla_{\mathbf{R}} \langle n(\mathbf{R})|\right) \tilde{\Lambda} \left(\nabla_{\mathbf{R}} | m(\mathbf{R})\rangle\right). \tag{8.85}$$

Substituting this equation in (8.83), we find

$$\tilde{\varepsilon}^{nm}(\mathbf{R}) = \varepsilon(\mathbf{R}) \langle n(\mathbf{R})|m(\mathbf{R})\rangle + \frac{1}{2\mu} \left(\nabla_{\mathbf{R}} \langle n(\mathbf{R})|\right) \tilde{\Lambda} \left(\nabla_{\mathbf{R}} | m(\mathbf{R})\rangle\right). \tag{8.86}$$

In particular, the matrix elements $\tilde{\varepsilon}^{nm}(\mathbf{R})$ of $\tilde{\varepsilon}(\mathbf{R})$ only depend on the basis vectors $|n(\mathbf{R})\rangle$ with $n = 1, \cdots, \mathcal{N}$.

Next, we consider the gauge transformed matrix $\tilde{\varepsilon}'$, i.e., the matrix (8.86) calculated in terms of the transformed $|n'(\mathbf{R})\rangle$:

$$\tilde{\varepsilon}'^{n'm'}(\mathbf{R}) = \varepsilon(\mathbf{R}) \langle n'(\mathbf{R})|m'(\mathbf{R})\rangle + \frac{1}{2\mu} \left(\nabla_{\mathbf{R}} \langle n'(\mathbf{R})|\right) \tilde{\Lambda} \left(\nabla_{\mathbf{R}} | m'(\mathbf{R})\rangle\right). \tag{8.87}$$

We can express these matrix elements in terms of the original matrix elements $\tilde{\varepsilon}^{nm}$ as we have done for the $\mathbf{A}'$ in (8.73). We first introduce

$$\tilde{\tilde{\varepsilon}}'^{n'm'} := -\left(\nabla_{\mathbf{R}} \langle n'(\mathbf{R})|\right) \tilde{\Lambda} \left(\nabla_{\mathbf{R}} | m'(\mathbf{R})\rangle\right), \tag{8.88}$$

and use (8.71) to calculate

$$\tilde{\tilde{\varepsilon}}'^{n'm'} = -\nabla_{\mathbf{R}} \left(u^{nn'}(\mathbf{R}) \langle n(\mathbf{R})|\right) \tilde{\Lambda} \nabla_{\mathbf{R}} \left(|m(\mathbf{R})\rangle u^{\dagger mm'}(\mathbf{R})\right)$$
$$= -\Big[u^{n'n} \left(\nabla_{\mathbf{R}} \langle n(\mathbf{R})|\right) \tilde{\Lambda} \left(\nabla_{\mathbf{R}} |m(\mathbf{R})\rangle\right) u^{\dagger mm'}$$
$$+ \left(\nabla_{\mathbf{R}} u^{n'n}\right) \langle n| \tilde{\Lambda} \left(\nabla_{\mathbf{R}} |m\rangle\right) u^{\dagger mm'}$$
$$+ u^{n'n} \left(\nabla_{\mathbf{R}} \langle n|\right) \tilde{\Lambda} |m\rangle \left(\nabla_{\mathbf{R}} u^{\dagger mm'}\right)$$
$$+ \left(\nabla_{\mathbf{R}} u^{n'n}\right) \langle n| \tilde{\Lambda} |m\rangle \left(\nabla_{\mathbf{R}} u^{\dagger mm'}\right)\Big].$$

The last three terms on the right-hand side of this equation are zero because $\tilde{\Lambda}$ is orthogonal to the $|n\rangle$ and $|m\rangle$ that occur in the above sums (and in the sum in (8.71)). Hence,

$$\tilde{\tilde{\varepsilon}}'^{n'm'} = -\sum_{n,m=1}^{\mathcal{N}} u^{n'n} \left(\nabla_{\mathbf{R}} \langle n(\mathbf{R})|\right) \tilde{\Lambda} \left(\nabla_{\mathbf{R}} | m(\mathbf{R})\rangle\right) u^{\dagger mm'},$$
$$= -\sum_{n,m=1}^{\mathcal{N}} u^{n'n} \sum_{l=\mathcal{N}+1}^{\infty} \left(\nabla_{\mathbf{R}} \langle n|\right) l\rangle \langle l| \left(\nabla_{\mathbf{R}} |m\rangle\right) u^{\dagger mm'}, \tag{8.89}$$

which means

$$\tilde{\varepsilon}'^{n'm'} = \mathcal{U}^{n'n}\tilde{\varepsilon}^{nm}\mathcal{U}^{\dagger mm'}. \tag{8.90}$$

The first term in (8.82) fulfills a similar relation trivially. Therefore,

$$\tilde{\varepsilon}'^{n'm'}(\mathbf{R}) = \mathcal{U}^{n'n}(\mathbf{R})\tilde{\varepsilon}^{nm}(\mathbf{R})\,\mathcal{U}^{\dagger mm'}(\mathbf{R}). \tag{8.91}$$

This is the analog of (8.73) and (8.75) for the matrix $\tilde{\varepsilon}$. It can be written in matrix notation in the form

$$\tilde{\varepsilon}'(\mathbf{R}) = \mathcal{U}(\mathbf{R})\tilde{\varepsilon}(\mathbf{R})\,\mathcal{U}^{\dagger}(\mathbf{R}).$$

Substituting this relation in (8.77), we find

$$\left(-\frac{1}{2\mu}\mathbf{D}'\mathbf{D}' + \tilde{\varepsilon}'(\mathbf{R})\right)\psi'^{E} = E\psi'^{E}. \tag{8.92}$$

Therewith we have shown that the Schrödinger equation (8.62) or (8.65) of the slow motion is independent of the choice of the basis vectors of $\mathcal{H}^{\text{fast}}$. It has the same form in terms of both the eigenvectors $|n(\mathbf{R})\rangle$ and the gauge transformed eigenvectors $|n'(\mathbf{R})\rangle$. Note that this is only the case if (8.66) holds.

Our analysis shows that in the case of an $\mathcal{N}$-fold degenerate electronic energy value we have a $U(\mathcal{N})$ gauge theory for the slow collective (nuclear) motion which leaves the Schrödinger equation unaltered. In this gauge theory the matter field is given by the so-called nuclear wave function (8.52) and the gauge potential is given by the Mead–Berry vector potential (8.47). This gauge theory is in general non-trivial.

In the case where $\mathcal{N} = 1$ the operator matrices D_i^{nm} and H_{eff}^{nm} become just operators. We call n_A the value of the electronic quantum number which is associated with the eigenvalue E. The wave functions $\langle \mathbf{R}, n | \psi^E \rangle = \psi_n^E(\mathbf{R})$ in (8.52) and (8.54) fulfill (8.61) as in the old Born–Oppenheimer approximation. For $n = n_A$ the Schrödinger equation (8.62) becomes

$$\left[\frac{-1}{2\mu}\mathbf{D}\cdot\mathbf{D} + \tilde{\varepsilon}(\mathbf{R})\right]\psi^{E}(\mathbf{R}) = E\psi^{E}(\mathbf{R}),$$

$$\left[\frac{-1}{2\mu}\left(\frac{\nabla_{\mathbf{R}}}{i} - \mathbf{A}\right)^{2} + \tilde{\varepsilon}(\mathbf{R})\right]\psi^{E}(\mathbf{R}) = E\psi^{E}(\mathbf{R}), \tag{8.93}$$

where we have defined

$$\mathbf{A} := \mathbf{A}^{n_A n_A} \tag{8.94}$$

$$\mathbf{D} := \mathbf{D}^{n_A n_A}, \tag{8.95}$$

$$\psi^E := \psi_{n_A}^E \tag{8.96}$$

$$\tilde{\varepsilon}(\mathbf{R}) := \varepsilon_{n_A}(\mathbf{R}) + \frac{1}{2\mu}\sum_{l\neq n_A}\mathbf{A}^{n_A l}(\mathbf{R})\cdot\mathbf{A}^{l n_A}(\mathbf{R}) \tag{8.97}$$

$$= \varepsilon_{n_A}(\mathbf{R})$$

$$+ \frac{1}{2\mu}\left(\nabla_{\mathbf{R}}\langle n_A(\mathbf{R})|\right)\cdot\left(\nabla_{\mathbf{R}}|n_A(\mathbf{R})\rangle\right) - \frac{1}{2\mu}\mathbf{A}(\mathbf{R})^2. \tag{8.98}$$

8.4 The Gauge Theory of Molecular Physics

Equation (8.93) is the new Schrödinger equation for the slow collective degrees of freedom that correspond to the one electronic state with quantum numbers $n = n_A$. This new Schrödinger equation uses the same assumption for the approximation of the wave function (8.60), (8.61) as the conventional Born–Oppenheimer equation but it does not ignore the **R** dependence of the eigenvectors $|n(\mathbf{R})\rangle$. As a consequence, this new Schrödinger equation differs from the conventional Born–Oppenheimer approximation (8.59) by two additional contributions. The first is the Mead–Berry potential **A** which is induced by the fast electronic motion. Since the curvature, which for $\mathcal{N} = 1$ has the form

$$F_{ij} = F_{ij}^{n_A n_A} = \partial_i A_j - \partial_j A_i, \tag{8.99}$$

is in general non-zero, there is no transformation by which $\mathbf{A} = \mathbf{A}^{n_A n_A}$ can be gauged away. This induced vector potential can make an important contribution and change the results of the conventional Born–Oppenheimer approximation. Its first contribution is to modify the momentum operators. The new "effective" momentum operator $\frac{1}{i}\mathbf{D}$ satisfies the new commutation relation

$$\left[\frac{1}{i}D_i, \frac{1}{i}D_j\right] = iF_{ij}, \tag{8.100}$$

which is a special case of (8.76). The second new contribution is the modification of the scalar potential by the second term in (8.97). In the following section, we will show that for a diatomic molecule this additional term in the scalar potential just contributes an irrelevant constant. The solutions of the equation (8.93) have to satisfy the following auxiliary conditions which follow from (8.62) for $n \neq n_A$,

$$\sum_l \mathbf{D}^{nl} \cdot \mathbf{D}^{l n_A} \psi^E(\mathbf{R}) = 0. \tag{8.101}$$

In the case that will be of interest to us in the following section, $\varepsilon_n(\mathbf{R})$ is doubly degenerate ($\mathcal{N} = 2$, $\varepsilon_n = \varepsilon_{n_+} = \varepsilon_{n_-}$) with the two states corresponding to electrons orbiting (or spinning) clockwise and counterclockwise around the internuclear axis (doubly degenerate Λ levels). Therefore, we have a 2×2 matrix equation given by (for $n = n_+, n_-$)

$$\sum_{m=n_+,n_-}\left[\sum_{l=n_+,n_-} \frac{-1}{2\mu}\mathbf{D}^{nl} \cdot \mathbf{D}^{lm} + \tilde{\varepsilon}^{nm}(\mathbf{R})\right] \psi_m^E(\mathbf{R}) = E\, \psi_n^E(\mathbf{R}), \tag{8.102}$$

where

$$\tilde{\varepsilon}^{nm}(\mathbf{R}) = \varepsilon_n(\mathbf{R})\delta^{nm} + \frac{1}{2\mu}\sum_{l \neq n_+, n_-} \mathbf{A}^{nl} \cdot \mathbf{A}^{lm}. \tag{8.103}$$

The covariant momenta are now 2×2 matrices whose commutator gives a non-zero field strength (curvature). We will evaluate these quantities in the following section for the example of doubly degenerate Λ-levels of diatomic molecules.

8.5 The Electronic States of Diatomic Molecule

In order to apply the new Born–Oppenheimer method to the diatomic molecule we have to discuss the properties of the electronic quantum numbers n in more details.[20]

The classification of the electronic states of diatomic molecules is in one respect similar to the classification of atomic energy levels, but in other respects it is quite different. The orbital angular momentum of the electrons is not a good quantum number, because in a diatomic molecule one does not have spherical symmetry but only axial symmetry about the internuclear axis $\mathbf{R} = \mathbf{R}_2 - \mathbf{R}_1$. We shall denote the unit vector in the direction of $\mathbf{R}$ by $\mathbf{e_R}$. The operator of electronic angular momentum squared, $\mathbf{L}_{\text{elec}}^2$, does not commute with the Hamiltonian h, whereas $\mathbf{L}_{\text{elec}} \cdot \mathbf{e_R}$ (and $(\mathbf{L}_{\text{elec}} \cdot \mathbf{e_R})^2$) commutes with h and is thus a good quantum number.

One, therefore, classifies the electronic states of a diatomic molecule according to the value of

$$\Lambda := +\sqrt{\text{eigenvalue of } (\mathbf{L} \cdot \mathbf{e_R})^2}. \tag{8.104}$$

This quantum number can take the values

$$\Lambda = 0, 1, 2, 3, \cdots, \tag{8.105}$$

and the corresponding molecular states are respectively called

$$\Sigma, \Pi, \Delta, \Phi \cdots, \tag{8.105'}$$

states (in analogy to the $S, P, D, F \ldots$ designation for atoms).

Since $\mathbf{L}_{\text{elec}} \cdot \mathbf{e_R}$ can have two eigenvalues, $+\Lambda$ and $-\Lambda$ for a given energy level (which depends only on Λ), the $\Pi, \Delta, \Phi \cdots$ states are doubly degenerate and the Σ states are non-degenerate. (The physical eigenstates are not eigenvectors of $\mathbf{L}_{\text{elec}} \cdot \mathbf{e_R}$ but eigenvectors of parity.[21] The physical states are thus linear combinations of two eigenvectors of $\mathbf{L}_{\text{elec}} \cdot \mathbf{e_R}$ with eigenvalue $+\Lambda$ and $-\Lambda$.)

The electron spin cannot be neglected in the classification of the electronic spectra of molecules. The electronic spin $\mathbf{S}_{\text{elec}}$ is the sum of the spins of each electron, the corresponding quantum number S can therefore be any integer or half-integer number depending on whether the number of electrons is even or odd.

The eigenvalue of $\mathbf{S}_{\text{elec}} \cdot \mathbf{e_R}$ is denoted by Σ and can take the following values

$$\Sigma = -S, \ -S+1, \ -S+2, \cdots, S. \tag{8.106}$$

---

[20] Readers familiar with this subject can omit this section. For more details consult [104, Vol. I].
[21] For a discussion of this point see [43, §V.4].

8.5 The Electronic States of Diatomic Molecule

Σ and Λ are not always well-defined separately. It can be that the spin–orbit coupling is so large that $\mathbf{L}_{\text{elec}}$ and $\mathbf{S}_{\text{elec}}$ are coupled to form the total electronic angular momentum $\mathbf{J} = \mathbf{L}_{\text{elec}} + \mathbf{S}_{\text{elec}}$ (this is called Hund's coupling case c). Then only

$$\Omega := \sqrt{\text{eigenvalue of } (\mathbf{J} \cdot \mathbf{e_R})^2} \tag{8.107}$$

is a good quantum number. We will consider the case where Λ and Σ separately are good quantum numbers (called Hund's coupling case a). Then

$$\Omega = |\Lambda + \Sigma|, \tag{8.108}$$

and for a fixed Λ (with $\Lambda \leq S$), Ω takes the $2S+1$ values

$$\Omega = \Lambda + S, \ \Lambda + S - 1, \cdots \Lambda - S. \tag{8.109}$$

The corresponding molecular states are denoted by $^{2S+1}(\Lambda)_\Omega$, where (Λ) stands for the symbols (8.105′) of the values (8.105) for Λ. Thus, e.g., the electronic states with $\Lambda = 1$ and $S = 1$ are denoted by $^3\Pi_0$, $^3\Pi_1$, $^3\Pi_2$ for the value $\Omega = 0 (\Sigma = -1)$, $\Omega = 1 (\Sigma = 0)$ and $\Omega = 2 (\Sigma = +1)$ respectively. They form the $^3\Pi$ multiplet. The quantum numbers $n_a \equiv {}^{2S+1}(\Lambda)_\Omega$ usually specify the electronic energy, i.e., the potential energy curves $\varepsilon_n(R)$ ($R = |\mathbf{R}|$). Sometimes, when the molecule has a center of inversion (homonuclear molecules like O_2), the additional quantum numbers g (gerade) and u (ungerade) appear with values $+1$ and -1 of this molecular fixed inversion operator (replacement of all electron coordinates by their negatives). Then the quantum numbers n_a which specify the potential energy curves $\varepsilon_n(R)$ are $n_a = {}^{2S+1}(\Lambda)_\Omega$, g or u: $\varepsilon_n(R) = \varepsilon_{n_a}(R)$.

In addition there are the degeneracy quantum numbers $n_b; n = (n_a; n_b)$, which the electronic energy does not depend upon. For the diatomic molecule this is the intrinsic parity with the eigenvalue $+1$ or -1 [43, §V.4]. The intrinsic parity is the eigenvalue π of the parity operator U_P (space fixed inversion operator) divided by $(-1)^j$, where j is the total angular momentum of the molecule, i.e., the angular momentum of the electronic and collective motions combined. The total angular momentum j is always (as a consequence of the isotropy of space) a good quantum number, and its component j_3 is also a good quantum number.

The total electronic angular momentum j_{el}, i.e., the eigenvalue $j_{\text{el}}(j_{\text{el}}+1)$ of $\mathbf{J}^2$, is not a good quantum number. The reason for this is that the electrons do not have spherical symmetry with respect to a center (like in atoms), but only axial symmetry with respect to the internuclear axis $\mathbf{e_R}$.

The basis vectors in the electronic space $\mathcal{H}^{\text{fast}}$ that depend on the internuclear vector $\mathbf{R}$ (as the slowly varying parameter) are thus labeled by the following quantum numbers

$$|n(\mathbf{R})\rangle = |n_a; n_b(\mathbf{R})\rangle = |\ {}^{2S+1}(\Lambda)_\Omega; \pm(\mathbf{R})\rangle. \tag{8.110}$$

The parameter $\mathbf{R}$ is the eigenvalue of $\hat{\mathbf{R}}$ and its components are given by $\mathbf{R} = (R\sin\theta\sin\varphi, R\sin\theta\cos\varphi, R\cos\theta)$.

The energy values $\varepsilon_{n_a}(R)$ depend upon $n_a = {}^{2S+1}(\Lambda)_\Omega$ and the degeneracy quantum number, $n_b = +1$ or -1, is the parity or intrinsic parity. Except for $\Lambda = 0$ (Σ states), when the intrinsic parity is either $+1$ or -1, the energy values have double degeneracy. This means both parity values $+$ and $-$ occur. Instead of using the parity as degeneracy quantum number, one can use the eigenvalue $k = +\Omega$ or $k = -\Omega$ of the operator $\mathbf{e_R} \cdot \mathbf{J}$. This means instead of choosing a basis consisting of the parity eigenvectors, a basis of eigenvectors of the operator $\mathbf{e_R} \cdot \mathbf{J}$ can be chosen. We denote these eigenvectors as $|n_a, k(R, \theta, \varphi)\rangle$ (θ and φ are the polar coordinates of the internuclear axis). For brevity we drop the labels n_a

$$\mathbf{e_R} \cdot \mathbf{J}|k(R, \theta, \varphi)\rangle = k|k(R, \theta, \varphi)\rangle. \tag{8.111}$$

The parity eigenvectors are linear combinations of these vectors

$$\begin{aligned}|{}^{2S+1}(\Lambda)^\pm_\Omega(R,\theta,\varphi)\rangle &:= |\pi = \pm; (R,\theta,\varphi)\rangle, \\ &= \frac{1}{\sqrt{2}}[|k{=}\Omega\ (R,\theta,\varphi)\rangle \pm |k{=}{-}\Omega\ (R,\theta,\varphi)\rangle]. \end{aligned} \tag{8.112}$$

We omit all the labels ${}^{2S+1}(\Lambda)^\pm_\Omega$ which are fixed in a given multiplet and we choose the notation $k = \pm\Omega = \pm|\Sigma + \Lambda|$. In general Ω can take any of its possible values: $(\frac{1}{2}, 1, \frac{3}{2}, 2\ldots)$; for a given electronic energy level it has a fixed value. In order to have a concrete example we choose $\Sigma = 0$ and $\Lambda = 1$ (i.e., $(\Lambda) = \Pi$), then $k = \pm 1$. The value $k = \Omega$ corresponds to an electron rotating counterclockwise around the internuclear axis and $k = -\Omega$ corresponds to a clockwise rotating electron. The diatomic molecule is thus a realization of the model of a dumbbell with flywheel [43, §V.4].

8.6 The Monopole of the Diatomic Molecule

The vectors $|k(R, \theta, \varphi)\rangle$ of (8.111) are familiar from Chap. 3. They are analogs of the vectors $|k; R\rangle$ of (3.11) which were eigenvectors of a Hamiltonian describing a magnetic moment in a rotating magnetic field. The vectors (8.111) are electronic energy eigenvectors but the eigenvalue k is now not an eigenvalue connected with the electronic energy $\varepsilon_{n_a}(R)$ as was the case in Chap. 3. Otherwise, the mathematics is the same as in Chap. 3 and also the physics is related. In place of the precessing external magnetic field we have the magnetic field in the direction of the internuclear axis created by the rapidly rotating electronic charges. In place of the magnetic moment of the spinning quantum system in Chap. 3 we have here the magnetic moment connected with the angular momentum of the electrons. Thus, in a certain sense, the diatomic dumbbell molecule with flywheel is the quantum analog of the

8.6 The Monopole of the Diatomic Molecule

quantum system in a slowly changing classical environment as discussed in Chaps. 2 and 3. It is a quantum system in a slowly changing quantum environment. The results here will, therefore, be reminiscent of the results we had previously obtained in Chaps. 2 and 3.

The calculation of the Mead–Berry connection proceeds in the same way as in Chap. 3, the only difference being that k now takes the two values $\pm\Omega$ and we obtain 2×2 matrices for the non-Abelian connections. Hence, the subject of the previous chapter, specially Sect. 7.2 is directly relevant to our discussion.

We start with the vector $|k(R, \mathbf{e}_3)\rangle$ which is an eigenvector of $\mathbf{e}_\mathbf{R} \cdot \mathbf{J}$ when the internuclear axis points along the space fixed 3-axis $\mathbf{e}_3$.

$$\mathbf{e}_3 \cdot \mathbf{J} |k(R, \mathbf{e}_3)\rangle = J_3 |k(R, \mathbf{e}_3)\rangle = k |k(R, \mathbf{e}_3)\rangle. \tag{8.113}$$

Note that $|k(R, \mathbf{e}_3)\rangle$ does not have to be, and in general is not, an eigenvector of $\mathbf{J}^2$ (because the total electronic angular momentum is not a good quantum number). It may only be an eigenvector of the electronic angular momentum component J_3. Starting with the vector $|k(R, \mathbf{e}_3)\rangle$, we define – as in Chap. 3 – a vector-valued function of (θ, φ) by

$$|k(R, \theta, \varphi)\rangle := U(\theta, \varphi)|k(R, \mathbf{e}_3)\rangle = e^{-i\varphi J_3} e^{-i\theta J_2} e^{i\varphi J_3} |k(R, \mathbf{e}_3)\rangle. \tag{8.114}$$

This defines a unique vector for all (θ, φ) *except* at $\theta = \pi$ (south pole). The other parameterization is given by

$$|k(R, \theta, \varphi)\rangle' := e^{-i\varphi J_3} e^{-i\theta J_2} e^{-i\varphi J_3} |k(R, \mathbf{e}_3)\rangle = e^{-i2k\varphi} |k(R, \theta, \varphi)\rangle, \tag{8.115}$$

which defines a unique vector for all (θ, φ) *except* at $\theta = 0$ (north pole).

With these vectors we can evaluate the Mead–Berry connection one-form (8.47):

$$\breve{A}^{k'k} = A_i^{k'k} dx^i = A_R^{k'k} dR + A_\theta^{k'k} d\theta + A_\varphi^{k'k} d\varphi = i\langle k'(R, \theta, \varphi)| \, d \, |k(R, \theta, \varphi)\rangle \tag{8.116}$$

the spherical components of which are given by

$$A_R^{k'k} = i\langle k'(R, \theta, \varphi)| \frac{\partial}{\partial R} |k(R, \theta, \varphi)\rangle, \tag{8.117}$$

$$A_\theta^{k'k} = i\langle k'(R, \theta, \varphi)| \frac{\partial}{\partial \theta} |k(R, \theta, \varphi)\rangle, \tag{8.118}$$

$$A_\varphi^{k'k} = i\langle k'(R, \theta, \varphi)| \frac{\partial}{\partial \varphi} |k(R, \theta, \varphi)\rangle. \tag{8.119}$$

The $A_i^{k'k}$ (with $i = R, \theta, \varphi$) are 2×2 matrices ($\breve{A}$ is a matrix-valued one-form). The calculation was done in Chap. 3 for $i = \theta, \varphi$ with the results (3.39) and (3.40), which we write here as

$$A_\theta^{k'k} = e^{-i\varphi k'} e^{i\varphi k} \langle k'(R, \mathbf{e}_3) | J_2 | k(R, \mathbf{e}_3) \rangle, \tag{8.120}$$

$$A_\varphi^{k'k} = \langle k'(R, \mathbf{e}_3) | [-\sin\theta \, (\cos\varphi \, J_1 \\ + \sin\varphi \, J_2) + (\cos\theta - 1) \, J_3] | k(R, \mathbf{e}_3) \rangle. \tag{8.121}$$

If the $|k(R, \mathbf{e}_3)\rangle$ were eigenvectors of $\mathbf{J}^2$, i.e., if they were basis vectors of an irreducible representation of $SO(3)_{J_i}$, the matrix elements in (8.120) and (8.121) would be well-known.[22] But even if the $|k(R, \mathbf{e}_3)\rangle$ are not $SO(3)$-basis vectors, the matrix elements are easily calculated (from the $SO(2)_{J_3}$-vector operator property of J_1 and J_2). If $|k - k'| \neq 1$ (i.e., if $|k| \equiv \Omega \neq \frac{1}{2}$) one obtains (for $\theta \neq \pi$)

$$A_\theta^{k'k} = 0, \tag{8.122}$$

$$A_\varphi^{k'k} = \begin{array}{c} k' = +\Omega \\ k' = -\Omega \end{array} \begin{bmatrix} k = +\Omega & k = -\Omega \\ -\Omega(1-\cos\theta) & 0 \\ 0 & +\Omega(1-\cos\theta) \end{bmatrix}$$
$$= -\Omega(1-\cos\theta) \sigma_3^{k'k}. \tag{8.123}$$

Using the parameterization (8.115) one calculates in the same way (for $\theta \neq 0$)

$$A'^{k'k}_\theta = 0, \tag{8.124}$$

$$A'^{k'k}_\varphi = k \, \delta_{k'k}(1 + \cos\theta) = \Omega \, (1 + \cos\theta) \, \sigma_3^{k'k}. \tag{8.125}$$

For $|k| \equiv \Omega = \frac{1}{2}$ the calculation is a little more complicated because for $|k' - k| = 1$ the operators J_1 and J_2 will have non-zero off-diagonal matrix elements. One can show (Problem 8.5) that $\check{A}$ can be expressed in terms of a real parameter κ which is defined by the matrix element $\langle k(R, \mathbf{e}_3) | J_2 | k(R, \mathbf{e}_3) \rangle$ as

$$\langle k(R, \mathbf{e}_3) | J_2 | k(R, \mathbf{e}_3) \rangle =: \frac{\kappa}{2} \sigma_2^{k'k}. \tag{8.126}$$

The result in terms of this arbitrary parameter is

$$A_\theta^{k'k} = \begin{array}{c} \\ k' = +\frac{1}{2} \\ k' = -\frac{1}{2} \end{array} \begin{pmatrix} k = +\frac{1}{2} & k = -\frac{1}{2} \\ 0 & \frac{-ie^{-i\varphi}}{2} \kappa \\ \frac{ie^{i\varphi}}{2} \kappa & 0 \end{pmatrix}, \tag{8.127}$$

---

[22] See for example [43, §III.3].

8.6 The Monopole of the Diatomic Molecule

$$A^{k'k}_\varphi = \begin{pmatrix} & k = +\frac{1}{2} & k = -\frac{1}{2} \\ k' = +\frac{1}{2} & \dfrac{-(1-\cos\theta)}{2} & \dfrac{-\sin\theta e^{-i\varphi}}{2}\kappa \\ k' = -\frac{1}{2} & \dfrac{-\sin\theta e^{i\varphi}}{2}\kappa & \dfrac{(1-\cos\theta)}{2} \end{pmatrix}. \tag{8.128}$$

As for $A^{k'k}_R$ we have

$$\begin{aligned} A^{k'k}_R &= i\langle k'(R,\mathbf{e}_3)|e^{-i\varphi J_3}\frac{\partial}{\partial R}e^{i\varphi J_3}|k(R,\mathbf{e}_3)\rangle \\ &= ie^{i\varphi(k-k')}\langle k'(R,\mathbf{e}_3)|\partial_R|k(R,\mathbf{e}_3)\rangle \\ &= i\langle k'(R,\mathbf{e}_3)|\partial_R|k(R,\mathbf{e}_3)\rangle, \end{aligned} \tag{8.129}$$

where $\partial_R = e^{-i\varphi J_3}\partial_R e^{i\varphi J_3}$ is used. In order that the second and third expressions are compatible, we must have $A^{k'k}_R = 0$ if $k \neq k'$. If we choose $|k(R,\mathbf{e}_3)\rangle$ to be real, we have $A^{kk}_R = 0$, therefore, the radial component of the vector potential can be taken to be zero:

$$A^{k'k}_R = 0. \tag{8.130}$$

In order to express the vector $\mathbf{A}$ in terms of its spherical orthonormal components we identify $\mathbf{A} = A_R\mathbf{e}_R + A_\theta\mathbf{e}_\theta + A_\varphi\mathbf{e}_\varphi$ with the one-form $\breve{A} = a_R dR + a_\theta R d\theta + a_\varphi R\sin\theta d\varphi$. By comparison with (8.122) and (8.123) we see that

$$a^{k'k}_R = 0, \tag{8.131}$$

$$a^{k'k}_\theta = 0, \tag{8.132}$$

$$a^{k'k}_\phi = \frac{-\Omega(1-\cos\theta)}{R\sin\theta}\sigma_3^{k'k}. \tag{8.133}$$

From the connection one-form – either in the gauge given by (8.122) and (8.123) or in the gauge given by (8.124) and (8.125) – we can calculate the curvature two-form (5.101):

$$\begin{aligned} \breve{F}^{k'k} &= \frac{1}{2}F^{k'k}_{ij}dx^i\wedge dx^j \\ &= F^{k'k}_{R\theta}dR\wedge d\theta + F^{k'k}_{R\varphi}dR\wedge d\varphi + F^{k'k}_{\theta\varphi}d\theta\wedge d\varphi, \end{aligned} \tag{8.134}$$

where $F_{R\theta}$, $F_{R\varphi}$, $F_{\theta\varphi}$ are the covariant spherical components given in terms of the matrices A_θ and A_φ by

$$F_{\theta\varphi} = \partial_\theta A_\varphi - \partial_\varphi A_\theta - i[A_\theta, A_\varphi]. \tag{8.135}$$

It is easily seen that all other components of $\check{F}$ except for $F_{\theta\varphi} = -F_{\varphi\theta}$ are zero. The $F_{\theta\varphi}$ component of the curvature calculated from (8.122) and (8.123) (or (8.124) and (8.125)) is the matrix (for $\Omega \neq \frac{1}{2}$)

$$F_{\theta\phi}^{k'k} = -\Omega \sin\theta\, \sigma_3^{k'k}. \tag{8.136a}$$

Thus the curvature two-form for our case is:

$$\check{F}^{k'k} = F_{\theta\varphi}^{k'k}\, d\theta \wedge d\varphi = -\Omega \sin\theta\, \sigma_3^{k'k} d\theta \wedge d\varphi. \tag{8.136b}$$

For the case $\Omega = \frac{1}{2}$ a separate calculation shows that (Problem 8.4).

$$F_{\theta\varphi}^{k'k} = -\frac{1}{2}(1-\kappa^2)\sin\theta\, \sigma_3^{k'k}. \tag{8.137}$$

The Mead–Berry connection and curvature that we have obtained for the diatomic molecule with fixed angular momentum component k along the internuclear axis is given by the same expressions as the ones we obtained also in Chap. 3. The difference between the molecule and the quantum system of Chap. 3 is that the fast electron state that we consider here is not a pure state described by a one-dimensional subspace of the space of electron states, but a mixture described by a two-dimensional subspace spanned by the vectors $|k(z,\theta,\varphi)\rangle$ with $k = \Omega$ and $k = -\Omega$ [43, §II.4]. As a consequence of this, the connection $\check{A}$ (or A) and curvature $\check{F}$ (or F) are 2×2 matrices.

Instead of using the basis vectors $|k(R,\theta,\varphi)\rangle$ for this two-dimensional space, we can use the parity basis vectors $|\pi = \pm;\ (\theta\varphi)\rangle$ of (8.112). Then $\check{F}$ is no longer diagonal but has the form

$$F^{\pi'\pi} = \sum_{k,k'=\pm\Omega} \langle \pi'|k'\rangle F^{k'k}\langle k|\pi\rangle, \tag{8.138}$$

where $\langle k|\pi\rangle$ is the transition matrix of the transformation (8.112):

$$\langle k|\pi\rangle = \begin{array}{c} \\ k=+\Omega \\ k=-\Omega \end{array} \overset{\begin{array}{cc}\pi=+ & \pi=-\end{array}}{\begin{pmatrix} \frac{1}{\sqrt{2}} & \frac{1}{\sqrt{2}} \\ \frac{1}{\sqrt{2}} & -\frac{1}{\sqrt{2}} \end{pmatrix}}. \tag{8.139}$$

Performing the transformation (8.138), one finds

$$F_{\theta\varphi}^{\pi'\pi} = -\Omega \sin\theta\, \sigma_1^{\pi'\pi} = -\Omega \sin\theta \begin{pmatrix} 0 & 1 \\ 1 & 0 \end{pmatrix}. \tag{8.140}$$

8.6 The Monopole of the Diatomic Molecule

The electronic energy eigenstate is a mixture of electron states rotating clockwise and counterclockwise around the internuclear axis. Equivalently the state can be considered a mixture of positive and negative parity eigenstates.

In the derivation of (8.136a) the commutator on the right-hand side of (8.135) is zero (because A_θ is zero). However, in the derivation of (8.137) this commutator gives a non-zero contribution. For $\Omega \neq \frac{1}{2}$, $\breve{F}$ is always different from zero (except in the trivial case $\Omega = 0$). For $\Omega = \frac{1}{2}$, the curvature $\breve{F}$ can vanish when $\kappa = 1$. According to the definition of κ given by (8.126), the case $\kappa = 1$ means that the $\{|k(R, \mathbf{e}_3)\rangle; k = \frac{1}{2}, -\frac{1}{2}\}$ span the irreducible representation space of the rotation group generated by the electron angular momentum operators J_1, J_2 and J_3 with electron angular momentum $\frac{1}{2}$.

Since the parameter space of the gauge theory that we are considering is three dimensional, we can also introduce the (contravariant matrix-valued) vector field (strength):

$$\mathbf{F} := \frac{1}{2} \epsilon_{ijk} \breve{F}_{jk} \cdot \frac{\partial}{\partial x^i}.$$

In order to express $\mathbf{F}$ in terms of $\mathbf{R}$, we use in (8.136b) the orthonormal spherical components $\hat{F}_{\theta\varphi}, \hat{F}_{R\theta}, \hat{F}_{R\varphi}$:[23]

$$\breve{F} = \hat{F}_{R\theta} dR \wedge (R d\theta) + \hat{F}_{R\varphi} dR \wedge (R \sin\theta d\varphi) + \hat{F}_{\theta\varphi} (R d\theta) \wedge (R \sin\theta d\varphi). \tag{8.141}$$

Comparing this with (8.136b) we obtain:

$$\hat{F}_{\theta\varphi} = -\Omega \frac{1}{R^2} \sigma_3, \qquad \hat{F}_{R\theta} = 0, \qquad \hat{F}_{R\varphi} = 0$$

or

$$\mathbf{F} = -\Omega \sigma_3 \frac{\mathbf{R}}{R^3}. \tag{8.142}$$

The vector potential $\mathbf{A}$ and the field strength $\mathbf{F}$ are identical with the vector potential and the field strength of a charge–monopole system [41], if the quantum number k (component of electronic angular momentum along the internuclear axis) is replaced by $-\frac{eg}{4\pi}$ where e is the electric charge and g the magnetic monopole strength. This is most apparent from the vector form of $\breve{F}$, i.e., (8.142). Equation (8.142) is the standard expression for the magnetic field of a monopole if $\Omega \to -\frac{eg}{4\pi}$ [68, 74, 87].

Thus we have derived the monopole vector potential and field from the rapid motion of the electrons around the internuclear axis of a diatomic molecule. This "magnetic monopole" is of course not of the electromagnetic kind as originally envisaged by Dirac [74] but has its origin in the fast motion of the electrons. This monopole is characterized by the motion constant k,

---

[23] Note that here the parameter space of the gauge theory is $\mathbb{R}^3$. Hence we can and do identify the parameter space and its tangent spaces. This is necessary for being able to relate the direction of $\mathbf{F}$ to the position vector $\mathbf{R} \in \mathbb{R}^3$.

in place of the electromagnetic constants (e, g). These "motional monopoles" of the diatomic dumbbell with flywheel always come in pairs with opposite motion constants $|k|$ and $-|k|$ or, equivalently, in pairs with opposite parity. The parity eigenstates are linear combinations (8.112) of these monopoles.

We shall now discuss the effective Hamiltonian of our system. The gauge-covariant momenta, defined by

$$\boldsymbol{\Pi}^{k'k} := \frac{1}{i}\mathbf{D}^{k'k} = \delta^{k'k}\mathbf{P} - \mathbf{A}^{k'k}, \qquad (8.143)$$

are 2×2 matrices like the $\mathbf{A}$ and $\mathbf{F}$ whose rows and columns are labeled by the eigenvalues k' and k of the operator $\mathbf{e_R} \cdot \mathbf{J}$. The effective Hamiltonian of (8.62) is also a 2×2 matrix:

$$H_{\text{eff}} = \frac{1}{2\mu}\boldsymbol{\Pi}^2 + \tilde{\varepsilon}(\mathbf{R}), \qquad (8.144)$$

with

$$\tilde{\varepsilon}^{k'k}(R) = \varepsilon_\Omega(R)\delta^{k'k} + \frac{1}{2\mu}\sum_{l \neq \pm \Omega} \mathbf{A}^{k'l} \cdot \mathbf{A}^{lk}. \qquad (8.145)$$

The multiplication of the $\boldsymbol{\Pi}$'s is matrix multiplication (there is also a sum over the three vector components in (8.144)).

The matrix elements of the 2×2 matrices Π_i $(i = 1, 2, 3)$ are operators in the space of the slow motion. For the case $|k| \neq \frac{1}{2}$ the matrices Π_i are diagonal but their matrix elements are non-commuting operators. For $|k| = \frac{1}{2}$ their matrix elements are non-commuting operators, and from (8.127) and (8.128) we see that they are non-diagonal matrices as well. In both cases the operator matrices Π_i fulfill the commutation relations (5.71), which in our case (8.136a) and (8.137) are given by

$$[\Pi_i, \Pi_j] = -i\varepsilon_{ijl}\frac{R_l}{R^3}\Omega\begin{pmatrix} 1 & 0 \\ 0 & -1 \end{pmatrix}, \quad |k| \neq \frac{1}{2}, \qquad (8.146)$$

and

$$[\Pi_i, \Pi_j] = -i\varepsilon_{ijl}\frac{R_l}{R^3}\frac{1}{2}(1-\kappa^2)\begin{pmatrix} 1 & 0 \\ 0 & -1 \end{pmatrix}, \quad |k| = \frac{1}{2}. \qquad (8.147)$$

If the matrix elements of the 2×2 matrices Π_i are labeled not by the quantum numbers k but by the quantum numbers π (eigenvalues of intrinsic parity) then their commutation relations are similar to (8.146) and (8.147) with σ_3 on the right-hand side replaced by σ_1 as in (8.140). If we restrict ourselves to a subspace of the space of electronic states with a fixed value for the operator $\mathbf{e_R} \cdot \mathbf{J}$ (e.g., to $k' = k = |k|$), then the Π_i are just operators and not operator matrices. Their commutation relations are (for $|k| \neq \frac{1}{2}$)

$$[\Pi_i, \Pi_j] = -i\epsilon_{ijl}\frac{R_l}{R^3}k, \qquad (8.148)$$

8.6 The Monopole of the Diatomic Molecule

which are the standard commutation relations for a charge–monopole system.

We now consider the scalar potential given by (8.145). The first term $\varepsilon_\Omega(R)$, which is the scalar potential for the diatomic molecule in a definite electronic energy level (the Born–Oppenheimer eigenvalue), is specified by the value of Ω and the other fixed electronic quantum numbers $^{2S+1}(\Lambda)$.

We shall now sketch the calculation of the second term in (8.145). The result is given in (8.157) below. By using the same techniques that were used in the evaluation of (8.85) and (8.98) one obtains

$$\sum_{l\neq\pm\Omega} \mathbf{A}^{k'l} \cdot \mathbf{A}^{lk} = \left(\nabla_\mathbf{R} \langle k'(R,\theta,\varphi)|\right) \nabla_\mathbf{R} |k(R,\theta,\varphi)\rangle - \sum_{l=\pm\Omega} \mathbf{A}^{k'l} \cdot \mathbf{A}^{lk}. \quad (8.149)$$

The first term is calculated using the following result, which can be obtained by a straightforward calculation from (3.33) and (3.34):

$$\nabla_\mathbf{R} U^\dagger(\theta,\varphi) \cdot \nabla_\mathbf{R} U(\theta,\varphi) = \frac{1}{R^2} \frac{\partial U^\dagger}{\partial \theta} \frac{\partial U}{\partial \theta} + \frac{1}{R^2 \sin^2\theta} \frac{\partial U^\dagger}{\partial \varphi} \frac{\partial U}{\partial \varphi}$$

$$= \frac{1}{R^2} \left[J_1^2 + J_2^2 - \frac{1-\cos\theta}{\sin^2\theta} J_3^2 \right] \quad (8.150)$$

$$+ \frac{1-\cos\theta}{R^2 \sin\theta} \{J_3, J_1\cos\varphi + J_2\sin\varphi\},$$

where { , } denotes the anticommutator of two operators. The second term of (8.149), which involves a summation over the two matrix indices $l = \pm|k|$ only, is calculated from (3.35) and (3.36):

$$\sum_{l=\pm\Omega} \mathbf{A}^{k'l} \mathbf{A}^{lk} = \frac{1}{R^2} \sum_{l=\pm\Omega} \left(j_1^{k'l} j_1^{lk} + j_2^{k'l} j_2^{lk} + j_3^{k'l} j_3^{lk} \frac{1-\cos\theta}{\sin^2\theta} \right), \quad (8.151)$$

where j_i^{lk} are the 2×2 matrices

$$j_i^{lk} = \langle l, (R, \mathbf{e}_3) | J_i | k, (R, \mathbf{e}_3) \rangle; \qquad l, k = +\Omega \text{ or } -\Omega. \quad (8.152)$$

Note that the calculation of the matrix elements of the operators $J_i \cdot J_j$ in (8.150) involves a summation over all infinite electronic quantum numbers, not just $l = \pm\Omega$:

$$\langle k', (R, \mathbf{e}_3) | J_i J_j | k, (R, \mathbf{e}_3) \rangle$$
$$= \sum_l \langle k', (R, \mathbf{e}_3) | J_i | l, (R, \mathbf{e}_3) \rangle \langle l, (R, \mathbf{e}_3) | J_j | k, (R, \mathbf{e}_3) \rangle.$$

Also note that the vectors $|k, (R, \mathbf{e}_3)\rangle$ are not eigenvectors of $\mathbf{J}^2$ (not basis vectors of an $SO(3)_{J_i}$ representation).

We first consider the case $\Omega \neq \frac{1}{2}$. The matrix elements of the anticommutator in (8.150) are zero, the matrix elements of the first term are not. Taking the matrix elements of (8.150) and inserting this and (8.151) into (8.149) one obtains

$$\sum_{l \neq \pm \Omega} \mathbf{A}^{k'l} \cdot \mathbf{A}^{lk} = \left(\frac{\partial}{\partial R} \langle k'(R, \mathbf{e}_3)| \right) \frac{\partial}{\partial R} |k(R, \mathbf{e}_3)\rangle$$
$$+ R^{-2} \langle k'(R, \mathbf{e}_3)| \left(J_1^2 + J_2^2 \right) |k(R, \mathbf{e}_3)\rangle. \quad (8.153)$$

The matrix elements on the right-hand side cannot be explicitly evaluated using the matrix elements of the generators of $SO(3)$ ($\langle k'(R, \mathbf{e}_3)|\mathbf{J}^2|k(R, \mathbf{e}_3)\rangle$ is not $j_{\mathrm{el}}(j_{\mathrm{el}}+1)$). However, they can be expressed in terms of an $SO(2)_{J_3}$ reduced matrix element Δ of the $SO(2)_{J_3}$-scalar operator $J_1^2 + J_2^2$ (using the Wigner–Eckart theorem for $SO(2)$) and the $SO(2)$ Clebsch–Gordan coefficient $\delta_{k'k}$):

$$\langle k'(R, \mathbf{e}_3)|(J_1^2 + J_2^2)|k(R, \mathbf{e}_3)\rangle = \Delta \, \delta^{k'k}. \quad (8.154)$$

The factor Δ can certainly be different for different values of the (fixed) quantum number $\Omega = |k|$; $\Delta = \Delta(\Omega)$. We also obtain

$$\left(\frac{\partial}{\partial R} \langle k'(R, \mathbf{e}_3)| \right) \frac{\partial}{\partial R} |k(R, \mathbf{e}_3)\rangle = \Xi \delta^{k'k}. \quad (8.155)$$

Thus, we can write the second term of (8.145) as

$$\frac{1}{2\mu} \sum_{l \neq \pm \Omega} \mathbf{A}^{k'l} \cdot \mathbf{A}^{lk} = \frac{1}{2\mu R^2} \Delta \, \delta^{k'k} + \frac{1}{2\mu} \Xi \delta^{kk}. \quad (8.156)$$

For the case $\Omega = \frac{1}{2}$, the matrix elements of (8.152) are given by the Pauli matrices, $j_i = \frac{\kappa}{2} \sigma_i, i = 1, 2$. A separate calculation shows that in this case Δ in (8.156) has to be replaced by $\Delta + \frac{\kappa^2}{2}$ where κ is the constant introduced in (8.126).

Equation (8.156) shows that the term with Δ is an inverse-square potential originating from an inverse-cube repulsive force centered at $R = 0$ [33]. For future reference we rewrite (8.145) (for both cases $|k| \neq \frac{1}{2}$ and $|k| = \frac{1}{2}$) as

$$\tilde{\varepsilon}^{k'k}(R) = \tilde{\varepsilon}_\Omega(R) \delta^{k'k}. \quad (8.157)$$

where

$$\tilde{\varepsilon}_\Omega(R) = \varepsilon_\Omega(R) + \frac{1}{2\mu R^2} \Delta + \frac{1}{2\mu} \Xi. \quad (8.158)$$

Therewith the effective Hamiltonian of the diatomic dumbbell with flywheel can be given by

$$H_{\mathrm{eff}} = \frac{1}{2\mu} \Pi^2 + \frac{1}{2\mu R^2} \Delta + \frac{1}{2\mu} \Xi + \varepsilon_\Omega(R). \quad (8.159)$$

If the last two terms in (8.159) were zero, i.e., $\tilde{\varepsilon}_\Omega(R) = 0$ then, the Hamiltonian would be the Hamiltonian of a charge–monopole system:

8.6 The Monopole of the Diatomic Molecule

$$H = H^{\text{mon}} = \frac{1}{2\mu}\boldsymbol{\Pi}^2 . \tag{8.160}$$

The radial motion of the charge–monopole system is free. If $\tilde{\varepsilon}_\Omega(R)$ were a Coulomb potential, then (8.159) would be the Hamiltonian of a dyon system (two mass points with charge and magnetic monopole strength). In our case of the diatomic molecule (dumbbell with flywheel) the effective Hamiltonian (8.159) has the same induced vector potential as the monopole or dyon system but the induced scalar potential is given by (8.158) where $\tilde{\varepsilon}_\Omega(R)$ is close to an oscillator potential. Thus the system has the same angular motion as that of a magnetic monopole but in addition it has a radial motion which is oscillatory.

We now investigate the property of the effective Hamiltonian for the "slow" system with the electronic potential energy curve $\tilde{\varepsilon}_{n_a}(R) = \tilde{\varepsilon}_\Omega(R)$. To $\tilde{\varepsilon}_\Omega(R)$ belong the two electronic quantum numbers $k = +\Omega, k = -\Omega$, and the electronic energy eigenspace $\mathcal{H}_{\tilde{\varepsilon}_\Omega} \subset \mathcal{H}^{\text{fast}}$ is two dimensional. The subspace $\mathcal{H}^{\text{slow}} \otimes \mathcal{H}_{\tilde{\varepsilon}_\Omega} \subset \mathcal{H} = \mathcal{H}^{\text{slow}} \otimes \mathcal{H}^{\text{fast}}$ is thus the direct sum

$$\mathcal{H}^{\text{slow}} \otimes \mathcal{H}_{\tilde{\varepsilon}_\Omega} \equiv \mathcal{H}^\Omega = \mathcal{H}^{k=\Omega} \oplus \mathcal{H}^{k=-\Omega}. \tag{8.161}$$

In this space the effective Hamiltonian

$$H^\Omega_{\text{eff}} = \langle k'(R, \mathbf{e_R})|H|k(R, \mathbf{e_R})\rangle = \frac{1}{2\mu}\boldsymbol{\Pi}^2 - \tilde{\varepsilon}_\Omega(R)$$

is a 2×2 matrix. Similarly the slow observables $\hat{\mathbf{R}}$ (whose eigenvalues are the parameters $\mathbf{R}$) and the momenta $\boldsymbol{\Pi} = \mathbf{P} - \mathbf{A}(\hat{\mathbf{R}})$ are also 2×2 (but not necessarily diagonal) matrices. Their elements are operators in $\mathcal{H}^\Omega$.

The commutation relations of R_i with R_j, and R_i with Π_j are as usual

$$[R_i, R_j] = 0, \qquad [\Pi_i, R_j] = i\delta_{ij}. \tag{8.162}$$

The commutation relations of the Π_i are the anomalous commutation relations (8.147).

Next we want to construct the angular momentum operator of the dumbbell molecule with flywheel. This is different from the angular momentum operator $\mathbf{J} = \mathbf{L}_{\text{elec}} + \mathbf{S}_{\text{elec}}$ which we discussed above (and in Chap. 3) and which was the angular momentum of the electronic motion. The angular momentum that we want to construct now is the angular momentum of the whole system, fast and slow. It is therefore the generator of rotations. As a consequence of the isotropy of space, this operator $\mathcal{J}_i$ must commute with the Hamiltonian of the molecule,

$$[H, \mathcal{J}_i] = 0. \tag{8.163}$$

Further it must fulfill the standard commutation relations:

$$[\mathcal{J}_i, \mathcal{J}_j] = i\epsilon_{ijl}\mathcal{J}_l, \tag{8.164}$$

and the position R_i and momentum Π_i operators must be vector operators[24] with respect to the angular momentum:

$$[\mathcal{J}_i, R_j] = i\epsilon_{ijl} R_l, \quad (8.165)$$
$$[\mathcal{J}_i, \Pi_j] = i\epsilon_{ijl} \Pi_l. \quad (8.166)$$

We want to construct $\mathcal{J}_i$ in terms of the position operator $\mathbf{R}$, the momentum operator $\mathbf{\Pi}$, and the other physical quantities that we constructed for the collective motion. We expect that this angular momentum will not be the operator $\mathbf{R} \times \mathbf{\Pi}$ (or the operator $\mathbf{R} \times \mathbf{P}$), because these operators have a zero component of angular momentum along the internuclear axis, $\mathbf{e_R}$. We know that the molecule in the state characterized by Ω has a non-zero component of angular momentum along the internuclear axis due to the fast intrinsic motion of the electrons (the flywheel motion)

$$\mathbf{e_R} \cdot \mathcal{J} = k = \pm\Omega = \mathbf{e_R} \cdot \mathbf{J}. \quad (8.167)$$

From the theory of Dirac's magnetic monopole, we would expect that

$$\mathcal{J}_i = \epsilon_{ijl} R_j \Pi_l - \frac{1}{2}\epsilon_{mnl} R_i R_m \check{F}_{nl}, \quad (8.168)$$

is the right expression for the angular momentum operator. That means the operator

$$\mathcal{J}_i^{(k)} = \epsilon_{ijl} R_j \Pi_l^{(k)} + k\frac{R_i}{R}, \quad (8.169)$$

is the angular momentum operator in each subspace $\mathcal{H}^k$. It can be shown by direct calculation that this operator indeed fulfills the conditions (8.162)–(8.166). From the above results we conclude that the dynamics of a diatomic molecule in a doubly degenerate Λ level is the same as that of the hypothetical charge–monopole system, except that the radial interaction is not free as in (8.160), but given by the scalar potential $\tilde{\varepsilon}_\Omega(R)$ in (8.159).

To derive the properties of this physical system (diatomic molecule in a fixed electronic state) we separate the radial and angular motion. From (8.168) we obtain by a straightforward calculation using the commutation relations (8.148) and (8.162)

$$\mathcal{J}^2 = \mathbf{R}^2 \mathbf{\Pi}^2 + i\mathbf{R} \cdot \mathbf{\Pi} - (\mathbf{R} \cdot \mathbf{\Pi})^2 + k^2. \quad (8.170)$$

Since the radial component of $\mathbf{A}$ ($a_R = \mathbf{e_R} \cdot \mathbf{A}$) is equal to zero (8.131), we have

$$\mathbf{R} \cdot \mathbf{\Pi} = \mathbf{R} \cdot \mathbf{P}. \quad (8.171)$$

Thus we can replace the $\mathbf{\Pi}$ in the second and third term on the right-hand side of (8.170) by $\mathbf{P}$. Then we obtain from (8.170)

---

[24] For more details see [43, §V.3].

8.6 The Monopole of the Diatomic Molecule

$$\Pi^2 = \frac{1}{R^2}(\mathcal{J}^2 - k^2) + \frac{1}{R^2}\left((\mathbf{R}\cdot\mathbf{P})^2 - i\mathbf{R}\cdot\mathbf{P}\right). \tag{8.172}$$

The last term on the right-hand side is the square of the radial momentum operator [43, §VII.2]:

$$P_{\rm rad} = \frac{1}{2}\left\{\frac{R_i}{R}, P_i\right\}. \tag{8.173}$$

$P_{\rm rad}$ is the component of $\mathbf{P}$ along the direction $\mathbf{R}$. The anticommutator $\{\frac{R_i}{R}, P_i\} = \frac{R_i}{R}P_i + P_i\frac{R_i}{R}$ takes into account that R_i and P_i are noncommuting operators. $P_{\rm rad}$ is the operator conjugate to the radius operator R ($[R, P_{\rm rad}] = i1$), and $\frac{P_{\rm rad}^2}{2\mu}$ is the operator for the kinetic energy of the radial motion (μ is the reduced mass of the two atoms). Taking the square of (8.173) gives

$$P_{\rm rad}^2 = \frac{1}{R^2}\left((\mathbf{R}\cdot\mathbf{P})^2 - i\frac{1}{R^2}(\mathbf{R}\cdot\mathbf{P})\right). \tag{8.174}$$

From this we see that (8.172) can be written (after dividing by 2μ) as

$$\frac{1}{2\mu}\Pi^2 = \frac{1}{2\mu R^2}(\mathcal{J}^2 - k^2) + \frac{1}{2\mu}P_{\rm rad}^2. \tag{8.175}$$

If we insert this into (8.159) we obtain for the effective Hamiltonian

$$H_{\rm eff} = \frac{1}{2\mu R^2}(\mathcal{J}^2 - k^2) + \frac{1}{2\mu}P_{\rm rad}^2 + \tilde{\varepsilon}_\Omega(R). \tag{8.176}$$

This Hamiltonian is best understood if we compare it with the two standard Hamiltonians connected with magnetic monopoles. This is illustrated in Fig. 8.4.

The charge–monopole system consists of a mass m_1 with magnetic monopole strength g and another mass m_2 with electric charge e. Its Hamiltonian is given by (8.160) with $k = -\frac{eg}{4\pi}$. Using (8.172) this can be expressed as:

$$H^{\rm mon} = \frac{1}{2\mu R^2}\left(\mathcal{J}^2 - k^2\right) + \frac{1}{2\mu}P_{\rm rad}^2. \tag{8.177}$$

The dyon system [229] consists of two particles of masses m_1 and m_2 each of which has electric charge, e_1 and e_2 respectively, and magnetic monopole strength g_1 and g_2 respectively. The Hamiltonian is given by

$$H^{\rm dyon} = \frac{1}{2\mu R^2}\left(\mathcal{J}^2 - k^2\right) + \frac{1}{2\mu}P_{\rm rad}^2 + \frac{1}{4\pi}\left(e_1 e_2 + g_1 g_2\right)\frac{1}{R}, \tag{8.178}$$

with

$$k = -\frac{e_1 g_2 - e_2 g_1}{4\pi}.$$

This dyon Hamiltonian differs from the monopole Hamiltonian (8.177) by the Coulomb potential term for the radial motion; for the charge monopole system the radial motion is free.

8. The Gauge Theory of Molecular Physics

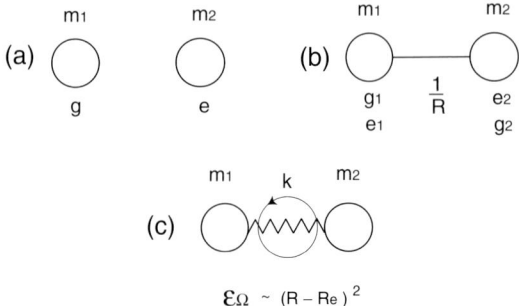

Fig. 8.4. Comparison of physical systems of two mass points m_1, and m_2 with magnetic monopole vector potentials. (**a**) is the charge–monopole system with electric charge e and magnetic monopole strength g with no force in the radial direction. (**b**) is the dyon system consisting of two mass – points m_i with electric charge e_i and magnetic monopole strength g_i and with a force in the radial direction given by the $\frac{1}{R}$ Coulomb potential. (**c**) is the vibrating dumbbell with flywheel. In contrast to system (**a**) and (**b**) the monopole potential is not of electromagnetic origin but caused by the fast flywheel motion with fixed angular momentum component k around the dumbbell axis. The radial interaction is not free but is given by the Born–Oppenheimer potential curve $\varepsilon_\Omega(R)$ which is approximated by an oscillator potential.

The diatomic dumbbell with flywheel is also a physical system that combines rotational motion, given (approximately) by the first term of (8.176), and radial motion given by the term $\frac{1}{2\mu} P_{\text{rad}}^2 + \tilde{\varepsilon}_\Omega(R)$. The rotational motion is again the same as for the monopole and dyon system, except that the k is not of electromagnetic origin but comes from the fast flywheel motion. The radial motion is now different and given by the induced scalar potential $\tilde{\varepsilon}_\Omega(R)$ which is often (approximately) oscillatory. We will now discuss the properties for the molecule in detail.

The scalar potential $\tilde{\varepsilon}_\Omega$, Fig. 8.1, has a minimum at the position R_e and is approximated by an oscillator potential given by

$$\tilde{\varepsilon}_\Omega \approx \frac{f}{2} X^2 + cX^3 + \cdots, \quad \text{with} \quad X := R - R_e. \tag{8.179}$$

The instantaneous moment of inertia is defined by $I = \mu R^2$, and is therefore written as

$$I = \mu R^2 = \mu(R_e + X)^2 = \mu R_e^2 + 2\mu R_e X + \mu X^2, \tag{8.180}$$

where X (as well as R) is the operator conjugate to P_{rad}. Introducing the constant $w = \sqrt{\frac{f}{\mu}}$, we define the new operators

$$q := \sqrt{\mu w}\, X, \quad p := \frac{1}{\sqrt{\mu w}} P_{\text{rad}}, \tag{8.181}$$

8.6 The Monopole of the Diatomic Molecule

and so
$$I = I_e(1 + \frac{1}{\sqrt{\mu w}\, R_e} q)^2, \qquad I_e = \mu R_e^2, \tag{8.182}$$

and
$$I^{-1} = \frac{1}{I_e}(1 - \gamma q + \frac{3}{4}\gamma^2 q^2 - \cdots), \qquad \gamma = \frac{2}{\sqrt{\mu w}\, R_e}. \tag{8.183}$$

Inserting this result into (8.176) we obtain
$$H_{\text{eff}} = \frac{1}{2I_e}(\mathcal{J}^2 - k^2) + \frac{w}{2}(p^2 + q^2) + H', \tag{8.184}$$

where H' is treated as a perturbation and given by
$$H' = -\frac{\gamma q}{2I_e}(\mathcal{J}^2 - k^2) + \frac{3\gamma^2 q^2}{8 I_e}(\mathcal{J}^2 - k^2) - \cdots. \tag{8.185}$$

We will not discuss this correction term H' any further.

The effective Hamiltonian for the slow motion of the diatomic molecule in a doubly degenerate Λ-level is thus given by the sum of the Hamiltonian for a radial vibrator
$$H_{\text{rad vib}} = \frac{w}{2}(p^2 + q^2), \tag{8.186}$$

and a rotator
$$H_{\text{rot}} = \frac{1}{2I_e}\left(\mathcal{J}^2 - (\mathbf{e}_z \cdot \mathcal{J})^2\right) = \frac{1}{2I_e}(\mathcal{J}^2 - k^2). \tag{8.187}$$

Thus we derived that the collective motion of the diatomic molecule is to a first approximation that of a vibrating rotator.[25] The operators $\mathcal{J}_i$ and R_i generate the enveloping algebra of E_3.[26] The irreducible representation spaces[27] $\mathcal{H}(k,R)$ of E_3 are characterized by two numbers k and R. These two numbers are related to the eigenvalues of the Casimir operators $R_i\mathcal{J}_i = Rk$ and $R_iR_i = R^2$, of which R^2 can take any real value and k can only be integer or half-integer (which we already derived by completely different means in Chap. 3 (3.52), and Chap. 6 (6.50)). $\mathcal{H}(k,R)$ is the space of physical states of the rotational motion (E_3 is the spectrum generating group of the rotator). The spectrum of (8.187) is given by the property of the space $\mathcal{H}(k,R)$.

---

[25] For a further discussion of this point see [43, §II.5]. The eigenvalues of the radial vibrator Hamiltonian (8.186) and the rotator (with flywheel and therefore symmetrical top) are given by Equations (II.3.30) and (V.4.55) of [43], respectively.

[26] E_3 is the Lie group consisting of rotations, reflections, and translations in the Euclidean space $\mathbb{R}^3$. Briefly, E_3 is the group of isometries of $\mathbb{R}^3$ with Euclidean metric.

[27] The irreducible representation spaces of E_3 are derived in the Appendix to Sect. V.3 of [43] and are applied to describe the spectrum of the rotator (and the symmetric top) in Sect. V.4 of this reference.

Since the energy levels are defined only up to an arbitrary constant, the k^2 and Δ terms in (8.187) can be ignored as fixed constants. For $k = 0$ one obtains from (8.187) the ordinary rotator spectrum

$$E_{\text{rot}} = \frac{1}{2I_e} j(j+1), \qquad j = 0, 1, 2, 3, \cdots . \tag{8.188}$$

For $k^2 > 0$ one obtains the spectrum

$$E_{\text{rot}} = \frac{1}{2I_e} \left(j(j+1) - k^2 \right) \qquad j = k, k+1, k+2, \cdots . \tag{8.189}$$

Equation (8.184) with (8.186) and (8.187) is the standard result, including the flywheel term $(\mathbf{e_R} \cdot \boldsymbol{\mathcal{J}})^2$ in H_{rot}, which is usually obtained by phenomenological arguments from the assumption that the electronic angular momentum $\mathbf{J}$ is pointed along the internuclear axis $\mathbf{R}$, ($\mathbf{J} = \pm \Omega \mathbf{e_R}$). We have shown here that it can be derived (without any further phenomenological assumption) directly from the Schrödinger equation if one does not make the drastic assumption of the Born–Oppenheimer approximation (which is that $\mathbf{R}$ is a fixed value). If the collective coordinates and momenta are treated as quantum mechanical observables, the effective Hamiltonian for the slow motion retains terms which can be expressed in terms of a non-Abelian gauge potential. The induced gauge potential has its origin in the fast (electronic) motion and has the same form as the Mead–Berry connection. If the fast motion is the flywheel motion (of the electrons) about the body axis, the Mead–Berry connection is identical with the vector potential of Dirac's hypothetical monopole – once the electromagnetic constants $\frac{eg}{4\pi}$ are replaced by the motion constant $|k|$. As a consequence, the effective Hamiltonian of the slow motion is governed, in addition to the radial potential, by a monopole Hamiltonian. This monopole Hamiltonian leads to the experimentally well-known rotator spectrum (8.184) with two-fold degeneracy (Λ-type doubling). These well-known results are thus a manifestation of the Berry phase.

Although we used the diatomic molecule as an example, our arguments made very little use of the fact that the complicated system was a molecule. If we have a quantum system that we can visualize as a bead which can slide along a rod that rotates slowly in space, we will come to the same conclusions: The rapid rotation of the bead about the rod induces into the dynamics of the slow motion of the bead along the rotating rod a vector potential which is of the same form as that of a magnetic monopole [153]. Quantum systems of this kind exist in many areas of physics. Thus, even though magnetic monopoles of the electromagnetic kind probably do not exist, physical systems with the same dynamics as that of magnetic monopoles are abundant. These physical systems are the slow "parts" of a complicated physical system in which the fast "part" is a rapid rotation.

Problems

8.1 Construct the quantum mechanical kinetic energy operator $\hat{T}_{\text{nuclei}}$ for (8.10) (consult [77]).

a) First, rewrite T_{nuclei} in (8.10) using the total angular momenta

$$J_\alpha = \frac{\partial T_{\text{nuclei}}}{\partial \omega_\alpha} = \sum_\beta I_{\alpha\beta}\omega_\beta + \sum_k \xi_k^\alpha \dot{Q}_k \qquad (8.190)$$

and the momenta conjugate to internal coordinates Q_k

$$P_k = \frac{\partial T_{\text{nuclei}}}{\partial \dot{Q}_k} = \dot{Q}_k + \sum_\beta \omega_\beta \xi_k^\beta. \qquad (8.191)$$

The result will be

$$2T_{\text{nuclei}} = \sum_{\alpha,\beta} \mu_{\alpha\beta}(J_\alpha - \pi_\alpha)(J_\beta - \pi_\beta) + \sum_k P_k^2, \qquad (8.192)$$

where π_α are the vibrational angular momenta defined as

$$\pi_\alpha := \sum_k \xi_k^\alpha P_k, \qquad (8.193)$$

and $\mu_{\alpha\beta}$ are elements of the matrix that relates $(J_\alpha - \pi_\alpha)$ to ω_β,

$$\sum_\alpha \mu_{\beta\alpha}(J_\alpha - \pi_\alpha) = \omega_\beta. \qquad (8.194)$$

Because $J_\alpha - \pi_\alpha$ are given by

$$J_\alpha - \pi_\alpha = \sum_\beta (I_{\alpha\beta} - \xi_k^\alpha \xi_k^\beta)\omega_\beta, \qquad (8.195)$$

the matrix $(\mu_{\alpha\beta})$ is actually the inverse of the matrix $(I_{\alpha\beta} - \xi_k^\alpha \xi_k^\beta)$.

b) Suppose that the quantum mechanical kinetic energy operator corresponding to the classical kinetic energy

$$2T_{\text{nuclei}} = \sum_{i,j} g^{ij} p_i p_j, \qquad (8.196)$$

is given by

$$2\hat{T}_{\text{nuclei}} = g^{\frac{1}{4}} \sum_{i,j} \hat{p}_i g^{-\frac{1}{2}} g^{ij} \hat{p}_j g^{\frac{1}{4}}, \qquad (8.197)$$

where g is the determinant of the matrix g^{ij} and the volume element for integration is $dq_1 dq_2 \cdots dq_{3N_n-3}$. Show that if we choose $p_1 = J_x - $

π_x, $p_2 = J_y - \pi_y$, $p_3 = J_z - \pi_3$, $p_4 = Q_1, \cdots$, and $p_{3N_n-3} = Q_{3N_n-6}$, then $\hat{T}_{\text{nuclei}}$ takes the form

$$2\hat{T}_{\text{nuclei}} = \mu^{\frac{1}{4}} \sum_{\alpha,\beta}(J_\alpha - \pi_\alpha)\mu_{\alpha\beta}\mu^{-\frac{1}{2}}(J_\beta - \pi_\beta)\mu^{\frac{1}{4}}$$
$$+ \mu^{\frac{1}{4}} \sum_k P_k \mu^{-\frac{1}{2}} P_k \mu^{\frac{1}{4}}. \tag{8.198}$$

Note that this expression may be further simplified to yield

$$2\hat{T}_{\text{nuclei}} = \sum_{\alpha,\beta} \mu_{\alpha\beta}(J_\alpha - \pi_\alpha)(J_\beta - \pi_\beta) + \sum_k P_k^2$$
$$- \frac{\hbar^2}{4} \sum_\alpha \mu_{\alpha\alpha}. \tag{8.199}$$

Although this final form is very simple, no simple derivation has been found so far [265].[28]

8.2 Show that the gauge transformations

$$\psi_n^E(\mathbf{R}) \to \psi'^E_{n'}(\mathbf{R}) = \sum_{l=1}^{\mathcal{N}} \mathcal{U}^{n'l}(\mathbf{R})\psi_l(\mathbf{R}),$$

of the collective wave function leave the $\mathcal{N} \times \mathcal{N}$-matrix Schrödinger equation for $\psi_n^E(\mathbf{R})$ unchanged. Hint: Use the fact that the total wave function $\psi^E(\mathbf{R},\mathbf{r}) = \sum_{n=1}^{\mathcal{N}} \phi_n(\mathbf{R},\mathbf{r})\psi_n^E(\mathbf{R})$ should not be affected by this gauge transformation.

8.3 Show that if one allows for the gauge transformations of molecular physics the set of (in general infinite-dimensional) unitary transformations of the basis vectors of $\mathcal{H}^{\text{fast}}$:

$$|n(\mathbf{R})\rangle \to |n'(\mathbf{R})\rangle = \sum_l |l(\mathbf{R})\rangle \mathcal{U}^{\dagger ln'}(\mathbf{R}),$$

with the sum extending over all values of the electronic quantum numbers $\sum_l |l(\mathbf{R})\rangle\langle l(\mathbf{R})| = 1^{\text{fast}}$, then one obtains a pure gauge theory: $F_{ij}^{nm} = 0$.

8.4 According to Problem 8.3 the gauge theory corresponding to $\mathbf{A}^{nm} = i\langle n(\mathbf{R})|\nabla_\mathbf{R}|m(\mathbf{R})\rangle$ is trivial if a complete electronic basis is employed. Show that indeed $\mathbf{A}^{nm}$ is a pure gauge:

---

[28] A gauge structure in a coupled motion of rotation and vibration is reviewed in [157].

$$\mathbf{A}^{nm}(\mathbf{R}) = i \sum_l \mathcal{U}^{\dagger nl}(\mathbf{R}) \left(\nabla_{\mathbf{R}} \mathcal{U}^{lm}(\mathbf{R}) \right).$$

Obtain a unitary matrix $\mathcal{U}^{nm}(\mathbf{R})$ which gives this $\mathbf{A}^{nm}$.

8.5 a) Calculate the Mead–Berry connection

$$A_\theta^{k'k} = i \langle k'(R,\theta,\varphi)| \frac{\partial}{\partial \theta} |k(R,\theta,\varphi)\rangle,$$

$$A_\varphi^{k'k} = i k' \langle (R,\theta,\varphi)| \frac{\partial}{\partial \varphi} |k(R,\theta,\varphi)\rangle,$$

for the diatomic molecule with $\Omega = |k| = \frac{1}{2}$ (component of angular momentum along the internuclear axis) and express it in terms of the real parameter κ defined by $\langle k'(R,\mathbf{e}_3)|J_2|k(R,\mathbf{e}_3)\rangle = \frac{\kappa}{2} \sigma_2^{k'k}$ (σ_i = Pauli matrix).

b) Show that the Mead–Berry curvature for this case is given (in terms of its covariant spherical components) by

$$F_{\theta\varphi}^{k'k} = -\frac{1}{2}(1-k^2)\sin\theta\, \sigma_3^{k'k}.$$

9. Crossing of Potential Energy Surfaces and the Molecular Aharonov–Bohm Effect

9.1 Introduction

In the preceding chapter we have developed a general gauge theory of molecular physics. In this chapter, we shall examine problems associated with the conical intersections of potential energy surfaces.

Conical intersections are singularities of the Born–Oppenheimer approximation in the sense that at the location of a conical intersection the derivatives of electronic wave functions with respect to nuclear coordinates are not well-defined. The most conspicuous property of conical intersections is the sign-change of the electronic wave function which occurs if it is transported around a conical intersection of potential energy surfaces [105, 158]. This sign-change alters the boundary condition of the Schrödinger equation for the nuclear motion.

The first attempt to solve the nuclear motion in the presence of a conical intersection involved a realization of the sign-change boundary condition by attaching an additional phase factor to the wave function [203]. This is the earliest example of a *singular gauge transformation* that transforms a single-valued wave function to a multi-valued wave function. Later, the sign-change was taken into account by introducing a gauge potential, which created a delta function fictitious magnetic flux at the conical intersection. At this stage the gauge structure of the Born–Oppenheimer approximation became clear and this sign-change effect was named the *molecular Aharonov–Bohm effect* [165, 167].

The conical intersections found in early studies were associated with *Jahn–Teller problems* where conical intersections arise due to the symmetries of the states. This was the reason for the misunderstanding that conical intersections and the sign-change of electronic wave functions were exceptional properties solely relevant to Jahn–Teller systems. Later it became clear that this was not true. Indeed we now know numerous examples of conical intersections whose position has no particular symmetry [278].

Potential surfaces usually involve more than two parameters. In this case, the associated conical intersections are not isolated but form lines or higher-dimensional geometric objects in the parameter space of the system.

A line of conical intersections resembles the Dirac string – a line of singularities attached to the Dirac monopole [74] – in two ways:

1. The amplitude of the wave function is zero at the singular points.
2. A fictitious magnetic flux exists through the string of singularities.

Because of these similarities, a line of conical intersections may be called a *molecular Dirac string*.

There are also marked differences between the original and molecular Dirac strings:

1. The molecular Dirac strings are arranged in such a way that the corresponding monopoles are not observable.
2. The magnetic flux through the molecular Dirac strings does not necessarily guarantee the single-valuedness of the wave function. Instead they often give rise to the sign-change boundary conditions.
3. The molecular Dirac strings are only relevant to low-energy processes with energy lower than that of the conical intersections.

In this chapter, we explore some of the properties of conical intersections of potential energy surfaces in molecular systems.

9.2 Crossing of Potential Energy Surfaces

In the quantum description of a system in the approximation in which one considers only a finite-dimensional subspace of the Hilbert space, the Hamiltonian takes the form of a Hermitian matrix. If the Hamiltonian depends on a set of parameters, then so does the corresponding Hermitian matrix. In molecular systems, the electronic Hamiltonian depends on internal coordinates $\mathbf{Q}$. The eigenvalues of the electronic Hamiltonian determine the potential energy surfaces. We shall denote the latter by $\varepsilon_n(\mathbf{Q})$. The crossing of potential energy surfaces is nothing but a degeneracy of the energy eigenvalues.

We shall begin our description of conical intersections of potential energy surfaces by calculating the number of parameters that are necessary to have degenerate eigenvalues [261].

Consider an electronic Hamiltonian $h(\mathbf{Q})$ that depends on $\mathbf{Q}$, and express $h(\mathbf{Q})$ using a parameter-dependent basis $\{|n(\mathbf{Q}_0)\rangle : i = 1, \cdots, \mathcal{N}\}$, where $\mathbf{Q}_0$ is a fixed point in the parameter space. This yields for the matrix elements of h:

$$h_{nm}(\mathbf{Q}) = \langle n(\mathbf{Q}_0)|h(\mathbf{Q})|m(\mathbf{Q}_0)\rangle. \tag{9.1}$$

Because the matrix (h_{nm}) is Hermitian, it can be diagonalized by an $\mathcal{N} \times \mathcal{N}$ unitary matrix $\mathcal{U}$. We can express the matrix elements (9.1) in terms of the eigenvalues ε_k of (h_{nm}) and the matrix elements of $\mathcal{U}$ according to

$$h_{nm}(\mathbf{Q}) = \sum_{k=1}^{\mathcal{N}} \varepsilon_k(\mathbf{Q}) \mathcal{U}_{nk}(\mathbf{Q}) \mathcal{U}^*_{mk}(\mathbf{Q}). \tag{9.2}$$

We can use this expression to deduce the number of the independent parameters characterizing (h_{nm}).

We have $\mathcal{N}^2$ independent parameters describing the unitary matrix $\mathcal{U}$ and f independent parameters corresponding to the number of distinct eigenvalues ε_k, $k = 1, \cdots, \mathcal{N}$. This yields $\mathcal{N}^2 + f$ parameters from which we should subtract the squares of the dimension of the degeneracy subspaces. To see this suppose that the i-th eigenvalue is g_i-fold degenerate. Because a $g_i \times g_i$ diagonal matrix with the common diagonal element ε_i is invariant under transformation by any $g_i \times g_i$ unitary matrix, we must subtract g_i^2 from the sum $\mathcal{N}^2 + f$. Repeating this for all the eigenvalues, we find a total of

$$\mathcal{N}^2 + f - \sum_{k=1}^{f} g_i^2 \qquad (9.3)$$

independent parameters. According to this equation, the number of independent parameters at a point in the parameter space where all the eigenvalues are distinct is $\mathcal{N}^2$, as we have $g_1 = \cdots = g_{\mathcal{N}} = 1$ and $f = \mathcal{N}$.

Now, suppose that we move slightly away from the above-mentioned point such that one of the eigenvalues (say ε_1) becomes doubly degenerate. At this point in the parameter space, the number of independent parameters is $\mathcal{N}^2 - 3$, because in this case $f = \mathcal{N} - 1$, $g_1 = 2$, $g_2 = \cdots = g_{\mathcal{N}-1} = 1$. This exercise shows that moving from a point where eigenvalues are all different to a point where one of them becomes doubly degenerate causes the loss of $\mathcal{N}^2 - (\mathcal{N}^2 - 3) = 3$ degrees of freedom. This loss of three degrees of freedom may be interpreted in terms of the existence of a set of three conditions that are necessary for maintaining the double degeneracy. Because three parameters (in our case internal coordinates) are required to satisfy the three double-degeneracy conditions, a minimum of three parameters are needed in order to have a degeneracy of an eigenvalue. In other words, the parameter space must be at least three-dimensional.

A molecule with N_n nuclei has $3N_n - 6$ internal coordinates ($3N_n - 5$ for linear molecules). In particular, a diatomic molecule has only one parameter; it is two parameters short of having two potential surfaces cross (this is known as the "non-crossing rule"). A triatomic molecule has three parameters and crossings are possible. In general, crossings of potential energy surfaces are possible for molecules with more than two constituent atoms. As we shall explain below, for the case of Kramers doublets, however, this condition must be modified.

It often happens that the molecular Hamiltonian is real.[1] Then, we can choose a basis in which its matrix representation is a real symmetric ma-

---

[1] This means that in the position representation h may be represented by a real differential operator (i.e., $h^* = h$). More generally, there is a unitary transformation of the Hilbert space that maps h into a real differential operator. Equivalently, there is a complete set of eigenvectors of h whose eigenfunctions are real-valued functions. A Hamiltonian is real if and only if it has time-reversal

trix. In this case, the number of parameters required for the occurrence of a potential energy surface crossing is reduced. This is because an $\mathcal{N} \times \mathcal{N}$ real symmetric matrix can be diagonalized by an $\mathcal{N} \times \mathcal{N}$ orthogonal matrix which has $\mathcal{N}(\mathcal{N}-1)/2$ independent elements. Thus the number of independent parameters is

$$\mathcal{N}(\mathcal{N}-1)/2 + f - \sum_{k=1}^{f} g_i(g_i - 1)/2. \tag{9.4}$$

For example in order for a double degeneracy to occur, one needs two independent parameters. This is the case for a molecule with more than two constituent atoms.

9.3 Conical Intersections and Sign-Change of Wave Functions

In this section we examine wave functions near a conical intersection. In the following, we assume that the Hamiltonian is real. For a real Hamiltonian, a potential energy surface crossing can occur, only if there are more than two internal coordinates. Suppose that such a crossing occurs at a point in the two-dimensional internal coordinate space parameterized by (Q_1, Q_2). As shown in Fig. 9.1, if the energy gap between the two potential energy surfaces varies linearly with the distance from the crossing point along both the Q_1 and Q_2-directions, the intersection forms a diabolic cone called a *conical intersection*.

Within the framework of the Born-Oppenheimer approximation, as seen from (8.58), the *non-adiabatic transition* from one electronic state $|m(\mathbf{R})\rangle$ to another $|n(\mathbf{R})\rangle$ is governed by the term

$$\frac{\hbar}{\mu}\mathbf{A}^{nm}\cdot\mathbf{P} = \frac{i\hbar}{\mu}\langle n(\mathbf{R})|\nabla_\mathbf{R}|m(\mathbf{R})\rangle\cdot\mathbf{P}, \tag{9.5}$$

where we have restored $\hbar$ and assumed that $\mathbf{A}^{nm}$ with $n \neq m$ are so small that their square may be neglected.[2] Using the identity

$$\langle n(\mathbf{R})|[h, \nabla_\mathbf{R}]|m(\mathbf{R})\rangle = [\varepsilon_n(\mathbf{R}) - \varepsilon_m(\mathbf{R})]\langle n(\mathbf{R})|\nabla_\mathbf{R}|m(\mathbf{R})\rangle$$
$$= \langle n(\mathbf{R})|(\nabla_\mathbf{R} V)|m(\mathbf{R})\rangle, \tag{9.6}$$

---

symmetry, i.e., $\hat{T}h\hat{T} = h$ where $\hat{T}$ is the anti-unitary operator representing the time-reversal transformation. A real Hermitian Hamiltonian admits a complete set of eigenvectors $|n\rangle$ that are time-reversal invariant, i.e., $\hat{T}|n\rangle = |n\rangle$. In the position representation $\hat{T}$ maps a wave function to its complex conjugate. Therefore, $|n\rangle$ is represented by a real-valued wave function and is therefore said to be real.

[2] As we shall explain in the following, this assumption is not valid near a conical intersection.

9.3 Conical Intersections and Sign-Change of Wave Functions

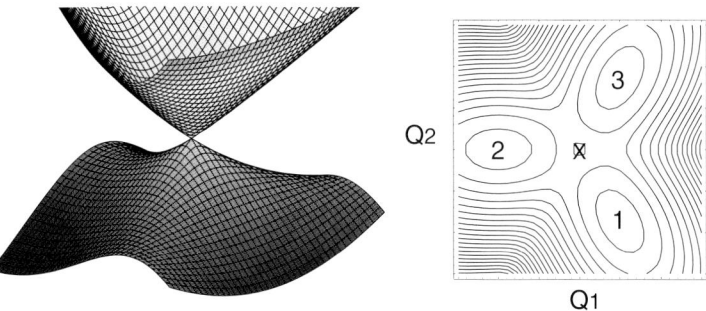

Fig. 9.1. Conical intersection of potential energy surfaces in the $E \otimes e$ Jahn–Teller system given by (9.57). Left: Two potential surfaces intersecting linearly, with respect to the change of Q_1 and Q_2 coordinates, and forming a *conical intersection*. Right: The contour plot of the lower surface. X indicates the point of crossing. 1, 2, and 3 denote three equivalent potential minima symmetrically placed around the point of crossing.

we can estimate the magnitude of $|\mathbf{A}^{nm}|$ as

$$|\mathbf{A}^{nm}| = \frac{|\langle n(\mathbf{R})|(\nabla_{\mathbf{R}} V)|m(\mathbf{R})\rangle|}{|\varepsilon_m(\mathbf{R}) - \varepsilon_n(\mathbf{R})|}, \tag{9.7}$$

where

$$h = \frac{1}{m_e}\mathbf{p}^2 + V(\mathbf{r}, \mathbf{R}) \tag{9.8}$$

is the electronic Hamiltonian parameterized by $\mathbf{R}$. The estimate (9.7) shows that if $V(\mathbf{R})$ is a slowly changing function of $\mathbf{R}$ which is usually the case, then $|\mathbf{A}^{nm}|$ is small except at the points where $|\varepsilon_m(\mathbf{R}) - \varepsilon_n(\mathbf{R})|$ tends to zero.

In order to quantify what we mean by "small" $|\mathbf{A}^{nm}|$, we introduce the "kinetic energy" at $\mathbf{R}$ on the potential surface $\varepsilon_n(\mathbf{R})$:

$$\frac{\hbar^2 k_n^2(\mathbf{R})}{2\mu} = E - \varepsilon_n(\mathbf{R}). \tag{9.9}$$

Then, the adiabaticity condition is equivalent to the requirement that the ratio of the magnitudes of (9.5) and (9.9) is negligibly small, i.e.,

$$1 \gg \frac{\frac{\hbar}{\mu}|\mathbf{A}^{nm}|P_{\text{nucl}}}{\frac{\hbar^2}{2\mu}|k_n^2(\mathbf{R}) - k_m^2(\mathbf{R})|} = \frac{\hbar v_{\text{nucl}}/\delta R}{|\varepsilon_n(\mathbf{R}) - \varepsilon_m(\mathbf{R})|}. \tag{9.10}$$

Here P_{nucl} and v_{nucl} represent the magnitudes of nuclear momentum and velocity, respectively, and $\delta R := |\mathbf{A}^{nm}|^{-1}$ is the characteristic length for the non-adiabatic transition. Note that the quantity on the right-hand side of (9.10) is dimensionless. Its inverse, namely $|\varepsilon_n(\mathbf{R}) - \varepsilon_m(\mathbf{R})|\delta R/(\hbar v_{\text{nucl}})$, is called the Massey parameter [62] which describes the effectiveness of an inelastic collision (an excitation in our case). Noting that the collision time $\tau := \delta R/V_{\text{nucl}}$ is related to the energy uncertainty $\hbar/\tau$, we find that the condition for ineffectiveness of the excitation with excitation energy $\Delta E = |\varepsilon_n(\mathbf{R}) - \varepsilon_m(\mathbf{R})|$ has the form

$$\hbar/\tau = \hbar V_{\text{nucl}}/\delta R \ll \Delta E = |\varepsilon_n(\mathbf{R}) - \varepsilon_m(\mathbf{R})|. \tag{9.11}$$

This is equivalent to the adiabaticity condition (9.10).[3]

In the following, we consider a pair of potential energy surfaces with a conical intersection at $\mathbf{R}_0$, and let $|1(\mathbf{R})\rangle$ and $|2(\mathbf{R})\rangle$ be the corresponding orthonormal and single-valued eigenvectors. We assume that in the vicinity of $\mathbf{R}_0$ we can neglect the effects of other potential energy surfaces, i.e., the condition (9.10) is satisfied except for $n, m = 1, 2$. Hence near $\mathbf{R}_0$ we can express the total wave functions in terms of the wave functions associated with $|1(\mathbf{R}_0)\rangle$ and $|2(\mathbf{R}_0)\rangle$. Since the Hamiltonian is real, we can assume without loss of generality that these wave functions are real-valued.

Denoting by h_{ij} (with $i, j = 1, 2$) the matrix elements of the electronic Hamiltonian h in this basis, we can identify h with the matrix

$$\begin{pmatrix} h_{11} & h_{12} \\ h_{21} & h_{22} \end{pmatrix}. \tag{9.12}$$

Alternatively, we introduce the basis vectors

$$|\phi\rangle := \frac{1}{\sqrt{2}}(|1(\mathbf{R}_0)\rangle + i|2(\mathbf{R}_0)\rangle),$$

$$|\phi^*\rangle := \frac{1}{\sqrt{2}}(|1(\mathbf{R}_0)\rangle - i|2(\mathbf{R}_0)\rangle),$$

In this basis h takes the form

$$\begin{pmatrix} V_{11} & V_{12} \\ V_{21} & V_{22} \end{pmatrix}, \tag{9.13}$$

where the matrix elements V_{ij} and h_{ij} are related according to

$$V_{11} = (h_{11} + h_{22})/2 + \Im\{h_{12}\}, \tag{9.14}$$
$$V_{12} = V_{21}^* = (h_{11} - h_{22})/2 - i\Re\{h_{12}\}, \tag{9.15}$$
$$V_{22} = (h_{11} + h_{22})/2 - \Im\{h_{12}\}, \tag{9.16}$$

and $\Re\{h_{12}\}$ and $\Im\{h_{12}\}$ stand for the real and imaginary parts of h_{12}, respectively. Note that for a real Hamiltonian h, $\Im\{h_{12}\} = 0$ and $V_{11} = V_{22}$.

The potential energy surfaces are determined by the eigenvalues of either of the matrices (9.12) or (9.13),

---

[3] For a recent development on non-adiabatic effects, consult [187].

9.3 Conical Intersections and Sign-Change of Wave Functions

$$\varepsilon_{\pm} = \frac{h_{11} + h_{22}}{2} \pm \sqrt{\left(\frac{h_{11} - h_{22}}{2}\right)^2 + |h_{12}|^2}$$

$$= \frac{V_{11} + V_{22}}{2} \pm \sqrt{\left(\frac{V_{11} - V_{22}}{2}\right)^2 + |V_{12}|^2}. \tag{9.17}$$

The electronic wave functions corresponding to ε_+ and ε_- are, respectively, given by

$$\psi_+ = \cos\frac{\alpha}{2} e^{-i\beta/2}\phi + \sin\frac{\alpha}{2} e^{i\beta/2}\phi^*, \tag{9.18}$$

and

$$\psi_- = -\sin\frac{\alpha}{2} e^{-i\beta/2}\phi + \cos\frac{\alpha}{2} e^{i\beta/2}\phi^*, \tag{9.19}$$

where α and β are defined according to

$$\cos\alpha := \frac{V_{11} - V_{22}}{\sqrt{(V_{11} - V_{22})^2 + 4|V_{12}|^2}}, \tag{9.20}$$

$$V_{12} =: |V_{12}|e^{i\beta}. \tag{9.21}$$

Note that for a real h, $V_{11} = V_{22}$ and $\cos\alpha = 0$.

Having obtained the expression for the electronic wave functions, we can compute the corresponding geometric phases. Let us consider the phase change after a circular transport along a loop in the nuclear coordinate space. Because of the single-valuedness of V_{12}, the difference of the values of the phase angle β at the beginning and the end of the loop must be a multiple of 2π. Denoting these values by β_1 and β_2, we have

$$\beta_2 - \beta_1 = 2n\pi, \tag{9.22}$$

where n is an integer. If n happens to be odd ψ_- and ψ_+ change sign after encircling the loop ($\beta \to \beta + 2\pi n$). Thus, they are double-valued. In order to calculate the geometric phase, we need to construct single-valued wave functions. We can change a double-valued wave function into a single-valued wave function by multiplying the former by a phase factor that depends only on the nuclear coordinates. For example, a single-valued wave function for the lower potential energy surface is given by

$$\psi = e^{i\beta/2}\psi_- = -\sin\frac{\alpha}{2}\phi + \cos\frac{\alpha}{2}e^{i\beta}\phi^*. \tag{9.23}$$

The same phase transformation on φ_+ yields a single-valued electronic wave function for ε_+.

Now we are in a position to calculate the Mead–Berry vector and induced scalar potentials using the single-valued wave function ψ. A straightforward calculations yields the following expressions for the components

$A_j := i\langle\psi|\frac{\partial}{\partial Q_j}|\psi\rangle$ of the Mead–Berry vector potential $\mathbf{A}$ and the induced scalar potential $\tilde{\varepsilon}_-$:

$$A_j = -\frac{1}{2}\frac{\partial \beta}{\partial Q_j}(1+\cos\alpha), \qquad (9.24)$$

$$\tilde{\varepsilon}_- = \varepsilon_- + \frac{1}{8}(\nabla_\mathbf{Q}\beta)^2, \qquad (9.25)$$

where

$$\nabla_\mathbf{Q} := \left(\frac{\partial}{\partial Q_1}, \ldots, \frac{\partial}{\partial Q_{3N_n-6}}\right). \qquad (9.26)$$

and the nuclear kinetic energy operator is supposed to have the form

$$T_{\text{nuclei}} = -\frac{1}{2}\nabla_\mathbf{Q}^2. \qquad (9.27)$$

In view of (9.24), the Mead–Berry connection one-form is given by

$$A := \sum_j A_j dQ_j = -\frac{1}{2}(1+\cos\alpha)d\beta. \qquad (9.28)$$

The Mead–Berry vector potential (respectively connection) can be split into two parts [140], namely a *topological part*:

$$A_i^{\text{top}} = -\frac{1}{2}\frac{\partial \beta}{\partial Q_i}, \qquad (9.29)$$

and a *magnetic part*:

$$A_i^{\text{mag}} = -\frac{1}{2}\frac{\partial \beta}{\partial Q_i}\cos\alpha. \qquad (9.30)$$

The topological part gives rise to a geometric phase which depends solely on the topological character of the trajectory traced by the nuclear coordinates in the parameter space, i.e., it is a *topological phase*. The corresponding phase angle is calculated as

$$\oint \sum_{i=1}^{3N_n-6} A_i^{\text{top}} dQ_i = -\frac{1}{2}\oint d\beta = -\frac{1}{2}(\beta_2 - \beta_1) = -n\pi. \qquad (9.31)$$

This phase angle is non-zero provided that the closed path in the complex V_{12} plane (which is obtained as the image of the path traced by the nuclear coordinates in the nuclear coordinate space) encircles the origin $V_{12} = 0$. According to (9.15), at the origin both $h_{11}-h_{22}$ and $\Re\{h_{12}\}$ vanish. Therefore the origin is a point of potential energy surface crossing provided that $\Im\{h_{12}\}$ is zero, i.e., h is real.

9.3 Conical Intersections and Sign-Change of Wave Functions

The topological phase only depends on the winding number of the image of the closed path in the nuclear coordinate space around the origin. It is $+1$ if the winding number is even and -1 if it is odd. This phase factor is not altered even if a different gauge is used to obtained the single-valued electronic wave functions. For example, we can construct a new set of single-valued wave functions by multiplying the double-valued wave functions $\psi_\pm$ by $e^{i(2\ell+1)\beta/2}$ where ℓ is a non-zero integer. This choice modifies the Mead–Berry vector potential, but does not change the topological phase factor.

The magnetic part (9.30) of the Mead–Berry vector potential is non-zero only if the Hamiltonian h is not real, i.e., the system has a broken time reversal symmetry. For a real h, $\cos\alpha = 0$ and the A_i^{mag} vanishes. In general, A_i^{mag} is not sensitive to the choice of gauge.

The components of the Mead–Berry gauge field strength (curvature two-form) corresponding to the (total) vector potential (9.28) is given by

$$F_{jk} = \frac{\partial A_k}{\partial Q_j} - \frac{\partial A_j}{\partial Q_k}$$
$$= \frac{\partial A_k^{\mathrm{mag}}}{\partial Q_j} - \frac{\partial A_j^{\mathrm{mag}}}{\partial Q_k}$$
$$= \frac{\sin\alpha}{2}\left(\frac{\partial \alpha}{\partial Q_j}\frac{\partial \beta}{\partial Q_k} - \frac{\partial \alpha}{\partial Q_j}\frac{\partial \beta}{\partial Q_k}\right). \tag{9.32}$$

Therefore, the Mead–Berry curvature two-form has the form

$$F := dA = \frac{1}{2}\sin\alpha\, d\alpha \wedge d\beta. \tag{9.33}$$

This equation can be obtained using (9.28) or by substituting (9.32) in

$$F = \frac{1}{2}F_{ij}dQ_i \wedge dQ_j. \tag{9.34}$$

Comparing the expression (9.33) for the Mead–Berry curvature two-form with (3.45) and using (9.20) and (9.21), we notice that (9.33) corresponds to the magnetic field of a monopole in three-dimensional (X, Y, Z) space, where X, Y, and Z are given by

$$X := \sqrt{(V_{11} - V_{22})^2/4 + |V_{12}|^2}\cos\beta \sin\alpha, \tag{9.35}$$
$$Y := \sqrt{(V_{11} - V_{22})^2/4 + |V_{12}|^2}\sin\beta \sin\alpha, \tag{9.36}$$
$$Z := \sqrt{(V_{11} - V_{22})^2/4 + |V_{12}|^2}\cos\alpha. \tag{9.37}$$

Now, in order to make our discussion more concrete, we consider a model which is suitable for describing certain types of molecules, e.g., Na$_3$ and Li$_3$. This is a model for a system of a doubly degenerate electronic state (E) and a doubly degenerate vibrational mode (E) in equilateral triangular configurations. As we explain in Appendix B, the point-group symmetry of such a system is D_{3h}, and this problem is called the $E \otimes e$ Jahn–Teller problem. If we treat the e mode as a two-dimensional isotropic harmonic oscillator and include only linear vibronic coupling, we obtain the following vibronic Hamiltonian from (B.52):

$$-\frac{1}{2}\left(\frac{\partial^2}{\partial Q_1^2} + \frac{\partial^2}{\partial Q_2^2}\right)\sigma_0 + \frac{\omega^2}{2}(Q_1^2 + Q_2^2)\sigma_0 + \begin{pmatrix} -kQ_1 & kQ_2 \\ kQ_2 & kQ_1 \end{pmatrix}, \quad (9.38)$$

where σ_0 is the 2×2 unit matrix, Q_1, Q_2 are two E symmetry normal coordinates, ω is the angular frequency of the oscillator, and k is the linear vibronic coupling constant. The 2×2 matrix Hamiltonian (9.38) acts on the two-dimensional space of electronic states.

Diagonalizing the vibronic Hamiltonian, we obtain the potential energy surfaces:

$$\varepsilon_\pm(\rho, \varphi) = \frac{1}{2}\omega^2\rho^2 \pm k\rho, \quad (9.39)$$

where $\rho := \sqrt{Q_1^2 + Q_2^2}$ and $\varphi := \tan^{-1}(Q_2/Q_1)$. Note that $\varepsilon_\pm$ do not depend on φ; in this approximation all distortions with the same value of ρ have the same energy. Figure 9.2 shows a series of isoenergetic configurations of a hypothetical X_3 molecule. This type of motion is called *pseudorotation*.

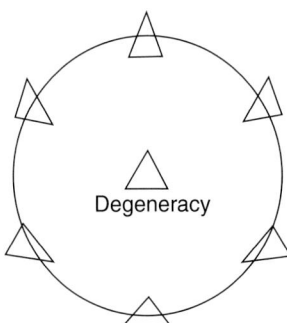

Fig. 9.2. Pseudorotation of a trimer. At the center is the equilateral triangle configuration, the point of electronic degeneracy with nuclear coordinate $\rho = 0$. Different shapes, at fixed $\rho \neq 0$ and variable φ, are shown arrayed around the origin. To first order in the vibronic coupling all are isoenergetic.

Comparing the expressions (9.38) and (9.13) for the Hamiltonian, we find

$$V_{11} = V_{22} = \frac{\omega^2}{2}\rho^2, \quad V_{12} = k\rho e^{i(\varphi+\pi)}. \quad (9.40)$$

Hence the two angles in (9.19) are $\alpha = \pi/2$ and $\beta = \varphi + \pi$, and the electronic wave function corresponding to $\varepsilon_-(\rho, \varphi)$ takes the form

9.3 Conical Intersections and Sign-Change of Wave Functions

$$\psi_- = \frac{i}{\sqrt{2}}(e^{-i\varphi/2}\phi - e^{i\varphi/2}\phi^*). \tag{9.41}$$

This wave function is *double-valued* with respect to encircling the origin ($\varphi = 0 \to 2\pi$). The single-valued wave function (9.23) is given by

$$\psi = \frac{i}{\sqrt{2}}(\phi - e^{i\varphi}\phi^*). \tag{9.42}$$

Next we calculate the Born–Huang term (the second term on the right-hand side of (9.25)),

$$\frac{1}{8}(\nabla_\mathbf{Q}\beta)^2 = \frac{1}{8\rho^2}. \tag{9.43}$$

This term which diverges at the crossing point ($\rho = 0$) is also called the *vibronic centrifugal term*. The reason for this terminology is that (9.43) diverges as the inverse square of distance from the conical intersection just like any other centrifugal potential. This behavior arises from the φ-dependence of ψ. Because of this divergence the amplitude of the nuclear wave function is zero at the conical intersection, i.e., conical intersections are inaccessible points for low-energy processes. Therefore, the effective parameter space for low-energy processes is multiply connected.

Having obtained the single-valued electronic wave functions, we can calculate the Mead–Berry connection one-form (9.42). The result is

$$A := i\langle\psi|d|\psi\rangle = -\frac{1}{2}d\varphi. \tag{9.44}$$

Expressing A in the components (Q_1, Q_2) and (φ, ρ),

$$A = A_1 dQ_1 + A_2 dQ_2 = A_\rho d\rho + A_\varphi d\varphi, \tag{9.45}$$

we obtain the corresponding components of the Mead–Berry connection one-form:

$$A_1 = -\frac{1}{2}\frac{\partial\varphi}{\partial Q_1} = \frac{Q_2}{2(Q_1^2 + Q_2^2)}, \tag{9.46}$$

$$A_2 = -\frac{1}{2}\frac{\partial\varphi}{\partial Q_2} = -\frac{Q_1}{2(Q_1^2 + Q_2^2)}, \tag{9.47}$$

and

$$A_\rho = -\frac{1}{2}\frac{\partial\varphi}{\partial\rho} = 0, \tag{9.48}$$

$$A_\varphi = -\frac{1}{2}\frac{\partial\varphi}{\partial\varphi} = -\frac{1}{2}. \tag{9.49}$$

The phase change of the wave function after an adiabatic transportation around the conical intersection is the line integral of the connection one-form,

$$\gamma(C) = \oint_C A = -\pi. \tag{9.50}$$

The single-valued wave function ψ acquires a phase angle $-\pi$, i.e., it changes sign, upon transportation around the point of degeneracy. This is similar to parallel transportation on a Möbius strip, in the sense that transporting once around changes the sign of the wave function, while transporting twice around gives no sign-change.

As we discussed in the preceding sections, the Mead–Berry connection (vector potential) may be viewed as the vector potential for a fictitious magnetic field which has a non-zero flux only in the conical intersection. It is this fictitious magnetic field that causes the extra phase factor – the sign-change of the wave function. This phase factor effects the interference pattern of the system. Figure 9.3 shows the change of the interference pattern caused by the fictitious magnetic field in the wave packet propagation with and without the Mead–Berry vector potential. The nuclear wave packet picks up an extra phase (-1), due to the Mead–Berry vector potential, yielding destructive interference in the diametrically opposite point to the center of the initial wave packet. In contrast, neglecting the Mead–Berry vector potential produces constructive interference.

We now calculate the energy spectrum associated with the pseudorotation. In view of (9.38), the effective nuclear Hamiltonian for the lower potential energy surface has the from

$$-\frac{1}{2}\frac{\partial^2}{\partial \rho^2} + \frac{1}{2\rho^2}\left(-i\frac{\partial}{\partial \varphi} - A_\varphi\right)^2 + \frac{\omega^2}{2}\rho^2 - k\rho + \frac{1}{8\rho^2}. \tag{9.51}$$

We obtain the Hamiltonian for the pseudorotational motion for φ by fixing ρ at an average value ρ_0. This yields

$$B\left(-i\frac{\partial}{\partial \varphi} - A_\varphi\right)^2 = B\left(-i\frac{\partial}{\partial \varphi} + \frac{1}{2}\right)^2, \tag{9.52}$$

where $B = 1/(2\rho_0^2)$ is the pseudorotational constant. This Hamiltonian leads to a ladder of states with energies Bj^2 where j is the pseudorotational angular momentum. Note that because $A_\varphi = -1/2$, $|j| = 1/2, 3/2, \ldots$. The levels with $|j| = 1/2, 5/2, \ldots$ are doubly degenerate with the degeneracy subspace spanned by the basis vectors having $j = \pm 1/2$, $j = \pm 5/2$, etc. The levels with $|j| = 3/2, 9/2, \ldots$, i.e., those with $2j$ equal to a multiple of 3, are accidentally degenerate. This means that one can lift the degeneracy by including higher-order vibronic coupling terms.

Next we discuss an interesting symmetry test for the geometric phase which was originally proposed in [99]. As we explained above, we label the states by the pseudorotational quantum number j which, in the presence of the geometric phase, is quantized as a half-odd-integer. We can also label the states according to their transformation properties under the action of

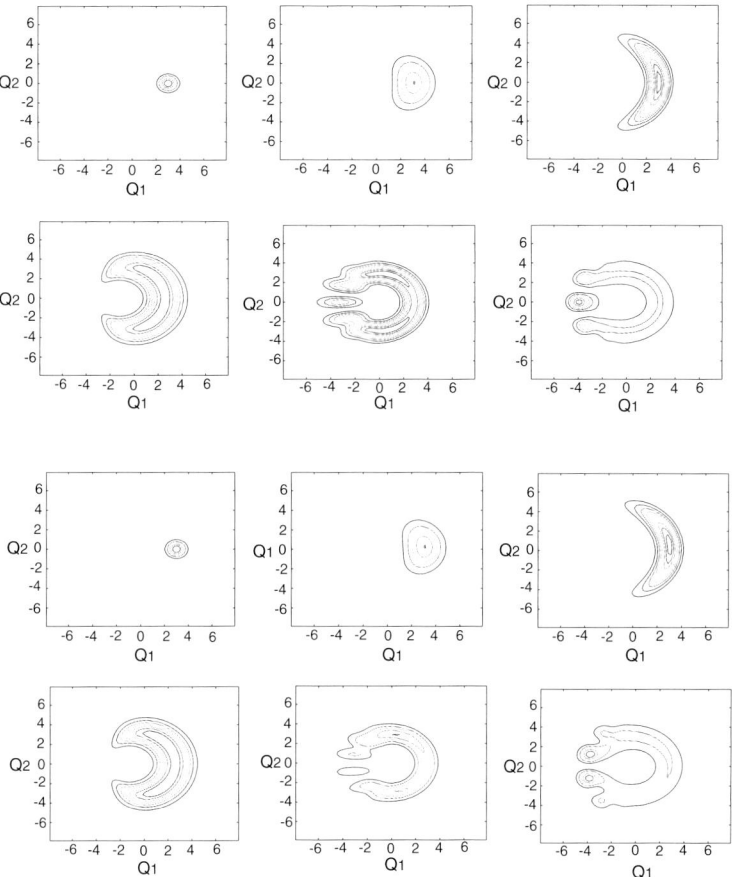

Fig. 9.3. Wave packets motion on the lower potential surface of the linear $E \otimes e$ problem. The initial wave packet is $f = \exp(-(Q_1-3)^2 - Q_2^2)$ and $|f|^2$ are plotted at $t = 0, 1.5, 3, 4.5, 6$, and 7.5. The upper (lower) six panels are results obtained by including (excluding) the vector potential. The linear vibronic coupling parameter k is 3 and the oscillator frequency ω is set to 1. Note that the interference of the wave packet in the opposite to the initial position is constructive for the result neglecting the vector potential, while it is destructive for that including the vector potential.

the point group of the molecule. For the triangular case which we consider here this is the group D_{3h} (see Appendix B for more details). This group has two inequivalent irreducible representations: the one-dimensional representation, which are labeled by A, and the two-dimensional representations which are labeled by E. The $j = 1/2$ state, which is doubly degenerate and thus transforms as the irreducible representation E and said to have symmetry E for short, is the ground state of the Hamiltonian. The states with $j = 3/2$, which transform as the irreducible representation A and said to have symmetry A, are the first excited states. If we exclude the geometric phase, the quantization of j would be integral; the ground state would have $j = 0$ (an A state) and the first excited states would have $j = 1$ and form a pair of E states. Therefore we can infer the presence of the geometric phase from the symmetries of the states alone. This conclusion is significant because the spectroscopic determination of the symmetry of a state can be performed in a model-independent way, as opposed to a spectral fitting procedure which inherently depends on a choice of the model. We shall present a further discussion of the relationship between the ground state symmetry and the conical intersection later in this chapter.

As we explained above conical intersections play an important role in the quantum mechanical description of a molecular. We shall next discuss how one can locate a conical intersection.

One way to determine the location of a conical intersection is to use the sign-change of the wave function [158].[4] For this purpose the *Pancharatnam connection* is most convenient [208, 224]. The Pancharatnam connection[5] defines the phase of the wave function at $\mathbf{R}_A$ in relation to that at $\mathbf{R}_B$, so as to maximize

$$||n(\mathbf{R}_A)\rangle + |n(\mathbf{R}_B)\rangle|^2 = 2 + 2\Re\{\langle n(\mathbf{R}_A)|n(\mathbf{R}_B)\rangle\}. \qquad (9.53)$$

For a real Hamiltonian $h(\mathbf{R})$, the wave functions associated with $|n(\mathbf{R}_A)\rangle$ and $|n(\mathbf{R}_B)\rangle$ can be chosen real. Then, the use of the Pancharatnam connection amounts to choosing the sign of $|n(\mathbf{R}_B)\rangle$ so that $\langle n(\mathbf{R}_A)|n(\mathbf{R}_B)\rangle$ is positive. In the limit $\mathbf{R}_B \to \mathbf{R}_A$, this connection is identified with

$$\mathbf{A}^{nn} = i\langle n(\mathbf{R}_A)|\nabla_{\mathbf{R}}|n(\mathbf{R}_B)\rangle = 0. \qquad (9.54)$$

A practical way of finding conical intersections is to take a closed loop in the nuclear coordinate space and transport the state vector along this loop using the Pancharatnam connection. This is done by taking N points along the loop[6] $\mathbf{R}(0), \mathbf{R}(1), \cdots, \mathbf{R}(N-1), \mathbf{R}(N) = \mathbf{R}(0)$. After a circular transportation using the Pancharatnam connection, one obtains the final state vector $|n(\mathbf{R}(N))\rangle$ which may differ from the initial state vector $|n(\mathbf{R}(0))\rangle$ by a phase factor, although $\mathbf{R}(0)$ and $\mathbf{R}(N)$ coincide. If an odd number of conical intersections exist within the loop, the sign-change should occur. In this case, one can find the position of the conical intersections by shrinking the loop systematically [83].

---

[4] Another efficient method is to find zeros of $(\varepsilon_n(\mathbf{Q}) - \varepsilon_{n+1}(\mathbf{Q}))^2$ and check if they are really conical intersections. See also [279] for a more elaborate method.

[5] "Connection" here means a tool to compare phases of two wave functions at different points in the parameter space. Readers who are interested in the relation between the Pancharatnam connection and the usual mathematical definition, i.e., an assignment of a horizontal subspace, should consult [224].

[6] This can be as few as three points.

As is manifested by the sign-change of the wave function, the existence of a conical intersection is a topological property that is resilient against small perturbations. This is shown in a very simple way: If a small perturbation is applied, the potential part of the Hamiltonian (9.38) is modified to

$$\begin{pmatrix} -kQ_1 + a & kQ_2 + b \\ kQ_2 + b & kQ_1 + c \end{pmatrix}, \tag{9.55}$$

where $a, b, c,$ and d are perturbation parameters. This perturbation leads to a shift of the location of the conical intersection to $(Q_1, Q_2) = (c-a)/2k, -b/k)$, but it does not eliminate it.

Next, we consider the case where one more nuclear coordinate (Q_3) is present, and study its effect on the position of a conical intersection. Suppose that a conical intersection exists at a certain point in the Q_1-Q_2 plane for a fixed value of Q_3, say $Q_3 = 0$. Now move Q_3 slightly away from zero. This leads to a small perturbation of the electronic Hamiltonian corresponding to its original value for $Q_3 = 0$. This perturbation in turn shifts the position of the conical intersection in the Q_1-Q_2 plane. If we keep changing Q_3, the position of the conical intersection will also move in the Q_1-Q_2 plane. By connecting the position of the conical intersection for different values of Q_3, we obtain a trajectory of conical intersections in Q_1-Q_2-Q_3 space. This is actually a demonstration of a theorem stating that conical intersections are not isolated but exist as parts of a curve or a higher-dimensional geometrical object when the dimension of the parameter space is greater than two [83].

9.4 Conical Intersections in Jahn–Teller Systems

Jahn–Teller systems involve a variety of conical intersections. Therefore, they provide a suitable testing ground for studying geometric phases. In this section we study some of the conical intersections occurring in Jahn–Teller systems.

$E \otimes e$ system

Setting $Q_\epsilon = Q_1$, $Q_\theta = Q_2$, $\omega = 1$, $W_E^{e \times e} = g/2$, $V_E = k$, and all other parameters to zero in (B.52), we obtain the following $E \otimes e$ Jahn–Teller vibronic Hamiltonian:

$$-\frac{1}{2}\left(\frac{\partial^2}{\partial Q_1^2} + \frac{\partial^2}{\partial Q_2^2}\right)\sigma_0 + \frac{1}{2}(Q_1^2 + Q_2^2)\sigma_0$$
$$+ \begin{pmatrix} \frac{g}{2}(Q_1^2 - Q_2^2) - kQ_1 & gQ_1Q_2 + kQ_2 \\ gQ_1Q_2 + kQ_2 & \frac{g}{2}(Q_2^2 - Q_1^2) + kQ_1 \end{pmatrix}. \tag{9.56}$$

Next, we diagonalize the potential energy part of this Hamiltonian. This yields the corresponding potential energy surfaces, namely

$$\varepsilon_\pm = \frac{1}{2}\rho^2 \pm \sqrt{k^2\rho^2 + 2kg\rho^3 \cos(3\varphi) + \frac{1}{4}g^2\rho^4}. \qquad (9.57)$$

From this we find for the location of the conical intersections: $(Q_1, Q_2)=(0,0)$ and $(Q_1, Q_2) = (\frac{2k}{3g} \cos \frac{2\pi n}{3}, \frac{2k}{3g} \sin \frac{2\pi n}{3})$, with $n = 0, 1, 2$.

The $E \otimes e$ Jahn–Teller system arises in modeling triatomic molecules. Figure 9.4(a) shows the three normal coordinates that are often used as internal coordinates for triatomic systems. Electronic states are doubly degenerate (E states) at equilateral triangular configurations (D_{3h} symmetry points). The degeneracy is lifted by displacement of E symmetry normal modes Q_1 and Q_2 but survives the changes of the totally symmetric normal coordinate Q_3. Because a shift of Q_3 does not destroy the D_{3h} symmetry, conical intersections occur along the line $(Q_1, Q_2, Q_3) = (0, 0, q)$. As seen in Fig. 9.1, the lower potential energy surface has three equivalent minima around the equilateral triangular position $(Q_1, Q_2) = (0, 0)$.

The $E \otimes e$ system also arises in vibronic problems of impurity ions in the cubic symmetry site of crystals, e.g., Cu^{2+} in MgO, and octahedrally coordinated ions. Figure B.2 shows the important normal vibrational modes for octahedral ion complexes.

As we explained in the preceding section, a small perturbation of the system does not destroy conical intersections. In the following we present an interesting demonstration of this using the H_3 + H system. The triatomic molecule H_3 is an $E \otimes e$ Jahn–Teller system. Adding another H atom to the H_3 system produces a perturbation that displaces the conical intersections of the H_3 system. Fig. 9.4(b) shows the resulting displacement of the conical intersection due to a movement of the additional H atom along a line $(Q_1, Q_2, Q_6) = (0, 0.2a_0, q)$, where a_0 is the Bohr radius and q is the parameter for the line. The conical intersection moves from its original position: $(Q_1, Q_2) = (0, 0)$ to $(Q_1, Q_2) = (0, Q_2 \neq 0)$.

In analogy with the Dirac string, a Dirac-monopole-like singularity occurs, if a line of conical intersection is terminated at a point. This does not happen in molecular systems. Lines of conical intersections are arranged in a way that monopole singularities do not show up. In fact, conical intersections disappear in pairs. This is known as the pair annihilation of conical intersections. Figure 9.4(c) shows pair annihilation of conical intersections in the H_3 + H system where one can see the approaching, merging, and disappearance of a pair of conical intersections.

$T \otimes t_2$ system

The $T \otimes t_2$ Jahn–Teller system is a simplified version of the $T \otimes (t_2 + e)$ system. By setting $\omega_T = 1$, $V_T = -k$, $W_E^{t \times t} = g/2$, $Q_\xi = Q_1$, $Q_\eta = Q_2$, $Q_\zeta = Q_3$, and all the other parameters to zero in (B.54), we find the matrix Hamiltonian:

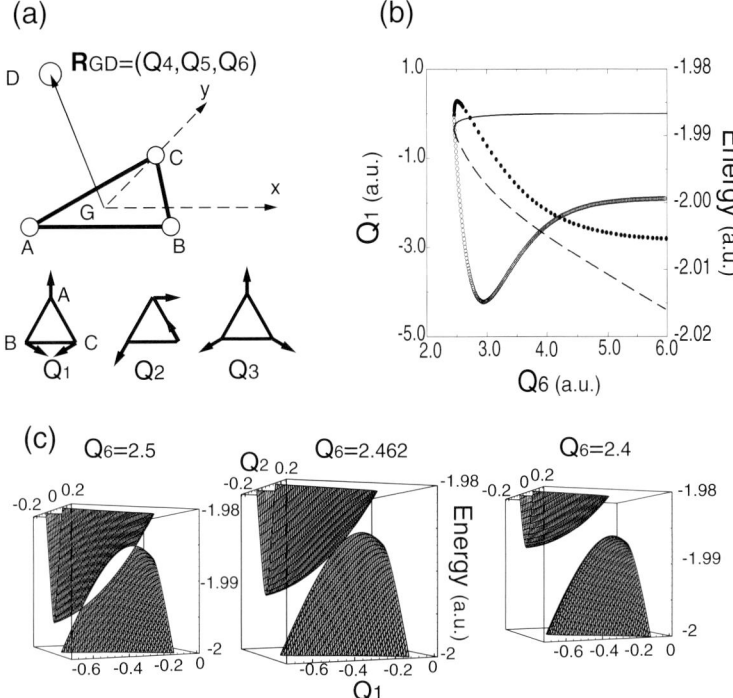

Fig. 9.4. Coordinates for the H$_3$ + H system (**a**), lines of conical intersection (**b**), and a pair annihilation of conical intersections (**c**). (**a**) Q_1, Q_2 are the E, and Q_3 is the A normal vibrational modes. Q_4, Q_5, Q_6 are Cartesian components of the vector from the center of mass of ABC to D. (**b**) The atom D moves along the straight line parallel to the z-axis $(Q_4, Q_5, Q_6) = (0, 0.2, q)$ (a.u.). The solid (dashed) line describes Q_3 values for the lines of conical intersections with the corresponding energy shown by filled (open) circles. Here the potential energy surfaces are calculated by the simple valence bond method [83]. (**c**) Three-dimensional plots of the lowest two potential surfaces, showing pair annihilation of conical intersections in the Q_1-Q_2 plane with increasing Q_6 value. In the left figure ($Q_6 = 2.5$), a pair of conical intersections are seen. They merge into one in the middle ($Q_6 = 2.462$), and are lifted in the right ($Q_6 = 2.4$).

$$-\frac{1}{2}\sum_{i=1}^{3}\frac{\partial^2}{\partial Q_i^2}\lambda_0 + \frac{1}{2}(Q_1^2+Q_2^2+Q_3^2)\lambda_0$$

$$+\begin{pmatrix} \frac{g}{2}(-2Q_1^2+Q_2^2+Q_3^2) & kQ_3 & kQ_2 \\ kQ_3 & \frac{g}{2}(Q_1^2-2Q_2^2+Q_3^2) & kQ_1 \\ kQ_2 & kQ_1 & \frac{g}{2}(Q_1^2+Q_2^2-2Q_3^2) \end{pmatrix}. \quad (9.58)$$

At $(Q_1, Q_2, Q_3) = (q, q, q), (q, -q, -q), (-q, q, -q)$, and $(-q, -q, q)$, the potential energy surfaces have the form

$$\varepsilon = \begin{cases} \frac{3}{2}q^2 - kq & \text{(two fold)} \\ \frac{3}{2}q^2 + 2kq. \end{cases} \quad (9.59)$$

This expression shows that for $k > 0$ four lines of conical intersection occur between the first and second lowest potential energy surfaces. They start from $(Q_1, Q_2, Q_3) = (0, 0, 0)$ and extend in the directions of $(1, 1, 1)$, $(1, -1, -1)$, $(-1, 1, -1)$, and $(-1, -1, 1)$, respectively (Fig. 9.5). These lines of conical intersections continue in the opposite directions, i.e., $(-1, -1, -1)$, $(-1, 1, 1)$, $(1, -1, 1)$, and $(1, 1, -1)$, as well. However now they are intersections between the second and third potential energy surfaces. Note that they are arranged in such a way that the monopole singularities do not arise.

The boundary conditions arising from these lines of conical intersection cause the ground state of the $T \otimes t_2$ problem to be a triply degenerate T symmetry state [100, 204]. We will present a more detailed discussion of this point in a following section.

At $(q, q, \frac{2k}{3g})$, $(q, -q, -\frac{2k}{3g})$, $(q, \frac{2k}{3g}, q)$ $(q, -\frac{2k}{3g}, -q)$, $(\frac{2k}{3g}, q, q)$, and $(-\frac{2k}{3g}, q, -q)$, the potential energy surfaces are given by

$$\varepsilon = \begin{cases} \frac{2-g}{2}q^2 + \frac{1-2g}{2}\left(\frac{2k}{3g}\right)^2 & \text{(two fold)} \\ \frac{1+g}{2}q^2 + \frac{1+4g}{2}\left(\frac{2k}{3g}\right)^2. \end{cases} \quad (9.60)$$

In view of (9.60), we see that there are six additional lines of conical intersection between the first and second lowest potential energy surfaces, if $g > 1/2$. They occur along the edges of the tetrahedron with vertices $(\frac{2k}{3g}, \frac{2k}{3g}, \frac{2k}{3g})$, $(\frac{2k}{3g}, -\frac{2k}{3g}, -\frac{2k}{3g})$, $(-\frac{2k}{3g}, \frac{2k}{3g}, -\frac{2k}{3g})$, and $(-\frac{2k}{3g}, -\frac{2k}{3g}, \frac{2k}{3g})$. It is remarkable that these additional conical intersections can change the ground

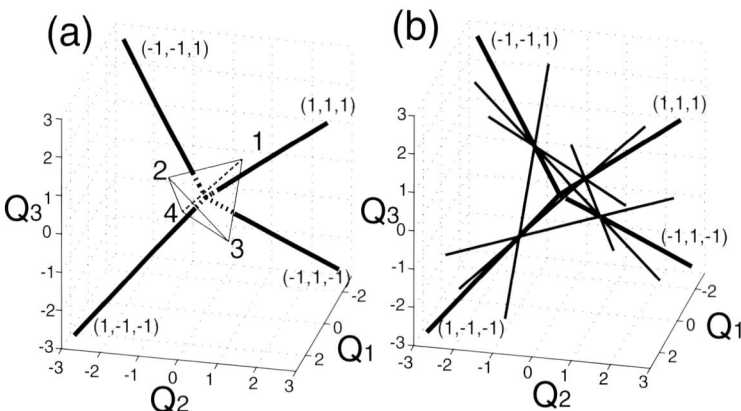

Fig. 9.5. Lines of conical intersection in the $T \otimes t_2$ Jahn–Teller system. (**a**) Four lines of conical intersection that penetrate faces of the tetrahedron formed by connecting trigonal minima of the lowest potential energy surface. (**b**) Six additional lines of conical intersection (*thin lines*) that arise if the quadratic coupling parameter g is larger than $1/2$. The original four lines of conical intersections in (**a**) are also shown (*thick lines*).

state symmetry from T to (the non-degenerate) A. This is done by modifying the geometric phase in a rather complicated way [137].

9.5 Symmetry of the Ground State in Jahn–Teller Systems

The geometric phase induced by conical intersections determines the symmetry and degeneracy of the ground state in Jahn–Teller systems. In this section we examine the relationship between the ground state symmetry and the geometric phase in *tunneling splitting* problems in Jahn–Teller systems.

The tunneling splitting is the energy splitting of localized states in r-equivalent minima, which are r-fold degenerate if the barriers between the minima are infinite but split when they are finite. If the geometric phase effect does not exist, the ground state will be a totally symmetric non-degenerate state, because the kinetic energy of the vibration is smallest for this state due to the least number of nodes in the vibrational wave function. However, the sign-change boundary condition imposed by circular transport of the electronic wave function around a conical intersection may not allow for the ground vibrational wave function to be nodeless, yielding a non-totally symmetric ground state. In the following we examine this problem by employing a simple model Hamiltonian for the $E \otimes e$ and $T \otimes t_2$ Jahn–Teller systems.

We first consider the $E \otimes e$ system. In the small quadratic coupling case, the source of the geometric phase is a single conical intersection located at the highest symmetry point. There are three equivalent minima symmetrically placed around this point (Fig. 9.1). If the depth of the minima is sufficiently large, near each minimum the wave function can be approximated by

$$\Psi^{(i)}(\mathbf{r}, \mathbf{R}) = \psi(\mathbf{r}; \mathbf{R})\Phi^{(i)}(\mathbf{R}), \tag{9.61}$$

where ψ is the electronic wave function for the potential energy surface, and $\Phi^{(i)}(\mathbf{R})$ is the ground vibrational wave function localized at the i-th minimum.

If we ignore the geometric phase effect, we obtain the following model Hamiltonian for tunneling splitting:

$$H_{TS} = -t \sum_{\langle i,j \rangle} \left(a_i^\dagger a_j + a_j^\dagger a_i \right) + \epsilon \sum_i a_i^\dagger a_i, \tag{9.62}$$

where $t \geq 0$ is the hopping matrix element given by

$$-t = \langle \Psi^{(i)} | H | \Psi^{(j)} \rangle, \tag{9.63}$$

ϵ is the site energy,

$$\epsilon = \langle \Psi^{(i)} | H | \Psi^{(i)} \rangle, \tag{9.64}$$

the three equivalent minima are labeled by 1, 2 , and 3, and the sum $\langle i, j \rangle$ is taken over pairs of minima, $\langle 1, 2 \rangle$, $\langle 2, 3 \rangle$, and $\langle 1, 3 \rangle$.

We now include the geometric phase effect in the Hamiltonian.

We do not impose global sign-change boundary conditions (using antiperiodic basis wave functions), because this involves multi-valued wave functions that are difficult to handle. As we demonstrated in Chap. 6, the basic mathematical structure underlying the geometric phase is the theory of fiber bundles where there is no need for multi-valued functions. Instead, one describes the wave function with a global section of a vector bundle (in this case a line bundle). As shown in (5.84), a global section may be expressed in terms of locally defined scalar functions – the local components – in a set of basis local sections (5.83) which are defined on local charts of the bundle.[7] The latter are glued together using the transition functions to yield the global (topological) structure of the bundle. In the intersection of two local charts, a global section may be described using two sets of basis local sections. Clearly a global section (5.84) has different local components in the basis local sections associated with different charts. In practice, one identifies a global section with its local components which can take different values at a single point if one uses different basis local sections. A global section is therefore determined by a collection of local components which are single-valued scalar functions in their local charts. But at the intersections of the local charts, it corresponds to more than one set of local components with different values. It is in this sense that a global section plays a similar role as a multi-valued function.

For the systems we consider here, the vector bundle is a line bundle. Hence, there is a single basis local section in each chart, and a global section is determined by a single local component. The local component associated with a chart labeled by j is the wave function $\Phi_0^{(j)}$ given below. The chart j is a simply-connected open region where $\Phi_0^{(j)}$ is non-zero. Because the Hamiltonian is real, the Mead–Berry connection one-form has vanishing curvature in each chart. Nevertheless, the geometric phase which is the holonomy of this connection may be different from unity. This is an indication of the non-trivial topology of the line bundle. As one performs a parallel transportation along a loop going around a conical intersection one has to use more than one local chart. In changing charts at their intersections one uses the transition functions of the fiber bundle which encode the topological information about the bundle. The geometric phase, therefore, includes the information about the transition functions that connect the basis local sections.

The local components $\Phi_0^{(j)}$ are the eigenfunctions of the approximate vibrational Hamiltonian near the local minimum $\mathbf{R}_i$, i.e., they are vibrational wave functions satisfying

---

[7] For details see Sect. 5.6.

9.5 Symmetry of the Ground State in Jahn–Teller Systems

$$\left[\frac{1}{2\mu}\left(\frac{1}{i}\frac{\partial}{\partial \mathbf{R}} - \mathbf{A}\right)^2 + V_i(\mathbf{R})\right]\Phi^{(i)}(\mathbf{R}) = E\Phi^{(i)}(\mathbf{R}), \tag{9.65}$$

where $V_i(\mathbf{R})$ is an approximate potential that mimics the potential surface around $\mathbf{R}_i$. We require that V_i contains an infinite barrier preventing circular motion around the conical intersection. The Mead–Berry vector potential $\mathbf{A}$ is given by

$$\mathbf{A} = i\int d\mathbf{r}\,\psi^*(\mathbf{r};\mathbf{R})\frac{\partial}{\partial \mathbf{R}}\psi(\mathbf{r};\mathbf{R}). \tag{9.66}$$

Next we perform the following gauge transformation in each chart

$$\Phi^{(i)}(\mathbf{R}) \to \tilde{\Phi}^{(i)}(\mathbf{R}) := e^{i\int_{\mathbf{R}_i}^{\mathbf{R}} \mathbf{A}\cdot d\mathbf{R}}\Phi^{(i)}(\mathbf{R}), \tag{9.67}$$

where $\mathbf{R}_i$ is an arbitrary point of the chart in which $\Phi^{(i)}(\mathbf{R})$ is defined and the line integral is evaluated along a curve C connecting $\mathbf{R}_i$ to $\mathbf{R}$. This transformation removes the gauge potential $\mathbf{A}$ from the expression for the vibrational Hamiltonian in (9.65) in the sense that the Schrödinger equation satisfied by the transformed wave function has the form

$$\left[\frac{1}{2\mu}\left(\frac{1}{i}\frac{\partial}{\partial \mathbf{R}}\right)^2 + V_i(\mathbf{R})\right]\tilde{\Phi}^{(i)}(\mathbf{R}) = E\tilde{\Phi}^{(i)}(\mathbf{R}). \tag{9.68}$$

In the presence of a conical intersection, the gauge transformation (9.67) is singular, i.e., it maps a single-valued wave function to a double-valued wave function.[8] If the curve C is a loop that encircles the conical intersection once, the gauge transformation (9.67) causes the wave function to change sign because

$$\exp(i\oint_C \mathbf{A}\cdot d\mathbf{R}) = -1. \tag{9.69}$$

Therefore, the transformed wave function $\tilde{\Phi}^{(i)}$ is double-valued. In the present case, however, the vibrational wave function $\Phi^{(i)}(\mathbf{R})$ is defined only in the vicinity of $\mathbf{R}_i$. Paths that go around the conical intersection are prohibited due to the infinite barrier in V_i. Therefore, the gauge-transformed wave function $\tilde{\Phi}^{(i)}(\mathbf{R})$ is also single-valued.

After the gauge transformation (9.67), our basis vibronic wave functions are given by

$$\tilde{\Psi}^{(i)}(\mathbf{r},\mathbf{R}) = \psi(\mathbf{r};\mathbf{R})\tilde{\Phi}^{(i)}(\mathbf{R}). \tag{9.70}$$

The Hamiltonian with eigenfunctions $\tilde{\Psi}^{(i)}$ is

---

[8] Strictly speaking a singular gauge transformation is not a true gauge transformation, because by definition one requires a gauge transformation to be single-valued.

9. Molecular Aharonov–Bohm Effect

$$H_{TS} = -\sum_{\langle i,j \rangle} \left(\tilde{t}_{ij} a_i^\dagger a_j + \tilde{t}_{ij}^* a_j^\dagger a_i \right) + \epsilon \sum_i a_i^\dagger a_i \quad (9.71)$$

where the transformed matrix elements are given by

$$-\tilde{t}_{ij} = \langle \tilde{\Psi}^{(i)} | H | \tilde{\Psi}^{(j)} \rangle = -t \exp(i \int_{\mathbf{R}_j}^{\mathbf{R}_i} \mathbf{A} \cdot d\mathbf{R}). \quad (9.72)$$

In the above analysis, the choice of gauge is not unique [167], but the vector potential

$$\mathbf{A} = \frac{3}{2} \nabla \theta \quad (9.73)$$

that preserves the symmetry of the system is preferred. Here θ is the angle around the conical intersection. Making this choice, we find for the transfer matrix elements

$$\tilde{t}_{ij} = \tilde{t} = -t \leq 0. \quad (9.74)$$

The symmetry condition that justifies the choice (9.73) for the vector potential follows from the requirement that each vibronic state vector should belong to an irreducible representation of the symmetry group of the original Hamiltonian for the Jahn–Teller system, which exhibits C_3 symmetry for the rotation described by θ. This requirement is satisfied only if the model Hamiltonian (9.71) has the same symmetry as the original one. We also recall that the global sign-change arises from the vector potential that is defined locally through the single-valued electronic wave function $\psi(\mathbf{r}; \mathbf{R})$. Therefore, we may regard a particular choice of the vector potential as a particular choice of the phase of the single-valued electronic wave function.

By diagonalizing the Hamiltonian H_{TS}, we obtain the following vibronic energies,

$$E = \begin{cases} \tilde{t} + \epsilon & \text{(two fold)} \\ -2\tilde{t} + \epsilon. \end{cases} \quad (9.75)$$

Since the hopping matrix element $\tilde{t}$ is negative, the vibronic ground state is doubly degenerate with energy $\tilde{t} + \epsilon = -t + \epsilon$.

In the same manner, the model Hamiltonian for the $T \otimes t_2$ problem with only linear vibronic coupling is given by

$$H_{TS} = -\tilde{t} \sum_{\langle i,j \rangle} \left(a_i^\dagger a_j + a_j^\dagger a_i \right) + \epsilon \sum_i a_i^\dagger a_i, \quad (9.76)$$

where $i = 1, 2, 3$, and 4 label the four equivalent trigonal minima. In this case only four lines of conical intersections are important. These are depicted in Fig. 9.5.

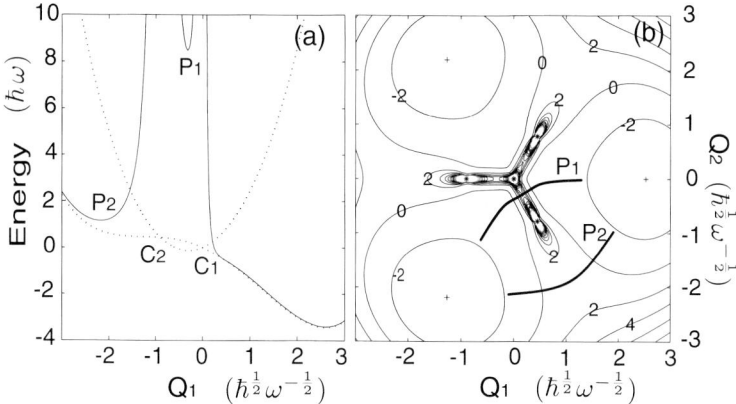

Fig. 9.6. Potential energy surface for the $E \otimes e$ system for $k = 0.9$, $g = 2$, and $f = 0.05$. (a) Cross-section of the potential energy surface with the Born–Huang term (solid line) and that without the Born–Huang term (dotted line) at $Q_2 = 0$. C_1 and C_2 indicate conical intersections, while P_1 and P_2 indicate two saddle points. (b) Contour Plot of the potential surface with the Born–Huang term. $*$ and $+$ indicate a conical intersection and a minimum, respectively. Two steepest decent tunneling paths P_1 and P_2 are shown by thick lines.

Vibronic energies are calculated as

$$E = \begin{cases} \tilde{t} + \epsilon & \text{(three fold)} \\ -3\tilde{t} + \epsilon. \end{cases} \tag{9.77}$$

Similarly to the case of the $E \otimes e$ system, the hopping matrix element $\tilde{t}$ is chosen as $\tilde{t} = -t$. Indeed each triangular face of the tetrahedron having the four minima as vertices is pierced by a line of conical intersection (Fig. 9.5(a)). This makes each triangular face resemble the triangle of three minima in the $E \otimes e$ problem. The ground state turns out to be triply degenerate with energy $-t + \epsilon$.

For a sufficiently large quadratic coupling parameter, additional conical intersections come into play. In the $T \otimes t_2$ case, six additional lines of conical intersections (Fig. 9.5(b)) turn out to be important. In the following, we examine the $E \otimes e$ Hamiltonian (9.56) with large quadratic coupling parameter [136]. We add the fourth-order term

$$f\rho^4 \sigma_0 \tag{9.78}$$

to the Hamiltonian so that even for the cases where $g > 1$ a bound ground state is obtained.

Figure 9.6 shows a potential energy surface and its cross-section for a large quadratic coupling case. Note that the potential energy surface diverges at conical intersections due to the Born–Huang term. Also shown are two important tunneling paths that connect nearby potential minima. One (P_1)

goes through the two conical intersections (one at the origin and the other on the barrier between the minima) and the other (P_2) goes around them.

We now examine the relationship between the ground state symmetry and the change of the dominant tunneling path, P_1 or P_2. For this purpose we employ the semiclassical approximation, which can be considered as a stationary phase approximation to the path integral [191]. The stationary-phase path is called a *semiclassical path*, which is calculated by the ordinary classical equation of motion in a classically allowed region, and which is obtained by the classical equation of motion in imaginary time with the inverted potential in a classically forbidden region.

In the semiclassical approximation, a bound state is associated with a closed semiclassical path. P_1 forms a semiclassical path that encircles a conical intersection located at the origin, while P_2 forms a semiclassical path encircling all four conical intersections. The former path gives rise to a change of the phase angle by π, i.e., it corresponds to the sign-change boundary condition, while the latter changes the phase angle by 4π. Then, it is expected that the change of the ground state symmetry will be related to the change of the dominant semiclassical loop between the two that have different boundary conditions. In order to confirm this assumption, we examine the correlation of the ground state symmetry and the dominant tunneling path change.

Figure 9.7(a) shows the absolute values of the energy difference between the lowest A and E states for the quadratic coupling parameter $g = 2.0, 1.9, 1.8$ with varying k between 0.5 and 1. The fourth-order coupling parameter is fixed at $f = 0.05$. A triply degenerate ground state occurs at the degeneracy point of A and E states. They are found at $(k,g)=(0.82, 2.0), (0.68, 1.9), (0.54, 1.8)$. The ground state has an E symmetry, if k is smaller than the values for triple-degeneracy.

The tunneling rate along P_1 or P_2 is estimated using the simple one-dimensional WKB approximation,

$$\exp(-S_i) = \exp\left(\int_{P_i} \sqrt{2(V-E)}dq\right), \quad i = 1, 2 \qquad (9.79)$$

where V is the potential surface including the Born–Huang term and E is the ground state energy.

Figure 9.7(b) shows the graph of $|\exp(S_1 - S_2) - 1|$ versus k with different g values. This figure also shows the correlation of the triply degenerate points $E_A = E_E$ and the equi-tunneling rate points $S_1 = S_2$. A qualitatively good correlation between $E_A = E_E$ and $S_1 = S_2$ points is convincing evidence that the change of the ground state symmetry is due to the change of the dominant tunneling path. This is despite the fact that the present estimate of the tunneling rates which uses the one-dimensional WKB approximation is not very accurate.

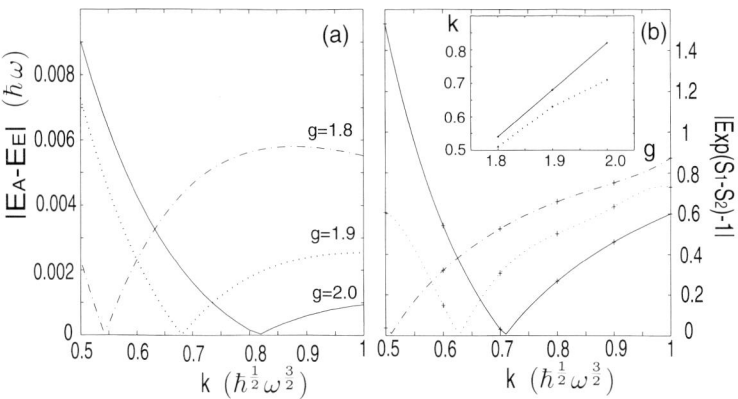

Fig. 9.7. (a) The absolute values of the energy difference $|E_A - E_E|$ between the lowest A and E levels, and (b) the ratios of the tunneling rates via P_1 and P_2 plotted as $|\exp(S_1 - S_2) - 1|$ as functions of k for fixed $f = 0.05$ and $g = 2.0, 1.9, 1.8$. Calculated points are indicated by $+$ and interpolated curves for $g = 2.0, 1.9$, and 1.8 are depicted by solid, dotted, and dash-dotted lines, respectively. Inset: The g versus k values at the crossing points $E_A - E_E$ (solid line) and at $S_1 = S_2$ (dotted line).

9.6 Geometric Phase in Two Kramers Doublet Systems

In this section we extend the two-state problem such as that for an $E \otimes e$ Jahn–Teller system by including spin–orbit interactions. When a two-electronic-state system has a half-odd integral total angular momentum, the system exhibits Kramers degeneracies. A Kramers degeneracy is the degeneracy of a time-reversal conjugate pair of states. A two-state system consisting of two Kramers doublets is described by a 4×4 matrix Hamiltonian.

In the following we assume that the two electronic wave functions, which we denote by u and v, are real-valued. We also introduce the complex-valued wave function

$$\phi = \frac{1}{\sqrt{2}}(u + iv), \tag{9.80}$$

and two half-integral spin wave functions α, β, which are related to one another through the time-reversal operation $\hat{T}$ according to

$$\hat{T}\alpha = \beta,$$
$$\hat{T}\beta = -\alpha. \tag{9.81}$$

These are also chosen to be eigenfunctions of the z-component of the spin operator,

$$\hat{S}_z \alpha = m_s \alpha,$$
$$\hat{S}_z \beta = -m_s \beta. \tag{9.82}$$

Employing $\{\phi\alpha, \phi^*\beta, \phi^*\alpha, \phi\beta\}$ as a basis, we obtain the following matrix Hamiltonian

$$-\frac{1}{2}\sum_n \frac{\partial^2}{\partial Q_n^2} + \begin{pmatrix} X & 0 & U & V \\ 0 & X & -V^* & U^* \\ U^* & -V & Y & 0 \\ V^* & U & 0 & Y \end{pmatrix}, \tag{9.83}$$

where the matrix elements are given by

$$\begin{aligned}
\langle\phi\alpha|H|\phi^*\beta\rangle &= \langle\phi^*\alpha|H|\phi\beta\rangle = 0, \\
\langle\phi\alpha|H|\phi\alpha\rangle &= \langle\phi^*\beta|H|\phi^*\beta\rangle = X, \\
\langle\phi^*\alpha|H|\phi^*\alpha\rangle &= \langle\phi\beta|H|\phi\beta\rangle = Y, \\
\langle\phi\alpha|H|\phi^*\alpha\rangle &= \langle\phi^*\beta|H|\phi\beta\rangle^* = U, \\
\langle\phi\alpha|H|\phi\beta\rangle &= -\langle\phi^*\beta|H|\phi^*\alpha\rangle^* = V.
\end{aligned} \tag{9.84}$$

Here the Hamiltonian H itself is assumed to be real, i.e., time-reversal invariant: $\hat{T}H\hat{T}^{-1} = H$.

Next we obtain the potential energy surfaces by diagonalizing the potential part of (9.83). This yields

$$W_\pm = \frac{1}{2}(X+Y)^2 \pm \frac{1}{2}\sqrt{(X-Y)^2 + 4|U|^2 + 4|V|^2}, \tag{9.85}$$

where each surface is doubly degenerate. The adiabatic electronic basis functions for W_- are given by

$$\begin{aligned}
\Psi_1^- &= \sin\frac{\theta}{2}\phi^*\beta + \cos\frac{\theta}{2}\cos\varphi e^{-i\chi}\phi^*\alpha - \cos\frac{\theta}{2}\sin\varphi e^{-i\eta}\phi\beta, \\
\Psi_2^- &= \sin\frac{\theta}{2}\phi\alpha - \cos\frac{\theta}{2}\sin\varphi e^{i\eta}\phi^*\alpha - \cos\frac{\theta}{2}\cos\varphi e^{i\chi}\phi\beta,
\end{aligned} \tag{9.86}$$

and those for W_+ are

$$\begin{aligned}
\Psi_1^+ &= \cos\frac{\theta}{2}\phi^*\beta - \sin\frac{\theta}{2}\sin\varphi e^{-i\chi}\phi^*\alpha + \sin\frac{\theta}{2}\cos\varphi e^{-i\eta}\phi\beta, \\
\Psi_2^+ &= \cos\frac{\theta}{2}\phi\alpha + \sin\frac{\theta}{2}\cos\varphi e^{i\eta}\phi^*\alpha + \sin\frac{\theta}{2}\sin\varphi e^{i\chi}\phi\beta,
\end{aligned} \tag{9.87}$$

where θ, φ, η, and χ are defined by

$$\begin{aligned}
\cos\theta &:= \frac{X-Y}{\sqrt{(X-Y)^2 + 4|U|^2 + 4|V|^2}}, \\
\cos\varphi &:= \frac{|V|}{\sqrt{|U|^2 + |V|^2}}, \\
U &=: |U|e^{i\eta}, \\
V &=: |V|e^{i\chi}.
\end{aligned} \tag{9.88}$$

9.6 Geometric Phase in Two Kramers Doublet Systems

The matrix elements of the matrix-valued connection one-form A,

$$A_{jk} = i\langle \Phi_j^- | d | \Phi_k^- \rangle \tag{9.89}$$

are calculated as

$$A_{11} = -A_{22} = \cos^2 \frac{\theta}{2} \left(\cos^2 \varphi \, d\chi + \sin^2 \varphi \, d\eta \right),$$

$$A_{21} = A_{12}^* = -\cos^2 \frac{\theta}{2} e^{i(-\eta+\chi)} \left[id\varphi + \sin \varphi \cos \varphi \, d(\eta - \chi) \right], \tag{9.90}$$

and the matrix elements of the Born–Huang term, namely

$$B_{jk} = \frac{1}{2} \sum_n \langle \frac{\partial}{\partial Q_n} \Phi_j^- | \frac{\partial}{\partial Q_n} \Phi_k^- \rangle - \frac{1}{2} \sum_n \sum_{l=1}^{2} A_{jl}^n A_{lk}^n, \tag{9.91}$$

are given by

$$B_{11} = B_{22}$$
$$= \frac{1}{8} \sum_n \left[(\partial_n \theta)^2 + \sin^2 \theta \left((\partial_n \varphi)^2 + \cos^2 \varphi (\partial_n \chi)^2 + \sin^2 \varphi (\partial_n \eta)^2 \right) \right],$$
$$B_{12} = B_{21} = 0, \tag{9.92}$$

where we have used the notation:

$$A_{jk}^n = i\langle \Phi_j^- | \frac{\partial}{\partial Q_n} | \Phi_k^- \rangle. \tag{9.93}$$

In the present problem the vector potential (connection) A and the geometric phase,

$$\mathcal{P} \exp(i \oint_C A), \tag{9.94}$$

are 2×2 matrices. Because A is Hermitian and traceless, the geometric phase (9.94) is an $SU(2)$ matrix.

In order to make the above model more concrete, we may assume the following form for the spin–orbit interaction:

$$H_{\text{SO}} = \sum_i \lambda_i \hat{S}_i \hat{L}_i \tag{9.95}$$

where $\hat{S}_i$ and $\hat{L}_i$ are the i-th component of the total spin and total orbital angular momentum, respectively, and the Hamiltonian H is expressed as

$$H = H_0 + H_{\text{SO}}, \tag{9.96}$$

with H_0 being real. Then, the matrix elements X, Y, U, and V are given by

$$\begin{aligned}
X &= \langle\phi|H_0|\phi\rangle + m_s\lambda_3\langle\phi|\hat{L}_3|\phi\rangle, \\
Y &= \langle\phi|H_0|\phi\rangle - m_s\lambda_3\langle\phi|\hat{L}_3|\phi\rangle, \\
U &= \langle\phi|H_0|\phi^*\rangle, \\
V &= m_s\delta_{1/2,m_s}(\lambda_1\langle\phi|\hat{L}_1|\phi\rangle - i\lambda_2\langle\phi|\hat{L}_2|\phi\rangle).
\end{aligned} \quad (9.97)$$

If $m_s \neq 1/2$, we have $V = 0$ and the 4×4 matrix Hamiltonian is reduced to two 2×2 matrix Hamiltonians, both of which are equivalent to (9.12). In this case, because of the spin–orbit coupling, $\Im\{h_{12}\} \neq 0$ and $V_{11} \neq V_{22}$. Therefore, in view of (9.17) and (9.30), the potential energy surface crossing is lifted; a non-zero A^{mag} arises; and the geometric phase ceases to be topological. As a result the phase change of the wave function is not given by a simple sign-change.

9.7 Adiabatic–Diabatic Transformation

In practical calculations for molecular systems, it is convenient to take a representation in which the non-adiabatic transition terms appear as a scalar potential rather than a vector one. This latter representation is called the *diabatic representation*, which has been extensively used in one-parameter non-adiabatic processes. The transformation from the *adiabatic representation* (i.e., the ordinary Born–Oppenheimer approximation) to the diabatic representation is performed by the unitary transformation (8.69) which transforms the vector potential according to (8.73). In the diabatic representation,

$$0 = \mathbf{A}'^{n'm'}(\mathbf{R}) = \mathcal{U}^{n'n}\mathbf{A}^{nm}(\mathbf{R})\,\mathcal{U}^{\dagger mm'} + i\mathcal{U}^{n'n}\left(\nabla_{\mathbf{R}}\mathcal{U}^{\dagger nm'}(\mathbf{R})\right). \quad (9.98)$$

This is equivalent to

$$\mathbf{A}^{nm}(\mathbf{R})\,\mathcal{U}^{\dagger mm'} + i\left(\nabla_{\mathbf{R}}\mathcal{U}^{\dagger nm'}(\mathbf{R})\right) = 0, \quad (9.99)$$

and

$$A\mathcal{U}^{\dagger} + id\mathcal{U}^{\dagger} = 0, \quad (9.100)$$

where the (n, m) element of the matrix-valued connection one-form A is given by

$$A^{nm} = \langle n|d|m\rangle = \sum_j \langle n|\frac{\partial}{\partial R_j}|m\rangle dR_j = \sum_j A_j^{nm} dR_j \quad (9.101)$$

with A_j^{nm} being the j-th component of $\mathbf{A}^{nm}$, and the (n, m) element of $d\mathcal{U}^{\dagger}$ is understood as

$$d\mathcal{U}^{\dagger nm} = \sum_j \left(\frac{\partial}{\partial R_j}\mathcal{U}^{\dagger nm}\right) dR_j. \quad (9.102)$$

9.7 Adiabatic–Diabatic Transformation

From (9.99), we obtain

$$\frac{\partial}{\partial R_j} \mathcal{U}^{\dagger nm'} = i A_j^{nm} \mathcal{U}^{\dagger mm'},$$

$$\frac{\partial}{\partial R_k} \mathcal{U}^{\dagger nm'} = i A_k^{nm} \mathcal{U}^{\dagger mm'}. \qquad (9.103)$$

In order that the above equations are integrable, we should have

$$\left(\frac{\partial}{\partial R_k}\frac{\partial}{\partial R_j} - \frac{\partial}{\partial R_j}\frac{\partial}{\partial R_k}\right)\mathcal{U}^{\dagger nm'}$$

$$= i\sum_m \left(\frac{\partial}{\partial R_k}A_j^{nm} - \frac{\partial}{\partial R_j}A_k^{nm} - i[A_k, A_j]^{nm}\right)\mathcal{U}^{\dagger mm'}$$

$$= i\sum_m F_{kj}^{nm} \mathcal{U}^{\dagger mm'} = 0. \qquad (9.104)$$

This condition can also be easily obtained using (9.100),

$$0 = d(A\mathcal{U}^\dagger + id\mathcal{U}^\dagger)$$

$$= \sum_{j,k} \left(\partial_j A_k dR_j \wedge dR_k \mathcal{U}^\dagger + A_k dR_k \wedge \partial_j \mathcal{U}^\dagger dR_j\right)$$

$$= \sum_{j,k} \left(\frac{1}{2}(\partial_j A_k - \partial_k A_j)\mathcal{U}^\dagger dR_j \wedge dR_k + A_k dR_k \wedge (-i A_j dR_j)\mathcal{U}^\dagger\right)$$

$$= \sum_{j,k} \frac{1}{2}(\partial_j A_k - \partial_k A_j - i[A_j, A_k])\mathcal{U}^\dagger dR_j \wedge dR_k$$

$$= \frac{1}{2}\sum_{j,k} F_{jk} \mathcal{U}^\dagger dR_j \wedge dR_k = F. \qquad (9.105)$$

In view of this equation, the integrability condition is given by $F = 0$ [168].
As seen from (9.99) if $F = 0$, the vector potential is a pure gauge,

$$A_i^{nm} = -i\sum_l \left(\partial_i \mathcal{V}^{nl}(\mathbf{R})\right)\mathcal{V}^{\dagger lm}(\mathbf{R}) = i\sum_l \mathcal{V}^{nl}(\mathbf{R})\left(\partial_i \mathcal{V}^{\dagger lm}(\mathbf{R})\right). \qquad (9.106)$$

In this case it can be shown by a straightforward calculation (Problem 8.4) that the matrix $\mathcal{V}$ is given by

$$\mathcal{V}^{nl}(\mathbf{R}) = \langle n(\mathbf{R})|l(\mathbf{R}_0)\rangle, \qquad (9.107)$$

where $\mathbf{R}_0$ is a *fixed* value of the parameter $\mathbf{R}$.

In the gauge where the vector potential vanishes, the scalar potential takes the form

$$\tilde{\varepsilon}^{n'm'} = \sum_{m,n} \mathcal{V}^{n'n} \tilde{\varepsilon}^{nm} \mathcal{V}^{\dagger mm'} = \sum_l \mathcal{V}^{n'l}\tilde{\varepsilon}_l \mathcal{V}^{\dagger lm'}. \qquad (9.108)$$

Therefore in this gauge $\tilde{\varepsilon}$ which is diagonal in the original gauge is no longer diagonal, and the Schrödinger equation (8.77) has the form

$$\sum_{m'}\left[-\frac{\nabla_\mathbf{R}^2}{2\mu}\delta^{n'm'}+\sum_l \mathcal{U}^{n'l}(\mathbf{R})\varepsilon_l(\mathbf{R})\mathcal{U}^{\dagger lm'}(\mathbf{R})\right]\psi'^E_{m'}(\mathbf{R})=E\psi'^E_{n'}(\mathbf{R}).\tag{9.109}$$

This is the diabatic representation.

In the approximation in which one only includes a finite number of basis vectors, the condition $F = 0$ is not usually satisfied (except for diatomic systems where the number of parameter is one). One must however note that even for the cases where $F = 0$ the effective parameter space for the low-energy processes may be multiply connected and the molecular Aharonov–Bohm effect may still arise.

The solution $\mathcal{U}^\dagger$ to (9.100) is given by

$$\mathcal{U}^\dagger = \mathcal{P}\exp\left(i\int_C A\right),\tag{9.110}$$

where $\mathcal{P}$ denotes the path-ordered product. In general, the adiabatic–diabatic transformation is possible only in a path-dependent way. A global adiabatic–diabatic transformation does not usually exist.

10. Experimental Detection of Geometric Phases I: Quantum Systems in Classical Environments

10.1 Introduction

In the previous chapters of this book we have shown that quantum states undergoing either adiabatic or exact time evolution can acquire phase factors that reflect the geometry of the spaces in which they evolve. Knowledge of these phases can be helpful in understanding the way in which a system is embedded in an environment.

In this chapter we will discuss in some detail experiments that detect the geometric phase, both in its original adiabatic formulation, due to Berry, and its generalization due to Aharonov and Anandan. The experiments presented in this chapter demonstrate that the geometric phase is observable and controllable, and that its description by Berry and by Aharonov and Anandan provides a remarkably simple way to view the evolution of quantum states in terms of the geometry of the parameter space or projective Hilbert space.

10.2 The Spin Berry Phase Controlled by Magnetic Fields

10.2.1 Spins in Magnetic Fields: The Laboratory Frame

The paradigm for the geometric phase, which we have discussed in previous chapters, is of a particle with spin in a slowly rotating magnetic field. As a particularly direct realization of this case, we will explain the measurement of the polarization of a beam of neutrons before and after interaction with a helical magnetic field [37]. The phase shift of the resulting polarization contains information about the geometric phase.

The initial condition of the experiment is a beam of neutrons, traveling at 500 m/sec. Neutrons have spin $J = 1/2$, and in the beam they were polarized to the extent of 97%. This means that the magnetic moments were almost perfectly oriented in a chosen direction. The polarized beam was then sent through a tube which had a helical coil wound on it. When a current was passed through this coil, a magnetic field was produced which varied in space as

$$\mathbf{B}(z) = B\cos\left(\frac{2\pi z}{L}\right)\mathbf{e}_1 + B\sin\left(\frac{2\pi z}{L}\right)\mathbf{e}_2, \tag{10.1}$$

where L is the length of the tube, the coordinate origin is set to be at the entrance to the tube, and the beam propagates in the $\mathbf{e}_3$ direction. The apparatus is sketched in Fig. 10.1. After propagating through the field, the polarization of the beam was analyzed.

The Hamiltonian for the neutrons in this experiment is

$$H = \frac{p^2}{2m} - \mathbf{m}\cdot\mathbf{B}, \tag{10.2}$$

where p and m are the neutron momentum and mass, respectively, and $\mathbf{m}$ the magnetic dipole moment of the neutron. The forces on the neutron due to the magnetic interaction (and gravity) are very weak, so that during the time the particle spends in the field there is negligible deviation from the original trajectory. Therefore the particle position is given by

$$\mathbf{r}(t) = vt\,\mathbf{e}_3, \tag{10.3}$$

with $v = 500$ m/sec in the present experiment. The magnetic field (10.1) can then be re-parameterized by time, resulting in

$$\mathbf{B}(t) = B\cos(\omega t)\mathbf{e}_1 + B\sin(\omega t)\mathbf{e}_2, \tag{10.4}$$

with $\omega = 2\pi v/L$. For the experimental parameters used, $\omega/2\pi = 1250$ Hz.

In the center-of-mass frame of the neutron, then, the magnetic dipole moment interacts with a magnetic field that traces one complete revolution around the origin. The strength of this field is controlled, and can be made weak or strong. Additionally, an axial field ($B_3\mathbf{e}_3$) could be added, although we will consider mainly the $B_3 = 0$ case. The experimental question probed by Bitter and Dubbers can be formulated as follows: given that the neutrons are injected into the field in some well-defined state of polarization, what is the intensity of the beam after it has left the field and gone through an analyzer

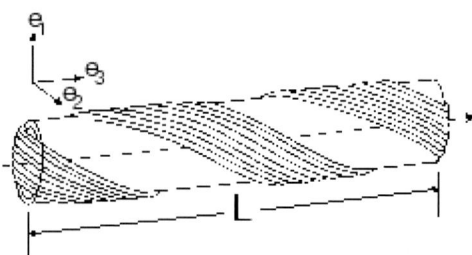

Fig. 10.1. Magnetic field geometry and coordinate system used in neutron spin rotation experiments. Reprinted with permission of the American Physical Society [37].

10.2 The Spin Berry Phase Controlled by Magnetic Fields

oriented in some specified direction (possibly different from the direction of polarization of the incoming beam)? We can guess intuitively what the answer will be. Consider the case of an initial polarization along $\mathbf{e}_1$ (thus aligned with the initial direction of the magnetic field). For zero applied field, the neutrons emerge with their polarization unchanged. For a very strong field, the neutron magnetic dipole, and hence its spin, should be locked along the field, and will again emerge polarized along the original direction. For intermediate cases, for which the field amplitude is comparable to the rate of change of the field direction, the spins undergo more complicated, but calculable, evolution. In particular, we will see that the final polarization oscillates as the ratio of the rate of field reorientation to field amplitude changes.

We can solve this problem exactly, using the theoretical treatment developed in Chap. 3. There, the time evolution operator for a geometry like this one was found to be (3.80)

$$U^\dagger(t) = e^{-i\omega t J_3} e^{-i\Omega t \mathbf{e} \cdot \mathbf{J}}, \qquad (10.5)$$

with

$$\mathbf{e} = \frac{b}{\Omega}\left(\cos\theta - \frac{\omega}{b}\right)\mathbf{e}_3 + \frac{b}{\Omega}\sin\theta\, \mathbf{e}_1 \qquad (10.6)$$

and

$$\Omega = b\sqrt{1 - 2\frac{\omega}{b}\cos\theta + \left(\frac{\omega}{b}\right)^2} \qquad (10.7)$$

[(3.72) and (3.73)]. Here θ is the angle between the magnetic field and $\mathbf{e}_3$, and $b = \gamma B$ is the field strength times the gyromagnetic ratio of the neutrons (this product gives the precession frequency of the neutrons in the magnetic field). The gyromagnetic ratio itself is just $\gamma = -ge/2mc$ [see (3.1)]. We will denote the density operator describing the state of the neutrons at the beginning of the flight tube as $\rho(0)$; upon emerging from the tube it is $\rho(T)$, where T is the flight time through the tube. The equation of motion for the density operator is the Liouville–von Neumann equation,

$$i\frac{d\rho}{dt} = [H, \rho], \qquad (10.8)$$

which is analogous to the Schrödinger equation. The same propagator $U(t)$ that solves the Schrödinger equation solves the Liouville–von Neumann equation, in the form

$$\rho(T) = U^\dagger(T)\rho(0)U(T). \qquad (10.9)$$

For neutrons initially polarized along $\mathbf{e}_1$, the appropriate density operator is, in the usual $|j,m\rangle = |1/2, \pm 1/2\rangle$ basis,

$$\rho(0) = |\mathbf{e}_1\rangle\langle\mathbf{e}_1| = \begin{bmatrix} 1/2 & 1/2 \\ 1/2 & 1/2 \end{bmatrix} = \frac{1}{2}\mathbf{1} + J_1, \qquad (10.10)$$

with

$$|\mathbf{e}_1\rangle = \frac{1}{\sqrt{2}}\begin{pmatrix}1\\1\end{pmatrix}. \tag{10.11}$$

The observable we wish to compute is the spin polarization of the beam in some arbitrary direction as it leaves the magnetic field, given that it entered the field polarized along some possibly different direction. We denote this observable $P(\mathbf{e}_\alpha, \mathbf{e}_\beta)$, for initial direction $\mathbf{e}_\alpha$ and final direction $\mathbf{e}_\beta$. This observable is proportional to the expectation value of the angular momentum in the final direction, which we can write as $\text{Tr}[(\mathbf{e}_\beta \cdot \mathbf{J})\rho(T)]$. Let us first consider the special case of both the initial neutron polarization and the final analyzer in the $\mathbf{e}_1$ direction. Then the observable is given by

$$P(\mathbf{e}_1, \mathbf{e}_1) = \text{Tr}[J_1 \rho(T)] = \text{Tr}[J_1 U^\dagger(T) \rho(0) U(T)], \tag{10.12}$$

with $\rho(0)$ given by (10.10) and $U^\dagger(T)$ by (10.5). Any convenient basis can be used to evaluate the trace; the greatest simplification is achieved by using the eigenstates of $U^\dagger(T)$, which for this type of Hamiltonian have already been derived (3.96), and denoted $|k; \tilde{\theta}, 0\rangle$ the angle $\tilde{\theta}$ orients the vector $\mathbf{e}$, and was defined in (3.75). The trace is evaluated as

$$\begin{aligned}
&\text{Tr}[J_1 U^\dagger(T) \rho(0) U(T)] \\
&= \sum_k \langle k; \tilde{\theta}, 0 | J_1 U^\dagger(T) \rho(0) U(T) | k; \tilde{\theta}, 0 \rangle \\
&= \sum_{k,k'} \langle k; \tilde{\theta}, 0 | J_1 U^\dagger(T) | k'; \tilde{\theta}, 0 \rangle \langle k'; \tilde{\theta}, 0 | \rho(0) U(T) | k; \tilde{\theta}, 0 \rangle \\
&= \sum_{k,k'} e^{i(\alpha_k - \alpha_{k'})} \langle k; \tilde{\theta}, 0 | J_1 | k'; \tilde{\theta}, 0 \rangle \langle k'; \tilde{\theta}, 0 | \rho(0) | k; \tilde{\theta}, 0 \rangle,
\end{aligned} \tag{10.13}$$

where in the last equality we have used the fact that the basis is an eigenbasis of $U^\dagger(T)$, (3.97). Now by using (3.94), which stated

$$|k; \tilde{\theta}, 0\rangle = e^{-i\tilde{\theta} J_2} |k; \mathbf{e}_3\rangle,$$

along with (10.10), we can write

$$\text{Tr}[J_1 U^\dagger(T) \rho(0) U(T)]$$
$$= \sum_{k,k'} e^{i(\alpha_k - \alpha_{k'})} \langle k | e^{i\tilde{\theta} J_2} J_1 e^{-i\tilde{\theta} J_2} | k' \rangle \langle k' | e^{i\tilde{\theta} J_2} (\frac{1}{2}\mathbf{1} + J_1) e^{-i\tilde{\theta} J_2} | k \rangle, \tag{10.14}$$

where we have used the shorthand $|k; \mathbf{e}_3\rangle = |k\rangle$. Application of the rotation operators to J_1 and $\mathbf{1}$, and summation over the states $k, k' = \pm 1/2$, are left as an exercise (Problem 10.1). The result is

$$P(\mathbf{e}_1, \mathbf{e}_1) = \frac{1}{4}\left[1 - \cos 2\tilde{\theta} + (1 + \cos 2\tilde{\theta})\cos(\alpha_{1/2} - \alpha_{-1/2})\right], \tag{10.15}$$

10.2 The Spin Berry Phase Controlled by Magnetic Fields

where $\alpha_{\pm 1/2}$ gives the total phases of these states after one cycle of the evolution. Note that the observable depends on the *difference* of phases between the two states. From (3.97) we find

$$\alpha_{\pm 1/2} = \pm\pi(1 + \Omega/\omega), \tag{10.16}$$
$$\alpha_{1/2} - \alpha_{-1/2} = 2\pi + 2\pi\Omega/\omega. \tag{10.17}$$

The observable is thus

$$P(\mathbf{e}_1, \mathbf{e}_1) = \frac{1}{4}\left[1 - \cos 2\tilde{\theta} + (1 + \cos 2\tilde{\theta})\cos\left(\frac{2\pi\Omega}{\omega}\right)\right]. \tag{10.18}$$

When we consider the special case $\theta = \pi/2$, (10.18) simplifies to

$$P(\mathbf{e}_1, \mathbf{e}_1) = \frac{1}{2}\frac{\cos\left[2\pi\sqrt{1 + (b/\omega)^2}\right] + (b/\omega)^2}{1 + (b/\omega)^2}. \tag{10.19}$$

The function $P(\mathbf{e}_1; \mathbf{e}_1)$ is plotted in Fig. 10.2. We see that indeed for the zero field case ($b = 0$), the polarization of the neutrons is unchanged, and for $b/\omega \gg 1$ the polarization regains its initial value. Particularly interesting are the oscillations. These give phase information about the system as it evolves in progressively stronger fields (that is, increasingly adiabatic conditions). From (10.19) we can read off that the phase is determined by $2\pi\sqrt{1 + (b/\omega)^2}$; 2π must be subtracted from this, to ensure that the phase is zero as $b \to 0$, that is, as the magnetic field is turned off. The total phase Φ is then

$$\Phi = 2\pi\sqrt{1 + (b/\omega)^2} - 2\pi = 2\pi\frac{b}{\omega}\sqrt{1 + (\omega/b)^2} - 2\pi. \tag{10.20}$$

The adiabatic limit in this problem corresponds to the direction of the magnetic field changing slowly compared to its amplitude, or, more precisely, to

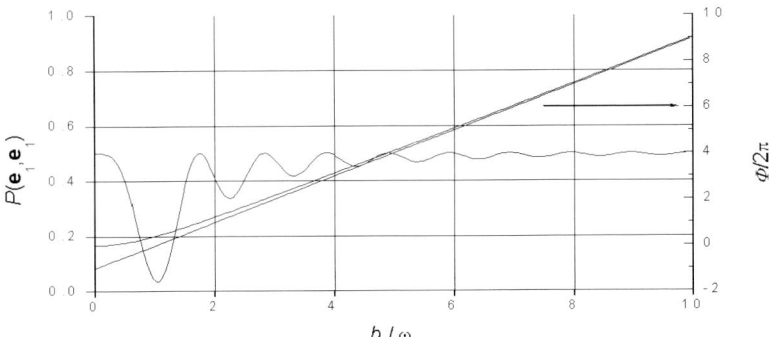

Fig. 10.2. Polarization along x after a single rotation of a magnetic field, given an initial polarization along x, as a function of b/ω. The right axis shows the phase in units of 2π; the curved line is the exact result, while the straight line is the adiabatic result, valid at large b/ω.

the precession frequency of the spin around the field. Mathematically, this condition is $\omega/b \ll 1$, yielding for the phase

$$\Phi \approx 2\pi b/\omega - 2\pi = bT - 2\pi. \tag{10.21}$$

Both the exact and adiabatic phases (10.20) and (10.21) are plotted in Fig. 10.2. We can identify the time-dependent term bT in (10.21) with the dynamical phase; it reflects the length of time that the neutrons experienced the field. The second term, 2π, is independent of time, and depends only on the shape of the path followed.

The experiment as described (i.e., with $\theta = \pi/2$) can only give a single value for the geometric phase. To explore more fully the geometric nature of the phase it is necessary to examine a wider class of paths, and this was done in the present case by including a static magnetic field along $\mathbf{e}_3$. The paths so generated are still conical, but with $\theta \leq \pi/2$. To analyze the phase in this case we must return to the exact result for $P(\mathbf{e}_1, \mathbf{e}_1)$ (10.18). As in the $\theta = \pi/2$ case, here too the terms in $\tilde{\theta}$ do not contribute to the oscillations as ω/b is changed, and so the entire phase behavior is determined by

$$\Phi = \frac{2\pi\Omega}{\omega} - 2\pi. \tag{10.22}$$

We have again subtracted 2π to give the correct behavior in the limit of small b. Upon substituting (10.7) and expanding to lowest order in ω/b, the adiabatic limit, the phase becomes

$$\Phi \approx \frac{2\pi b}{\omega} - 2\pi(1 \pm \cos\theta), \tag{10.23}$$

where the two choices in sign of the $\cos\theta$ term reflect the two choices in the sense of the rotating magnetic field. Geometrically, the phase is thus sensitive to the solid angle $2\pi(1 - \cos\theta)$ enclosed by the path, or $4\pi - 2\pi(1 - \cos\theta)$, the solid angle enclosed by the path viewed the other way around. Varying the axial magnetic field B_3 varies θ, and Bitter and Dubbers demonstrated that the phase of the neutron spin polarization indeed shows the dependence on θ predicted by (10.23).

This experiment provides a very clean demonstration of the presence of the geometric phase, using the paradigmatic Hamiltonian. We note again that the experiment is sensitive to the difference in phases between the two evolving states (10.15), rather than the phases themselves. The present experiment is not a true interferometry experiment, in which a state is compared against an unchanging reference; here two evolving states are probed as a function of time. The difference in their evolutions contains the difference in their geometric phases, which can be extracted. Later we will discuss a truly interferometric measurement of the geometric phase of a single state.

We make one final, and important, comment on the Bitter and Dubbers experiment: while we have analyzed this experiment in terms of $P(\mathbf{e}_1; \mathbf{e}_1)$, this function is merely the most convenient of the many possible polarization choices. It is convenient because for it, the state of the neutrons at $t = 0$ is indeed an eigenstate of the Hamiltonian at $t = 0$, and thus has the simplest geometric interpretation. Other choices, such as $P(\mathbf{e}_2; \mathbf{e}_1)$ and $P(\mathbf{e}_3; \mathbf{e}_3)$ show exactly the same phase behavior, and are in fact the functions Bitter and Dubbers present in their paper. In particular $P(\mathbf{e}_2; \mathbf{e}_1)$ is easier to measure experimentally, because it does not include an offset term – the polarization oscillates about zero, not about a b-dependent value that tends to a non-zero quantity in the adiabatic limit (see Problem 10.2). The fact that the phase shifts can be calculated exactly in all these cases illustrates an important aspect of the geometric phase: it is possible to calculate correctly the outcome of experiments, without knowledge of the geometry; however, like symmetry, taking geometry into account leads to deeper understanding, and an explanation for why the system depends in the way it does on different parameters.

10.2.2 Spins in Magnetic Fields: The Rotating Frame

We turn now to a second exploration of the standard example of the geometric phase: again, it is a spin in a time-varying magnetic field. This time, however, the experiment is carried out not in the laboratory frame but in a rotating frame, using nuclear magnetic resonance spectroscopy (NMR). Briefly, spins in a liquid are immersed in a magnetic field, and transitions are excited with a circularly polarized magnetic field. The detected magnetization acquires phase at a constant rate, which yields, in the frequency domain, a shift of the spectral features. This shift is a signature of the geometric phase for properly chosen experimental parameters. The experiment is due to Suter *et al.* [238].

The experiment begins with mutually uncoupled spins in a strong, static external magnetic field. The kinetic energy contributes nothing significant to the spin dynamics in this case, so we can take the Hamiltonian to be

$$h = -\mathbf{m} \cdot \mathbf{B}, \tag{10.24}$$

where $\mathbf{m}$ is the magnetic dipole moment of the spin. Choosing the external field to be in the $\mathbf{e}_3$ direction, and using the fact that $\mathbf{m} = \gamma \mathbf{J}$, with again $\gamma = -ge/2mc$ the gyromagnetic ratio and $\mathbf{J}$ the angular momentum, we obtain

$$h = -\gamma B_3 J_3. \tag{10.25}$$

The combination γB_3 has units of frequency, and is called the Larmor frequency (denoted ω_L). Note that this definition differs by a factor of $-1/g$ from the one given in Chap. 3; however, the present definition is the convention in magnetic resonance experiments. In typical NMR spectrometers, with magnetic fields in the range of about 2.5 T to 14 T, $\omega_L/2\pi$ for protons is 100–600 MHz.

In the experiment of Suter *et al.* [238], the spins are initially in thermal equilibrium, and so the state of the system can be represented by the following density operator (see for example [217]):

$$\rho(0) = \frac{\exp(-\beta h)}{\text{Tr}[\exp(-\beta h)]} = \exp(-\beta h)/Z, \qquad (10.26)$$

with $\beta = 1/k_B T$ and k_B the Boltzmann constant. Now, the spacing between energy levels in the Hamiltonian (10.25) is just $\hbar\omega_L$, and at all but the lowest temperatures this splitting is small compared to $k_B T$. Our above estimate for ω_L, in fact, leads to $\beta\hbar\omega_L \approx 10^{-5}$ at 300 K. It is therefore appropriate to expand the density operator in powers of $\beta\hbar\omega_L$ around 0, which is the high-temperature limit. Substituting (10.25) into (10.26) and expanding to first order in β, we find

$$\rho(0) = \frac{1 + \beta\omega_L J_3}{Z}. \qquad (10.27)$$

The constant operator $\mathbf{1}/Z$ and the numerical constant $\beta\omega_L/Z$ contribute nothing significant to the dynamics, and so we may ignore them with no essential loss of information about the system. The useful approximation to the initial density operator in the high-temperature limit is therefore

$$\rho(0) = J_3. \qquad (10.28)$$

At first sight this density operator is surprising because its trace is zero, not unity. This is only because we've dropped all the constants in (10.27). A density operator like (10.28) is interpreted as conveying information not about the equilibrium state of the system, but about deviations from equilibrium [1].

We can probe the system by perturbing it with a second magnetic field, and measuring the response of the spins.[1] This strategy is typically accomplished by applying an oscillating field perpendicular to the static field, yielding a total field of

$$\mathbf{B} = B_3 \mathbf{e}_3 + 2B_1 \cos(\omega_{RF} t)\mathbf{e}_1. \qquad (10.29)$$

Here the perturbing field has amplitude $2B_1 \ll B_3$, and frequency ω_{RF}. The frequency ω_{RF} is chosen to be close to ω_L, leading to resonant excitation of the spins. Classically, the field $\mathbf{B}$ produces a torque $\mathbf{m} \times \mathbf{B}$ on the magnetic dipole moments, leading to precession around the total applied field. The dynamics of the density operator (10.28) are identical in form to the classical result. In either case, the precession is clockwise or counter-clockwise, depending on the sign of γ. Therefore, only one of the two circularly polarized components of $2B_1 \cos(\omega_{RF} t)$ will be resonant, the other being off-resonant by roughly $2\omega_L$, because it is rotating in the wrong direction. This component thus has negligible effect. The effective total magnetic field (assuming, without loss of generality, $\gamma > 0$) is therefore

---

[1] This is the standard approach of time-domain nuclear magnetic resonance.

10.2 The Spin Berry Phase Controlled by Magnetic Fields

$$\mathbf{B} = B_1 \cos(\omega_{\text{RF}} t)\mathbf{e}_1 + B_1 \sin(\omega_{\text{RF}} t)\mathbf{e}_2 + B_3 \mathbf{e}_3. \quad (10.30)$$

Detection of the system response to this perturbation is accomplished, macroscopically, by detecting the changing magnetization; but because the magnetization is ultimately proportional to the system angular momentum, again through the gyromagnetic ratio γ, it is easiest to think microscopically of the observables as just the angular momentum operators themselves. The NMR measurement, therefore, amounts to a perturbation of $\rho(0) = J_3$ followed by determination of $\langle J_1 \rangle$, $\langle J_2 \rangle$, and $\langle J_3 \rangle$.

In complex spin systems with mutual couplings, there are small additional contributions to the magnetic field $\mathbf{B}$ due to chemical bonding, etc.; the NMR spectrum indicates these as frequency shifts, and this forms the basis of NMR as a probe of molecular and solid state structure. In the present case, we consider a very simple spin system, but use control of the magnetic fields to provide a very sensitive laboratory for manipulation and detection. In either case, the major contribution to the dynamics of the system is the precession around the large static field $B_3\mathbf{e}_3$. This precession is present in all NMR experiments, and so can be removed with no loss of information content. If the magnetization could be detected in a frame rotating around $\mathbf{e}_3$ at or near ω_{L}, this fast precession would be effectively canceled. Generating such a rotating frame is in fact not difficult; it is only necessary to mix the signal with a reference field oscillating at the "detector frequency" ω_{D} near ω_{L}. The desired signal appears at the difference frequency $\omega_{\text{L}} - \omega_{\text{D}}$. This technology is used in almost all NMR spectrometers, and will be implicit in our discussion below. It is described in detail in [84]. Conventionally, $\omega_{\text{D}} = \omega_{\text{RF}}$, making the excitation appear stationary in the detector frame, but this is not necessary.

Let us now combine these various ideas. The uncoupled spins in the sample, described initially by $\rho(0) = J_3$, are subjected to a time-varying field of the form

$$\mathbf{B} = B_1 \cos(\omega_{\text{RF}} t)\mathbf{e}_1 + B_1 \sin(\omega_{\text{RF}} t)\mathbf{e}_2 + B_3 \mathbf{e}_3. \quad (10.31)$$

Given this field, the spin Hamiltonian (10.24) is

$$h = -\omega_{\text{L}} J_3 - \omega_1 J_1 \cos(\omega_{\text{RF}} t) - \omega_1 J_2 \sin(\omega_{\text{RF}} t), \quad (10.32)$$

with $\omega_1 = \gamma B_1$. The detector frame is effectively rotating around $\mathbf{e}_3$ at frequency ω_{D}. To transform from the laboratory frame to this rotating frame it is necessary to apply the rotation operator

$$\tilde{U}(t) = \exp(iJ_3 \omega_{\text{D}} t). \quad (10.33)$$

The dynamics of the density operator viewed from the rotating frame are related to those in the laboratory frame, which are determined by the Liouville–von Neumann equation. By following steps analogous to the derivation of (3.67) (see Problem 10.3) one can show that in the rotating frame the equation of motion is

234   10. Experimental Detection of Geometric Phases I

$$i\frac{d\tilde{\rho}}{dt} = [\tilde{h} - i\tilde{U}\dot{\tilde{U}}^\dagger, \tilde{\rho}], \tag{10.34}$$

where

$$\tilde{\rho} = \tilde{U}\rho\tilde{U}^\dagger, \tag{10.35}$$
$$\tilde{h} = \tilde{U}h\tilde{U}^\dagger. \tag{10.36}$$

Using these definitions and (10.32) and (10.33) we calculate the effective Hamiltonian as

$$\tilde{h} - i\tilde{U}\dot{\tilde{U}} = -(\omega_\text{L} - \omega_\text{D})J_3 - \omega_1 J_1 \cos[(\omega_\text{RF} - \omega_\text{D})t] \tag{10.37}$$
$$- \omega_1 J_2 \sin[(\omega_\text{RF} - \omega_\text{D})t].$$

The dynamics of the spin viewed in this frame is thus the dynamics of a spin subjected to a magnetic field with components

$$B_1 = \frac{\omega_1}{\gamma}\cos[(\omega_\text{RF} - \omega_\text{D})t], \tag{10.38}$$

$$B_2 = \frac{\omega_1}{\gamma}\sin[(\omega_\text{RF} - \omega_\text{D})t], \tag{10.39}$$

$$B_3 = (\omega_\text{L} - \omega_\text{D})/\gamma, \tag{10.40}$$

and total strength

$$B = \left[(\omega_\text{L} - \omega_\text{D})^2 + \omega_1^2\right]^{1/2}/\gamma. \tag{10.41}$$

The effective magnetic field traces out a conical path, and so this problem is equivalent to our canonical example of geometric phase as outlined in (3.4). Through judicious choices of the excitation and detector frequencies, and the amplitude of the excitation field, we can generate adiabatic evolutions with a variety of paths. A selection of paths used in NMR experiments on geometric phases is shown in Fig. 10.3.

The evolution of the density operator in the detector frame is given again by the formalism described previously,

$$\tilde{\rho}(t) = U^\dagger(t)\tilde{\rho}(0)U(t), \tag{10.42}$$

with

$$U^\dagger(t) = e^{-i\omega t J_3} e^{-i\Omega t \mathbf{e} \cdot \mathbf{J}}. \tag{10.43}$$

Here and for the remainder of this section we will always be working in the detector frame, so all quantities refer to its coordinate system. The definitions of Ω and $\mathbf{e}$ are as in (10.6) and (10.7) (and are expressed in the detector frame). The parameter b is the total field strength in frequency units, and so in the present problem has the form

$$b = \left[(\omega_\text{L} - \omega_\text{D})^2 + \omega_1^2\right]^{1/2}. \tag{10.44}$$

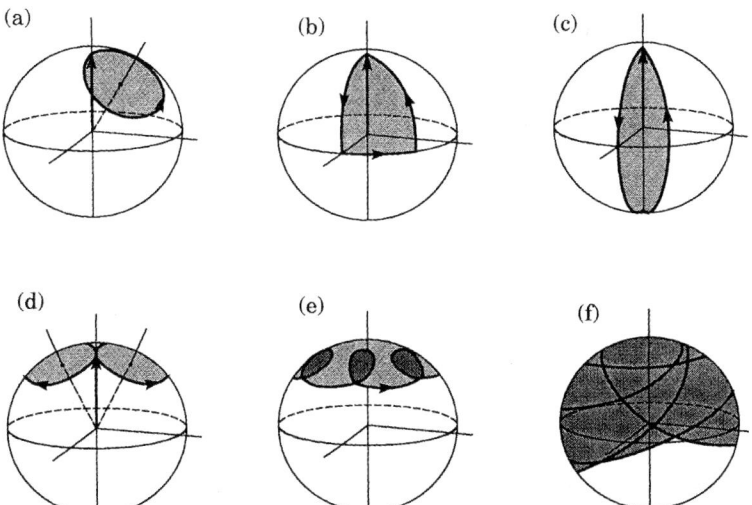

Fig. 10.3. Paths used in NMR experiments on geometric phases. (**a**) shows a simple rotation about a fixed axis. (**b**) and (**c**) show a wedge and a slice; these have been used in studies of the Aharonov–Anandan phase. (**d**) shows a figure-eight, which would comprise the simplest way to study non-Abelian phases. (**e**) and (**f**) show variants of the path traced by a double rotor, a device which has been used to detect the effects of non-Abelian phases. Reprinted with permission of Annual Reviews, Inc. [286].

The parameter θ defining the angle between the total field and the $\mathbf{e}_3$ direction is given by

$$\cos\theta = \frac{\omega_L - \omega_D}{b}, \tag{10.45}$$

$$\sin\theta = \frac{\omega_1}{b}. \tag{10.46}$$

Finally, the angular frequency ω of the rotating field is expressed by

$$\omega = \omega_{RF} - \omega_D. \tag{10.47}$$

Applying this evolution to $\tilde{\rho}(0) = J_3$ yields (Problem 10.4)

$$\tilde{\rho}(t) = \sin\tilde{\theta} J_1 \left[\cos\tilde{\theta}(1 - \cos\Omega t)\cos\omega t - \sin\Omega t \sin\omega t\right] \tag{10.48}$$
$$+ \sin\tilde{\theta} J_2 \left[\cos\tilde{\theta}(1 - \cos\Omega t)\sin\omega t + \sin\Omega t \cos\omega t\right]$$
$$+ J_3 \left[\cos^2\tilde{\theta} + \sin^2\tilde{\theta}\cos\Omega t\right],$$

where the angle $\tilde{\theta}$ is defined in (3.75). The evolving density operator (10.49) looks a good deal more complicated than the evolutions calculated for the same geometry in Chap. 3, e.g., (3.104). The reason is that we made no effort in the present case to use as initial condition an eigenstate of the Hamiltonian at time zero, that is, a vector like (3.94). Moreover, as we discuss next, we will not detect the density operator along a preferred direction either. The result will be a signal that includes information on the geometric component of the evolution, but also additional, more routine information about the evolving magnetization. While these features complicate the interpretation (mildly), they are present because the implementation of the experiment in this form is much simpler.

From $\tilde{\rho}(t)$ we can calculate the detected signal. Due to the proportionality through the gyromagnetic ratio γ, measurement of the signal (the magnetization) is essentially a determination of the angular momentum components. Typical NMR spectrometers can only measure the components orthogonal to the large, static magnetic field, namely, the transverse components J_1 and J_2. The magnetization is measured, as noted above, by mixing it with a reference oscillator at ω, and measuring the difference frequency. This is done with both the reference oscillator, and the reference phase-shifted by 90°. The result is a phase-sensitive determination, capable of distinguishing positive ($> \omega_D$) from negative ($< \omega_D$) frequencies in the rotating frame (this procedure is known as quadrature detection [84]). Mathematically, the detected observables are acquired in the form $\langle J_1 \rangle(t) + i \langle J_2 \rangle(t)$, where $\langle J_i \rangle(t) = \text{Tr}[J_i \tilde{\rho}(t)]$. For the signal we thus obtain

$$\begin{aligned}\langle J_1\rangle(t) + i\langle J_2\rangle(t) &= \frac{1}{2}\sin\tilde{\theta}\cos\tilde{\theta}(1 - \cos\Omega t)e^{i\omega t} + \frac{i}{2}\sin\tilde{\theta}\sin\Omega t\, e^{i\omega t}\\ &= \frac{1}{2}\sin\tilde{\theta}\cos\tilde{\theta}e^{i\omega t} + \frac{1}{4}\sin\tilde{\theta}(1-\cos\tilde{\theta})e^{i(\omega+\Omega)t}\\ &\quad - \frac{1}{4}\sin\tilde{\theta}(1+\cos\tilde{\theta})e^{i(\omega-\Omega)t}.\end{aligned} \quad (10.49)$$

This signal has components with phase shifts of ωt, and $(\omega \pm \Omega)t$. Since ω and Ω are constants, the phase is being acquired at a constant rate. Therefore, in the frequency domain, the Fourier transform of the signal consists of three resonances[2] at ω and $\omega \pm \Omega$. Note that after one cycle ($t = 2\pi/\omega$) the phase acquired by the $(\omega + \Omega)t$ component is $2\pi(1 + \Omega/\omega)$, which is just the phase difference between the two evolving states (10.17). To see this, calculate $\langle J_1 \rangle(T) + i \langle J_2 \rangle(T)$ using the basis of eigenvectors of $U^\dagger(T)$, as in the previous section (10.13). Therefore, encoded within the frequency shift of this component is the phase difference, including the geometric phase, of the two spin states. Using our previous definitions we can write this frequency as

$$\Omega + \omega = b\left(1 - 2\frac{\omega}{b}\cos\theta + \frac{\omega^2}{b^2}\right)^{1/2} + \omega, \quad (10.50)$$

---

[2] For experimental reasons, the data published from this experiment are presented in absolute value, doubling the apparent number of peaks in the spectrum [238].

10.2 The Spin Berry Phase Controlled by Magnetic Fields

which, in the adiabatic limit $\omega/b \ll 1$, becomes

$$\Omega + \omega \to b + \omega(1 - \cos\theta). \tag{10.51}$$

The phase itself is thus $bt + \omega(1 - \cos\theta)t$, which, per cycle ($t = 2\pi/\omega$) yields $2\pi b/\omega + 2\pi(1 - \cos\theta)$. The first term is dynamic: it depends on how fast the effective field completes a circuit. The second term, $2\pi(1 - \cos\theta)$, depends only on the shape of the conical path, through θ, and not on any details of how the path is traversed.

The experiment performed by Suter et al. consisted of measuring NMR spectra of uncoupled spins (the protons of water in a dilute solution, with deuterated acetone as solvent) under the conditions outlined above, for a variety of values of $\omega = \omega_{RF} - \omega_D$. The shift of the feature at $\Omega + \omega$ was measured and plotted as a function of ω, and the linear regime identified in the $\omega \to 0$ limit. The slope in this limit is just $1 - \cos\theta$ (10.51), yielding the geometric phase divided by 2π. As in the example discussed in the previous section, here again due to the non-interferometric nature of the experiment the measured quantity reflects the difference in phase between the two states. Representative data are shown in Fig. 10.4. Finally, by varying the angle θ (by varying either the detector offset or the excitation field strength) and carrying out the same sequence of measurements, these researchers were able to demonstrate the predicted solid-angle dependence of the phase shift (Fig. 10.5), thus confirming its geometrical nature.

10.2.3 Adiabatic Reorientation in Zero Field

The above sections have detailed several different experiments which match the original example of the geometric phase, namely, a spin in a slowly rotating magnetic field. This example is merely one particularly concrete realization of the general case; it should be clear that any slow reorientation of a quantization axis can give rise to an adiabatic geometric phase. In the present section we discuss one regime in which the general case has been explored in some detail, namely a nuclear quadrupole moment in a slowly changing electric field gradient. We draw on the experimental work of [254] and [285], and the theoretical treatments of [282] and [21].

Nuclei with spin greater than $1/2$ may have a non-spherical charge distribution, which implies that they possess a non-zero electric quadrupole moment. This moment couples to local electric field gradients, giving rise to splittings of the nuclear energy levels. The study of these splittings is referred to as nuclear quadrupole resonance spectroscopy (NQR).[3] In order to express the Hamiltonian of the nucleus in the presence of a field gradient it is easiest to use the Wigner–Eckart theorem to express the multipole moment operators in terms of spin operators. Here we merely state the result [231], expressed in the principal axis frame of the gradient tensor:

---

[3] NQR is discussed for example in [72]

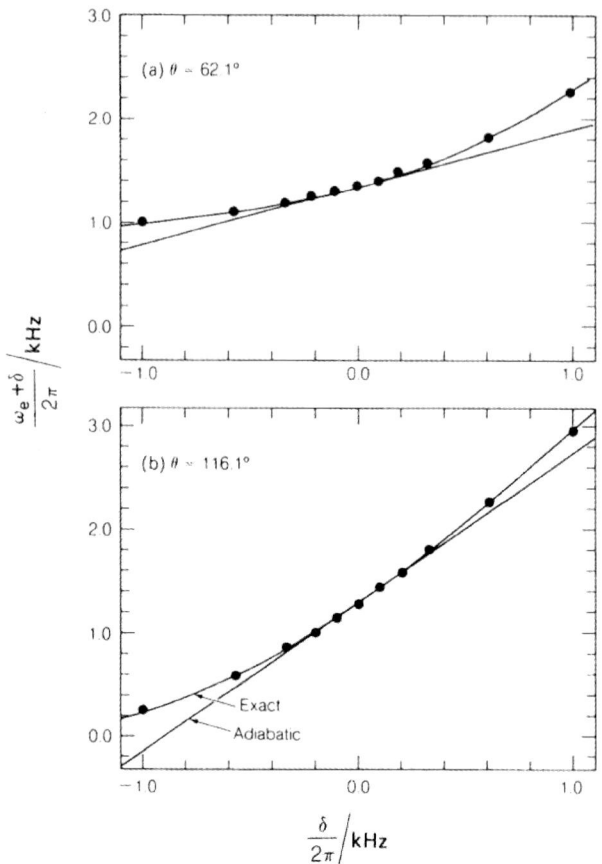

Fig. 10.4. Frequency shift as a function of detector offset in the experiment of Suter et al. 1987. Reprinted with permission of Taylor and Francis Group [238].

$$h = \frac{eQ}{4J(2J-1)} \left[V_{33}(3J_3^2 - J^2) + (V_{11} - V_{22})(J_1^2 - J_2^2) \right], \quad (10.52)$$

where eQ is the quadrupole moment of the nucleus under consideration, and $V_{ii} = \partial^2 V/\partial x_i^2$, the second derivative of the potential, is the component of the electric field gradient in direction $\mathbf{e}_i$. It is conventional in NQR to label the axes such that $V_{33} \geq V_{11} \geq V_{22}$, and define

$$eq = V_{33} \quad (10.53)$$

and

$$\eta = \frac{V_{11} - V_{22}}{V_{33}}. \quad (10.54)$$

Then the Hamiltonian can be written in the form

10.2 The Spin Berry Phase Controlled by Magnetic Fields

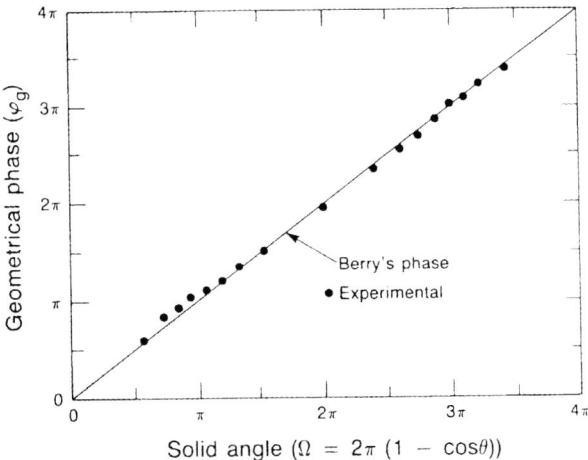

Fig. 10.5. Dependence of the geometric phase on solid angle, as determined in the rotating frame experiment of Suter et al. 1987. Reprinted with permission of Taylor and Francis Group [238].

$$h = \frac{e^2qQ}{4J(2J-1)}\left[(3J_3^2 - J^2) + \eta(J_1^2 - J_2^2)\right]. \tag{10.55}$$

This Hamiltonian is invariant with respect to time-reversal; for half-odd-integer J, this fact implies that the eigenvalues will occur in degenerate pairs. In particular, consider $J = 3/2$. The states split into two manifolds, $m_J = \pm 3/2$ and $m_J = \pm 1/2$, with a splitting between them of

$$\omega_Q/2\pi = \frac{e^2qQ}{2h}\sqrt{1+\eta^2/3}. \tag{10.56}$$

In the following we will consider in detail NQR experiments on the ^{35}Cl nucleus in crystalline sodium chlorite, NaClO$_3$; for this example, $J = 3/2$ and $\omega_Q/2\pi \approx 30$ MHz. In this case the resonance can be easily excited and detected with a standard NMR spectrometer (less the magnet!), provided the sample is shielded from all magnetic fields.

The Hamiltonian of (10.55) depends in general on five parameters: e^2qQ/h, η, and three Euler angles specifying the orientation of the principal axis frame in some laboratory fixed frame. The first two of these parameters are set by the choice of sample, in an experiment, and thus are hard to vary. We will take e^2qQ/h as fixed, and in fact take $\eta = 0$; this choice is consistent with the experiments we will describe, and does not substantially diminish the variety of phenomena observable in this system. With these restrictions we need only two parameters to describe the system, namely a polar and an azimuthal angle to orient the (now axially symmetric) quadrupole in the laboratory frame. Whereas the original example of the geometric phase could be thought of in terms of the Hamiltonian

$$h(t) = b\hat{R} \cdot J, \tag{10.57}$$

the present case can now be expressed as

$$h(t) = (b\hat{R} \cdot J)^2, \tag{10.58}$$

with

$$b^2 = \frac{3e^2 qQ/h}{4J(2J-1)} \tag{10.59}$$

(ignoring the additive constant).

This example is qualitatively different from our previous ones, because here the adiabatically evolving subspace is of dimension two, rather than one. Adiabatic evolution of a multi-dimensional eigenspace gives rise in general to a non-Abelian geometric phase, as discussed in detail in Chap. 7. Since the eigenspace in the present problem is of dimension two, the various phase factors are elements of the group $U(2)$. We write the geometric phase in this case as (see (7.52))

$$U^{\text{geometric}} = \mathcal{P}\exp\left(i\oint A\right). \tag{10.60}$$

In the above expression $\mathcal{P}$ is the path-ordering operator, and A the connection one-form. Because the eigenspace is two-dimensional, the "phase factor" generated by A is an element of $U(2)$, hence a 2×2 matrix. Matrix elements of A have the form (7.53)

$$A^{k,k'} = \sum_{R=\theta,\varphi} \langle k;\theta,\varphi | i\partial/\partial R | k';\theta,\varphi\rangle. \tag{10.61}$$

For the NQR examples of interest, the Hamiltonian is parameterized through the two angles θ and φ, which serve to express its orientation in the laboratory frame. We thus have

$$h = e^{-i\varphi J_3} e^{-i\theta J_2} e^{i\varphi J_3} h_0 e^{-i\varphi J_3} e^{i\theta J_2} e^{i\varphi J_3}, \tag{10.62}$$

where

$$h_0 = \frac{e^2 qQ}{4J(2J-1)}(3J_3^2 - J^2). \tag{10.63}$$

For the Hamiltonian (10.62) the instantaneous eigenstates are

$$|k;\theta,\varphi\rangle = e^{-i\varphi J_3} e^{-i\theta J_2} e^{i\varphi J_3} |k;0,0\rangle. \tag{10.64}$$

The apparent "extra" rotation around $\mathbf{e}_3$ in (10.62) and (10.64) is important, as it guarantees that the states will be well-defined at $\theta = 0$. Using these states we can calculate the matrix elements of the connection one-form:

$$\begin{aligned} A_\theta^{k,k'} &= \langle k;\theta,\varphi | i\partial/\partial\theta | k';\theta,\varphi\rangle \\ &= \langle k;0,0 | J_2 \cos\varphi - J_1 \sin\varphi | k';0,0\rangle, \end{aligned} \tag{10.65}$$

and

$$A_\varphi^{k,k'} = \langle k; \theta, \varphi | i\partial/\partial\varphi | k'; \theta, \varphi \rangle$$
$$= \langle k; 0, 0 | J_3 \cos\theta - J_1 \sin\theta \cos\varphi - J_2 \sin\theta \sin\varphi | k'; 0, 0 \rangle. \quad (10.66)$$

Matrix elements of both A_θ and A_φ can in principle cause mixing between the degenerate manifolds, because for example $\langle 3/2 | J_1 | 1/2 \rangle \neq 0$, and $\langle 3/2 | J_2 | 1/2 \rangle \neq 0$. The physical meaning of the adiabatic approximation in this case is that such inter-manifold mixing is negligible. Therefore, the matrix elements are to be evaluated only within the degenerate manifolds. Thus, for $J = 3/2$, for example, within the $k = \pm 3/2$ block we have only an Abelian connection, given by

$$A^{\pm 3/2} = \begin{bmatrix} \frac{3}{2}\cos\theta\, d\varphi & 0 \\ 0 & -\frac{3}{2}\cos\theta\, d\varphi \end{bmatrix}. \quad (10.67)$$

Within the $k = \pm 1/2$ block, however, we have the possibility for non-Abelian phases (Problem 10.5):

$$A^{\pm 1/2} = \begin{bmatrix} \frac{1}{2}\cos\theta\, d\varphi & -\sin\theta e^{-i\varphi}\, d\varphi - i\, e^{-i\varphi}\, d\theta \\ -\sin\theta e^{i\varphi}\, d\varphi + i\, e^{i\varphi}\, d\theta & -\frac{1}{2}\cos\theta\, d\varphi \end{bmatrix}. \quad (10.68)$$

The signature of the non-Abelian nature of the connection one-form of the $k = \pm 1/2$ manifold in this example is not the presence of off-diagonal terms, but rather the dependence on changes in both parameters, that is, $d\theta$ and $d\varphi$. If either term is absent, the resulting connection samples only a one-parameter subgroup, which is necessarily Abelian. From an experimental point of view, what happens in this case is that the eigenvalues of A do not collide for any values of the parameters over the path. Thus for this case there exists some set of states that don't mix during the evolution, but independently acquire phase, as in our earlier examples of spins in magnetic fields. Such a case is thus a realization of Berry's (Abelian) formulation of the geometric phase, and has been studied experimentally by Tycko [254].

In Tycko's experiment a single crystal of $NaClO_3$ was mounted in a rotatable sample holder such that a cleavage plane was perpendicular to the rotation axis; this direction is designated $\mathbf{e}_3$ in the laboratory frame. The crystal structure of $NaClO_3$ has four formula units in the unit cell, and its symmetry is that of the cubic space group $P2_13$. The structure consists of ClO_3^- trigonal pyramids, charge-balanced by sodium cations. The principal axis of the electric field gradient at each chlorine atom is collinear with the three-fold rotation axis of the ClO_3^- ion; the four ions in the unit cell are oriented such that the principal axes are parallel to body diagonals of the cubic cell. The crystal cleaves along the cell axes. Therefore, by orienting the crystal with a cleavage plane perpendicular to the rotation axis, it is assured that the four chlorine nuclei will appear identical, with the major axes of their local electric field gradients all at an angle θ with respect to $\mathbf{e}_3$, where

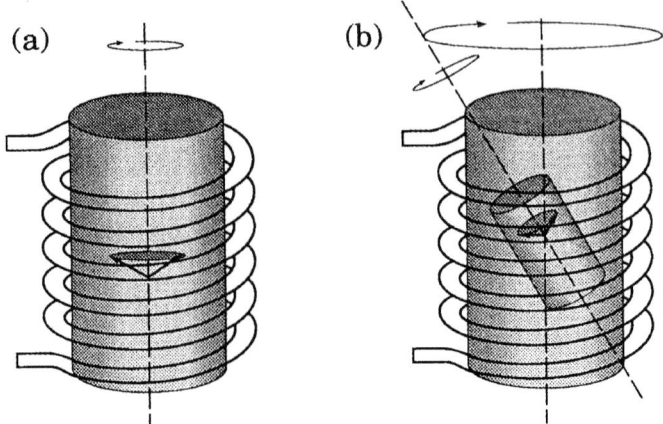

Fig. 10.6. Experimental configurations used in NQR experiments to detect geometric phases. (**a**) Sample spinning about a stationary axis yields a probe of the Abelian (Berry) phase. (**b**) Spinning about a moving axis is sensitive to non-Abelian phases. Reprinted with permission of Annual Reviews, Inc. [286].

$\cos\theta = 1/\sqrt{3}$. A tuned coil is wound around the sample holder, collinear with the sample rotation axis (Fig. 10.6a).

Consider first a stationary sample. Without loss of generality we take the $\mathbf{e}_3$ direction of the chlorine principal axis system to lie in the $\mathbf{e}_1$–$\mathbf{e}_3$ plane of the laboratory frame (so $\varphi = 0$), inclined at an angle $\theta = \arccos 1/\sqrt{3}$ with respect to the laboratory $\mathbf{e}_3$ axis. Dropping all constant terms from (10.63) we write the Hamiltonian in the principal axis frame as

$$\tilde{h} = \frac{\omega_Q}{2} J_3^2, \qquad (10.69)$$

where $\omega_Q = e^2 qQ/2$. In the laboratory frame the Hamiltonian is (10.62)

$$h = \frac{\omega_Q}{2} e^{-i\theta J_2} J_3^2 e^{i\theta J_2} = \frac{\omega_Q}{2} (J_3 \cos\theta + J_1 \sin\theta)^2. \qquad (10.70)$$

The density operator at thermal equilibrium for the system is found as in (10.27) and (10.28), but using the Hamiltonian (10.69); the result is $\tilde{\rho} = J_3^2$ in the principal axis frame, and

$$\rho = (J_3 \cos\theta + J_1 \sin\theta)^2 \qquad (10.71)$$

in the laboratory frame. A radio-frequency pulse rotates this operator around $\mathbf{e}_3$ (the coil axis), preparing a non-equilibrium state the evolution of which will be observed. By using the analog of a $\pi/2$ pulse (rotation by $\pi/2$), the initial density operator is

$$\rho(0) = (J_3 \cos\theta - J_2 \sin\theta)^2. \qquad (10.72)$$

10.2 The Spin Berry Phase Controlled by Magnetic Fields

This density operator evolves under the action of (10.69), producing oscillating magnetization that is detected by the coil. The evolution operator is expressed in the principal axis frame as $\tilde{U}^\dagger(t) = \exp(-i\omega_Q t J_3^2/2)$. The observable is the magnetization collinear with the coil, which is proportional to J_3 (in the principal axis frame this operator is $J_3 \cos\theta + J_1 \sin\theta$). The signal $S(t)$ is given by

$$S(t) = \text{Tr}[J_3 U^\dagger(t)\rho(0)U(t)]. \tag{10.73}$$

Calculating the signal is easiest when all quantities are converted to the principal axis frame, for there $U(t)$ has a particularly simple form. The result, after lengthy algebra and dropping the time-independent component arising from the $J_3 \cos\theta$ term of the observable in the principal axis frame, is

$$S(t) = \frac{4}{3} \sin \omega_Q t. \tag{10.74}$$

Therefore the spectrum consists of a single resonance at frequency $\omega_Q/2\pi$, and due to the structure of the NaClO$_3$ crystal all four chlorine nuclei in the unit cell contribute equally to this result.

If, simultaneously, the sample is made to rotate about $\mathbf{e}_3$, then additionally the states acquire phase as described by the connection one-forms above. For the present experiment, $d\theta = 0$ and $d\varphi = \omega\, dt$. Adiabaticity is ensured when the rotation frequency, that is, the rate of change of the quantization axis direction, is small compared to the characteristic system evolution frequency. In the present case this requires that $\omega/2\pi \ll 30$ MHz, a condition that would be difficult to violate with a mechanical rotor. Tycko used $\omega/2\pi = 1$–4 kHz. In the laboratory frame the Hamiltonian is therefore

$$h = \frac{\omega_Q}{2} e^{-i\omega t J_3} e^{-i\theta J_2} e^{i\omega t J_3} J_3^2 e^{-i\omega t J_3} e^{i\theta J_2} e^{i\omega t J_3}$$

$$= \frac{\omega_Q}{2}(J_3 \cos\theta + J_1 \sin\theta \cos\omega t + J_2 \sin\theta \sin\omega t)^2. \tag{10.75}$$

The initial density operator is again given by (10.72), and the observable is J_3. The signal can be calculated from $\text{Tr}[J_3 U^\dagger(t)\rho(0)U(t)]$ as usual, but the algebra is lengthy and obscures understanding.

The action of this time-dependent Hamiltonian is made more apparent by transforming to a frame moving with it, as follows:

$$\tilde{h} = \tilde{U}h\tilde{U}^\dagger = e^{i\theta J_2} e^{i\varphi J_3} h\, e^{-i\varphi J_3} e^{-i\theta J_2} = \frac{\omega_Q}{2} J_3^2, \tag{10.76}$$

where $\tilde{U} = e^{i\theta J_2} e^{i\varphi J_3}$. In this moving frame the generator of time-evolution will be (3.68)

$$\tilde{h} - i\tilde{U}\dot{\tilde{U}}^\dagger = \frac{\omega_Q}{2} J_3^2 - \omega(J_3 \cos\theta - J_1 \sin\theta). \tag{10.77}$$

This effective Hamiltonian is time-independent. Actually, if we were to undo also the "extra" rotation about $\mathbf{e}_3$ in (10.62), the resulting operator would be time-dependent. This approach is formally correct, but very cumbersome. We defer discussion of this point until after we have calculated the spectrum of the rotating quadrupole, using for the time being the states

$$\tilde{U}|k;\theta,\varphi\rangle = e^{i\theta J_2}e^{i\varphi J_3}|k;\theta,\varphi\rangle = |k;0,0\rangle, \quad (10.78)$$

rather than (10.64) above. In this basis the initial density operator and observable are

$$\tilde{\rho}(0) = \tilde{U}\rho(0)\tilde{U}^\dagger = (J_3\cos^2\theta - J_1\cos\theta\sin\theta - J_2\sin\theta)^2, \quad (10.79)$$

$$\tilde{U}J_3\tilde{U}^\dagger = J_3\cos\theta - J_1\sin\theta. \quad (10.80)$$

We can calculate the spectrum by working either in the frequency domain, or the time domain. In the frequency domain we use the static effective Hamiltonian (10.77), find its eigenvalues and eigenstates, and compute the matrix elements of the observable in the eigenstate basis, to obtain the intensities of the resonances. The differences between the eigenvalues are the energies of the resonances. In the time domain, we integrate the diagonal elements of the propagator, in its eigenbasis, including the contributions from the connection one-form, to get the phase accumulation on each eigenstate as a function of time. The two approaches are equivalent, provided in the frequency domain we apply first-order static perturbation theory. This neglects mixing between the levels induced by the motion, precisely as does the adiabatic approximation in the time domain.

We will work in the time domain. The propagator includes the dynamical part $\exp(-iJ_3^2\omega_Q t/2)$, and the geometric contribution discussed above, $\exp(i\int A)$, where in our present simplified basis, using the orientation of the NaClO$_3$ crystal,

$$A = \begin{bmatrix} \frac{\sqrt{3}}{2}\omega & 0 & 0 & 0 \\ 0 & \frac{1}{2\sqrt{3}}\omega & -\sqrt{\frac{2}{3}}\omega & 0 \\ 0 & -\sqrt{\frac{2}{3}}\omega & -\frac{1}{2\sqrt{3}}\omega & 0 \\ 0 & 0 & 0 & -\frac{\sqrt{3}}{2}\omega \end{bmatrix}. \quad (10.81)$$

Here the columns are in the order: $|3/2\rangle$, $|1/2\rangle$, $|-1/2\rangle$, $|-3/2\rangle$. Again, there is no coupling between states in the $\pm 3/2$ block and the $\pm 1/2$ block, due to the adiabatic nature of the motion, but the $\pm 1/2$ states are mutually coupled.

Combining the dynamical evolution with the geometric contribution shows that the $|\pm 3/2\rangle$ states evolve in time as

$$\exp\left(-i\int_0^t \frac{9}{8}\omega_Q \mp \frac{\sqrt{3}}{2}\omega \, dt'\right)|\pm 3/2\rangle = e^{-i(\frac{9}{8}\omega_Q \mp \frac{\sqrt{3}}{2}\omega)t}|\pm 3/2\rangle. \quad (10.82)$$

For the $\pm 1/2$ block we must diagonalize the connection one-form, (10.81). The eigenvalues are $\pm\frac{\sqrt{3}}{2}\omega$, and the eigenvectors are

10.2 The Spin Berry Phase Controlled by Magnetic Fields

$$|\psi_+\rangle = \cos\frac{\tilde{\theta}}{2}|1/2\rangle + \sin\frac{\tilde{\theta}}{2}|-1/2\rangle, \tag{10.83}$$

$$|\psi_-\rangle = -\sin\frac{\tilde{\theta}}{2}|1/2\rangle + \cos\frac{\tilde{\theta}}{2}|-1/2\rangle. \tag{10.84}$$

Here $\tan\tilde{\theta} = 2\tan\theta$, so $\cos\tilde{\theta}/2 = \sqrt{2/3}$ and $\sin\tilde{\theta}/2 = \sqrt{1/3}$. The states $|\psi_\pm\rangle$ evolve as

$$\exp\left(-i\int_0^t \frac{1}{8}\omega_Q \mp \frac{\sqrt{3}}{2}\omega\, dt'\right)|\psi_\pm\rangle = e^{-i(\frac{1}{8}\omega_Q \mp \frac{\sqrt{3}}{2}\omega)t}|\psi_\pm\rangle. \tag{10.85}$$

As noted above, the observable in this frame is $J_3\cos\theta - J_1\sin\theta$. Only the term in J_1 couples the $|\pm 3/2\rangle$ states to the $|\psi_\pm\rangle$ states. The observed signal thus includes components oscillating at the difference frequencies between the pairs of states, with amplitudes given by the matrix elements of J_1. There are four components:

$$\omega_Q - \sqrt{3}\omega, \quad |\langle 3/2|J_1|\psi_-\rangle|^2 = 1/4; \tag{10.86}$$
$$\omega_Q, \quad |\langle 3/2|J_1|\psi_+\rangle|^2 = 1/2; \tag{10.87}$$
$$\omega_Q, \quad |\langle -3/2|J_1|\psi_-\rangle|^2 = 1/2; \tag{10.88}$$
$$\omega_Q + \sqrt{3}\omega, \quad |\langle -3/2|J_1|\psi_+\rangle|^2 = 1/4. \tag{10.89}$$

The predicted spectrum, then, consists of four resonances: two at $\pm\sqrt{3}\omega$, relative to the original, unshifted peak, with intensity $1/4$, and two unshifted peaks with relative intensity $1/2$. Overall the spectrum has the appearance of a triplet, with peak intensities 1:4:1. This spectrum is precisely what was observed by Tycko, see Fig. 10.7.

Readers familiar with Magic Angle Spinning (MAS) and related techniques from high-field NMR may find the spectra in Fig. 10.7 reminiscent of the spinning side-bands typically observed in experiments in which the sample orientation is modulated. The present case is different, however. Note that the new features are separated by $\sqrt{3}\omega$ from the central peak, rather than an integer multiple of ω; furthermore, the intensities of these features are constant for all spinning speeds. In fact, MAS and other high-field sample spinning experiments in NMR should also show geometric effects, but theory and simulations suggest that even in favorable cases the effects will be extremely small [88].

We now return to the issue we raised above, that of using the simplified basis (10.78) rather than the formally proper (10.64) [282]. One problem with the simple basis is its ambiguity at $\theta = 0$ and $\theta = \pi$; this ambiguity is corrected in the proper basis, but since the experiment under analysis avoids this point (in the present experiment, θ is always given by $\cos\theta = 1/\sqrt{3}$) we are justified in using the simpler point of view. More problematic is an artifactual winding number that can appear. Since both $\theta = 0$ and $\theta = \pi$ must be avoided in the simplified basis, paths that wind around the sphere multiple

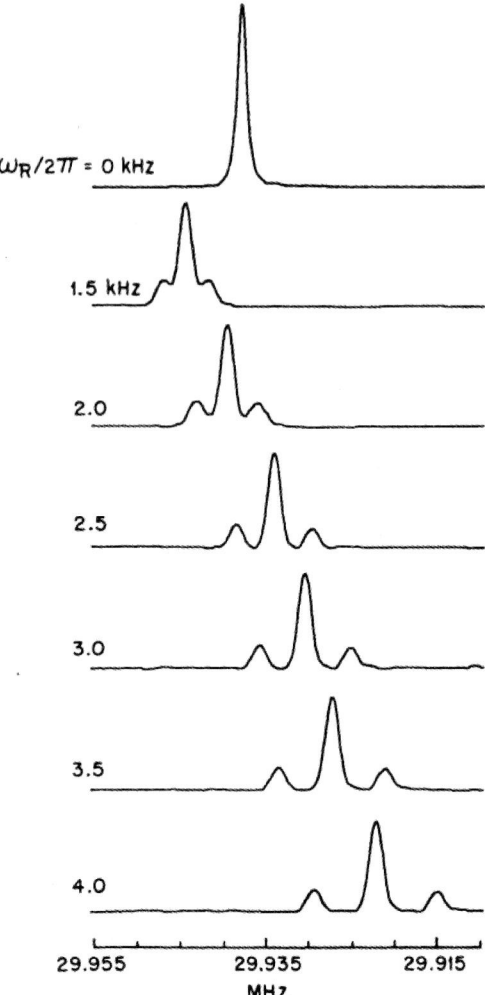

Fig. 10.7. NQR spectra of NaClO$_3$ rotating at different speeds. Reprinted with permission of the American Physical Society [254].

times induce factors of -1 for each circumnavigation; this occurs because the simplified basis is double-valued. The proper basis is single-valued, due to the inclusion of the additional $\exp(iJ_3\varphi)$ operator. In the present case, however, since we are calculating phase *differences* these extra factors do not appear, and we can ignore them in order to provide a simpler discussion.

We close this section by noting an experiment that is sensitive to the non-Abelian part of the connection one-form derived above [285]. As we noted, the non-Abelian phase arises from variation of both θ and φ, which is experimen-

10.2 The Spin Berry Phase Controlled by Magnetic Fields

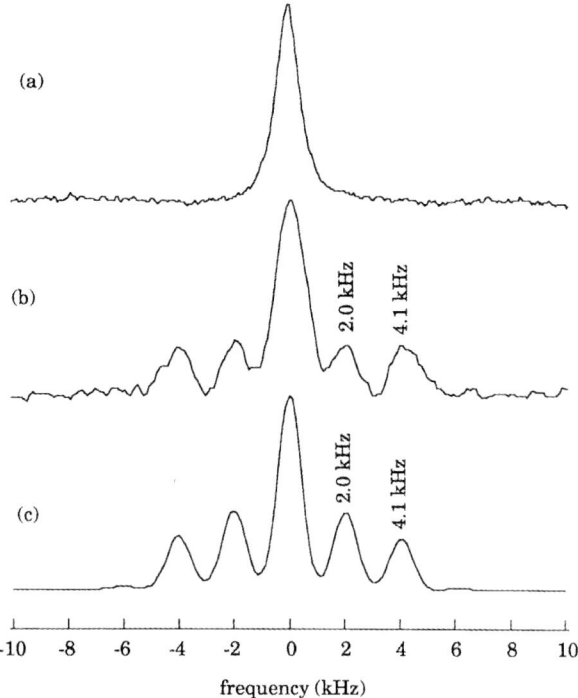

Fig. 10.8. Simulated and experimental NQR spectra of $NaClO_3$ undergoing double rotation. Reprinted with permission of the American Physical Society [285].

tally difficult. Nonetheless, a device (called a double rotor) has been invented in the laboratory of Professor A. Pines which accomplishes this, by rotating a sample holder which is itself rotating; the device is sketched in Fig. 10.6(b). Using this device in a straightforward extension of the experiment of Tycko, one of the authors and colleagues acquired NQR spectra of $NaClO_3$ undergoing double rotation. The trajectory followed by the quantization axis in this experiment is shown in Fig. 10.3(f); both θ and φ change over this path, and it does not lead to a simple analytic form for the acquired phase matrix. Nonetheless, solutions to the Schrödinger equation can be calculated for this experiment, using both exact and adiabatic time evolution; the NQR spectrum predicted on the basis of the connection one-forms (10.67) and (10.68) is in good agreement with the measured spectrum (Fig. 10.8).

While this experiment represents the most direct verification to date of the prediction of non-Abelian effects in adiabatic systems, it is not without its limitations. To demonstrate fully the non-Abelian nature, the ideal experiment, as noted in [282], would subject a system to two different closed paths, and measure the acquired phase matrix; then the paths would be traversed in the reverse order, and one would (presumably) find that the acquired matrix is different. Changing the order of the paths is the key to demonstrating the non-Abelian nature of the connection one-form; in the experiment described above, this was not done, for reasons of experimental difficulty; the experiment shows a result that is fully consistent with the predicted non-Abelian connection, but does not fully explore its geometry.

10.3 Observation of the Aharonov–Anandan Phase Through the Cyclic Evolution of Quantum States

The previous examples in this chapter have illustrated the geometric phase in the spirit originally proposed by Berry, namely a Hamiltonian which is driven through a cycle by means of its dependence on external parameters. As we have noted previously, this picture has been generalized in a deep way by Aharonov and Anandan [7], who showed that the more fundamental object to consider is a family of quantum states (density operators), as opposed to the previously considered family of Hamiltonians. This approach is more fundamental because it removes the restriction of adiabaticity; one merely requires the evolution of a density operator, by whatever means, through a closed cycle. As shown in [7], and recapitulated above, the state vector associated with the density operator acquires a geometric phase, which agrees in the adiabatic limit with the phase defined by Berry. Suter *et al.* [239] have performed an elegant experiment that measures the geometric phase of the density operator, which we discuss in detail in this section.

The experiment discussed here is different in spirit from those above in that it is a true interferometric experiment, and as such is a particularly direct test of the geometric view of quantum evolution. It is a complex experiment, and we build it up in stages. Consider first a two-level system described by a density operator ρ. We will consider only pure states, so we can write $\rho = |\psi\rangle\langle\psi|$, with $|\psi\rangle = c_1|\phi_1\rangle + c_2|\phi_2\rangle$. Because the Hilbert space is of dimension two, the state space is the projective Hilbert space $\mathcal{P}(\mathcal{H}) = \mathbb{C}P^1 = S^2$, as discussed in Chap. 4. Now suppose that initially the system is described by $\rho(0)$, and that it evolves through the space of density operators, eventually returning to its initial state at time T, that is, $\rho(T) = \rho(0)$. As we saw earlier, a state vector associated with this density operator acquires a geometric phase

$$\gamma(T) = \oint i\langle\phi|d|\phi\rangle, \tag{10.90}$$

where $|\phi\rangle$ is the single-valued vector related to $|\psi\rangle$ by (4.25). But in the two-level system this phase is unobservable, as we noted in the introduction to this chapter.

To observe the phase we must couple the two-level system to an environment, to destroy its isolation. This was done by Suter, Mueller, and Pines by considering a three-level system, in which a two-level subset could be manipulated; the result is observed through its effect on the third level. The

10.3 Observation of the Aharonov–Anandan Phase

relevant density operator takes the form

$$\rho = \begin{bmatrix} \rho_{11} & \rho_{12} & 0 \\ \rho_{21} & & \\ & (\rho^{2-3}) & \\ 0 & & \end{bmatrix}, \qquad (10.91)$$

where levels 2 and 3 form the density operator to be manipulated, and level 1 the level that will probe the result. Thus suppose some non-zero matrix elements ρ_{12} and ρ_{21} are created, say by selectively exciting the transition between levels 1 and 2. This coherence can be detected at some time T later, which establishes a phase reference. In a second experiment, the same coherence can be prepared, but during the time T the ρ^{2-3} subdensity operator for levels 2 and 3 can be driven through a cycle, by using selective excitation. At time T this subdensity operator has returned to its initial state, but because it has a level in common (level 2) with the ρ^{1-2} subsystem, the coherence in that system will have evolved differently from in the reference experiment. The phase shift between the two signals reflects the extra phase accumulated by the evolution of ρ^{2-3}.

It proves convenient to describe the three-level system as a collection of two-level systems, by assigning to each pair of states a set of three fictitious spin-1/2 operators [79]. Thus for states r and s define operators $J_1^{(rs)}$, $J_2^{(rs)}$, and $J_3^{(rs)}$ such that

$$J_1^{(rs)} = \frac{1}{2}\left(|r\rangle\langle s| + |s\rangle\langle r|\right), \qquad (10.92)$$

$$J_2^{(rs)} = \frac{i}{2}\left(-|r\rangle\langle s| + |s\rangle\langle r|\right), \qquad (10.93)$$

$$J_3^{(rs)} = \frac{1}{2}\left(|r\rangle\langle s| - |s\rangle\langle r|\right). \qquad (10.94)$$

These operators satisfy the usual angular momentum algebra

$$[J_i^{(rs)}, J_j^{(rs)}] = i\epsilon_{ijk} J_k^{(rs)}; \quad i,j,k = 1,2,3. \qquad (10.95)$$

Operators of this type related to two disjoint two-level systems commute:

$$[J_i^{(rs)}, J_j^{(tu)}] = 0. \qquad (10.96)$$

If, however, the pair of systems shares a level in common, the commutation rules are more complicated:

$$[J_1^{(rt)}, J_1^{(st)}] = [J_2^{(rt)}, J_2^{(st)}] = \frac{i}{2} J_2^{(rs)}, \qquad (10.97)$$

$$[J_3^{(rt)}, J_3^{(st)}] = 0, \qquad (10.98)$$

$$[J_1^{(rt)}, J_2^{(st)}] = \frac{i}{2} J_1^{(rs)}, \qquad (10.99)$$

$$[J_1^{(rt)}, J_3^{(st)}] = -\frac{i}{2} J_2^{(rt)}, \qquad (10.100)$$

$$[J_2^{(rt)}, J_3^{(st)}] = \frac{i}{2} J_1^{(rt)}. \qquad (10.101)$$

From the commutation rules the effects of selective rotations on the different subsystems can be calculated. Within a subsystem we have the familiar result

$$e^{-iJ_i^{(rs)}\varphi} J_j^{(rs)} e^{iJ_i^{(rs)}\varphi} = J_j^{(rs)} \cos\varphi + J_k^{(rs)} \sin\varphi. \qquad (10.102)$$

Rotations involving pairs sharing one state, however, take a more complicated form, for example

$$e^{-iJ_1^{(rs)}\varphi} J_1^{(st)} e^{iJ_1^{(rs)}\varphi} = J_1^{(st)} \cos\varphi/2 + J_2^{(rt)} \sin\varphi/2. \qquad (10.103)$$

To generate a three-level system, methylene chloride (CH_2Cl_2) was dissolved in a liquid crystal solvent. The protons of CH_2Cl_2 are coupled by their magnetic dipole interaction; in a neat liquid this interaction is not apparent in the spectrum, because its isotropic average is zero. However, when the molecule is held at a fixed orientation, as in a liquid crystal environment, the dipole term is observable. The four levels of the two coupled protons consist of three symmetric states (the triplet) and one antisymmetric state (the singlet). Singlet–triplet transitions are forbidden, and so the protons of oriented methylene chloride provide a pseudo-three-level system. Due to the dipolar coupling, the two allowed transitions in this manifold are separated by some 4 kHz, well within the range of selective excitation experiments on modern NMR spectrometers.

The system is probed with NMR techniques similar to those described above in Sect. 10.2.2. The experiment begins with the system at equilibrium. Following arguments given above, (10.28), the effective density operator is

$$\rho(0) = J_3 = J_3^{(1)} + J_3^{(2)} + J_3^{(3)}, \qquad (10.104)$$

where system (1) consists of states 1 and 2; system (2), states 2 and 3; and system (3), states 1 and 3. If we selectively excite the $1-2$ transition with a $\pi/2$ pulse, the resulting density operator is

$$e^{iJ_1^{(1)}\pi/2} J_3 e^{-iJ_1^{(1)}\pi/2} = J_2^{(1)} + J_3^{(2)} + J_3^{(3)}. \qquad (10.105)$$

The component $J_2^{(1)}$ evolves under the influence of the static external field; we can detect it in the rotating frame, as described above, in which it appears static. Its phase at any time T provides the phase reference for the experiment.

10.3 Observation of the Aharonov–Anandan Phase

Suppose now the density operator of (10.105) is subjected to an additional radio-frequency field, at resonance with the transition between levels 2 and 3, and of sufficient strength and duration to effect a 2π rotation in this manifold. Using the commutators for the fictitious spin-1/2 operators it is straightforward to check that

$$e^{2\pi i J_1^{(2)}} J_3^{(2)} e^{-2\pi i J_1^{(2)}} = J_3^{(2)}, \qquad (10.106)$$

$$e^{2\pi i J_1^{(2)}} J_3^{(3)} e^{-2\pi i J_1^{(2)}} = J_3^{(3)}, \qquad (10.107)$$

$$e^{2\pi i J_1^{(2)}} J_2^{(1)} e^{-2\pi i J_1^{(2)}} = -J_2^{(1)}. \qquad (10.108)$$

What is startling here is that the rotation on the $2-3$ transition *inverts* the phase of the $1-2$ coherence. Since the geometry of the projective space for a two-level system is a sphere, and the 2π rotation described above is just a great circle on this sphere, we expect from the work of Aharonov and Anandan that the geometric phase in this case should be $-\frac{1}{2}\Omega(\mathcal{C}) = -\pi$, precisely as observed in the experiment. Here as usual $\Omega(\mathcal{C})$ is the solid angle subtended by the path in the space of density operators.

The path described above was implemented by Vaughan and co-workers to investigate the spinor behavior of a two-level system [233]. Suter, Mueller, and Pines have fully explored the geometry of this case by implementing a

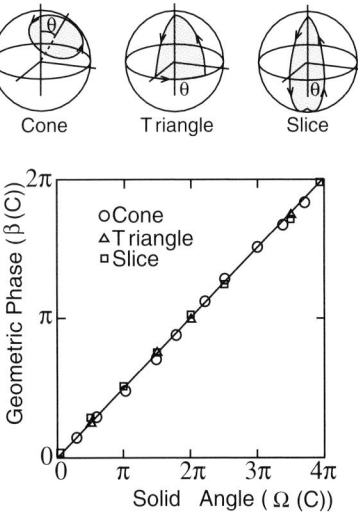

Fig. 10.9. (a) Paths in the space of density operators used to measure the Aharonov–Anandan geometric phase. (b) Measured phase shifts, as a function of solid angle enclosed by the paths. Note that a solid angle of 2π yields a measured phase shift of π in this experiment, due to its interferometric nature (in contrast to the experiments discussed earlier in this chapter). Reprinted with permission of the American Physical Society [239].

variety of paths, enclosing a variety of solid angles – the relation between the geometric phase and solid angle, predicted by Aharonov and Anandan, is confirmed in all respects (Fig. 10.9). This experiment stands as the most complete study to date of the geometric effects discussed in this volume: it is interferometric in nature, and so does not measure phase differences between pairs of states, and does not depend on extrapolating to the adiabatic limit from energy spectra. Moreover, it implements many different paths, and so shows clearly the dependence on solid angle and independence of path shape.

Problems

10.1) Derive (10.18) from (10.14) and (3.104).

10.2) Show that $P(\mathbf{e}_2, \mathbf{e}_1)$, that is, $\text{Tr}[J_2 \rho(T)]$, is given by

$$P(\mathbf{e}_2, \mathbf{e}_1) = \frac{1}{2} \cos \tilde{\theta} \sin \frac{2\pi \Omega}{\omega},$$

using the approach of Problem 10.1. Plot this function for $\theta = \pi/2$ and increasing b, and compare to a similar plot of (10.19).

10.3) Suppose a system is described by the Hamiltonian H, so its density operator evolves in time according to

$$i\frac{d\rho}{dt} = [H, \rho].$$

For $\rho = |\psi\rangle\langle\psi|$ (pure state) and $\rho = \sum_i w_i |\psi_i\rangle\langle\psi_i|$, with the w_i's classical probabilities (mixed state), show that under a unitary transformation $\tilde{U}(t)$ the correct transformed equation of motion is

$$i\frac{d\tilde{\rho}}{dt} = [\tilde{H} - i\tilde{U}\dot{\tilde{U}}^\dagger, \tilde{\rho}],$$

where

$$\tilde{\rho} = \tilde{U}\rho\tilde{U}^\dagger,$$
$$\tilde{H} = \tilde{U}H\tilde{U}^\dagger.$$

10.4) Derive (10.49). This is easiest to do by thinking of the initial density operator $\tilde{\rho}(0) = J_3$ as a component of a spherical tensor, and using the usual transformation rule for such objects:

$$T_{l,m} \xrightarrow{\text{rotation}} \sum_{m'} T_{l,m'} D^l_{m,m'},$$

where the tensor T is of rank l with components m. For the rotation matrix elements D, you can use either Wigner rotation matrix elements or the functions U [257]. Note that J_3 is the $m = 0$ component of a rank-1 spherical tensor.

10.5) Derive the expressions for the non-Abelian matrix elements, (10.65) and (10.66), and their values for the $J = 3/2$ nuclear quadrupole resonance problem, (10.67) and (10.68).

11. Experimental Detection of Geometric Phases II: Quantum Systems in Quantum Environments

11.1 Introduction

In the previous chapter we described a variety of experiments that explore geometric phases under controlled conditions, in other words, when the (classical) environment can be manipulated by the experimenter. In this case the exact solution and adiabatic solutions are both known, thus, the experimental challenge is whether we can control the quantum state so that the geometric phase effect can be observable.

In the present chapter we deal with cases where the environment itself is also a quantum system by examining 1) the interaction between internal and external rotations; 2) the interaction between electronic and rotational motions; 3) the interaction between electronic and vibrational motions.

For realistic quantum many-body systems such as molecules, exact solutions are not usually obtained. The most accurate theoretical results are often obtained from the Born–Oppenheimer approximation. This approximation, when it is employed correctly, introduces gauge potential terms, which produce fictitious magnetic flux at crossing points of potential energy surfaces (PESs). Although the fictitious magnetic flux is a consequence of the careful application of the Born–Oppenheimer approximation, its existence is nevertheless surprising, and it causes the wave functions to be multi-valued. Naturally, it is very important to verify experimentally whether such gauge potentials are really necessary ingredients of the theory. A number of experiments have been performed to observe the geometric phase effect arising from the fictitious magnetic flux. Such experiments are difficult, and comparison between theory and experiment has been rather controversial. Recently, however, experiments that demonstrate the unambiguous manifestation of the fictitious magnetic field effect have been performed using high-resolution spectroscopy of alkali trimers [122, 259].

Previous to these high-resolution spectroscopic studies of alkali trimers, the geometric effect had also been taken into account in order to explain the EPR spectra of vibronic states of impurity ions in crystals [99]. We also discuss these experiments in this chapter.

11.2 Internal Rotors Coupled to External Rotors

In order to motivate the later, more complex examples, we begin with a very simple system, that of two disks connected on an axis by a light, rigid rod. Each disk is free to rotate about the connecting rod, and we exclude any interaction between the disks. This model represents a highly simplified view of molecules like ethane (C_2H_6) or methanol (CH_3OH), that have two groups that are more or less free to rotate with respect to one another, as discussed by [196] and [129] (see also [251]). The entire molecule may rotate about its long axis. We will consider only the motions about this long axis. As we shall see, a description of the internal motion (the relative rotation of the two disks) requires knowledge of the state of the environment. This ingredient is added either as a boundary condition on the wave function of the internal degree of freedom, or as a gauge potential. In either case, it has an effect on the observed spectra.

Let us now analyze the model. We take the moment of inertia of each disk to be I, and the long axis to lie in the $\mathbf{e}_3$ direction. The classical Hamiltonian is then

$$H = \frac{J_{1,3}^2}{2I} + \frac{J_{2,3}^2}{2I}, \tag{11.1}$$

where $J_{i,3}$ is the component of angular momentum of disk i in direction $\mathbf{e}_3$. In this description the two disks are treated as independent entities, as indeed they are. We take the coordinates of the two disks to be φ_1 and φ_2, which are just the azimuthal angles of each disk. Let us now transform to an internal coordinate $\varphi_{\text{int}} = \varphi_2 - \varphi_1$, that describes the relative orientation of the two disks, and an external coordinate $\Phi = (\varphi_1 + \varphi_2)/2$. The canonical transformation is effected by a generating function of type F_2 (see [89])

$$F_2(\varphi_1, \varphi_2, \tilde{J}_1, \tilde{J}_2) = \frac{\tilde{J}_1}{2}(\varphi_1 + \varphi_2) + \tilde{J}_2(\varphi_2 - \varphi_1). \tag{11.2}$$

Here $\tilde{J}_1$ and $\tilde{J}_2$ are the momenta conjugate to the new variables Φ and φ_{int} respectively. In terms of the original momenta we have

$$\tilde{J}_1 = J_1 + J_2, \tag{11.3}$$

$$\tilde{J}_2 = \frac{1}{2}(J_2 - J_1). \tag{11.4}$$

The new momenta have the simple interpretation as the total angular momentum and the relative angular momentum. The Hamiltonian in these coordinates takes the form

$$H = \frac{\tilde{J}_1^2}{2I_{\text{ext}}} + \frac{\tilde{J}_2^2}{2I_{\text{int}}}, \tag{11.5}$$

where $I_{\text{ext}} = 2I$ is the moment of inertia of the entire system, and $I_{\text{int}} = I/2$ is the moment of inertia of the relative motion. In this coordinate system

11.2 Internal Rotors Coupled to External Rotors

there is no obvious interaction term between the internal degree of freedom (φ_{int}) and the environment (the overall rotation).

Let us now consider the quantum states of this Hamiltonian. In spite of the simple form of the Hamiltonian in these coordinates, the effect of the environment on the internal motion is not absent. It appears in the boundary condition on the internal wave function, as we now show. In the original coordinate system the energy eigenfunction is clearly

$$\hat{\psi}(\varphi_1, \varphi_2) = \frac{1}{2\pi} e^{im_1\varphi_1} e^{im_2\varphi_2}, \tag{11.6}$$

with m_1 and m_2 integers. This wave function is single-valued with respect to transformations of the sort $\varphi_i \to \varphi_i + 2\pi n$, for integer n. In the $(\Phi, \varphi_{\text{int}})$ coordinate system we have a similar form for the wave function:

$$\tilde{\psi}(\Phi, \varphi_{\text{int}}) = \frac{1}{2\pi} e^{iK\Phi} e^{i\sigma\varphi_{\text{int}}}. \tag{11.7}$$

We require this wave function to be invariant under the same transformations as the original wave function, since these just correspond to rotations of both parts of the system through 2π. Substituting in for φ_{int} and Φ we find

$$e^{iK\Phi} e^{i\sigma\varphi_{\text{int}}} = e^{iK\Phi} e^{i\sigma\varphi_{\text{int}}} e^{2\pi i [K(n_1+n_2)/2 + \sigma(n_1-n_2)]}. \tag{11.8}$$

We therefore require that $K(n_1 + n_2)/2 + \sigma(n_1 - n_2)$ be an integer, which is accomplished if K is itself an integer, and $\sigma = m - K/2$, with integer m. The wave function thus takes the modified form

$$\tilde{\psi}(\Phi, \varphi_{\text{int}}) = \frac{1}{2\pi} e^{iK\Phi} e^{im\varphi_{\text{int}}} e^{-iK\varphi_{\text{int}}/2}, \tag{11.9}$$

in which we interpret $e^{iK\Phi}$ as the wave function for the overall rotation about the long axis of the molecule, with quantum number K, and $e^{im\varphi_{\text{int}}} e^{-iK\varphi_{\text{int}}/2}$ as the wave function of the internal motion. The boundary condition on the internal degree of freedom is now given by

$$e^{im(\varphi_{\text{int}}+2\pi)} e^{-iK(\varphi_{\text{int}}+2\pi)/2} = e^{im\varphi_{\text{int}}} e^{-iK\varphi_{\text{int}}/2} e^{-i\pi K}. \tag{11.10}$$

Evidently the state of the environment is manifested on the internal degree of freedom, regardless of the apparent lack of coupling between the two in the Hamiltonian!

The coupling between the system and environment in this problem is not merely bookkeeping. It is easy to compute the energies of the internal degree of freedom, using the Hamiltonian $H_{\text{int}} = \tilde{J}_2^2/2I_{\text{int}}$ and (11.9); these are

$$E_{\text{int}} = \frac{\hbar^2}{2I_{\text{int}}} (m - K/2)^2. \tag{11.11}$$

We see that the ladder of states of the internal rotor is actually affected by the value of K, the overall rotational state of the system. The coupling between the internal and external degrees of freedom arises, of course, from the factor $e^{-iK\varphi_{\text{int}}/2}$ in the internal wave function. Suppose we attempt to remove this term, by applying the transformation $e^{iK\varphi_{\text{int}}/2}$ to the internal wave function. This procedure yields

$$e^{im\varphi_{\text{int}}}e^{-iK\varphi_{\text{int}}/2} \to e^{iK\varphi_{\text{int}}/2}e^{im\varphi_{\text{int}}}e^{-iK\varphi_{\text{int}}/2} = e^{im\varphi_{\text{int}}}, \quad (11.12)$$

yielding a single-valued set of wave functions. Of course, we must simultaneously transform the momentum operator in the same way, yielding

$$\tilde{J}_2 \to e^{iK\varphi_{\text{int}}/2}\tilde{J}_2 e^{-iK\varphi_{\text{int}}/2}$$
$$= e^{iK\varphi_{\text{int}}/2}\left(-i\frac{\partial}{\partial\varphi_{\text{int}}}\right)e^{-iK\varphi_{\text{int}}/2}$$
$$= \tilde{J}_2 - K/2. \quad (11.13)$$

The transformed Hamiltonian is thus

$$H_{\text{int}} = \frac{(\tilde{J}_2 - K/2)^2}{2I_{\text{int}}}, \quad (11.14)$$

in which we see that the internal degree of freedom experiences a gauge potential $K/2$, determined by the state of its environment.

Thus this extremely simple model illustrates all the features of effective gauge interactions in molecular systems: an internal degree of freedom that is never completely decoupled from its environment, and the coupling that affects the energy levels of the internal system, through either boundary conditions or a gauge potential. A variety of real molecules fit this simple picture, and fit even better the geometrically similar model of two rotors with a hindering torsional barrier to internal rotation [251].

The rotational spectrum of the above model does not show the interaction between the external and internal rotation. This is so because the model describes a true symmetric top, the electric dipole of which must lie along the internal rotation axis. The resulting $\Delta K = 0$, $\Delta m = 0$ selection rules would result in a simple spectrum in which the internal state of the molecule cannot be changed. Certain asymmetric rotors, such as nitromethane, CH_3NO_2, have nearly free internal rotation, and a generalization of the above picture has been applied to their microwave spectra [241, 251]. Features of the nitromethane spectrum illustrate the points we raised above. Consider for example the $J = 2 \leftarrow J = 1$ transition [241]. The $\Delta K = 0$, $\Delta m = 0$ selection rules are nearly good for this molecule, even though it is an asymmetric rotor. The transitions with $K = 0$, $m = 0$ and $K = \pm 1$, $m = 0$ differ in frequency by some 3.6 MHz. Even though the internal energy depends on K and m, and so is different in these two cases, if the molecule were symmetric the transition frequencies would be the same. Therefore, we can think of the 3.6 MHz difference as an estimate of the effect of the rotor asymmetry. Now consider the $K = 0$, $m = \pm 2$ and $K = \pm 1$, $m = \pm 2$ transitions. The first of these two is 0.9 MHz lower in frequency than the second. Were there no interaction between the internal and external degrees of freedom, these two transitions would have also been split by 3.6 MHz. Because of the complicated nature of the asymmetric rotor spectrum it is difficult to provide a transparent interpretation for these differences; but it is clear nevertheless that the internal rotation is coupled to the external degrees of freedom.

11.3 Electronic–Rotational Coupling

The interaction between an internal *electronic* angular momentum and an external angular momentum fits more closely the theme of this book than does the case discussed above. Because of the low effective mass of the electronic motion relative to the nuclear motion, one can describe the two degrees of freedom within the adiabatic approximation. This case thus provides a clear molecular instance of a geometric phase. Our aim in this section is to present several cases in which the consequences of the geometric phases have been observed experimentally.

The interaction between the rotational angular momentum and the electronic angular momentum in a diatomic molecule is the well-studied phenomenon known as Λ-doubling. We described this in some detail previously (Section 8.6) – here we remind the reader that the interaction is between the component Λ of the electronic angular momentum along the molecular axis, and the rotational angular momentum of the nuclei. The electronic problem is a two-state problem, the two states corresponding to projections $\pm\Lambda$ onto the internuclear axis. The operators that make up the gauge potential are linear combinations of the angular momentum operators (8.120) and (8.121). For $|\Lambda| \neq 1/2$ there are no off-diagonal matrix elements. The effective gauge potential is Abelian, and generates an effective monopole field that interacts with the nuclear motion. The result is a modified rotational Hamiltonian [174]

$$\hat{H} = B(\tilde{J}^2 - \Lambda^2), \qquad (11.15)$$

where B is the rotational constant of the molecule and $\tilde{J}$ includes both the nuclear and electronic angular momentum (J and Λ respectively). The component of $\tilde{J}$ in the $\mathbf{e}_3$ direction is quantized to $\tilde{J}_3 = J \pm \Lambda$. Because the degeneracy of the Λ levels is not lifted by the diagonal gauge potential, this potential causes no additional splittings to appear in the spectrum.[1] The ladder of levels for $\Lambda > 1/2$ is thus the same as that for $\Lambda = 0$, aside from starting at $J = \Lambda$ rather than $J = 0$ of course (8.188) and (8.189).

When $|\Lambda| = 1/2$, however, there are off-diagonal matrix elements of the gauge potential and a true non-Abelian gauge interaction is present. Strictly speaking Λ cannot equal $1/2$, since it is an orbital angular momentum; this

---
[1] Actually, splittings of $\Lambda = 1$ levels are often observed, but these are due to higher-order interactions [142].

case is realized in practice in Hund's case (*a*) molecules, for which both the spin and orbital angular momentum projections onto the molecular axis are conserved quantities. The resulting total projection is denoted Ω, and so such a case for a diatomic molecule might better be called "Ω-doubling". The degeneracy of the levels is lifted immediately by a term proportional to $J + 1/2$ [142]. This happens essentially because in this case the gauge potential is not constrained to lie always along the internuclear axis, but can point in any direction. Thus in particular it can lift the degeneracy of the Λ levels.

There are numerous experimental examples of the above interactions [142, 185, 251]. The ZnH molecule provides an especially clear case, because its spectroscopy has been studied for both the $^2\Pi_{1/2}$ and $^2\Pi_{3/2}$ states [142, 185]. The former shows the non-Abelian gauge-induced splitting, the latter does not. These splittings are shown as a function of J in Fig. 11.1.

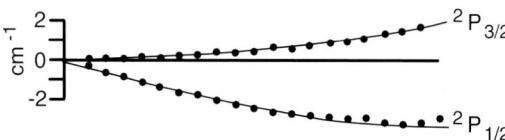

Fig. 11.1. Splitting due to Λ-doubling in the non-Abelian $|\Lambda| = 1/2$ case of the $^2\Pi_{1/2}$ state of ZnH (bottom curve) and the Abelian, $|\Lambda| = 3/2$ case ($^2\Pi_{3/2}$). The theoretical prediction of a linear dependence on $\tilde{J}$ is shown for the non-Abelian case, as well as the lack of splitting in the Abelian case, until higher-order effects appear at high J.

11.4 Vibronic Problems in Jahn–Teller Systems

In the previous section, electronic and rotational angular momenta provided the coupled quantum systems. If instead the electronic degrees of freedom are coupled to vibrational angular momenta, the coupling is termed vibronic. In typical cases the coupling is linear in the vibrational coordinates, to lowest order, and such cases are termed Jahn–Teller systems. These systems are characterized by PESs that intersect as a function of nuclear displacements, and hence in the Born–Oppenheimer framework include fictitious magnetic flux originating at the crossings.

One of the observable consequences of the fictitious magnetic flux in Jahn–Teller systems is the change of the ground state symmetry. For example, the ground state symmetry of the $E \otimes e$ Jahn–Teller system can be shown to be E (that is, doubly degenerate) when the fictitious magnetic flux situated at the highest symmetry point is properly included; the symmetry is A (non-degenerate) if it is ignored; see Sect. 9.5 and [99] for detailed discussion of this point. In this section we describe some experimental results which confirm that the ground vibronic state of the $E \otimes e$ Jahn–Teller system is indeed E and thus verify the necessity of the fictitious magnetic flux as an ingredient of the theory.

11.4.1 Transition Metal Ions in Crystals

The verification of the doubly degenerate ground state in the $E \otimes e$ Jahn–Teller system was first realized in electron paramagnetic resonance (EPR) experiments of impurity ions in crystals.

Zeeman splittings of vibronic states can be measured in an EPR experiment. The magnetic resonance spectrum of a Jahn–Teller system is usually treated with the following effective Hamiltonian [34]:

$$H = \sum_{\Gamma,\gamma} G_{\Gamma\gamma} \mathbf{C}_{\Gamma\gamma}, \quad (11.16)$$

where $G_{\Gamma\gamma}$ is an electronic operator that transforms as the γ component of the Γ representation, and $\mathbf{C}_{\Gamma\gamma}$ are the matrices composed of the Clebsch–Gordan coefficients (see Appendix B). For a cubic system, for example, the electronic operators are given by

$$G_{A_1} = g_1 \beta_e \mathcal{H} \cdot \mathbf{S} + a_1 \mathbf{I} \cdot \mathbf{S}$$
$$G_{E\theta} = \frac{1}{2} g_2 \beta_e (3\mathcal{H}_z S_z - \mathcal{H} \cdot \mathbf{S}) + \frac{1}{2} a_2 (3 I_z S_z - \mathbf{I} \cdot \mathbf{S})$$
$$G_{E\epsilon} = \frac{\sqrt{3}}{2} g_2 \beta_e (\mathcal{H}_x S_x - \mathcal{H}_y S_y) + \frac{\sqrt{3}}{2} a_2 (I_x S_x - I_y S_y) \quad (11.17)$$

where β_e is the electronic Bohr magneton. The applied magnetic field and electron spin are given by $\mathcal{H}$ and $\mathbf{S}$, respectively. Note that the A_1, $E\theta$, and $E\epsilon$ components transform as $x^2 + y^2 + z^2$, $3z^2 - r^2$, and $x^2 - y^2$, respectively (see the character table for O_h in Appendix B). From EPR experiments the parameters g_1, g_2, a_1, and a_2 are obtained.

An anomalous spectrum (nearly isotropic EPR spectrum), which is a dynamical manifestation of the Jahn–Teller instability for the orbital doublet ground state (E term electronic state), was first observed for Cu^{2+} in crystalline zinc fluorosilicate [39].[2] The spectrum obtained at 20 K is the superposition of the individual static Jahn–Teller distorted configurations of the Cu^{2+} complexes, and becomes isotropic due to the dynamical Jahn–Teller distortion at 90 K [38]. This temperature dependence is in accord with the predicted ordering of the vibronic states; the E state is the ground state, and the A state is the first excited state. The latter is populated only at high temperatures, giving rise to the isotropic spectrum.

---

[2] The theory for the Jahn–Teller effect for paramagnetic ions in solids was already developed in 1937 [258].

11. Experimental Detection of Geometric Phases II

An important consequence of the degenerate ground state is the reduction of parameters for electronic operators [34]. The reduction factor can be derived purely by group theoretical considerations, as follows. A ground state vibronic wave function is written as [see (B.33)]

$$\Psi_{\Gamma\gamma}(\mathbf{r},\mathbf{Q}) = \sum_{\Gamma_1,\gamma_1,\gamma'} \chi_{\Gamma_1\gamma_1}(\mathbf{Q})\psi_{\Gamma\gamma'}(\mathbf{r})\langle\Gamma_1\gamma_1\Gamma\gamma'|\Gamma\gamma\rangle, \quad (11.18)$$

where $\chi_{\Gamma\gamma}(\mathbf{Q})$ are vibrational wave functions, $\psi_{\Gamma\gamma}(\mathbf{r})$ are electronic wave functions, and $\langle\Gamma_1\gamma_1\Gamma\gamma'|\Gamma\gamma\rangle$ are Clebsch–Gordan (C-G) coefficients.

Using the above wave functions, vibronic matrix elements of an electronic operator $O_{\Gamma_2\gamma_2}$ are calculated as

$$\langle\Psi_{\Gamma\gamma}|O_{\bar{\Gamma}\bar{\gamma}}|\Psi_{\Gamma\gamma'}\rangle$$
$$= \langle\Psi_\Gamma||O_{\bar{\Gamma}}||\Psi_\Gamma\rangle\langle\Gamma\gamma|\Gamma\gamma'\bar{\Gamma}\bar{\gamma}\rangle$$
$$= \langle\psi_\Gamma||O_{\bar{\Gamma}}||\psi_\Gamma\rangle$$
$$\times \sum_{\Gamma_1,\gamma_1,\gamma_1',\gamma_1''} \langle\chi_{\Gamma_1}^2\rangle\langle\Gamma_1\gamma_1\Gamma\gamma_1'|\Gamma\gamma\rangle\langle\Gamma\gamma'|\Gamma\gamma_1''\Gamma_1\gamma_1\rangle\langle\Gamma\gamma_1''|\Gamma\gamma_1'\bar{\Gamma}\bar{\gamma}\rangle$$

(11.19)

where $\langle\psi_\Gamma||O_{\bar{\Gamma}}||\psi_\Gamma\rangle$ and $\langle\Psi_\Gamma||O_{\bar{\Gamma}}||\Psi_\Gamma\rangle$ are the reduced matrix elements (see B.36), and the vibrational functions are normalized as

$$\langle\chi_{A_1}^2\rangle + \langle\chi_{A_2}^2\rangle + \langle\chi_E^2\rangle = 1,$$
$$\langle\chi_{\Gamma_1\gamma_1}|\chi_{\Gamma_2\gamma_2}\rangle = \langle\chi_{\Gamma_1}^2\rangle\delta_{\Gamma_1\Gamma_2}\delta_{\gamma_1\gamma_2}. \quad (11.20)$$

The reduction factor for the operator that transforms as $\bar{\Gamma}$ is defined as

$$K(\bar{\Gamma}) := \frac{\langle\Psi_\Gamma||O_{\bar{\Gamma}}||\Psi_\Gamma\rangle}{\langle\psi_\Gamma||O_{\bar{\Gamma}}||\psi_\Gamma\rangle}$$
$$= \sum_{\Gamma_1,\gamma_1,\gamma_1',\gamma_1'',\gamma',\bar{\gamma}} \langle\chi_{\Gamma_1}^2\rangle\langle\Gamma_1\gamma_1\Gamma\gamma_1'|\Gamma\gamma\rangle$$
$$\times \langle\Gamma\gamma'|\Gamma\gamma_1''\Gamma_1\gamma_1\rangle\langle\Gamma\gamma_1''|\Gamma\gamma_1'\bar{\Gamma}\bar{\gamma}\rangle\langle\Gamma\gamma'\bar{\Gamma}\bar{\gamma}|\Gamma\gamma\rangle \quad (11.21)$$

where the orthogonality relation of the C-G coefficients (B.34) has been used.

The reduction factors for the E term vibronic state are calculated using the C-G coefficients in Table B.2 as

$$K(A_1) = \langle\chi_{A_1}^2\rangle + \langle\chi_{A_2}^2\rangle + \langle\chi_E^2\rangle = 1$$
$$K(A_2) = \langle\chi_{A_1}^2\rangle + \langle\chi_{A_2}^2\rangle - \langle\chi_E^2\rangle$$
$$K(E) = \langle\chi_{A_1}^2\rangle - \langle\chi_{A_2}^2\rangle. \quad (11.22)$$

Coupling parameters of electronic operators are then modified as

$$g_1 \to \tilde{g}_1 = g_1$$
$$g_2 \to \tilde{g}_2 = K(E)g_2$$
$$a_1 \to \tilde{a}_1 = a_1$$
$$a_2 \to \tilde{a}_2 = K(E)a_2 \qquad (11.23)$$

As seen from (11.22), coupling parameters for A_2 and E symmetries are reduced because $K(A_2)$ and $K(E)$ are less than one. On the other hand those transforming as A_1 are unaffected.

For real calculations the vibronic reduction factor is calculated using the following relation:

$$\frac{\langle \Psi_{\Gamma\gamma}|\mathbf{C}_{\bar{\Gamma}\bar{\gamma}}|\Psi_{\Gamma\gamma'}\rangle}{\langle \psi_{\Gamma\gamma}|\mathbf{C}_{\bar{\Gamma}\bar{\gamma}}|\psi_{\Gamma\gamma'}\rangle} = \frac{\langle \Psi_\Gamma||\mathbf{C}_{\bar{\Gamma}}||\Psi_{\Gamma\gamma'}\rangle\langle\Gamma\gamma|\Gamma\gamma'\bar{\Gamma}\bar{\gamma}\rangle}{\langle \psi_\Gamma||\mathbf{C}_{\bar{\Gamma}}||\psi_{\Gamma\gamma'}\rangle\langle\Gamma\gamma|\Gamma\gamma'\bar{\Gamma}\bar{\gamma}\rangle} = K(\bar{\Gamma}) \qquad (11.24)$$

where the vibronic wave functions $\Psi_{\Gamma\gamma}$ are numerically obtained by solving the vibronic problem, such as (B.52).

The theory of vibronic reduction in EPR spectra was first applied to explain the anomalously large reduction of the orbital contribution to the g-factor in the triplet ground state for Mn^+, Cr, Mn, and Fe^+ in silicon [97]. The reduction is anomalous because it is achieved without destroying the cubic symmetry. If the reduction were due to the covalent bonding between the ion and surrounding atoms, the symmetry would be lowered. A satisfactory explanation is given by the vibronic reduction due a dynamic Jahn–Teller effect.

In Table 11.1, experimental g-factors and central hyperfine parameters for transition metal ions (^2E) in cubic symmetry are tabulated. Their low-

Table 11.1. Experimental g-factors g_1, g_2, and hyperfine parameters a_1, a_2 for transition metal ions (^2E) in cubic symmetry, as defined in (11.17). Systems with low-temperature anisotropic EPR spectra due to zero-point ionic motion are tabulated [98]. For specific data, see [65–67, 206] for Cu^{2+}, and [106] for Sc^{2+}.

Ion	Host	T (K)	g-factor	Hyp. Int. 10^{-4} cm^{-1}
Cu^{2+} ($3d^9$)	MgO	77	$g = 2.192$ (isotropic)	$a = 19.0$
		1.2	$g_1 = 2.195$ $K(E)g_2 = 0.108$ $K(E) \simeq 0.5$	$a_1 + K(E)a_2 = 63$
Sc^{2+} ($3d^1$)	CaF_2	77	$g = 1.967$ (isotropic)	$a = 65.0$
		10	$g = 1.969$ (isotropic)	$a = 65.5$
		1.5	$g_1 = 1.973$ $K(E)g_2 = -0.022$ $K(E) = 0.75$	$a_1 = 65.0$ $K(E)a_2 = 24.5$
Sc^{2+} ($3d^1$)	SrF_2	10	$g=1.963$ (isotropic)	$a=67.0$
		1.5	$g_1=1.963$ $K(E)g_2 = -0.028$ $K(E) = 0.71$	$a_1 = 67.1$ $K(E)a_2 = 24.1$

temperature anisotropic EPR spectra are due to zero-point vibrational motion, which exhibits a cubic anisotropy consistent with E symmetry, even at liquid helium temperatures. This means that they are hopping (tunneling) around equivalent minima due to the zero-point vibrational motion. These dynamics have to be distinguished from the case for Cu^{2+} in zinc fluorosilicate. In this case the ion complex stays in one of the three equivalent minima at low temperatures, yielding an EPR spectrum that is a superposition of three axial spectra of static Jahn–Teller distortions.

Splitting or broadening of levels due to defects, impurities, and lattice strain usually obscure the above effects in real spectra. Nevertheless, the ground vibronic E and T states have consistently explained many observed EPR spectra of transition metal ions in crystals. Because the degenerate ground state requires the existence of the geometric phase (and conversely, see Sect. 9.5), the systems discussed above demonstrate the importance of taking into account the geometric phase effect to understand real systems.

11.4.2 Hydrocarbon Radicals

Along with transition metal compounds, degenerate and nearly degenerate electronic states are common in hydrocarbon radicals. In this section determination of the symmetry and ordering of vibronic states in an isolated Jahn–Teller hydrocarbon radical is discussed. We will take up the zero-kinetic-energy (ZEKE) pulsed-field ionization experiment of $C_6H_6^+$ [156]. In this experimental method, the pulsed-field ionized electron current from the Rydberg states of benzene prepared by lasers is measured. The more precise procedure is as follows: within the pulsed-field ionization and electron collection region of the spectrometer, a pulsed beam of benzene seeded in argon was crossed by two lasers. The first laser creates rovibrational states in the first excited electronic state (S_1) of neutral benzene. The second laser scans transitions from a specific S_1 rotational state to the Rydberg states, forming accessible rovibrational states of the cations. Very high Rydberg states with relatively long lifetimes exist near each ionization limit to the cation of the Rydberg states. The transition to these states by the second laser was detected by application of a delayed pulsed field followed by collection of the electrons produced in the ionization. Significantly, resolution of individual rovibronic states was achieved in this experiment.

The relevant Jahn–Teller active vibrational mode in the present system is the doubly-degenerate e_{2g} mode, which couples with the doubly-degenerate E_{1g} electronic states in the D_{6h} point group. The spectrum arising from the $E_{1g} \otimes e_{2g}$ coupling can be fitted with the standard Hamiltonian discussed in detail previously (9.56). In order to confirm the presence of the geometric phase in this case, it is again necessary to consider the order of the vibronic levels: degenerate ground state and two non-degenerate first excited states in the case of a geometric phase, and the reverse order if the geometric phase is absent. To make the symmetry assignment of the observed transitions, the rotational resolution must be used. For each of the two first excited states, sets of rotational resonances were observed, with only even nuclear angular momentum N for the first excited state, and only odd nuclear angular momentum N for the second. If the geometric phase were absent the first of these states would correspond to a degenerate vibronic level, but in the presence of the geometric phase these two states correspond to the two expected non-degenerate excited vibronic levels. For the $E_{1g} \otimes e_{2g}$ coupling, the non-degenerate levels could be of symmetry B_{1g} and B_{2g}. The rotational wave functions must satisfy Fermi–Dirac statistics, because they interchange identical protons (spin-1/2). It turns out that the spin-statistics requirement forces odd-N rotational lines in the B_{1g} states to be 13 times weaker than the even-N lines, while in the B_{2g} state the situation is reversed. This alternation is essentially what was observed experimentally, allowing the excited levels to be assigned definitely as B_{1g} and B_{2g}. This then proves the existence of the geometric phase in this case, because it verifies that the lowest excited vibronic states are non-degenerate.

11.4.3 Alkali Metal Trimers

Alkali metal trimers, such as Na_3 and Li_3, have been intensively studied to detect the geometric phase effect. Such molecules have nominally D_{3h} geometries (that is, equilateral triangular shapes), and have only two vibrational modes: totally symmetric a_1' and doubly degenerate e'. Due to their odd number of valence electrons they have many degenerate electronic states, including the ground state, and thus are a source of simple Jahn–Teller systems of $E \otimes e$ type. Because the geometric phase imposes half-odd-integral quantization in this system, the spectroscopy of alkali metal trimers should reflect the presence (or absence) of the geometric phase.

An early paper on the $B \leftarrow X$ transition in Na_3 was taken for some time as a clear demonstration of the geometric phase [73]. The vibronically resolved spectrum was fitted with good precision to the eigenvalues of the standard $E \otimes e$ Hamiltonian (9.56), with large linear coupling parameter k. This limit of the $E \otimes e$ system is called the adiabatic free rotor, and is described by a harmonic oscillation due to radial distortions and nearly free rotation about the conical intersection. The presence of the geometric phase is then manifested as the half-integral quantization of the rotation about the conical intersection, and a different ladder of energy levels is obtained as compared to the standard integral quantization. The fit of the $B \leftarrow X$ spectrum of Na_3 to this limit was taken as verification of the geometric phase in this system. Subsequently, however, detailed calculations showed that this spectral region is not dominated by the B state but includes substantial overlap (via a pseudo-Jahn–Teller mechanism) with the A state [169]. Thus, simple though this system is, it cannot be taken as a direct test of the presence of geometric phases.

With the aid of much higher resolution spectroscopy than was performed in [73], the spectroscopy of the A state of Na_3 itself proves to yield a good check of the geometric phase in alkali metal trimers [259].

The rovibronic (rotational + vibrational + electronic) transitions of Na_3 show hyperfine substructure which overlaps considerably even at resolution better than 50 MHz. In order to disentangle these dense and complicated spectra, the optical–optical double-resonance (OODR) technique was used. The experimental set-up is shown in Fig. 11.2. In this technique, the laser-induced fluorescence (LIF) and resonance two-photon ionization (RTPI) methods are combined as follows: the first continuous laser (pump laser) is fixed to a prominent spectral line of the LIF response. This laser is chopped at the frequency f, creating the modulation of the initial and final state populations accordingly, although with opposite signs. The second laser (probe laser) is then used for the subsequent RTPI. The signal of the RTPI is filtered by a lock-in amplifier tuned to the modulation frequency f. Therefore, the observed transitions are either those that start from the same level as the pump transition (primary resonances), or start from levels populated by fluorescence from the pumped level (secondary resonances). Note that primary and secondary resonances differ in sign. The reduction of the number of lines provided by this technique is significant and makes the assignment of the spectral lines considerably easier. However, accurate theoretical calculations are needed to assign transitions to spectral lines.

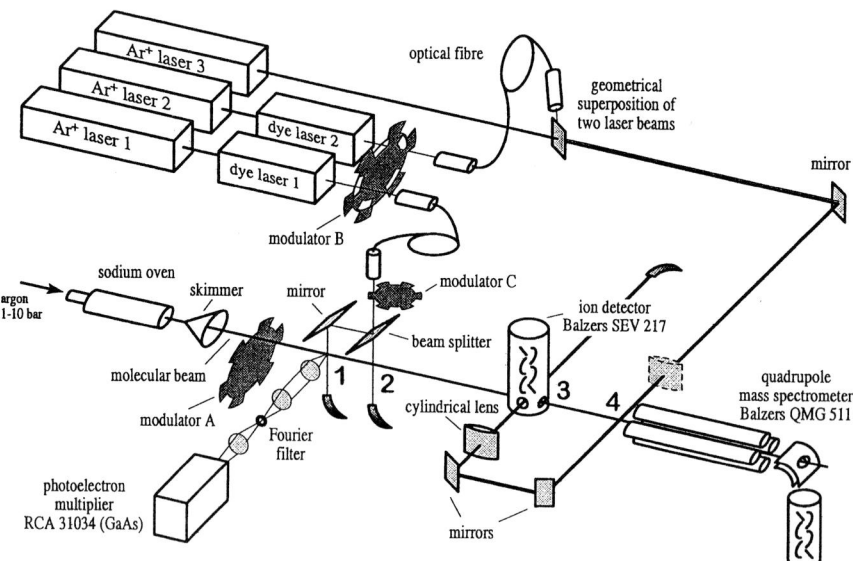

Fig. 11.2. Experimental setup for the high-resolution optical–optical double-resonance (OODR) spectroscopy for Na_3. Reprinted with permission from [52].

11.4 Vibronic Problems in Jahn–Teller Systems

The experimental results were compared with *ab initio* calculations. The potential energy surfaces (PESs) for X and A states were obtained by multi-reference configuration interaction calculations, including an effective core polarization potential by a large Gaussian-type orbital basis set including bond-center functions. From 109 calculated points, PESs in analytical form with standard deviation of $0.6\,\text{cm}^{-1}$ were constructed. Rovibrational (rotation + vibration) energy levels were obtained using these PESs.

We shall succinctly explain the calculational method for rovibrational levels on each PES in the following. The vibration–rotation Hamiltonian for three identical atoms A, B, and C with mass m on the potential surface V, is given by

$$\begin{aligned}
H &= -\frac{1}{2\mu}\left(\frac{\partial^2}{\partial \mathbf{R}^2} + \frac{\partial^2}{\partial \mathbf{r}^2}\right) + V \\
&= -\frac{1}{2\mu}\rho^{-5/2}\frac{\partial^2}{\partial \rho^2}\rho^{5/2} - \frac{1}{8\mu\rho^2}\left(\frac{\partial^2}{\partial \theta^2} + \cot\theta\frac{\partial}{\partial \theta} + \frac{1}{2(1+\cos\theta)}\frac{\partial^2}{\partial \theta^2} - \frac{\partial^2}{\partial \phi^2}\right) \\
&\quad + \frac{1}{\mu\rho^2}\left[\frac{J_x^2}{1-\cos(\theta/2)} + \frac{J_y^2}{1+\cos(\theta/2)} + \frac{J_z^2}{1+\cos(\theta)}\right] \\
&\quad + \frac{1}{\mu\rho^2}\frac{4\sin(\theta/2)}{1+\cos(\theta)}J_z\frac{1}{i}\frac{\partial}{\partial \phi} + \frac{1}{8\mu\rho^2} + V(\rho,\theta,\phi),
\end{aligned} \quad (11.25)$$

where $\mathbf{R} = s\mathbf{R}_{A,BC}$ and $\mathbf{r} = s^{-1}\mathbf{R}_{BC}$ are scaled Jacobi vectors ($s = 2^{1/2}3^{-1/4}$), and J_x, J_y, and J_z are three components of the total angular momentum $\mathbf{J}$, respectively. The scaled system mass μ is given by

$$\mu = \left(\frac{m_A m_B m_C}{m_{ABC}}\right)^{1/2} = \left(\frac{m^3}{3m}\right)^{1/2} = 3^{-1/2}m. \quad (11.26)$$

The Hamiltonian contains six variables: three Euler coordinates α, β, γ, and three internal coordinates, ρ, θ, ϕ, where the latter are related to lengths of three (unscaled) distances as

$$\begin{aligned}
|\mathbf{R}_{BC}| &= 3^{-1/4}\rho[1 - \cos(\theta/2)\cos(\phi + 2\pi/3)]^{1/2}, \\
|\mathbf{R}_{CA}| &= 3^{-1/4}\rho[1 - \cos(\theta/2)\cos(\phi + 4\pi/3)]^{1/2}, \\
|\mathbf{R}_{AB}| &= 3^{-1/4}\rho[1 - \cos(\theta/2)\cos(\phi)]^{1/2},
\end{aligned} \quad (11.27)$$

with $0 \leq \rho \leq \infty, 0 \leq \theta \leq \pi, 0 \leq \phi \leq 2\pi$. The hyper-radius

$$\rho := \sqrt{|\mathbf{r}|^2 + |\mathbf{R}|^2} \quad (11.28)$$

indicates the size of the system, θ describes the shape of the triangle (C_{2v} structure at $\theta = 0$ and D_{3h} structure at $\theta = \pi$), and ϕ is the pseudorotational angle around the conical intersection at the D_{3h} symmetry (equilateral triangular) points.

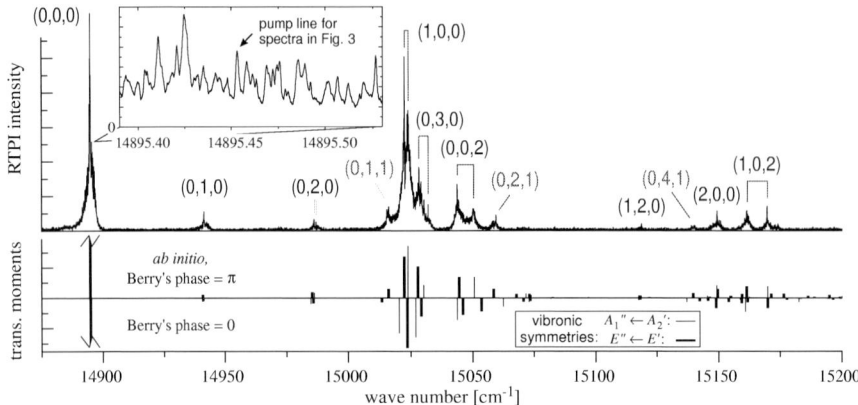

Fig. 11.3. Top: the $A \leftarrow X$ system of Na$_3$ and its assignment in terms of transitions $A(v_s, v_b, v_a) \leftarrow X(0,0,0)$. The brackets indicate the pseudorotational splittings. Bottom: absolute values of transition moments calculated by *ab initio* methods. Reprinted with permission from [53].

The Hamiltonian is diagonalized using the following basis functions:

$$\Psi^{JM} = \sum_{\Omega,t} F^J_{\Omega t}(\rho,\theta,\phi) D^J_{M\Omega}(\alpha,\beta,\gamma), \qquad (11.29)$$

where $D^J_{M\Omega}(\alpha,\beta,\gamma)$ is the Wigner rotational function, $F^J_{\Omega t}(\rho,\theta,\phi)$ is the internal function and t denotes labels for the internal state, collectively. The projection of the total angular momentum on the space-fixed z-axis, M, is a constant of the motion and is taken to be 0 without loss of generality.[3]

The internal wave function $F^J_{\Omega t}$ is a product of three primitive bases for the three variables, ρ, θ, and ϕ. The basis functions for ϕ are $\{\cos(m\phi), \sin(m\phi)\}$. The geometric phase effect is included here by using half-odd integers for m; these quantum numbers would be integers without the geometric phase effect. Thereby the sign-change boundary condition on $\phi \to \phi + 2\pi$ is imposed.

In Fig. 11.3 the observed OODR spectra and theoretical calculations are depicted. The order of vibronic states E and A changes depending on whether the geometric phase is included or not. The experimental results are well reproduced by the *ab initio* calculations including the geometric phase.

In Fig. 11.4, an experimental OODR spectrum obtained for the transition $A(1,0,0) \leftarrow X(0,0,0)$, for which the given pump-line involves levels with

---

[3] In order to construct the Hamiltonian matrix, the following parity-definite rotational functions are usually used;

$$D^{Jp}_{0\Omega} = \sqrt{\frac{1}{2(1+\delta_{\Omega 0})}} \left[D^J_{0\Omega} + (-1)^p D^J_{0-\Omega} \right], \qquad (11.30)$$

where $p = 0, 1$ is the parity.

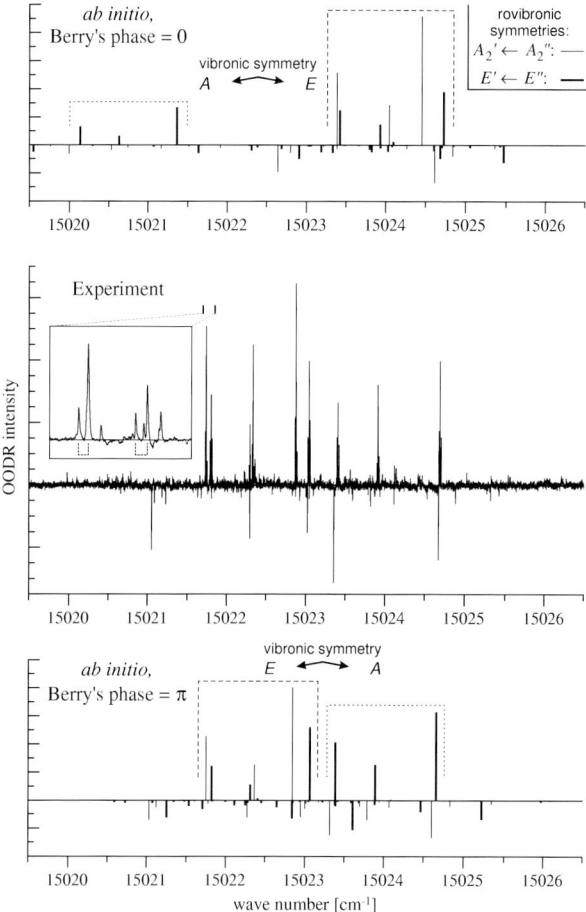

Fig. 11.4. OODR spectra in the band $A(1,0,0) \leftarrow X(0,0,0)$ for the pump line marked in Fig. 11.3. Calculations with geometric phases of 0 and π are compared with the experimental spectrum. Positive and negative signals are primary and secondary resonances, respectively. Reprinted with permission from [53].

$1 \leq N \leq 5$ (N is the angular momentum) is displayed. The agreement is much better for the calculation including the geometric phase. This demonstrates again that the fictitious magnetic field is a necessary ingredient of the Born–Oppenheimer approximation.

11.5 The Geometric Phase in Chemical Reactions

The presence of conical intersections in PESs of a system undergoing reactive scattering can also be expected to induce effects related to the geometric phase. The most extensively studied example is the reaction $H + H_2 \to H + H_2$ and isotopically substituted variants. If in the fully protonated case, the reagent and product H_2 are distinguishable if we monitor the conversion between the *ortho* H_2 and *para* H_2. For this system the conical intersection exists at 2.7 eV above the H_2 potential minimum. The minimal potential barrier for the reaction is at a collinear H-H-H geometry with 0.42 eV. After much effort, the first three-dimensional quantum calculations of the reactive scattering cross-section were performed in 1975 [225, 226]. The results exhibit remarkable quantum mechanical effects, including tunneling, resonances, and interferences.

Soon after this accomplishment, the geometric phase effect in this reaction, which had not been included in the above calculations, was pointed out [167]. At that time the importance of this effect was not fully appreciated by most of the researchers in this field. To be sure, however, at the low collision energies that were the object of primary attention the dominant reactive trajectories are those that go through the collinear saddle points. Contributions from the trajectories that encircle the conical intersection are negligible, and thus geometric effects in that early work were not large.

After decades of improvement in both experimental techniques and computational methods, it is now possible to do experiments and theoretical calculations at high collision energies, for which geometric phase effects may be important. Currently, however, the results are rather controversial as to the significance of the geometric phase in this reaction because of the subtlety and difficulty of the problem. Different groups have been producing somewhat different results and no consensus has been reached. We will however sketch the experimental and theoretical progress so far in the following.

Theoretical calculations are performed using the Hamiltonian (11.25) with the scattering boundary condition. The most accurate method to solve the Schrödinger equation is the coupled-channel method. In this method one coordinate is chosen as the propagation coordinate for scattering (usually the hyper-radius ρ) and the motion for the remaining coordinates is solved as bound state problems while treating ρ as a parameter.

The interval for the scattering coordinate ρ ($\rho_i \leq \rho \leq \rho_f$) is split into sectors, where the amplitude of the wave function is essentially zero at the initial ρ value, ρ_i. The tri-atomic system tends to atom plus diatom asymptotes (A + BC, B+CA, C+AB) at the final ρ value, ρ_f. In each sector the bound state wave functions are evaluated at the mid-point of the sector $\rho = \rho_\xi$. The bound state eigenfunctions Φ_t^J, which are called *surface functions*, are solutions to

$$H_s(\rho_\xi)\Phi_t^J = \mathcal{E}_t^J \Phi_t^J, \qquad (11.31)$$

where

$$H_s(\rho_\xi) = -\frac{8}{\mu\rho_\xi^2}\left(\frac{\partial^2}{\partial\theta^2} + \cot\theta\frac{\partial}{\partial\theta} + \frac{1}{2(1+\cos\theta)}\frac{\partial^2}{\partial\theta^2} - \frac{\partial^2}{\partial\phi^2}\right)$$
$$+ \frac{1}{\mu\rho_\xi^2}\left[\frac{J_x^2}{1-\cos(\theta/2)} + \frac{J_y^2}{1+\cos(\theta/2)} + \frac{J_z^2}{1+\cos(\theta)}\right]$$
$$+ \frac{1}{\mu\rho_\xi^2}\frac{4\sin(\theta/2)}{1+\cos(\theta)}J_z\frac{1}{i}\frac{\partial^2}{\partial\phi} + \frac{15}{8\mu\rho_\xi^2} + V(\rho_\xi,\theta,\phi). \tag{11.32}$$

The total wave function in the ξ-th sector is expressed using the surface functions as

$$\Psi_E^{JM} = \sum_t \rho^{-5/2}\psi_{Et}^J(\rho)\Phi_t^J(\alpha,\beta,\gamma,\theta,\phi;\rho_\xi). \tag{11.33}$$

Substituting (11.33) into the Schrödinger equation gives

$$H\Psi_E^{JM} = E\Psi_E^{JM}, \tag{11.34}$$

and by integrating out all coordinates except ρ, we obtain the coupled-channel equations

$$\frac{1}{2\mu}\left[\frac{\partial^2}{\partial\rho^2} + 2\mu E\right]\psi_{it}^J$$
$$= \sum_{t'}\langle\Phi_t^{JM}|\frac{\rho_\xi^2}{\rho^2}\left[\mathcal{E}_t^J(\rho_\xi) - V(\rho_\xi,\theta,\phi)\right] + V(\rho,\theta,\phi)|\Phi_{t'}^{JM}\rangle\psi_{Et'}^J. \tag{11.35}$$

These equations are solved from ρ_i to ρ_f. At $\rho = \rho_f$ the wave function is expressed as linear combinations of incident waves and outgoing waves, from which elements of the S-matrix are extracted.

The geometric phase is introduced in H_s. One way to include the geometric phase is to use double-valued basis functions for the pseudorotational coordinate as described in the previous section. Another way is to introduce the gauge potential in the Hamiltonian H_s. In the latter method, the gauge potential is introduced by modifying the surface function as

$$\Phi_t^J \to e^{il_A\eta/2}\Phi_t^J, \tag{11.36}$$

where l_A is an odd integer, and $\eta = \eta(\theta,\phi)$ is the angle around the conical intersection. For the case of three identical atoms η is simply given by $\eta = \phi$. But when some of the atoms are replaced by isotopes, η depends on both θ and ϕ. This is because the relations (11.27) are modified for general mass combinations as

$$|\mathbf{R}_{ij}| = 2^{-1/2}s_k\rho[1 - \cos(\theta/2)\cos(\phi + \epsilon_k)]^{1/2},$$
$$\epsilon_A = 2\tan^{-1}(m_B/\mu), \epsilon_B = -2\tan^{-1}(m_B/\mu), \epsilon_C = 0,$$
$$s_i = \left(\frac{m_{ABC}}{\mu}\frac{m_{ABC} - m_i}{m_{ABC}}\right)^{1/2}, \tag{11.37}$$

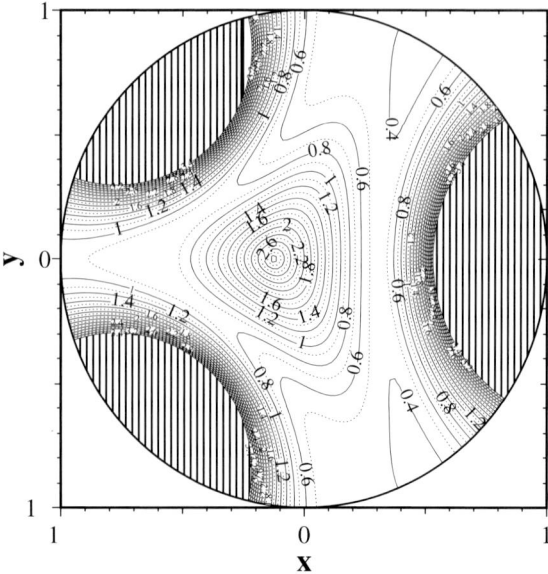

Fig. 11.5. Contour plot of the HD$_2$ PES with $\rho = 3.1 a_0$ (a_0: Bohr radius), where $x = \rho \sin\theta \cos\phi, y = \rho \sin\theta \sin\phi$. The contours start at 0.2 eV and end at 3.4 eV. The highest energy point near the origin is the conical intersection. (Courtesy of B. Kendrick.)

where (ijk) denotes $(ABC), (BCA)$, and (CAB). In this case the position of the conical intersection is shifted from the origin, as is shown in Fig. 11.5.

As a consequence of the replacement (11.36), the gauge potential

$$A_\theta = -\frac{l_A}{2}\frac{\partial \eta}{\partial \theta}$$
$$A_\phi = -\frac{l_A}{2}\frac{\partial \eta}{\partial \phi} \qquad (11.38)$$

is introduced in H_s. This replacement means also that the derivatives in the Hamiltonian are replaced by the "covariant" derivatives:

$$\frac{1}{i}\frac{\partial}{\partial \theta} \to \frac{1}{i}\frac{\partial}{\partial \theta} - A_\theta$$
$$\frac{1}{i}\frac{\partial}{\partial \phi} \to \frac{1}{i}\frac{\partial}{\partial \phi} - A_\phi. \qquad (11.39)$$

We note that introducing the geometric phase by including the gauge potential has several advantages over the method using a double-valued basis. First of all, multiple conical intersections are easily included, and their positions are immaterial. This method is also mathematically more tractable. Gauge invariance of the Hamiltonian ($\Phi_t^J \to e^{in\eta}\Phi_t^J$, n: integer) is apparent

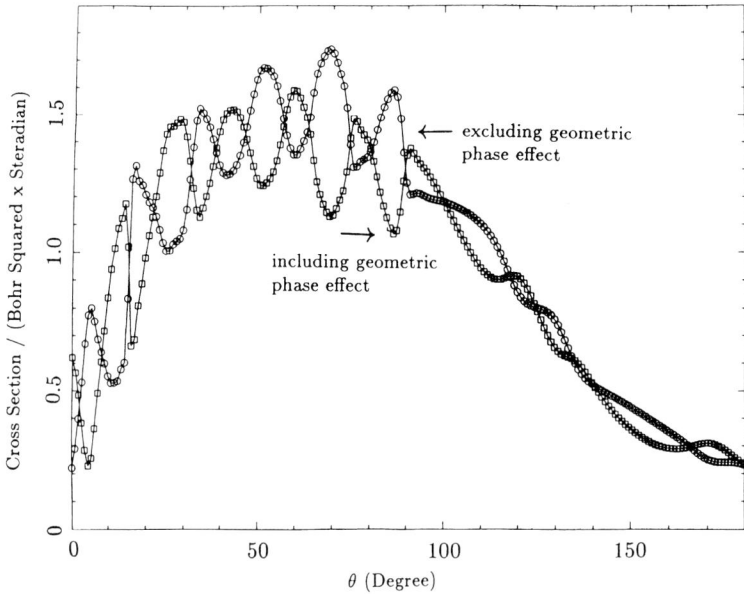

Fig. 11.6. Degeneracy-summed differential cross-sections for the H + H_2 ($v = 0, j = 0$) → H_2 ($v' = 0, j' = 2$) + H reaction, at a total energy of 0.7 eV, as a function of scattering angle. The squares (circles) correspond to the calculations with (without) geometric phase. Reprinted with permission of Elsevier Science BV [276].

and can be used as a check for computational accuracy, and the derivatives of the wave functions are well-defined at all points.

The results of the first theoretical calculations for H + H_2 → H_2 + H including the geometric phase effect are shown in Fig. 11.6 [276]. These results were obtained by employing a double-valued basis. They clearly show the significant change of oscillation pattern in the differential cross-sections. Comparison between experiment and theory was first done for the D+ H_2 → DH + H reaction (Fig. 11.7) [127, 144, 194, 274]. Agreement between the measured and calculated rates as a function of j' is rather poor if the geometric phase is ignored. For example, such a calculation predicts a maximum rate of reaction at $j' = 10$, while in the data the maximum occurs at $j' = 7$. Inclusion of the geometric phase effect shifts the predicted maximum to the correct value, and brings the predicted reaction rates into excellent agreement with experiment. However, this agreement may be due to accidental compensation of errors [123]. One source of error is that the PES used in this calculation is not of high quality in the high-energy region. Another is that the double-valued basis functions used in this work do not correctly represent pseudorotational angles, since they used ϕ for the pseudorotational

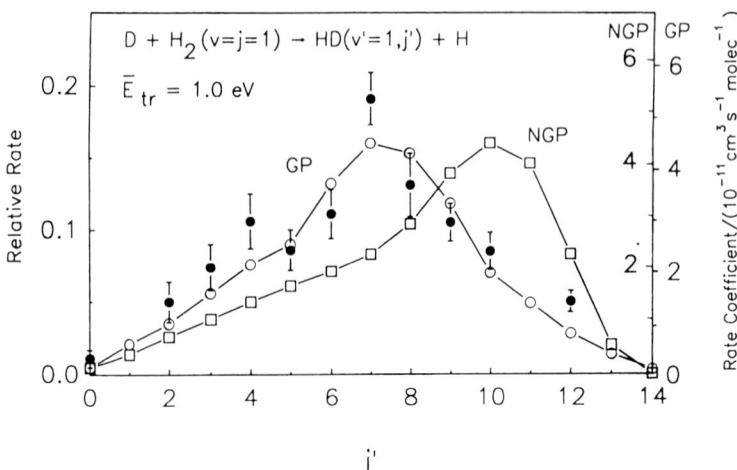

Fig. 11.7. Comparison of theoretical and experimental rates (products of initial relative velocities by integral cross-sections) as a function of rotational state quantum number j' for D + H$_2$ ($v = 1, j = 1$) → DH ($v' = 1, j'$) + H for an average initial relative translational energy $E_{tr} = 1.0$ eV (average total energy $E = 1.8$ eV). The solid circles represent experimental results. The open squares (circles) represent calculations with (without) the geometric phase. Reprinted with permission of Elsevier Science BV [144].

angle. Actually, the center of the pseudorotation is shifted in the DH$_2$ system compared with that in the H$_3$ system.

Comparison between theory and experiment is also made for the reaction H + D$_2$ → DH + D [125, 145, 228, 268–270, 275]. A typical experimental setup is depicted in Fig. 11.8. The results are again controversial. Early results indicated the importance of the geometric phase, but inaccuracy of the PES used in them became evident.

The most accurate calculations to date indicate that the geometric phase effect is negligible in the H + D$_2$ → DH + D reaction. The effect shows up in each partial wave, especially for low angular momentum at high collision energies (Fig. 11.9). But after summation over many partial waves, the effect is washed out. It seems further investigations are necessary to decide whether the geometric phase is important in the H + H$_2$ → H$_2$ + H reaction. On the theoretical side, the effect of the Born–Huang term and the non-adiabatic effect from the upper potential surface have not yet been carefully taken into account. Notably lacking is the experimental results of the H + o-H$_2$ → p-H$_2$ + H reaction that can be compared with theory. It is expected that the geometric phase effect should be largest in this reaction because a larger de Broglie wavelength is associated with a lighter hydrogen mass, which will make the quantum interference effect more evident.

11.5 The Geometric Phase in Chemical Reactions

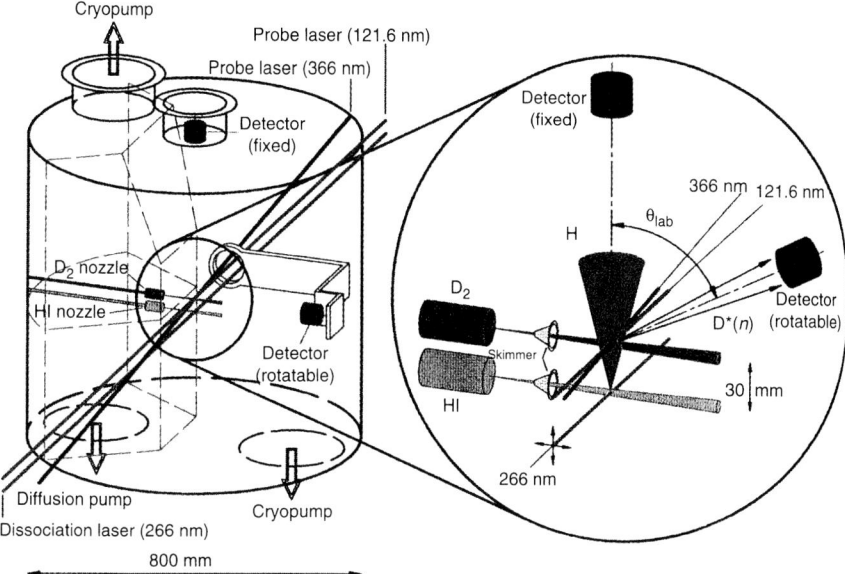

Fig. 11.8. A typical crossed molecular beam experimental set-up for the H + D_2 → HD + D reaction. Two parallel pulsed molecular beams, one with $o-D_2$ and the other with HI, are skimmed into the scattering and detection chamber. Hot H atoms are produced by the photolysis of the HI molecules. The product D atoms are excited into metastable Rydberg states, followed by field-ionization and detection with a particle multiplier. Reprinted with permission of the American Association for the Advancement of Science [228].

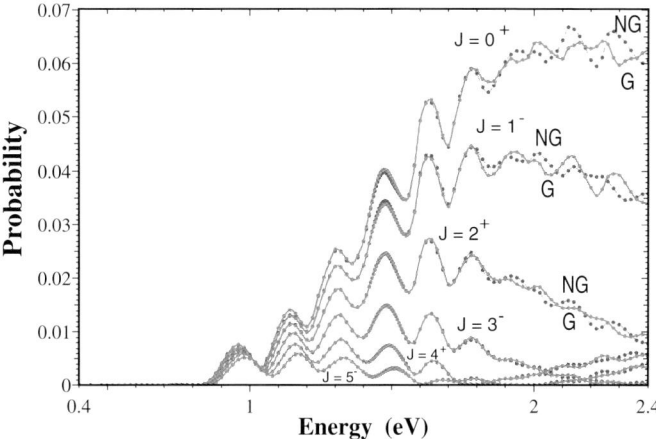

Fig. 11.9. Reaction probabilities for H + D_2 ($v = 0, j = 0$) → HD ($v' = 0, j' = 0$) + D for several values of J^p (J: total angular momentum, p: parity). The data indicated by G (NG) includes (does not include) the geometric phase. The BKMP2 potential surface is used. (Courtesy of B. Kendrick.)

12. Geometric Phase in Condensed Matter I: Bloch Bands

12.1 Introduction

The Schrödinger equation in a spatially periodic system has eigenstates in the form of Bloch waves, plane waves with amplitudes modulated periodically in space. We will consider in this chapter geometric phases associated with the Bloch waves. Crystalline solids are naturally occuring periodic systems in which an electron sees a periodic potential due to the nuclei and other electrons. Periodic systems also include artificial structures such as superlattices, networks of microwires, and arrays of quantum dots or antidots. A perfect periodic potential can also be produced on cold atoms by an interference pattern of laser beams of suitable frequency. Throughout this chapter, we will treat the periodic potentials as given and focus on the properties of particles in the Bloch states under external forces.

We will first introduce the Bloch theorem regarding the general structure of the eigenstates and energies in periodic potentials. We will also describe briefly some basic methods for the calculation of band structures of the eigenstates. We will then show how geometric phase arises in semiclassical dynamics and transport of particles in weak external fields. As we shall see, the concept of geometric phase is also very useful in our understanding of the behavior of Wannier functions. These are localized wave functions that can span the Hilbert space of the Bloch bands and that are extensively used in solid state physics. The remaining part of the chapter will be devoted to the discussion of insulators, where some energy bands of the Bloch states are completely filled while others are completely empty. There is no electric conduction by an electric field, however particle transport is still possible under adiabatic deformation of the potential. This is a phenomenon deeply rooted in the physics involving the geometric phase. Further applications of the geometric phase and adiabatic particle transport will also be discussed in connection with polarization calculations for crystal dielectrics.

12.2 Bloch Theory

12.2.1 One-Dimensional Case

We first consider the one-dimensional situation for simplicity where the Hamiltonian is given by

$$\hat{H} = \frac{1}{2m}\hat{P}^2 + V(x), \qquad (12.1)$$

and the potential is periodic, $V(x+a) = V(x)$, with a being the period known as the lattice constant. Without knowing the detailed shape of the potential within a *unit cell* (defined as any segment of length a), the translational symmetry alone can yield much information about the energy spectrum and eigenfunctions.

Consequence of Translational Symmetry. Introduce the translation operator by the lattice constant a,

$$\hat{T}(a) = e^{ia\hat{P}/\hbar}, \qquad (12.2)$$

which has the effect of $\hat{T}(a)\psi(x) = \psi(x+a)$ on any wave function. Periodicity of the potential implies that

$$\hat{T}\hat{H} = \hat{H}\hat{T}. \qquad (12.3)$$

Therefore, the energy eigenvectors can also be taken as the eigenvectors of the translation operator,

$$\hat{T}\psi(x) = \lambda\psi(x), \text{ or, } \psi(x+a) = \lambda\psi(x). \qquad (12.4)$$

For an infinite system ($-\infty < x < \infty$), λ should have unit magnitude, otherwise the wave function would diverge either at $x = -\infty$ or $x = +\infty$.[1] We can therefore write $\lambda = e^{ika}$, with k being real and called the *Bloch wave number*. This relation defines k only up to an additional constant of a multiple of $2\pi/a$. It is therefore sufficient to restrict k to an interval of length $2\pi/a$ called a *Brillouin zone*. It is customary to take this interval to be $(-\pi/a, \pi/a]$. This is called the *first Brillouin zone* or simply *the Brillouin zone*.

For each k, the eigenenergies and their eigenfunctions are obtained by solving the Schrödinger equation over a unit cell under the *Bloch condition*: $\psi(x+a) = e^{ika}\psi(x)$. For a given Hamiltonian involving a second-order differential operator, the Bloch condition is equivalent to the pair of equation: $\psi(a) = e^{ika}\psi(0)$ and $\psi'(a) = e^{ika}\psi'(0)$, where ' indicates a derivative with respect to x. Because the eigenvalue problem can be viewed as defined over the

---

[1] This result reflects the fact that $\hat{T}$ is a unitary operator whose eigenvalues must have unit modulus.

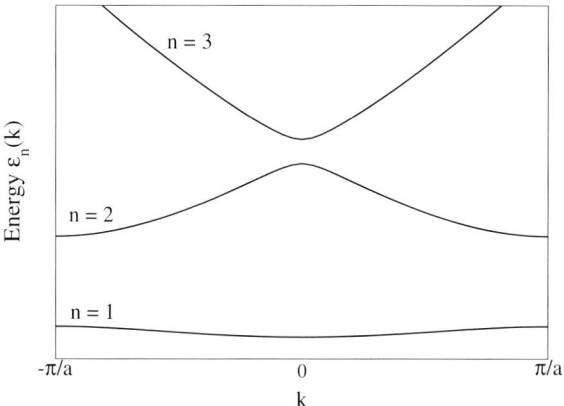

Fig. 12.1. Typical band structure in a one-dimensional periodic potential.

finite interval $0 < x < a$, the eigenenergies and their eigenfunctions form a discrete set $\{\epsilon_n(k),\ \psi_{nk}(x),\ n = 0, 1, 2, ...\}$. As k varies continuously over the Brillouin zone, the eigenenergies trace out a set of bands, called the Bloch bands. The wave functions $\{\psi_{nk}(x)\}$ are called the Bloch waves or Bloch states.

Additional Structures of the Bloch Bands. The Hamiltonian (12.1) has time-reversal symmetry, i.e., it is invariant under complex conjugation. This implies that $\psi_{nk}^*(x)$ satisfies the time-independent Schrödinger equation with the same eigenenergy $\epsilon_n(k)$. Since it also satisfies the Bloch condition for $\psi_{n(-k)}$, we conclude that $\epsilon_n(k) = \epsilon_n(-k)$. When the state vector ψ_{nk} for a given k is non-degenerate, which is generically the case (see below), $\psi_{n(-k)}$ is the same as ψ_{nk}^* except for a possible phase factor.[2]

Moreover, since the time-independent Schrödinger equation has at most two independent solutions with a given energy, different bands cannot overlap in energy, and each band must be monotonic on either side of $k = 0$. In the next section, we will see that the sense of monotonic behavior alternates from one band to the next. Therefore, if two bands accidentally touch, this must occur at the center or the edge of the Brillouin zone. However, if the Hamiltonian involves higher-order derivatives than the second, or if the system consists of coupled chains, or has a spin degree of freedom, there can be more than two eigenstates at the same energy, and the above results must be modified. Nevertheless, according to the Wigner–von Neumann theorem (9.2), different bands generically do not cross one another as k varies.

---

[2] In this chapter we usually identify state vectors ψ with the corresponding wave functions $\psi(x)$ in the position representation. In this representation the above-mentioned phase factor multiplies the wave function. But as it is associated with the state vector, it does not depend on x.

12.2.2 Three-Dimensional Case

The Bloch theory applies in higher dimensions quite similarly to the one-dimensional case discussed above. In three dimensions, a periodic potential is characterized by invariance under translations along the three basic lattice vectors:

$$V(\mathbf{r} + \mathbf{a}_j) = V(\mathbf{r}), j = 1, 2, 3. \tag{12.5}$$

The translation operators,

$$\hat{T}_j = e^{i\mathbf{a}_j \cdot \hat{\mathbf{P}}/\hbar}, \tag{12.6}$$

then commute with the Hamiltonian and with each other. The energy eigenstates can be chosen as the eigenstates of the translation operators with unimodular eigenvalues λ_j, which may be parameterized as $\lambda_j = e^{i\mathbf{k} \cdot \mathbf{a}_j}$ by the *Bloch wave vector* $\mathbf{k}$.

Brillouin Zone. The parameterization is unique if $\mathbf{k}$ is restricted to a Brillouin zone. This is determined in the following manner. We first define three so-called *reciprocal lattice vectors*:

$$\mathbf{G}_1 = \frac{2\pi \mathbf{a}_2 \times \mathbf{a}_3}{\mathbf{a}_1 \cdot \mathbf{a}_2 \times \mathbf{a}_3}, \tag{12.7}$$

$$\mathbf{G}_2 = \frac{2\pi \mathbf{a}_3 \times \mathbf{a}_1}{\mathbf{a}_2 \cdot \mathbf{a}_3 \times \mathbf{a}_1}, \tag{12.8}$$

and

$$\mathbf{G}_3 = \frac{2\pi \mathbf{a}_1 \times \mathbf{a}_2}{\mathbf{a}_3 \cdot \mathbf{a}_1 \times \mathbf{a}_2}. \tag{12.9}$$

Note that the last two equations may be obtained from the first by cyclic permutations of the indices 1, 2 and 3. A general reciprocal lattice vector $\mathbf{G}$ is defined as a linear combination of the above three vectors with integer coefficients. It is then easy to see that the exponentials $e^{i\mathbf{k} \cdot \mathbf{a}_j}$, which appear in the Bloch conditions for the wave functions, $\psi(\mathbf{r} + \mathbf{a}_j) = \exp(i\mathbf{k} \cdot \mathbf{a}_j)\psi(\mathbf{r})$, are invariant if $\mathbf{k}$ is translated by a reciprocal lattice vector. Therefore, we may choose a Brillouin zone to be the region defined by $\{\mathbf{k} = \alpha_1 \mathbf{G}_1 + \alpha_2 \mathbf{G}_2 + \alpha_3 \mathbf{G}_3, (-\frac{1}{2} < \alpha_j \leq \frac{1}{2})\}$. We note that this choice does not correspond to the commonly used first Brillouin zone, which is defined as the region of points that are closer to the origin, $\mathbf{G} = 0$, than any other reciprocal lattice points.

Structures of the Bloch Bands. For a periodic Hamiltonian with time-reversal symmetry the energy bands are degenerate between $\mathbf{k}$ and $-\mathbf{k}$. If the potential possesses rotational and inversion symmetries, the bands have similar symmetries in $\mathbf{k}$ space. Different bands may touch or cross at special $\mathbf{k}$ points that are invariant under symmetry operations. There are excellent textbooks that include a systematic treatment of this subject. The interested readers may consult [57] for a more detailed exposition.

In two dimensions, different bands do not generally touch at **k** points other than those special ones discussed above. But, according to the Wigner–von Neumann theorem (9.2), in three dimensions different bands can cross at isolated **k** points that are not related by symmetry.

12.2.3 Band Structure Calculation

The calculation of band structures for real electronic systems has been a major task of theoretical condensed matter physics where one needs to deal with the complications caused by Coulomb interactions between the electrons. In the following, we will confine our attention to the simpler question: What is the band structure for a given periodic potential? In general, we can employ numerical methods to solve the Schrödinger equation in a unit cell under the Bloch condition. We may also expand the wave function in a Fourier series, reduce the time-independent Schrödinger equation to the eigenvalue problem for the coefficients, and try to solve the latter. We will instead consider the two extreme situations, namely the very weak and very strong potentials, that admit analytic solutions.

Near-Free-Electron Regime. Because the free particle Hamiltonian is invariant under translation of the lattice vectors $\mathbf{R} = n_1\mathbf{a}_1 + n_2\mathbf{a}_2 + n_3\mathbf{a}_3$, its eigenstates can also be classified according to the Bloch theory. In fact, a plane wave state $e^{i\mathbf{q}\cdot\mathbf{r}}$ is a Bloch state of wave vector **k** satisfying $e^{i\mathbf{k}\cdot\mathbf{R}} = e^{i\mathbf{q}\cdot\mathbf{R}}$. In the Bloch theory, the wave vector **k** is restricted in a given Brillouin zone, and we therefore have $\mathbf{k} = \mathbf{q} - \mathbf{K}$, where $\mathbf{K} = m_1\mathbf{G}_1 + m_2\mathbf{G}_2 + m_3\mathbf{G}_3$ is a reciprocal lattice vector such that **k** lies in the Brillouin zone. The mapping of the plane wave vector **q** into the Bloch wave vector **k** is known as the *reduced zone scheme*. The Bloch band picture for the free particle is a very useful framework, if one wishes to consider the effect of a weak periodic potential by perturbation theory. Because the Bloch wave vector is conserved under the perturbation of a periodic potential, one only needs to consider coupling between states with the same **k**. Note also that the effect of perturbation becomes significant at the locations where the levels are degenerate or nearly degenerate.

In one dimension, the degeneracy occurs either at the center of the zone, $k = 0$, or its boundary, $k = \pm\frac{\pi}{a}$. These correspond to the plane waves with $q = \pm\frac{n\pi}{a}$. Degenerate perturbation theory gives an energy gap of $2|V_n|$ where

$$V_n = \int_0^a \frac{dx}{a} V(x) e^{in\frac{2\pi}{a}} \tag{12.10}$$

is the matrix element between the degenerate states. Therefore, the n-th gap between the Bloch bands equals twice the absolute value of the n-th Fourier component of the periodic potential. In case the Fourier component vanishes, one can still lift the degeneracy by taking into account higher-order perturbative corrections.

The situation in higher dimensions is similar, except that degeneracies can occur along lines or planes of the **k** space, and that the degree of degeneracy (multiplicity of the degenerate eigenvalue) can exceed two, i.e., more than two states can become degenerate.

Two plane waves become degenerate if their wave vectors have the same magnitude $|\mathbf{q}| = |\mathbf{q}'|$, and they correspond to the same Bloch wave vector if $\mathbf{q}' = \mathbf{q} - \mathbf{K}$ for some reciprocal lattice vector $\mathbf{K}$. These conditions imply $\mathbf{q} \cdot \mathbf{K} = \frac{1}{2}K^2$. This means that $\mathbf{q}$ lies in the plane perpendicular to $\mathbf{K}$ and at a distance of $|\mathbf{K}|/2$ from the origin. The latter is known as a *Bragg plane* in crystal diffraction. It is customary to choose the Brillouin zone as the region near the origin which is enclosed by all possible Bragg planes. Therefore, the boundary of the Brillouin zone is a surface of degeneracy in the free particle limit. For the lowest band this is the only place where degeneracy occurs. For higher bands degeneracy can also occur inside the Brillouin zone. Perturbation of the potential usually lifts such degeneracies, but partial degeneracy can still remain at some high-symmetry points where more than two plane waves are coupled. Group theory provides a very general framework for the determination of the pattern of degeneracy-lifting at such a point [57].

Tight-Binding Regime. For energies deep in the potential wells, the Bloch bands can be constructed out of the energy levels of each potential well. In the context of a solid state problem, one imagines that the periodic potential is made of a superposition of attractive atomic potentials centered at the lattice points $\{\mathbf{R}\}$,

$$V(\mathbf{r}) = \sum_{\mathbf{R}} V_a(\mathbf{r} - \mathbf{R}), \qquad (12.11)$$

each of which possesses a set of atomic orbitals. Here for simplicity we assume that there is one atom in each unit cell. If an atomic orbital $\psi_j(\mathbf{r} - \mathbf{R})$ is very tightly bound, in the sense that the wave function is nearly zero wherever the periodic potential differs appreciably from the atomic potential $V_a(\mathbf{r} - \mathbf{R})$, then it is also an approximate solution of the time-independent Schrödinger equation for the periodic potential. The set of atomic orbitals form a degenerate space of energy levels. We can use their linear combinations to construct more accurate approximate solutions. This is a standard procedure of degenerate perturbation theory. For the problem at hand, we have an additional guide for choosing appropriate linear combinations, namely the Bloch condition that leads to the solutions

$$|\psi_j(\mathbf{k})\rangle = \sum_{\mathbf{R}} e^{i\mathbf{k}\cdot\mathbf{R}} \psi_j(\mathbf{r} - \mathbf{R}). \qquad (12.12)$$

We can obtain the energy of the Bloch band formed this way by taking the expectation value of the Hamiltonian. This yields

$$\epsilon_j(\mathbf{k}) = \epsilon_j^0 + \frac{\sum_{\mathbf{R}} e^{i\mathbf{k}\cdot\mathbf{R}} \langle 0|\Delta V|\mathbf{R}\rangle}{\sum_{\mathbf{R}} e^{i\mathbf{k}\cdot\mathbf{R}} \langle 0|\mathbf{R}\rangle}, \qquad (12.13)$$

where ϵ_j^0 is the energy of the atomic orbital, $|\mathbf{R}\rangle$ denotes $\psi_j(\mathbf{r} - \mathbf{R})$, and ΔV is the difference between the periodic potential and the atomic potential at the origin. Conforming with the tight-binding assumption, we can neglect the overlap integrals beyond nearest neighbors. Hence,

$$\epsilon_j(\mathbf{k}) = \epsilon_j^0 + \langle 0|\Delta V|0\rangle + \sum_{\mathbf{R}}{}' \gamma(\mathbf{R}) e^{i\mathbf{k}\cdot\mathbf{R}}, \qquad (12.14)$$

where $\gamma(\mathbf{R}) = \langle 0|\Delta V|\mathbf{R}\rangle - \langle 0|\mathbf{R}\rangle\langle 0|\Delta V|0\rangle$, and the summation is restricted to the nearest neighbors only. Reality of the atomic potentials ensures that $\gamma(-\mathbf{R}) = \gamma(\mathbf{R})$. Therefore we can replace $e^{i\mathbf{k}\cdot\mathbf{R}}$ in the sum by $\cos\mathbf{k}\cdot\mathbf{R}$. Note that this is consistent with the general property of the $\mathbf{k}$-inversion-symmetry of the Bloch band.

The above treatment should be modified when the atomic orbital is degenerate (or nearly degenerate) with others within the same atomic potential; linear combinations should be sought in the larger degenerate space. The Bloch condition does not completely specify the combinations, but suggests the form

$$|\psi_j(\mathbf{k})\rangle = \sum_{\mathbf{R}} e^{i\mathbf{k}\cdot\mathbf{R}}(C_{j_1}\psi_{j_1}(\mathbf{r}-\mathbf{R}) + C_{j_2}\psi_{j_2}(\mathbf{r}-\mathbf{R}) + ...), \qquad (12.15)$$

where $\psi_{j_1}, \psi_{j_2}, \cdots$ are the degenerate atomic orbitals. The coefficients $C_{j_1}, C_{j_2}, \cdots$ can be determined by the condition of extremizing the energy, and obtained by solving finite matrix eigenvalue problems.

12.3 Semiclassical Dynamics

12.3.1 Equations of Motion

In this section, we will examine the dynamics of Bloch electrons driven by external electric $\mathbf{E}$ and magnetic $\mathbf{B}$ fields. We will assume that the fields are sufficiently weak so that the interband transitions, the so-called *Zener tunneling*, are negligible. We will consider a wave packet constructed from the state vectors in a given band, and study how its center of mass $(\mathbf{r}_c, \mathbf{k}_c)$, which we shall define below, moves in phase space. We will require that the spread in $\mathbf{k}$ of the wave packet be small compared to the Brillouin zone. Because of the uncertainty principle, the spread in $\mathbf{r}$ must then be much larger than the lattice constant. When the spatial variations of the external fields over the spatial spread of the wave packet are neglected, it is possible for the $\mathbf{r}$ motion of the wave packet to also depend locally on the fields. Under these conditions, we have the following pair of semiclassical equations of motion.

$$\hbar\dot{\mathbf{k}}_c = -e\mathbf{E}(\mathbf{r}_c) - e\dot{\mathbf{r}}_c \times \mathbf{B}(\mathbf{r}_c), \qquad (12.16)$$

and

$$\dot{\mathbf{r}}_c = \frac{\partial \epsilon_n(\mathbf{k}_c)}{\hbar \partial \mathbf{k}_c}. \qquad (12.17)$$

The first equation indicates that the rate of change of the crystal momentum is equal to the Lorentz force of the external fields. The second equation gives the group velocity in terms of the gradient of the band energy in $\mathbf{k}$ space. These equations constitute the foundation of dynamic and transport theory of Bloch electrons in weak external fields [20]. They have led to a number of very useful results in solid state physics. For example, one can use these equations to show that the filled bands do not contribute to the conductivity, that electrons near the bottom of a band behave like free electrons with an effective mass, and that an almost filled band may be adequately described in terms of virtual positively charged particles called holes.

We will show that, in general, to first order in the fields, the above velocity formula is incomplete. The correct formula has the following more complicated expression [237]

$$\dot{\mathbf{r}}_c = \frac{\partial E_n(\mathbf{k}_c, \mathbf{r}_c)}{\hbar \partial \mathbf{k}_c} - \dot{\mathbf{k}}_c \times \boldsymbol{\Omega}_n(\mathbf{k}_c). \qquad (12.18)$$

Apart from a modification of the band energy, an extra term previously known as an anomalous velocity [3, 120, 134] contributes to the $\dot{\mathbf{r}}_c$. This term turns out to be proportional to the Berry curvature of the Bloch states,

$$\boldsymbol{\Omega}_n(\mathbf{k}) = i \left\langle \frac{\partial u_n}{\partial \mathbf{k}} \middle| \times \middle| \frac{\partial u_n}{\partial \mathbf{k}} \right\rangle. \qquad (12.19)$$

Here $|u_n(\mathbf{k})\rangle = e^{-i\mathbf{k}\cdot\mathbf{r}}|\psi_n(\mathbf{k})\rangle$ is the periodic part of the Bloch function, and the cross-product and the inner product which is denoted by the Dirac brakets refer to the $\mathbf{k}$ vectors and wave functions, respectively.[3] Furthermore, the E_n in the general velocity formula differs from the band energy $\epsilon_n(\mathbf{k})$ by the magnetization energy $\frac{e}{2m}\mathbf{B}(\mathbf{r}_c) \cdot \mathbf{L}_n(\mathbf{k}_c)$. Here $\mathbf{L}_n$ is the angular momentum of the wave packet about its center which is expressed as [237]

$$\mathbf{L}_n(\mathbf{k}) = i\frac{m}{\hbar}\left\langle \frac{\partial u_n}{\partial \mathbf{k}} \middle| \times (\tilde{H} - \epsilon_n) \middle| \frac{\partial u_n}{\partial \mathbf{k}} \right\rangle, \qquad (12.20)$$

and $\tilde{H}$ is the unitarily transformed Hamiltonian $e^{-i\mathbf{k}\cdot\mathbf{r}}He^{i\mathbf{k}\cdot\mathbf{r}}$.

In summary, in order to execute the semiclassical dynamics correctly, one needs to know the dispersion of the Berry curvature and the angular momentum as well as the energy.

---

[3] Throughout this chapter, we adopt the convention that the inner products of u functions or their derivatives involve an integration over a unit cell. The normalization is $\langle u_n(\mathbf{k})|u_n(\mathbf{k})\rangle = V_c/(2\pi)^3$, which is consistent with the following normalization (over infinite real space) of the Bloch functions $\langle \psi_{n'}(\mathbf{k}')|\psi_n(\mathbf{k})\rangle = \delta_{nn'}\delta(\mathbf{k}-\mathbf{k}')$.

12.3.2 Symmetry Analysis

The semiclassical equations should be invariant under time-reversal transformation, spatial inversion, and the rotations that leave the unperturbed system invariant. These types of conditions that are imposed due to the presence of symmetries severely restrict the possible form of the Berry curvature $\boldsymbol{\Omega}_n$ and the angular momentum $\mathbf{L}_n$ as functions of $\mathbf{k}$.

Under time-reversal transformation, $\mathbf{k}_c$, $\dot{\mathbf{r}}_c$, and $\mathbf{B}$ change sign, while $\mathbf{r}_c$, $\dot{\mathbf{k}}_c$, and $\mathbf{E}$ are fixed. Under spatial inversion, $\mathbf{E}$, $\mathbf{r}_c$, $\mathbf{k}_c$ and their time-derivatives change sign, while $\mathbf{B}$ is fixed. Under pure rotations, all of these quantities behave as vectors. The Lorentz formula is invariant under all these operations. In contrast, the velocity formula is not generally invariant unless the unperturbed system has these symmetries. The presence of these symmetries has the following consequences. Firstly, if the unperturbed system has time-reversal symmetry, the symmetry condition on the velocity formula requires that $\epsilon(-\mathbf{k}) = \epsilon(\mathbf{k})$, $\boldsymbol{\Omega}_n(-\mathbf{k}) = -\boldsymbol{\Omega}_n(\mathbf{k})$, and $\mathbf{L}_n(-\mathbf{k}) = -\mathbf{L}_n(\mathbf{k})$. Secondly, if the unperturbed system is invariant under spatial inversion, the symmetry condition on the velocity formula gives rise to $\epsilon(-\mathbf{k}) = \epsilon(\mathbf{k})$, $\boldsymbol{\Omega}_n(-\mathbf{k}) = \boldsymbol{\Omega}_n(\mathbf{k})$, and $\mathbf{L}_n(-\mathbf{k}) = \mathbf{L}_n(\mathbf{k})$. Therefore, with either time-reversal or spatial-inversion symmetry, the band energy is an even function of $\mathbf{k}$, whereas the Berry curvature and the angular momentum behave differently in the two situations: odd in $\mathbf{k}$ with time-reversal symmetry and even in $\mathbf{k}$ with spatial-inversion symmetry. Finally, if the unperturbed system is invariant under a group of proper rotations, then the symmetry condition requires that $\epsilon(\mathbf{k})$ behaves like a scalar, while $\boldsymbol{\Omega}_n(\mathbf{k})$, and $\mathbf{L}_n(\mathbf{k})$ behave like vectors under such rotations.

The above analysis shows that for crystals with simultaneous time-reversal and spatial inversion symmetry both the Berry curvature and angular momentum vanish identically throughout the Brillouin zone. In this case the velocity formula reduces to the simple expression (12.17) which is usually quoted in standard textbooks. In most applications of solid state physics, the unperturbed system has time-reversal symmetry, and all elemental and many compound crystals also have spatial inversion symmetry. Therefore, the usual velocity formula is valid for a large number of physical situations. However, there are many important physical systems where these symmetries are not simultaneously present, and their proper description requires the use of the full velocity formula. For example, important crystals such as GaAs do not have spatial inversion symmetry. Moreover in the presence of ferro- or anti-ferro-magnetic ordering the crystal breaks time-reversal symmetry as well. Finally, when the external magnetic field is strong compared to some energy scales of interest, it is desirable to include the field or the major part of it into the unperturbed system [60, 61], breaking the time-reversal symmetry.

12.3.3 Derivation of the Semiclassical Formulas

We will now present a brief derivation of the semiclassical equations that uses the methods of Chang and Niu [61]. A derivation based on a more general setting is given by Sundaram and Niu [237]. A wave packet constructed from the Bloch states $|\psi_n(\mathbf{k})\rangle$ of a band may be written as

$$|W_0\rangle = \int_{BZ} d^3k\, w(\mathbf{k},t)|\psi_n(\mathbf{k})\rangle, \qquad (12.21)$$

where $|w(\mathbf{k},t)|^2$ should be sharply centered at $\mathbf{k}_c$, $\int_{BZ} d^3k\, \mathbf{k}|w(\mathbf{k},t)|^2 = \mathbf{k}_c$, and the phase of w must be chosen to satisfy $\langle W_0|\mathbf{r}|W_0\rangle = \mathbf{r}_c$. Using the relation

$$\langle\psi_n(\mathbf{k}')|\mathbf{r}|\psi_n(\mathbf{k})\rangle = -i\frac{\partial}{\partial \mathbf{k}}\delta(\mathbf{k}-\mathbf{k}') + \delta(\mathbf{k}-\mathbf{k}')\langle u_n(\mathbf{k})|i\frac{\partial}{\partial \mathbf{k}}|u_n(\mathbf{k})\rangle,$$

we may write the expectation value of the position vector $\mathbf{r}$ in the wave packet as

$$-\frac{\partial}{\partial \mathbf{k}_c}\arg w(\mathbf{k}_c,t) + A_n(\mathbf{k}_c) = \mathbf{r}_c, \qquad (12.22)$$

where $\arg w$ is the phase of w, and $A_n(\mathbf{k})$ is the Berry connection of the Bloch states

$$A_n(\mathbf{k}) := i\langle u_n(\mathbf{k})|\frac{\partial}{\partial \mathbf{k}}|u_n(\mathbf{k})\rangle. \qquad (12.23)$$

We note that if the basis functions of the wave packet were plane waves, then in the above formula for the position $\mathbf{r}_c$ the Berry connection term would be zero.

We can generate the dynamics of the wave packet using the following Lagrangian

$$L = \hbar\mathbf{k}_c \cdot \dot{\mathbf{r}}_c - e\dot{\mathbf{r}}_c \cdot \mathbf{A}(\mathbf{r}_c,t) + \hbar\dot{\mathbf{k}}_c \cdot A_n(\mathbf{k}_c) + e\phi(\mathbf{r}_c) - E_n(\mathbf{k}_c,\mathbf{r}_c), \qquad (12.24)$$

which we regard as a function of $(\mathbf{r}_c, \mathbf{k}_c)$ and their time derivatives. This Lagrangian is invariant up to a total time derivative under both the usual gauge transformation of the electromagnetic potentials and the transformation of the wave function by a $\mathbf{k}$-dependent phase. This is because the curl of both the vector potential and the Berry connection are well defined. Next, we show that this Lagrangian can actually be derived from the effective Lagrangian,

$$\langle W|i\hbar\frac{\partial}{\partial t}|W\rangle - \langle W|H|W\rangle, \qquad (12.25)$$

for the Schrödinger equation. We choose $|W\rangle = e^{-i(e/\hbar)\mathbf{A}(\mathbf{r}_c,t)\cdot\mathbf{r}}|W_0\rangle$, where $\mathbf{A}$ is the vector potential of the external fields. The exponential factor is used to locally gauge away the vector potential at the center of the wave packet, so that $\langle W|H|W\rangle = \langle W_0|H'|W_0\rangle$, where

$$H' = \frac{1}{2m}\{-i\hbar\nabla + e[\mathbf{A}(\mathbf{r},t) - \mathbf{A}(\mathbf{r}_c,t)]\}^2 + V(\mathbf{r}) - e\phi(\mathbf{r}), \quad (12.26)$$

and ϕ is the scalar potential of the external fields. Then, within the length scale of the wave packet, we may expand

$$H' = H_0 + \frac{e}{4m}\{\mathbf{B}(\mathbf{r}_c) \times (\mathbf{r} - \mathbf{r}_c) \cdot \mathbf{P} + H.c.\} - e\phi(\mathbf{r}). \quad (12.27)$$

The expectation value of H_0 in $|W_0\rangle$ is just $\epsilon_n(\mathbf{k}_c)$. The average of the last term contributes the term $e\phi(\mathbf{r}_c)$ to the Lagrangian, while the expectation value of the second term of H' is $\frac{e}{2m}\mathbf{B}(\mathbf{r}_c) \cdot \mathbf{L}_n(\mathbf{k}_c)$. This makes up the difference between E_n and ϵ_n in the Lagrangian. It vanishes for a wave packet of plane waves. Straightforward calculations also show that

$$\langle W|i\hbar\frac{\partial}{\partial t}|W\rangle = e\dot{\mathbf{A}} \cdot \mathbf{r}_c - \hbar\frac{\partial}{\partial t}\arg w(\mathbf{k}_c, t). \quad (12.28)$$

Using the expression (12.22) and neglecting the unimportant total time-derivative terms, we find that this gives the first three terms of the Lagrangian (12.24).

12.3.4 Time-Dependent Bands

The addition of the geometric phase curvature term makes the semiclassical equations of motion more symmetrical between the $\mathbf{k}$ and $\mathbf{r}$ variables. This term has a similar form as the magnetic Lorentz force. Therefore, one may identify it with a magnetic field in $\mathbf{k}$ space. This raises the question whether there is an analogous electric field in $\mathbf{k}$ space. The answer turns out to be affirmative [141].

In order to see this, we introduce a slow time dependence in the periodic potential, but suppose that the spatial periodicity is intact. We can still consider the wave-packet dynamics in a Bloch band, but the basis states are now explicitly time dependent. In the evaluation of the geometric phase connection for the wave packet, the following extra term arises because of the explicit time dependence of the basis states,

$$\chi_n(\mathbf{k}, t) = \langle u_n(\mathbf{k}, t)|i\frac{\partial}{\partial t}|u_n(\mathbf{k}, t)\rangle. \quad (12.29)$$

This term, which we should add to the Lagrangian, is called the *geometric scalar potential*. Note also that in this case the geometric vector potential,

$$\mathbf{A}_n(\mathbf{k}, t) = \langle u_n(\mathbf{k}, t)|i\frac{\partial}{\partial \mathbf{k}}|u_n(\mathbf{k}, t)\rangle, \quad (12.30)$$

becomes explicitly time dependent.

The addition of the geometric scalar potential and the explicit time dependence of the geometric vector potential modify the Euler–Lagrange equations for the semiclassical dynamics of the wave packet. The Lorentz force for the time derivative of **k** has the same form as before, but the expression for the velocity acquires an extra term, namely

$$\dot{\mathbf{r}}_c = \frac{\partial E_n(\mathbf{k}_c, \mathbf{r}_c)}{\hbar \partial \mathbf{k}_c} - \dot{\mathbf{k}}_c \times \boldsymbol{\Omega}_n(\mathbf{k}_c) - \boldsymbol{\Xi}_n(\mathbf{k}_c), \tag{12.31}$$

where $\boldsymbol{\Xi}_n = -\frac{\partial \mathbf{A}_n}{\partial t} - \frac{\partial \chi_n}{\partial \mathbf{k}}$. This extra velocity term is induced due to the time dependence of the band structure. It is responsible for the phenomenon of adiabatic particle transport and closely related to the geometric phase formulation of polarization in dielectric crystals.

12.4 Applications of Semiclassical Dynamics

12.4.1 Uniform DC Electric Field

In the absence of the magnetic field, the semiclassical motion is determined by

$$\hbar \dot{\mathbf{k}}_c = -e\mathbf{E}, \quad \text{and} \quad \dot{\mathbf{r}}_c = \frac{\partial \epsilon_n}{\hbar \partial \mathbf{k}} + e\mathbf{E} \times \boldsymbol{\Omega}_n(\mathbf{k}_c). \tag{12.32}$$

In a uniform and DC electric field, $\mathbf{k}_c$ moves in the direction of the field with a constant rate. Because of the periodic topology of the band structure, the Bloch electrons actually execute a cyclic motion in the Brillouin zone. These are known as Bloch oscillations. For example, consider the case where **E** is parallel to a reciprocal lattice vector **G**. Then the motion will be closed after $\mathbf{k}_c$ has moved a length of **G**. The velocity averaged over such a cycle is given by $e\mathbf{E}\times <\boldsymbol{\Omega}_n(\mathbf{k}_c)>$, where $<\ >$ indicates a time average. A Hall current would result if the average Berry curvature is non-zero, which generally occurs in the absence of time reversal symmetry.

The semiclassical quantization of the Bloch oscillations gives rise to the so-called *Wannier–Stark ladders*. In the following, we will work in the gauge where $\mathbf{A} = 0$ and $\phi = -\mathbf{E} \cdot \mathbf{r}$. The Lagrangian is then equal to $L = \hbar \mathbf{k}_c \cdot \dot{\mathbf{r}}_c + \hbar \dot{\mathbf{k}}_c \cdot \mathbf{A}_n(\mathbf{k}_c) + e\phi(\mathbf{r}_c) - \epsilon_n(\mathbf{k}_c)$. The generalized momenta for $\mathbf{r}_c$ and $\mathbf{k}_c$ are $\mathbf{P}_{\mathbf{r}_c} = \hbar \mathbf{k}_c$ and $\mathbf{P}_{\mathbf{k}_c} = \hbar \mathbf{A}_n$, respectively. The Hamiltonian for the semiclassical dynamics is $H_c = \epsilon_n(\mathbf{k}_c) - e\phi(\mathbf{r}_c)$. According to the semiclassical quantization rule, the classical action over a periodic orbit should be quantized as

$$\oint \mathbf{P}_{\mathbf{r}_c} \cdot d\mathbf{r}_c + \mathbf{P}_{\mathbf{k}_c} \cdot d\mathbf{k}_c = j + \nu/4\hbar, \tag{12.33}$$

where j is an integer and ν stands for the number of caustics (turning points) traversed by the orbit. The latter is called the *Maslov index*. Therefore, we have

$$\oint [\mathbf{A}_n(\mathbf{k}_c) - \mathbf{r}_c] \cdot d\mathbf{k}_c = 2\pi j. \tag{12.34}$$

The Maslov index for the Bloch oscillation is zero. Equations (12.32) and (12.34) indicate that the average position is quantized in the field direction in such a way that $\mathbf{E} \cdot <\mathbf{r}_c> = (\Gamma_n - 2\pi j)eE/G$. Here G is the magnitude of $\mathbf{G}$ and $\Gamma_n = \oint \mathbf{A}_n(\mathbf{k}_c) \cdot d\mathbf{k}_c$. Since the semiclassical motion is along a constant energy contour of H_c, the quantized energies are obtained by averaging H_c in time and using the above-mentioned quantization condition. This yields

$$E_j = <\epsilon_n(\mathbf{k}_c)> + (\Gamma_n - 2\pi j)eE/G. \tag{12.35}$$

Therefore, the band energy is quantized along the field direction with the level spacing $eE\ell$ where $\ell = 2\pi/G$ is effectively the lattice constant in the field direction. These equally spaced levels are known as a Wannier–Stark ladder. The Berry phase induced by going across a Brillouin zone was first discovered by Zak [281], although its role in the Wannier–Stark ladders was known earlier.

12.4.2 Uniform and Constant Magnetic Field

In the absence of the electric field, we can combine the semiclassical equations to yield

$$\hbar \dot{\mathbf{k}}_c = -\frac{e}{\hbar(1 - \frac{e}{\hbar}\mathbf{B} \cdot \mathbf{\Omega}_n)} \frac{\partial E_n(\mathbf{k}_c, \mathbf{r}_c)}{\partial \mathbf{k}_c} \times \mathbf{B}(\mathbf{r}_c). \tag{12.36}$$

Therefore, the Bloch wave vector moves on a plane perpendicular to the field and along a constant contour of the energy E_n known as a cyclotron orbit. According to the Lorentz force formula, there is a close relationship between the orbits in the $\mathbf{r}$ and $\mathbf{k}$ spaces as $\mathbf{k}_c(t) - \mathbf{k}_c(0) = \frac{e}{\hbar}[\mathbf{r}_c(t) - \mathbf{r}_c(0)] \times \mathbf{B}$.

The quantization of these orbits gives rise to Landau levels which are responsible for the de Haas–van Alphen effect and related phenomena [20]. In view of the expression for the Lagrangian, i.e.,

$$L = \hbar \mathbf{k}_c \cdot \dot{\mathbf{r}}_c - e\dot{\mathbf{r}}_c \cdot \mathbf{A}(\mathbf{r}_c) + \hbar \dot{\mathbf{k}}_c \cdot \mathbf{A}_n(\mathbf{k}_c) - E_n(\mathbf{k}_c, \mathbf{r}_c), \tag{12.37}$$

the canonical momenta conjugate to $\mathbf{r}_c$ and $\mathbf{k}_c$ are $\hbar \mathbf{k}_c - e\mathbf{A}$ and $\hbar \mathbf{A}_n$, respectively. The Hamiltonian for the semiclassical dynamics is simply $H_c = E_n(\mathbf{k}_c)$. According to the EBK formula for quantization (12.33), we find that

$$\frac{1}{2} \oint (\mathbf{r}_c \times \mathbf{B}) \cdot d\mathbf{r}_c = \left(j + \frac{1}{2} - \frac{\Gamma_n(C_j)}{2\pi}\right) \frac{h}{e}. \tag{12.38}$$

Therefore, apart from the corrections due to the Maslov index and the Berry phase, the magnetic flux through the cyclotron orbit is quantized in units of the flux quantum $\frac{h}{e}$. Using the relationship between the orbit in $\mathbf{r}$ and $\mathbf{k}$ spaces, we can write the quantization rule for the $\mathbf{k}$ space orbit area as

$$A_j = 2\pi \left(j + \frac{1}{2} - \frac{\Gamma_n(C_j)}{2\pi} \right) \frac{eB}{\hbar}. \tag{12.39}$$

Onsager [205] was the first to derive this formula without the Berry phase correction, Roth [221] found this correction without calling it a Berry phase (for the obvious reason that the latter was discovered much later), and Berry [32] realized that even the Maslov index is a kind of geometric phase.

12.4.3 Perpendicular Electric and Magnetic Fields

When both an electric and a magnetic field are present, the semiclassical equations become more complicated but are still tractable when the fields are constant and uniform. In this context an important concept is the drift velocity, $\mathbf{w} = \frac{E}{B}(\hat{\mathbf{E}} \times \hat{\mathbf{B}})$, with which a free electron gas would drift in perpendicular electric and magnetic fields. The semiclassical equations of motion may be combined to yield an equation identical to (12.36) except with E_n replaced by $\bar{E}_n(\mathbf{k}_c) = E_n(\mathbf{k}_c) - \hbar \mathbf{k}_c \cdot \mathbf{w}$. This shows that the orbits in $\mathbf{k}$ space are still perpendicular to the magnetic field and along a constant contour of the energy $\bar{E}_n$.

To determine the motion in real space, we consider the Lorentz equation (12.16). If the orbit is closed in $\mathbf{k}$ space, then taking the time average of both sides of this equation yields

$$<\dot{\mathbf{r}}_c> \times \mathbf{B} = -\mathbf{E}. \tag{12.40}$$

Therefore, on average $\mathbf{r}_c$ drifts with a rate given by $\mathbf{w}$, and there is no net motion along the electric field direction (though there might still be a net motion in the $\mathbf{B}$ direction). On the other hand, if the $\mathbf{k}$ orbit is open, then the average of $\dot{\mathbf{k}}_c$ cannot be taken as zero. The average velocity perpendicular to the $\mathbf{B}$ field will then be given by

$$<\dot{\mathbf{r}}_c> \times \mathbf{B} = -\mathbf{E} - \frac{\hbar}{e} <\dot{\mathbf{k}}_c>, \tag{12.41}$$

and there will be a component along the electric field direction in this case.

12.4.4 Transport

The semiclassical dynamics coupled with the Boltzmann equation provides a comprehensive theory of transport phenomena in solid state physics [20]. In this theory, the Bloch electrons are treated as classical particles described by a *distribution function* $g(\mathbf{k}, \mathbf{r}, t)$ in the phase space $(\mathbf{k}, \mathbf{r})$. The normalization of the distribution function is fixed by the requirement that $g\, d^3r d^3k/(2\pi)^3$ is the number of electrons in the phase space volume $d^3r d^3k$ about the point $(\mathbf{r}, \mathbf{k})$.

There are two sources that make g change in time: (1) drifting of $\mathbf{k}$ and $\mathbf{r}$ according to the semiclassical dynamics, and (2) sudden jumps of $\mathbf{k}$ due to

12.4 Applications of Semiclassical Dynamics

collision of the Bloch electrons with impurities, phonons, and other electrons. In the relaxation time approximation for the collision [20], the steady state distribution satisfies the following Boltzmann equation:

$$\dot{\mathbf{r}} \cdot \frac{\partial g}{\partial \mathbf{r}} + \dot{\mathbf{k}} \cdot \frac{\partial g}{\partial \mathbf{k}} = (f - g)/\tau(\mathbf{k}), \qquad (12.42)$$

where $\tau(\mathbf{k})$ is the transport relaxation time of the Bloch state and $f = 1/(e^{(\epsilon_n(\mathbf{k})-\mu)/k_B T} + 1)$ is the equilibrium distribution function. The relaxation time has to be provided by separate calculations of the problem of Bloch electron scattering. The solution of the Boltzmann equation, however, depends on $\tau(\mathbf{k})$ only for those states that are near the Fermi surface within an energy width of about $k_B T$.

In the absence of a magnetic field, we do not need to consider the modification of the band energy by the magnetic moment of the wave packet. We may solve (12.42) for the distribution function to first order in the electric field. This yields

$$g = f + \tau \frac{e}{\hbar} \mathbf{E} \cdot \frac{\partial \epsilon_n}{\partial \mathbf{k}} \frac{\partial f}{\partial \epsilon}, \qquad (12.43)$$

where we have assumed that the chemical potential and temperature are uniform in space. In view of the last factor of the second term on the right-hand side of this equation, the electric field alters the distribution only near the Fermi surface.

The electric current density is given by

$$\mathbf{J} = -e \int \frac{d^3 k}{(2\pi)^3} \dot{\mathbf{r}} g. \qquad (12.44)$$

To first order in the field, it may be expressed as the sum of two terms:

$$\mathbf{J} = -\mathbf{E} \times \frac{e^2}{\hbar} \int \frac{d^3 k}{(2\pi)^3} f(\epsilon_n) \mathbf{\Omega}_n(\mathbf{k}) - e^2 \tau \int \frac{d^3 k}{(2\pi)^3} \frac{\partial \epsilon_n}{\partial \mathbf{k}} \mathbf{E} \cdot \frac{\partial \epsilon_n}{\partial \mathbf{k}} \frac{\partial f}{\partial \epsilon}. \qquad (12.45)$$

The second term gives a symmetric conductivity tensor and depends only on the properties of the system near the Fermi energy. The first term comes from the anomalous velocity due to the Berry curvature and gives a Hall conductivity tensor which is antisymmetric.[4] For systems with time-reversal symmetry, the Berry curvature is antisymmetric in $\mathbf{k}$ and is averaged to zero with the equilibrium distribution which is symmetric in $\mathbf{k}$. In ferromagnetic and other magnetic materials where time-reversal symmetry is broken, one may have a non-zero Hall conductivity, called the anomalous Hall conductivity. This can be much larger than the ordinary Hall conductivity due to the Lorentz force of a magnetic field. Here spin–orbit interactions must be taken into account for a non-zero anomalous Hall conductivity, so that the time-reversal asymmetry in the spin degree of freedom leads to the modification of

---

[4] Such a term for the Hall conductivity can also be derived from a systematic linear response theory for transport [120, 247].

the spatial wave functions. Explicit calculations of the anomalous Hall conductivity have been done for ferromagnetic metals and semiconductors based on the above formula.

12.5 Wannier Functions

It is known in solid state physics that a filled Bloch band is inert to external fields. This can be revealed directly by the fact that the Hilbert space of a Bloch band can be spanned by a set of localized functions known as the *Wannier functions* [263]. Many important concepts such as polarization and adiabatic particle transport can be easily understood.

12.5.1 General Properties

The Wannier functions are defined as the Fourier transform of the Bloch functions:

$$W_n(\mathbf{r} - \mathbf{R}) := \sqrt{\frac{V_c}{(2\pi)^3}} \int_{Bz} d^3k \, e^{-i\mathbf{k}\cdot\mathbf{R}} \psi_{n\mathbf{k}}(\mathbf{r}), \quad (12.46)$$

where $\mathbf{R}$ is a lattice vector and the integral is over a Brillouin zone. The normalization of the Bloch function is such that $\langle \psi_{n'}(\mathbf{k}') | \psi_n(\mathbf{k}) \rangle = \delta_{nn'}\delta(\mathbf{k}-\mathbf{k}')$. There is still a freedom of multiplying the Bloch function by a $\mathbf{k}$-dependent phase factor. But in order for the Fourier transform to be well defined, the phase should be chosen in such a manner that the Bloch function is periodic in k. The fact that the Wannier function depends on position through $\mathbf{r}-\mathbf{R}$ follows from the translational property of the Bloch functions in space. The above expression can be inverted to give

$$\psi_{n\mathbf{k}}(\mathbf{r}) = \sqrt{\frac{V_c}{(2\pi)^3}} \sum_{\mathbf{R}} e^{i\mathbf{k}\cdot\mathbf{R}} W_n(\mathbf{r}-\mathbf{R}). \quad (12.47)$$

This is an indication of the completeness of the Wannier functions W_n for each Bloch band. It shows that a Bloch function expressible in terms of Wannier functions is naturally periodic in $\mathbf{k}$. Using the orthogonality and normalization of the Bloch functions, one can also show that the Wannier functions form a normalized and orthogonal set,

$$\int d^3r \, W^*_{n'}(\mathbf{r}-\mathbf{R}') W_n(\mathbf{r}-\mathbf{R}) = \delta_{nn'}\delta_{\mathbf{R}\mathbf{R}'}. \quad (12.48)$$

Next, we note that the projection operator for a given band has the following equivalent expressions:

$$P_n(\mathbf{r},\mathbf{r}') = \int_{Bz} d^3k \, \psi_{n\mathbf{k}}(\mathbf{r})\psi^*_{n\mathbf{k}}(\mathbf{r}') = \sum_{\mathbf{R}} W_n(\mathbf{r}-\mathbf{R})W^*_n(\mathbf{r}'-\mathbf{R}), \quad (12.49)$$

This, in particular, implies that the particle density for a filled band is the same as the sum of the probability densities of the Wannier functions for that band. Therefore, as far as the expectation values of physical observables are concerned, a filled band of Bloch states can be regarded as a filled set of Wannier states. In the next section, we will show that the flow of electrons in a filled band induced by an adiabatic change of the Hamiltonian can be entirely described by the shift of the position of a Wannier function, defined by

$$\langle W_n(\mathbf{R})|\mathbf{r}|W_n(\mathbf{R})\rangle = \mathbf{R} + \int_{Bz} d^3k\, A_n(\mathbf{k}). \quad (12.50)$$

This expression raises the question: How well is the position of a Wannier function defined? Under a phase change of the Bloch function $|\psi_n(\mathbf{k})\rangle \to e^{i\phi(\mathbf{k})}|\psi_n(\mathbf{k})\rangle$, the Berry connection changes by $-\frac{\partial \phi}{\partial \mathbf{k}}\frac{V_c}{(2\pi)^3}$. We required that the Bloch function be periodic in $\mathbf{k}$, so ϕ must be the sum of a periodic function of $\mathbf{k}$ and $\mathbf{k}\cdot\mathbf{R}'$ for some lattice vector $\mathbf{R}'$. This then leads to a change of the position of the Wannier function by a lattice vector. Therefore, the collection of Wannier centers is well defined, which makes it possible to attach physical significance to it.

If the system has (spatial) inversion symmetry, the Wannier centers must be on the lattice sites or shifted away from the lattice sites by half of a lattice vector. To see this, note that if $\mathbf{R}_0$ is the position of one Wannier function then $-\mathbf{R}_0$ must also be the position of another Wannier function. On the other hand, according to the discussions in the last paragraph, the positions of different Wannier functions can differ only by a lattice vector $\mathbf{R}$, implying $\mathbf{R}_0 - (-\mathbf{R}_0) = 2\mathbf{R}_0 = \mathbf{R}$ in particular. Therefore, the Wannier functions are either centered on the atoms (lattice site), on the bonds between the atoms, or at the interstitials. Precisely which situation occurs depends on the chemical nature of the band. For systems without inversion symmetry, the Wannier centers can assume other positions.

12.5.2 Localization Properties

Consider an isolated band that does not touch or cross other bands at each $\mathbf{k}$. The shape, size and localization properties of the Wannier functions are extremely sensitive to the phase of the Bloch function. The sheer possibility of constructing well-localized Wannier functions depends on whether the Chern number (see Sect. 6.5, where it is called the first Chern number), defined as the integral of the Berry curvature $\Omega_n(\mathbf{k})$ over the Brillouin zone, is zero [246].

This observation is based on the following argument. In order to have a well-localized Wannier function, we need a Bloch function which is a smooth and periodic function of $\mathbf{k}$. In this case, the Berry connection $A_n(\mathbf{k})$ is also smooth and periodic. This in turn implies that the Chern number, which is the integral of the curl of $A_n(\mathbf{k})$ over the Brillouin zone, must vanish.

In other words, if the Chern number is zero, then it is possible to choose the phase of the Bloch function in such a way that the Wannier functions are well localized. This is for example the case for systems with time-reversal symmetry. It turns out that the smallest Wannier function, with $<\mathbf{r}^2> - <\mathbf{r}>^2$ minimized among different phase choices of the Bloch functions, corresponds to a Berry connection satisfying $\frac{\partial}{\partial \mathbf{k}} A_n = 0$ (see [40]). This condition together with the periodicity requirement determine the Berry connection as

$$A_n(\mathbf{k}) = \mathbf{R}_w \frac{(2\pi)^3}{V_c} + \sum_{\mathbf{R} \neq 0} \frac{i\mathbf{R} \times \boldsymbol{\Omega}_n(\mathbf{R})}{R^2} e^{i\mathbf{k}\cdot\mathbf{R}}, \qquad (12.51)$$

where $\boldsymbol{\Omega}_n(\mathbf{R})$ is the Fourier transform of the Berry curvature and $\mathbf{R}_w$ is a vector independent of $\mathbf{k}$. As we showed in the last subsection, $\mathbf{R}_w$ is related to the position of the Wannier function, and is fixed up to a lattice vector.

Once the Berry connection is given, we can fix the phase of the Bloch wave function as follows. Let $e^{i\mathbf{k}\cdot\mathbf{r}}|u_n^0(\mathbf{k})\rangle$ be a given Bloch function which is k-periodic. A general wave function satisfying the same condition can be written as $|u_n(\mathbf{k})\rangle = e^{-i\theta(\mathbf{k})}|u_n^0(\mathbf{k})\rangle$, where the phase $\theta(\mathbf{k})$ should be the sum of a periodic function of $\mathbf{k}$ and $\mathbf{k} \cdot \mathbf{R}$ for some lattice vector $\mathbf{R}$. The periodic function can be determined (apart from a trivial constant) by comparing the Fourier components of both sides of the following relation that defines the Berry connection,

$$\frac{\partial \theta}{\partial \mathbf{k}} = A_n - \langle u_n^0|i\frac{\partial}{\partial \mathbf{k}}|u_n^0\rangle. \qquad (12.52)$$

Since the Berry curvature is a physical quantity, we expect it to be an analytic function of $\mathbf{k}$ with the same domain of analyticity as the band energy $\epsilon_n(\mathbf{k})$. The Bloch function determined in the above manner should also have the same analytic properties. The Wannier function, being a Fourier transform of a periodic and analytic wave function, must be exponentially localized.

When the Berry curvature is zero everywhere, as for a system with time-reversal and spatial inversion symmetries or for a general one-dimensional system, the most localized Wannier function requires that the Berry connection is given by a constant vector proportional to the position of a Wannier function [197]. In this case, there is an interesting physical interpretation of such Wannier functions: they are the eigenstates of the projected position vector $P_n \mathbf{r} P_n$ where P_n is the projection on a given band [126]. When the Berry curvature is non-zero such an interpretation does not exist, because the different components of the projected position vector do not commute.

When the system has time-reversal and inversion symmetry, the Wannier functions can be taken as real functions. Kohn has worked out explicit formulas for constructing the most localized Wannier functions in such cases [133]. Explicit numerical calculation of the Wannier functions have been performed more recently for the Kronig–Penney model [209]. From this work one can very much appreciate how a good choice of phase of the Bloch functions can lead to Wannier functions which are highly localized, while a bad choice makes them terribly spread.

12.6 Some Issues on Band Insulators

12.6.1 Quantized Adiabatic Particle Transport

Thouless [245] considered a one-dimensional system in a periodic potential at zero temperature and Fermi energy lying in a gap between the Bloch bands. He found that if the potential varies slowly in time and returns to itself after some time, the particle transport during the time cycle is an integer. Here the particle transport is defined as the quantum mechanical average of the number of particles transported across a section of the system in the given time.

The integer corresponding to the particle transport depends on particular variation of the potential in time. If the potential slides its position without changing its shape, we expect that the electrons simply follow the potential. If the potential shifts one spatial period in the time cycle, the particle transport should be equal to the number of filled Bloch bands discounting the spin degeneracy. This follows from the fact that there is on average one state per unit cell in each filled band.

In general the situation can be intuitively understood using the Wannier functions. We can view the electrons in a band as a set of localized charges centered at the position of the Wannier function. They are spaced apart by the lattice constant, and their locations are uniquely determined by the potential. These locations change continuously as the potential changes in time. When the potential restores its initial form in a time cycle, the set of the positions of the Wannier function must be the same as before, but the individual positions may have shifted by an integer multiple of the lattice constant. As we show below this integer gives the particle transport.

We employ the adiabatic approximation to evaluate the electron current to first order in the rate of change of the potential. We then integrate the current over a time cycle to find the particle transport. It is convenient to work with the basis vectors $|u_n(k,t)\rangle$ which are periodic in position over the lattice constant and are the instantaneous eigenstates (with energies $\epsilon_n(k,t)$) of the Hamiltonian,

$$H(k,t) = \frac{\hbar^2}{2m}\left(-i\frac{\partial}{\partial x} + k\right)^2 + V(x,t). \tag{12.53}$$

The velocity operator is then given by $\frac{\partial H(k,t)}{\hbar \partial k}$. We identify the domain of integration corresponding to the inner product of the wave functions with a unit cell, and in order to conform with the normalization of the Bloch states in the previous sections, we take the norm of $|u_n\rangle$ to be the lattice constant a divided by 2π. Apart from an unimportant overall phase factor and up to first order in the rate of change of the potential, the wave function is given by

$$|u_n\rangle - i\hbar \frac{2\pi}{a} \sum_{n'(\neq n)} \frac{|u_{n'}\rangle\langle u_{n'}|\frac{\partial u_n}{\partial t}\rangle}{\epsilon_n - \epsilon_{n'}}. \qquad (12.54)$$

The average velocity in a state of given k is found to first order as

$$v_n(k) = \frac{\partial \epsilon_n(k)}{\hbar \partial k} - i\hbar \left(\frac{2\pi}{a}\right)^2 \sum_{n'(\neq n)} \frac{\langle u_n|\frac{\partial H}{\hbar \partial k}|u_{n'}\rangle\langle u_{n'}|\frac{\partial u_n}{\partial t}\rangle}{\epsilon_n - \epsilon_{n'}} + \text{c.c.}, \qquad (12.55)$$

where c.c. denotes a term which is the complex conjugate of the term involving summation over n'. Using the fact that $\langle u_n|\frac{\partial H}{\hbar \partial k}|u_{n'}\rangle = (\epsilon_n - \epsilon'_n)\langle \frac{\partial u_n}{\partial k}|u_{n'}\rangle$ and the identity $\frac{2\pi}{a}\sum_{n'}|u_{n'}\rangle\langle u_{n'}| = 1$, we find that

$$v_n(k) = \frac{\partial \epsilon_n(k)}{\hbar \partial k} - i\frac{2\pi}{a}\left\{\left\langle\frac{\partial u_n}{\partial k}\bigg|\frac{\partial u_n}{\partial t}\right\rangle - \left\langle\frac{\partial u_n}{\partial t}\bigg|\frac{\partial u_n}{\partial k}\right\rangle\right\}. \qquad (12.56)$$

Upon integration over the Brilloiun zone, the zeroth order term given by the derivative of the band energy vanishes, and only the first order term survives.

We have thus derived the remarkable result that the average velocity in a band is equal to the k-integral of the Berry curvature,

$$\bar{v}_n = \int_{Bz} dk\, i\left\{\left\langle\frac{\partial u_n}{\partial t}\bigg|\frac{\partial u_n}{\partial k}\right\rangle - \left\langle\frac{\partial u_n}{\partial k}\bigg|\frac{\partial u_n}{\partial t}\right\rangle\right\}. \qquad (12.57)$$

Multiplying this expression with the average density $1/a$ gives the average particle current, i.e., the particle transport over a time cycle, as

$$c_n = \int dt \int_{Bz} dk\, \frac{i}{a}\left\{\left\langle\frac{\partial u_n}{\partial t}\bigg|\frac{\partial u_n}{\partial k}\right\rangle - \left\langle\frac{\partial u_n}{\partial k}\bigg|\frac{\partial u_n}{\partial t}\right\rangle\right\}. \qquad (12.58)$$

This is again the Chern number (called the first Chern number in Sect. 6.5), which takes integer values and characterizes the topological structure of the mapping from the parameter space (k, t) to the Bloch states.

Next, we show that the average velocity over a band is the same as the rate of change of the position of the Wannier function, so that the integer corresponding to the particle transport of a band coincides with the number of unit cells that a Wannier function has traversed during the time period.

We first note that the Berry curvature is invariant under k- and t-dependent phase transformations of the wave function. However, in order to make contact with the Wannier functions, we need to impose the k-periodicity on the Bloch function, so that $|u_n(k+2\pi/a, t)\rangle = e^{-i2\pi x/a}|u_n(k, t)\rangle$. We may write the Berry curvature as

$$i\left\{\frac{\partial}{\partial t}\left\langle u_n\bigg|\frac{\partial u_n}{\partial k}\right\rangle - \frac{\partial}{\partial k}\left\langle u_n\bigg|\frac{\partial u_n}{\partial t}\right\rangle\right\}. \qquad (12.59)$$

Upon integration over the Brilloiun zone, the second term vanishes because of the k-periodicity condition. Therefore, we find for the average velocity in the band

$$\bar{v}_n = \frac{\partial}{\partial t} \int dk\, i \left\langle u_n \left| \frac{\partial u_n}{\partial k} \right. \right\rangle, \qquad (12.60)$$

which is just the rate of change of the position of the Wannier function given in (12.50). We have thus reached the conclusion that the Wannier functions not only give the electron density for filled bands, they also describe the adiabatic flow of electrons in filled bands.

12.6.2 Polarization

In the simplest picture of a neutral dielectric, each negative charge is paired with a positive charge of equal magnitude. Such a pair may or may not have a dipole moment when averaged over time. Different dipoles orient in random directions with respect to one another such that the net polarization of the dielectric is zero. An external electric field may polarize the dielectric either by inducing new dipole moments or by re-orienting the existing dipoles towards the field direction. Spontaneous or forced structural changes of the material can also bring in a net polarization known as ferro electricity or piezoelectricity.

This simple picture breaks down for ionic crystals such as NaCl in which there is no unique or natural way of grouping the positive and negative ions into neutral units. The polarization obtained under different choices of grouping can make a difference of $e\mathbf{R}$ per unit cell where $\mathbf{R}$ is a lattice constant. We stress that $-e$ is the electronic charge, and we have assumed that the ions carry quantized charges. However, one should not be bothered too much by the ambiguity of the term $e\mathbf{R}$ if only the change of the polarization is of concern. This is because for any given grouping of the ions this term remains a constant during the process.

The classical ionic picture is only an ideal limit of real systems. The electron density is a continuous function of space, and there is no sharp division between different ions. If one "isolates" an ion by the surfaces of minimal electron density, then the ionic charge is not quantized, and it can change continuously as the configuration of the system changes. Ideal molecular solids do not exist either; there can always be some charge flowing in and out of the molecules because of the proximity of their locations. The most severe challenge comes from the so-called covalent solids, where the electrons near the Fermi energy reside in between the atoms.

One classic problem is, therefore, whether and how the electric polarization of a crystal dielectric may be defined in terms of its bulk properties so that the powerful machinery of Bloch theory may be utilized. Note that any attempt to define a dipole moment per unit cell from the electronic density distribution is destined to fail. Suppose one defines the dipole moment of a cell by integrating the charge density (including the nuclear charges) times $\mathbf{r}$ over the unit cell. This definition is independent of the origin of $\mathbf{r}$ because of the charge neutrality of the cell, but depends on the choice of the unit cell with respect to its shape and position. One can see this point very clearly by considering a one-dimensional problem. The above-defined dipole moment has the form $\int_{x_0}^{x_0+a} dx\, x \rho(x)$, where a is the lattice constant and $\rho(x)$ is the charge density. The dipole moment for the unit cell indeed depends on the position x_0, because its derivative with respect to x_0 is $\rho(x_0)a$. Also it does not help to average the cell's dipole moment over its position, because the result is always zero. Still taking the one-dimensional problem as an example, we introduce a (uniquely defined) periodic function $c(x)$ by requiring that its average be zero and $c'(x) = \rho(x)$. Then the cell's dipole moment can be written as $ac(x_0)$, which becomes zero when averaged over the cell's position x_0.

If one is concerned with the change of the polarization during a continuous physical process, then it is sufficient to know the time derivative of the polarization $\frac{\partial \mathbf{p}_c}{\partial t}$, [164, 218]. The latter is just the electric current induced by the adiabatic change of the Hamiltonian of the system and can be calculated using the method described in the last section. This observation has led King-Smith and Vanderbilt [124] to formulate the change of the polarization in terms of the Berry phase of the occupied Bloch states. They have further made the connection to Wannier function positions and proposed defining the polarization per unit cell as the dipole moment of the Wannier charge density [256],

$$\mathbf{p}_c = -e \sum_n \int d^3 r\, \mathbf{r} |W_n(\mathbf{r})|^2, \tag{12.61}$$

where the summation is over the occupied bands and the origin of $\mathbf{r}$ can be fixed, say, at the position of an ion. In this definition, one effectively maps a band insulator into a periodic array of localized distributions with truly quantized charges. This resembles an ideal ionic crystal except for the fact that the localized charges can overlap with one another. Similarly to the situation for an ideal ionic crystal, the dipole moment per unit cell defined by the Wannier functions has the ambiguity of a discrete term $e\mathbf{R}$, because the Wannier center position is defined only up to a lattice vector.

The polarization defined in the above paragraph is clearly a bulk quantity, and the Berry phase expression for the position of a Wannier function can be evaluated purely using the occupied Bloch states. This also provides a powerful quantitative method, because the Bloch functions are obtained routinely in modern band structure calculations. Alternatively, the position of the Wannier functions can be obtained directly using iterative procedures starting from localized atomic orbitals [256]. The latter method is particularly suitable for calculating polarizations induced by an electric field which breaks the translational symmetry of the system. Previously, people used to calculate polarizations either by using linear response theory, which involves occupied as well as unoccupied bands, or by using artificial supercells. Both these methods are computationally expensive and may be subject to truncation errors.

Problems

12.1) Derive the symmetry properties of the Berry curvature and orbital angular momentum directly from their definitions in terms of the wave functions for a non-degenerate band. It is useful to show the following properties of wave functions first.
   a) If the crystal has a symmetry under rotation or inversion, then $u_n(\mathbf{k}, \mathbf{r})$ and $u_n(\mathbf{k}', \mathbf{r}')$ can differ by at most a phase. Here the primed and unprimed vectors are related to one another by such a symmetry transformation.
   b) If the crystal has time-reversal symmetry, then $u_n(-\mathbf{k}, \mathbf{r})$ and $u_n^*(\mathbf{k}, \mathbf{r})$ can differ at most by a phase.

12.2) Show the gauge invariance of the semiclassical Lagrangian for the center of mass motion of a Bloch wave packet. Consider both the gauge freedom in the potentials of the external fields and in the choice of phase in the basis wave functions. Note that the total time-derivatives may be ignored.

12.3) Show that in the Bloch state representation the projected position operator in a band satisfies $P_n \mathbf{r} P_n \times P_n \mathbf{r} P_n = \delta(\mathbf{k} - \mathbf{k}') \mathbf{\Omega}_n(\mathbf{k})$. Note that as a result one can introduce a complete set of simultaneous eigenstates for the different components of the projected position operator, if and only if the Berry curvature is identically zero. Show that when this is possible the set of eigenstates are ordinary Wannier functions with smallest possible spread in space.

13. Geometric Phase in Condensed Matter II: The Quantum Hall Effect

13.1 Introduction

In this chapter we review the application of the geometric phase in the context of the quantum Hall effect, a striking quantum phenomenon involving two-dimensional electrons in strong magnetic fields and low temperatures. We will focus our attention on the case of the integer quantum Hall effect in this chapter, and show how the quantization of the Hall conductance observed in experiments may be explained as a topological invariant represented by the (first) Chern number.

We will first give an introduction to the basic phenomenology and present the ideal model of two-dimensional electrons in a magnetic field in order to set the stage for the discussion. We will then consider the problem with a periodic potential and show how the Hall conductance for a filled band may be written as a Chern number and is thus quantized. We will also present concrete results for a simple model, namely the so-called Harper model, explain the intricate band structures using semiclassical ideas, and calculate the Chern numbers. We will finally address the difficult issue of the quantization of Hall conductance in disordered systems and present a topological invariant theory of the quantum Hall effect.

For a more complete review of the quantum Hall effect, we refer the reader to the following books.

- R. E. Prange and S. M. Girvin (eds.): *The Quantum Hall Effect* (Springer-Verlag, New York 1987);
- G. Morandi: *Quantum Hall Effect–Topological Problems in Condensed Matter Physics* (Bibliopolis, New York 1988);
- M. Stone: *Quantum Hall Effect* (World Scientific, New York 1992);
- S. Das Sarma and A. Pinczuk (eds.): *Perspectives in Quantum Hall Effect* (Wiley, New York 1997).

13.2 Basics of the Quantum Hall Effect

13.2.1 The Hall Effect

Figure 13.1 depicts a block of conducting material. If we add a magnetic field B in the z-direction and pass a current I in the x-direction, then we can measure a voltage drop V in the y-direction. This phenomenon was discovered in 1878 by E. H. Hall and is called the Hall effect.

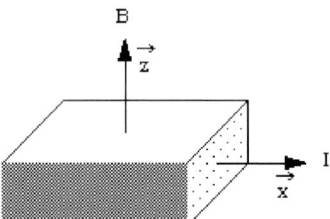

Fig. 13.1. Geometry of Hall effect measurement.

According to elementary physics, a charged particle moving in a magnetic field feels a Lorentz force $-e\mathbf{v} \times \mathbf{B}$ perpendicular to the velocity and the magnetic field. Initially the electrons bend their trajectories and move to a side boundary. As enough charges are accumulated on the side, a static electric field builds up that balances the Lorentz force. The current then keeps its originally intended direction of flow. The electric field gives rise to a potential drop in the transverse direction.

Consider a simpler situation: a gas of charged particles moving with a common velocity $v\mathbf{e}_x$. This corresponds to a current density $j_x = -env$. The Lorentz force that is to be balanced by the electric force $-eE_y\mathbf{e}_y$ is given by $evB\mathbf{e}_y$. Therefore $E_y = vB$ and the current density satisfies

$$j_x = -en\frac{E_y}{B}.$$

This equation suggests the definition $\rho_{xy} := E_y/j_x = -B/en$ which is known as the Hall resistivity. This quantity provides a simple and direct probe of the charge density of the current carriers. It is one of the most valuable tools in the experimental studies of various condensed matter systems.

13.2.2 The Quantum Hall Effect

The quantum Hall effect was discovered in 1980 by von Klitzing [260], and has since become an important research field in condensed matter physics. The sample used in von Klitzing's experiment was a semiconductor device called

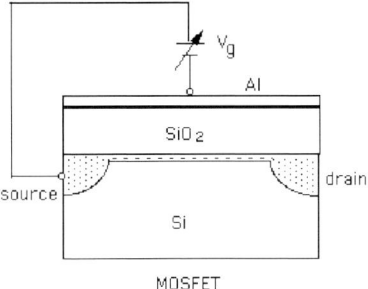

Fig. 13.2. MOSFET device housing a two-dimensional electron gas.

a MOSFET (metal oxide silicon field effect transistor, see Fig. 13.2). Under a certain gate voltage V_g, there is a thin (*inversion*) layer of electrons attracted to the interface between Si and SiO$_2$. The thickness of the inversion layer is about 10 nm. Therefore, it is so thin that the electronic degree of freedom in the perpendicular direction is completely frozen into its ground state, and the electronic motion is thus restricted to the transverse plane. The *two-dimensional density* of the electrons, defined as the number per unit area, may be changed by varying V_g.

In a quantum Hall effect experiment, the device is immersed in a strong magnetic field which is perpendicular to the two-dimensional electron layer. If one passes a DC current from the source to the drain, the current will go through the two-dimensional electron layer. As shown in Fig. 13.3, a voltage drop V_{ab} is measured between a and b, yielding a Hall resistance of

$$\rho_H = \frac{V_{ab}}{I}.$$

Ordinarily, ρ_H would depend smoothly on the magnetic field and the electron density. The striking discovery of von Klitzing was that at very low temperatures (a few K) and very high magnetic fields (a few tesla), ρ_H behaved as a staircase function of V_g with extremely flat plateaus, namely

$$\rho_{xy} = \frac{h}{e^2 p}$$

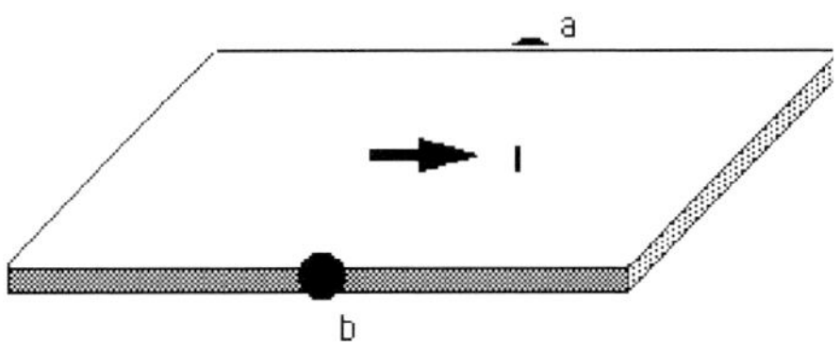

Fig. 13.3. Geometry for quantum Hall effect measurement.

where p is an integer. This phenomenon is called the quantum Hall effect, or more precisely the integer quantum Hall effect. The accuracy of these plateau values was less than six parts per million in the initial work. Later it has improved to the level of a few parts per billion [244]. Further improvement of this precision turned out to be impossible due to the lack of a more accurate resistance standard and more precise values of the fundamental constants. Indeed, in 1990, the quantized Hall resistance was officially adopted as a practical reference standard for resistance.

13.2.3 The Ideal Model

In order to understand the quantum Hall effect, we consider first the following idealized model: a non-interacting two-dimensional electron gas in the absence of impurities. Under a uniform and perpendicular magnetic field, the energy spectrum is quite different from the free electron case. This problem was solved by Landau several decades ago [148]. To arrive at a detailed physical picture and set the mathematical language, we will now provide a discussion of the Landau problem.

The magnetic field $\mathbf{B}$ enters the Schrödinger equation through the vector potential $\mathbf{A}$. As is well known the vector potential yields the magnetic field according to $\nabla \times \mathbf{A} = \mathbf{B}$, and it has the freedom of electromagnetic gauge transformations. In order to perform explicit calculations, we need to make a choice of gauge. A particularly convenient choice is the so-called Landau gauge, namely

$$A_x = 0, \quad A_y = Bx,$$

in which the time-independent Schrödinger equation reads

$$\frac{1}{2m}\left[\left(-i\hbar\frac{\partial}{\partial x}\right)^2 + \left(-i\hbar\frac{\partial}{\partial y} + eBx\right)^2\right]\psi = E\psi.$$

Because of the translational invariance in the y-direction, we can write

$$\psi = e^{iqy}\varphi(x),$$

where φ satisfies

$$\frac{1}{2m}\left[\left(-i\hbar\frac{\partial}{\partial x}\right)^2 + (\hbar q + eBx)^2\right]\varphi(x) = E\varphi(x).$$

This is the time-independent Schrödinger equation for a simple harmonic oscillator. The solutions are

$$E = \left(n + \frac{1}{2}\right)\hbar\omega_c =: E_n \qquad (13.1)$$

$$\varphi(x) = \varphi_n\left(x + \frac{\hbar q}{eB}\right), \qquad (13.2)$$

where $\omega_c = \frac{eB}{m}$ and $\varphi_n(x)$ is the n-th energy eigenfunction for the oscillator.

The energy spectrum consists of a series of equally spaced levels. These are called the *Landau levels*. Each level is highly degenerate as there are many energy eigenstates with different values of q but the same energy. The following argument shows that the number of these degenerate states for each level is proportional to the surface area of the electron gas. Consider a two-dimensional electron gas filling a rectangular area with sides aligned along the x- and y-directions and side length L_x and L_y, respectively. If we apply the periodic boundary condition in the y-direction: $\psi(y + L_y) = \psi(y)$, q is quantized as $q = \frac{2\pi}{L_y} j$, $j =$ integer. Since $-\frac{\hbar q}{eB}$ represents the center of the wave function in the x-direction, it is confined within the width L_x of the system in the x-direction. Therefore the allowed number of different q values is

$$\frac{L_x eB}{\hbar} \Big/ \frac{2\pi}{L_y} = (L_x L_y) \cdot \frac{eB}{h}.$$

In other words, each Landau level has $\frac{eB}{h}$ number of states per unit area. This result is accurate in the thermodynamic limit where $L_x, L_y \to \infty$.

If we add an electric field E_y in the y-direction, an extra term $-E_y t$ is added to A_y. Then the eigenfunction $\varphi(x)$ becomes $\varphi_n(x + \frac{\hbar q}{eB} - \frac{E_y}{B} t)$. This shows that the center of each eigenfunction moves with velocity $\frac{E_y}{B}$ in the x-direction. If one Landau level is filled, the current density is then

$$j_x = -e \cdot \frac{eB}{h} \cdot \frac{E_y}{B} = \frac{e^2}{h} E_y,$$

which corresponds to a Hall conductivity of

$$\sigma_{xy} = -\frac{e^2}{h}.$$

If p Landau levels are filled, the density increases p-fold and

$$\sigma_{xy} = -\frac{pe^2}{h}.$$

The Hall resistivity is then

$$\rho_{xy} = -\frac{h}{pe^2}.$$

In two dimensions, the Hall resistivity is the same as the Hall resistance. Also, the experimetalists usually just quote the magnitude of the measured quantities. Therefore, the above result corresponds to the experimentally observed quantization of the Hall resistance.

13.2.4 Corrections to Quantization

The above simplified model disregards many physical effects such as thermal fluctuations of the system and the presence of impurities. In what follows we will briefly discuss the consequences of these effects. We will first consider the thermal effect.

According to the above model, at finite temperature the Hall conductance should have the form

$$\sigma_H = \frac{e^2}{h} \sum_{n=0}^{\infty} \frac{1}{e^{(E_n - \mu)/kT} + 1}.$$

At absolute zero this is a staircase function of the chemical potential μ and the magnetic field. This is consistent with the experimental results. At finite but low temperatures the slope of the Hall conductance at the center of a plateau is given by

$$\frac{\partial \sigma_H}{\partial \mu} \simeq \frac{2}{kT} e^{-\hbar \omega_c / kT} \frac{e^2}{h}.$$

The quantizing condition may be written as

$$\frac{\partial \sigma_H}{\partial \mu} \cdot \hbar \omega_c = c \frac{e^2}{h}, \qquad c := 2 \frac{\hbar \omega_c}{kT} \exp(-\frac{\hbar \omega_c}{kT}) \ll 1.$$

If we take $c = 10^{-7}$, which is the figure found in the experiment, we obtain

$$\frac{\hbar \omega_c}{kT} > 40.$$

For Si heterostructures, the effective electron mass is given by $m = 0.19\, m_e$. Substituting this value in the expression for ω_c and taking $B = 10\,\text{T}$, we find $T < 2$ K which agrees with the value of the temperature adopted in the experiment.

Since $\omega_c \propto \frac{1}{m}$, the quantizing condition is more easily satisfied for smaller effective masses. This observation led Tsui *et al.* [253] to use GaAs heterojunctions, with an effective mass of $m \approx 0.068\, m_e$, to study the quantum Hall effect. This eventually gave rise to the discovery of the fractional quantum Hall effect which we shall discuss in the next chapter.

Next, we consider the effect of impurities. The calculation of the energy spectrum in the presence of impurities actually predates the discovery of the quantum Hall effect. The result was a broadening of the Landau levels. Later calculations showed that most of the states became localized except at the band centers where the states were extended. The experimental facts may therefore be interpreted as: (1) Changing V_g is equivalent to changing the chemical potential μ. (2) When μ lies in an energy gap or an energy range of localized states, called a *mobility gap*, the Hall conductance is quantized; σ_H jumps by a quantum $\frac{e^2}{h}$ when μ goes across a level of extended states.

There still remains the question of why σ_H has to be quantized when μ is away from a level of extended states. We will show in Sect. 13.4 that this can be explained in terms of the first Chern numbers characterizing a topological invariant of the states. Before discussing this interesting phenomenon, we will first consider the relatively simpler case in which the electrons are subject to a periodic potential.

13.3 Magnetic Bands in Periodic Potentials

13.3.1 Single-Band Approximation in a Weak Magnetic Field

In the presence of a magnetic field we need to deal with a spatially varying vector potential. Our strategy will be to work with a basis of localized states and to locally gauge away the vector potential. It turns out that this is possible for weak magnetic fields [162, 264].

Let $|W_n(\mathbf{R})\rangle$ be a Wannier state vector in the n-th band of the Hamiltonian H_0 in the absence of the vector potential. To simplify notation we choose to label the position of the Wannier functions, so that $|W_n(\mathbf{R})\rangle$ corresponds to a Wannier function in the n-th band centered at $\mathbf{r} = \mathbf{R}$. Next, we define a set of modified Wannier state vectors, namely

$$|\tilde{W}_n(\mathbf{R})\rangle = e^{-i\frac{e}{\hbar}\mathbf{A}(\mathbf{R})\cdot(\mathbf{r}-\mathbf{R})}|W_n(\mathbf{R})\rangle, \tag{13.3}$$

which form a basis of our band with a given n. Note that these states are not orthogonal. This does not, however, prevent them from serving as a basis. Obviously it would have been convenient if they were orthogonal. For a weak magnetic field, one can choose a vector potential with a slow variation in space. In this case, as we shall argue in the following, the modified Wannier functions form a nearly orthogonal basis.

The overlap between two modified Wannier state vectors associated with the same band is proportional to

$$\langle W_n(\mathbf{R}')|e^{i\frac{e}{\hbar}(\mathbf{A}(\mathbf{R}')-\mathbf{A}(\mathbf{R}))\cdot\mathbf{r}}|W_n(\mathbf{R})\rangle. \tag{13.4}$$

The overlap is essentially zero if $|\mathbf{R}' - \mathbf{R}|$ is much larger than the width Δr of the Wannier state vector $|W_n(\mathbf{R})\rangle$. Even if $|\mathbf{R}' - \mathbf{R}|$ is smaller than or comparable to the width, the overlap is still negligible provided that $|\mathbf{A}(\mathbf{R}') - \mathbf{A}(\mathbf{R})|$ is so small that the exponential in the above expression may be regarded as constant over a length scale of Δr. This condition is satisfied for sufficiently weak magnetic fields B where we can choose $\mathbf{A}$ to vary with a slope proportional to B. Quantitatively, we require that $eB(\Delta r)^2/\hbar \ll 1$.

Under the same condition, we can show that the modified Wannier functions from the same band form an invariant subspace of the Hamiltonian.[1] To see this we first write

$$\left[-i\hbar\frac{\partial}{\partial \mathbf{r}} + e\mathbf{A}(\mathbf{r})\right]|\tilde{W}_n(\mathbf{R})\rangle = e^{-i\frac{e}{\hbar}\mathbf{A}(\mathbf{R})\cdot(\mathbf{r}-\mathbf{R})}\left[-i\hbar\frac{\partial}{\partial \mathbf{r}}\right]|W_n(\mathbf{R})\rangle, \tag{13.5}$$

where we have neglected a term proportional to $\mathbf{A}(\mathbf{r}) - \mathbf{A}(\mathbf{R})$ relying on the assumption that the vector potential varies slowly over the length scale of the Wannier function. Next let H and H_0 denote the Hamiltonians with and without the vector potential, respectively. Then

---

[1] This means that the Hamiltonian operator maps them to their linear combinations.

$$H|\tilde{W}_n(\mathbf{R})\rangle = e^{-i\frac{e}{\hbar}\mathbf{A}(\mathbf{R})\cdot(\mathbf{r}-\mathbf{R})} H_0|W_n(\mathbf{R})\rangle. \tag{13.6}$$

Now using the expansion

$$H_0|W_n(\mathbf{R})\rangle = \sum_{\mathbf{R}'} h_n(\mathbf{R}'-\mathbf{R})|W_n(\mathbf{R}')\rangle, \tag{13.7}$$

we have

$$H|\tilde{W}_n(\mathbf{R})\rangle = \sum_{\mathbf{R}'} h_n(\mathbf{R}'-\mathbf{R})e^{i\frac{e}{\hbar}\{[\mathbf{A}(\mathbf{R}')-\mathbf{A}(\mathbf{R})]\cdot\mathbf{r}+\mathbf{A}(\mathbf{R})\cdot\mathbf{R}-\mathbf{A}(\mathbf{R}')\cdot\mathbf{R}'\}}|\tilde{W}_n(\mathbf{R}')\rangle. \tag{13.8}$$

Finally, since $h_n(\mathbf{R}'-\mathbf{R})$ is negligible when $|\mathbf{R}'-\mathbf{R}| \gg \Delta r$, we only need to consider the terms for which $|\mathbf{R}'-\mathbf{R}|$ is comparable with or smaller than Δr. We can then set $\mathbf{r} = (\mathbf{R}'+\mathbf{R})/2$ in the exponential on the right-hand side of (13.8), for the exponential is a slowly varying function of $\mathbf{r}$. This argument shows that a weak magnetic field does not induce mixing between different bands of the modified Wannier states.

In summary, under the action of the Hamiltonian the modified Wannier functions behave as

$$H|\tilde{W}_n(\mathbf{R})\rangle = \sum_{\mathbf{R}'} h_n(\mathbf{R}'-\mathbf{R})e^{i\frac{e}{2\hbar}[\mathbf{A}(\mathbf{R})+\mathbf{A}(\mathbf{R}')]\cdot(\mathbf{R}-\mathbf{R}')}|\tilde{W}_n(\mathbf{R}')\rangle. \tag{13.9}$$

The effect of the magnetic field enters through the phase factor, which is essentially the same as the Dirac phase introduced in (2.106). The scalar product in the phase factor can also be written as $-\int \mathbf{A}(\mathbf{r}) \cdot d\mathbf{r}$, where the path of the integral is a straight line from $\mathbf{R}$ to $\mathbf{R}'$. For an energy eigenstate vector of the band

$$|\psi\rangle = \sum_{\mathbf{R}} \psi_n(\mathbf{R})|\tilde{W}_n(\mathbf{R})\rangle, \tag{13.10}$$

where the coefficients $\psi_n(\mathbf{R})$ satisfy the following time-independent Schrödinger equation,

$$\sum_{\mathbf{R}'} h_n(\mathbf{R}'-\mathbf{R})\exp\left[i(e/\hbar)\int_{\mathbf{R}}^{\mathbf{R}'} \mathbf{A}(\mathbf{r})\cdot d\mathbf{r}\right]\psi_n(\mathbf{R}') = E\psi_n(\mathbf{R}), \tag{13.11}$$

where an extra phase, called the *Peierls phase* in this context, appears in the equation due to the modification of the Wannier function in (13.9).

Several remarks are in order:

(i) The existence of localized Wannier functions of a band in the absence of the vector potential is sufficient to ensure the decoupling of that band from the other bands in a weak magnetic field.

(ii) Going beyond the regime of weak magnetic field, one can establish the following results rigorously [193]. The energy gap between Bloch bands changes continuously with the magnetic field [25, 192]. This allows one to follow the evolution of an isolated Bloch band as the magnetic field is turned on (see Fig. 13.4 below). This band may have split into a very complicated structure (see the next section), but the gaps bounding this band do not disappear until the magnetic field reaches a critical value.

(iii) Before the critical field is reached, it is possible to construct a basis of orthogonal and exponentially localized Wannier functions for the band complex [51, 71]. Using this basis, we still have the Schrödinger equation (13.11) except that $h_n(\mathbf{R}' - \mathbf{R})$ is replaced by a function of the field (but not of the vector potential). The new hopping matrix element may be expanded (in the asymptotic sense) as a power series of the field. The leading term of this expansion is given by (13.11) [40, 133, 220].

A more elegant but equivalent way of writing the (implicit) Hamiltonian of (13.11) is

$$\epsilon_n\left(-i\frac{\partial}{\partial \mathbf{R}} + (e/\hbar)\mathbf{A}(\mathbf{R})\right). \tag{13.12}$$

This is defined as the operator obtained from the band energy $\epsilon_n(\mathbf{k})$ by replacing $\mathbf{k}$ with $-i\frac{\partial}{\partial \mathbf{R}} + (e/\hbar)\mathbf{A}(\mathbf{R})$. This is known as the *Peierls substitution*, [210]. The equivalence is realized once one uses the Fourier series of the band energy and notices that $\exp[i\mathbf{R}' \cdot (-i\frac{\partial}{\partial \mathbf{R}} + (e/\hbar)\mathbf{A}(\mathbf{R}))]$ is just the magnetic translation operator by a lattice vector $\mathbf{R}'$. This form of the Hamiltonian becomes extremely useful in situations where $\psi(\mathbf{R})$, called the *envelope wave function*, is expected to be slowly varying. In this case one may make a Taylor expansion of the band energy about $\mathbf{k} = 0$ and obtain an effective Hamiltonian of the form [162]

$$\epsilon_0 \pm \frac{\hbar^2}{2m^*}\left[-i\frac{\partial}{\partial \mathbf{R}} + (e/\hbar)\mathbf{A}(\mathbf{R})\right]^2 - e\phi(\mathbf{R}), \tag{13.13}$$

where ϵ_0 is the band energy at $\mathbf{k} = 0$, m^* is the effective mass defined by the curvature of the band (which is assumed to be isotropic), and the $+$ sign ($-$ sign) corresponds to $\mathbf{k} = 0$ being at a band minimum (maximum). The Landau levels are evenly spaced near such a maximum or minimum.

13.3.2 Harper's Equation and Hofstadter's Butterfly

We now study the effect of a uniform magnetic field $\mathbf{B} = B\mathbf{e}_z$ on a Bloch band by considering some examples. We will confine our attention to motion in the plane perpendicular to the magnetic field. For a three-dimensional crystal, the motion along the field is still described by Bloch waves. This is because one can choose a gauge with $A_z = 0$ so that the translational symmetry along the direction of the field is preserved. For a two-dimensional electron gas we assume that the motion in the perpendicular direction is frozen in its ground state.

We consider the simple case of a tight-binding band where only coupling to nearest neighbors is important. For a rectangular lattice, the four nearest neighbors of $\mathbf{R}$ are $\mathbf{R} \pm a\mathbf{e}_x$ and $\mathbf{R} \pm b\mathbf{e}_y$. We choose the gauge $\mathbf{A} = Bx\mathbf{e}_y$, so that the phases in the coupling along the x-direction is zero while those along the y-direction are given by $\pm \frac{eb}{\hbar}BX$. Here X is the x-component of $\mathbf{R}$. The time-dependent Schrödinger equation then becomes

$$\Delta_1 \left[\psi(\mathbf{R}+a\mathbf{e}_x) + \psi(\mathbf{R}-a\mathbf{e}_x)\right] + \Delta_2 \left[e^{i\frac{eb}{\hbar}BX}\psi(\mathbf{R}+b\mathbf{e}_y) + e^{-i\frac{eb}{\hbar}BX}\psi(\mathbf{R}-b\mathbf{e}_y)\right]$$
$$= E\psi(\mathbf{R}), \tag{13.14}$$

where $\Delta_{1,2}$ are the hopping matrix elements in the two directions, and we have chosen the center of the original band to be the zero of energy so that the $\mathbf{R}' = \mathbf{R}$ term in (13.11) is removed.

Equation (13.14) can be reduced to a one-dimensional time-independent Schrödinger equation, if we utilize the translational symmetry along the y-direction. Writing $\psi(\mathbf{R}) = e^{ik_y Y}C_m$, where $m = X/a$, we obtain the three-term difference equation

$$\Delta_1(C_{m+1} + C_{m-1}) + 2\Delta_2 \cos(2\pi\phi m + k_y b)C_m = EC_m, \tag{13.15}$$

where $\phi = eabB/h$ is the magnetic flux per unit cell divided by the flux quantum h/e. This equation is known as *Harper's equation* [102]. It is invariant under $\phi \to \phi + 1$ which reflects the periodical dependence of the system on the magnetic flux. Harper's equation also arises in the study of the Landau level splitting due to a weak periodic potential or a weak periodic modulation of the magnetic field [284].

The solution of Harper's equation strongly depends on whether ϕ is a rational or irrational number. If $\phi = p/q$ for some integers p and $q > 0$ with no common divisors (except ± 1), the spectrum of the original (parent) band is split into q subbands. This can be seen easily from the fact that Harper's equation is invariant under the translation $m \to m + q$. Imposing the Bloch condition $C_{m+q} = e^{ik_x aq}C_m$, we obtain a $q \times q$ matrix eigenvalue problem. Solving this problem we find q eigenvalues: $E_j(k_x, k_y)$, with $j = 1, 2, ..., q$, that depend on the wave numbers. As we shall see in the next chapter, these q bands are globally separated from one another except for even q where the two central subbands touch at $E = 0$.

In view of the way in which the wave numbers enter into the equations, we can deduce that k_x and $k_x + \frac{2\pi}{qa}$ (and similarly k_y and $k_y + \frac{2\pi}{b}$) are equivalent. We can use this observation to choose the Brillouin zone as $(0 \le k_x < \frac{2\pi}{qa}, 0 \le k_y < \frac{2\pi}{b})$. Furthermore, since shifting m by 1 in (13.15) is equivalent to adding $\frac{2\pi p}{qb}$ to k_y, we can infer that the band energy has an additional periodicity of $\frac{2\pi p}{qb}$ in k_y. The combined effect of these periodicities is that each energy band consists of q identical pieces in the Brillouin zone, one of which is $(0 \le k_x < \frac{2\pi}{qa}, 0 \le k_y < \frac{2\pi}{qb})$.

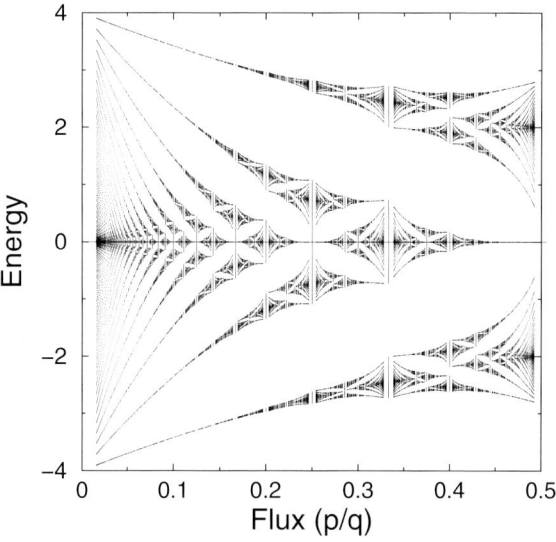

Fig. 13.4. The Hofstadter spectrum for the tight-binding model with square symmetry.

Figure 13.4 shows the energy spectrum of Harper's equation for the case $\Delta_1 = \Delta_2$. Only the rational values of ϕ with $q < 50$ are displayed. The result is a startling hierarchical structure known as *Hofstadter's butterfly* after its original creator [108].

The main skeleton of the spectral structure is given by the subbands for the cases $\phi = 1/q$. These have been systematically calculated and analyzed in [54]. The subbands correspond to Onsager's semiclassical energy levels of quantized cyclotron orbits in the Brillouin zones of the parent band. The widths of the subbands are results of level broadening due to coupling between degenerate orbits in different Brillouin zones [49, 132]. Those near the band edges are narrow because of large separations between the corresponding orbits in different zones. They correspond to the Landau levels in the effective mass approximation.

The finer structures in the Hofstadter butterfly may be understood as levels split from the main skeleton [27, 40, 59]. To understand this we will first present, in the next section, a general theory of magnetic Bloch bands and their dynamical properties based on a symmetry group and semiclassical analysis.

13.3.3 Magnetic Translations

It is a fundamental fact that the vector potential rather than the magnetic field enters in the basic laws of quantum mechanics. Therefore, even for a constant magnetic field which we will take along the z-direction, the translational symmetry of the Schrödinger equation must be broken in the (x-y) plane perpendicular to the field. The best one can do is to preserve the translational symmetry along one direction in this plane by making a convenient choice of gauge. This is what we did towards the end of the preceding section. In the following, in order to concentrate on the behavior of the vector potential, we will restrict our discussion to translations that leave the scalar potential invariant.

It turns out that, up to a gauge transformation, the Hamiltonian is still invariant under translations in the plane perpendicular to the magnetic field. This implies that the physical properties are not affected by lack of translational symmetry of the vector potential. This is clearly because they are gauge invariant. In order to utilize these physical symmetries in the solution of the Schrödinger equation, one introduces the so-called magnetic translation operators which commute with the Hamiltonian [51, 80, 102, 280]. A magnetic translation operator $T(\mathbf{R})$ is obtained by combining an ordinary translation with a gauge transformation according to

$$T(\mathbf{R}) = e^{i\mathbf{R}\cdot\boldsymbol{\kappa}/\hbar}, \tag{13.16}$$

where $\mathbf{R}$ is the displacement vector and $\boldsymbol{\kappa}$ is an operator with components

$$\kappa_x = -i\hbar\frac{\partial}{\partial x} + eA_x + eBy, \tag{13.17}$$

$$\kappa_y = -i\hbar\frac{\partial}{\partial y} + eA_y - eBx, \tag{13.18}$$

$$\kappa_z = -i\hbar\frac{\partial}{\partial z} + eA_z. \tag{13.19}$$

Our aim is to use the symmetry under magnetic translations to classify the eigenstates of the Hamiltonian. However, we have to deal with the fact that two magnetic translations do not generally commute unless the defining displacements form an area that encloses an integer number of magnetic flux quanta. This is a consequence of the identity

$$T(\mathbf{R})T(\mathbf{R}') = T(\mathbf{R}')T(\mathbf{R})e^{i\frac{e}{\hbar}\mathbf{B}\cdot\mathbf{R}'\times\mathbf{R}}, \tag{13.20}$$

which follows from the commutation relations $[\kappa_x, \kappa_y] = i\hbar eB$ and $[\kappa_x, \kappa_z] = [\kappa_y, \kappa_z] = 0$. In order to use the magnetic translational symmetry in classifying the energy eigenstates, we must find an Abelian subset of these translations, i.e., a subset consisting of mutually commuting elements.

To see the problem more clearly, we first consider the situation of two-dimensional space, where the field is perpendicular to the lattice plane in which the primitive lattice vectors are $\mathbf{a}$ and $\mathbf{b}$. If the unit cell area $\mathbf{a}\times\mathbf{b}$ is not an integer multiple of $\frac{h}{eB}$, we cannot diagonalize the Hamiltonian and the magnetic translations $T(\mathbf{a})$ and $T(\mathbf{b})$ simultaneously. What we did in the last section was in fact to use the commuting pair $T(q\mathbf{a})$ and $T(\mathbf{b})$ when the parameter $\phi = \frac{e}{\hbar}\mathbf{B}\cdot\mathbf{a}\times\mathbf{b}$ is a rational number $\frac{p}{q}$. Other choices are also possible, e.g., $T(\mathbf{a})$ and $T(q\mathbf{b})$, or in the case of a non-prime integer $q = q_1 q_2$, $T(q_1\mathbf{a})$ and $T(q_2\mathbf{b})$. Different choices give different but physically equivalent classifications of the eigenstates.

Using a commuting pair of magnetic translations such as $T(q\mathbf{a})$ and $T(\mathbf{b})$, we can define the magnetic Bloch states by the conditions

$$T(q\mathbf{a})|\psi(\mathbf{k})\rangle = e^{i\mathbf{k}\cdot q\mathbf{a}}|\psi(\mathbf{k})\rangle, \qquad (13.21)$$
$$T(\mathbf{b})|\psi(\mathbf{k})\rangle = e^{i\mathbf{k}\cdot\mathbf{b}}|\psi(\mathbf{k})\rangle.$$

Because the real-space unit cell is q times larger than ab, the magnetic Brillouin zone is q times smaller than that in the absence of the magnetic field. In the single-band approximation, this corresponds to folding the original Brillouin zone q times and splitting the parent band into q subbands with the total number of states conserved.

Furthermore, there is a q-fold degeneracy of the magnetic Brillouin zone. Let $|\psi(\mathbf{k})\rangle$ be an eigenvector of energy $\epsilon(\mathbf{k})$. Then $T(\mathbf{a})|\psi(\mathbf{k})\rangle$ is also an eigenvector with the same energy, because $T(\mathbf{a})$ commutes with the Hamiltonian. On the other hand, from the commutation relation $T(\mathbf{b})T(\mathbf{a}) = e^{i\frac{e}{\hbar}\mathbf{B}\cdot\mathbf{a}\times\mathbf{b}}T(\mathbf{a})T(\mathbf{b})$, we have

$$T(\mathbf{b})[T(\mathbf{a})|\psi(\mathbf{k})\rangle] = e^{i\frac{e}{\hbar}\mathbf{B}\cdot\mathbf{a}\times\mathbf{b}}e^{i\mathbf{k}\cdot\mathbf{b}}[T(\mathbf{a})|\psi(\mathbf{k})\rangle]. \qquad (13.22)$$

This means that $T(\mathbf{a})|\psi(\mathbf{k})\rangle$ is an eigenvector with wave vector $\mathbf{k}' = \mathbf{k} + \frac{e}{\hbar}\mathbf{B}\times\mathbf{a}$ with energy $\epsilon(\mathbf{k}') = \epsilon(\mathbf{k})$. Therefore, each energy subband should consist of q identical pieces, with the basic zone being q times smaller in each direction of the original Brillouin zone of the lattice.

It is easy to generalize the above results to a three-dimensional crystal when it is formed by repeating the two-dimensional lattice in the perpendicular direction with the lattice vector $\mathbf{c}$. In this case, we only need to append a third Bloch condition, namely $T(\mathbf{c})|\psi(\mathbf{k})\rangle = e^{i\mathbf{k}\cdot\mathbf{c}}|\psi(\mathbf{k})\rangle$, to (13.22). The results of the two-dimensional discussions are then valid for each k_z.

For a general three-dimensional crystal, we can find a sublattice with mutually commuting associated magnetic translations, if the cell areas contain rational multiples of the flux quantum. It turns out that such a rational magnetic field always coincides with the direction of a line of lattice vectors [143, 283]. We denote the smallest lattice vector in that direction by $\mathbf{c}$. It is also possible to find two more lattice vectors $\mathbf{a}$ and $\mathbf{b}$ to form a primitive (and right handed) set with $\mathbf{c}$. Then, the rationality condition reduces to the single requirement $\frac{e}{\hbar}\mathbf{B}\cdot\mathbf{a}\times\mathbf{b} = p/q$; the fluxes through other cell areas are zero because they are parallel to the field by construction. We can use the magnetic translations $T(q\mathbf{a})$, $T(\mathbf{b})$, and $T(\mathbf{c})$ (the latter being a simple translation) to define the magnetic Bloch states as before. Note however that now the lattice plane $\mathbf{a}\times\mathbf{b}$ is no longer perpendicular to the magnetic field.

The three-dimensional magnetic Brillouin zone is q times smaller than that in the absence of the field, and may be taken as the region spanned by $\mathbf{G}_a/q$, $\mathbf{G}_b$, and $\mathbf{G}_c$, where the $\mathbf{G}$'s are primitive vectors of the reciprocal lattice, i.e.,

$$\mathbf{G}_a := 2\pi \frac{\mathbf{b} \times \mathbf{c}}{\mathbf{a} \cdot \mathbf{b} \times \mathbf{c}}, \quad \mathbf{G}_b := 2\pi \frac{\mathbf{c} \times \mathbf{a}}{\mathbf{a} \cdot \mathbf{b} \times \mathbf{c}}, \quad \mathbf{G}_c := 2\pi \frac{\mathbf{a} \times \mathbf{b}}{\mathbf{a} \cdot \mathbf{b} \times \mathbf{c}}. \tag{13.23}$$

Therefore, the magnetic Brillouin zone is a prism defined by a base plane which is spanned by $\mathbf{G}_a/q$ and $\mathbf{G}_b$ and is perpendicular to the magnetic field, and the primitive vector $\mathbf{G}_c$ which in general points along an inclined direction with respect to the base plane.

13.3.4 Quantized Hall Conductivity

In the presence of a constant electric field $\mathbf{E}$, we can still use the magnetic translation symmetry to classify the states provided that the electric field is represented by a time-dependent vector potential. In this case, we have

$$|\psi(\mathbf{k}, t)\rangle = e^{i\mathbf{k} \cdot \mathbf{r}} |u(\mathbf{k}, t)\rangle, \tag{13.24}$$

where $|u(\mathbf{k}, t)\rangle$ is invariant under the magnetic translations $T(q\mathbf{a})$ and $T(\mathbf{b})$, and satisfies the following time-dependent Schrödinger equation.

$$i\hbar \frac{\partial}{\partial t} |u(\mathbf{k}, t)\rangle = H(\mathbf{k}(t)) |u(\mathbf{k}, t)\rangle, \tag{13.25}$$

where $\mathbf{k}(t) := \mathbf{k} - e\mathbf{E}t/\hbar$ and

$$H(\mathbf{k}) := \frac{1}{2m}(-i\hbar \nabla + \hbar \mathbf{k})^2 + V(\mathbf{r}). \tag{13.26}$$

These equations are valid for an arbitrary strength of the electric field.[2]

In the weak field limit, we may use the adiabatic approximation to find the solutions of (13.25) up to first order in the field strength. This yields

$$|u_n(\mathbf{k}(t))\rangle + ieE_x \sum_{n'(\neq n)} \frac{|u_{n'}(\mathbf{k}(t))\rangle \langle u_{n'}(\mathbf{k}(t))| \frac{\partial}{\partial k_x} |u_n(\mathbf{k}(t))\rangle}{\epsilon_n(\mathbf{k}(t)) - \epsilon_{n'}(\mathbf{k}(t))}, \tag{13.27}$$

where we have taken the field to be in the x-direction and used $|u_n(\mathbf{k})\rangle$ to denote an eigenstate of $H(\mathbf{k})$ of energy $\epsilon_n(\mathbf{k})$ in the n-th magnetic Bloch band. The y-component of the average velocity in such a state is given by

$$\bar{v}_y = \langle u_n | v_y | u_n \rangle + ieE_x \sum_{n' \neq n} \frac{\langle u_n | v_y | u_{n'} \rangle \langle u_{n'} | \frac{\partial}{\partial k_x} | u_n \rangle}{\epsilon_n - \epsilon_{n'}} + \text{c.c.}, \tag{13.28}$$

where c.c. stands for the complex conjugate of the term involving a summation over n'. We have encountered a similar expression in our discussion of

---

[2] Note the difference between $\mathbf{k}$ and $\mathbf{k}(t)$.

13.3 Magnetic Bands in Periodic Potentials

the adiabatic particle transport in the preceding chapter. We may express the first term in (13.28) in terms of the k derivative of the band energy using the relation $v_y = \frac{1}{\hbar}\frac{\partial H(\mathbf{k})}{\partial \mathbf{k}}$. Making use of the identities

$$\langle u_n | \frac{\partial H(\mathbf{k})}{\partial \mathbf{k}} | u_{n'} \rangle = (\epsilon_n - \epsilon_{n'}) \langle \frac{\partial u_n}{\partial \mathbf{k}} | u_{n'} \rangle$$

and $\sum_{n'} |u_{n'}\rangle\langle u_{n'}| = 1$ to simplify the second term, we then find

$$\bar{v}_y = \frac{1}{\hbar}\frac{\partial \epsilon_n}{\partial k_y} - i\frac{eE_x}{\hbar}\left[\left\langle \frac{\partial u_n}{\partial k_x} \Big| \frac{\partial u_n}{\partial k_y} \right\rangle - \left\langle \frac{\partial u_n}{\partial k_y} \Big| \frac{\partial u_n}{\partial k_x} \right\rangle\right]. \tag{13.29}$$

Therefore, under a weak electric field, the magnetic Bloch states undergo Bloch oscillations and also give rise to a Hall current proportional and perpendicular to the electric field. This confirms the predictions of the semiclassical theory which we will be dealing with in the next section.

For a filled band, the first term in the velocity formula sums to zero and only the Hall current survives. In two-dimensions, a $\mathbf{k}$ space area element contributes to the electron density as $\frac{dk_x dk_y}{(2\pi)^2}$, yielding a Hall current of $-e \int \frac{dk_x dk_y}{(2\pi)^2} \bar{v}_y$. The Hall conductivity, defined as the ratio of this current density and the electric field E_x, is therefore given by

$$\sigma_H = \frac{e^2}{h} \int \frac{i dk_x dk_y}{2\pi} \left[\left\langle \frac{\partial u_n}{\partial k_x} \Big| \frac{\partial u_n}{\partial k_y} \right\rangle - \left\langle \frac{\partial u_n}{\partial k_y} \Big| \frac{\partial u_n}{\partial k_x} \right\rangle\right]. \tag{13.30}$$

This formula was first derived by Thouless et al. [247] and has since been the source of inspiration for a large number of theoretical studies of the quantum Hall effect and other subjects. The integral on the right-hand side of (13.30), i.e.,

$$\sigma := \int \frac{i dk_x dk_y}{2\pi} \left[\left\langle \frac{\partial u_n}{\partial k_x} \Big| \frac{\partial u_n}{\partial k_y} \right\rangle - \left\langle \frac{\partial u_n}{\partial k_y} \Big| \frac{\partial u_n}{\partial k_x} \right\rangle\right]$$

is nothing but the Chern number which as we pointed out earlier takes integer values. It characterizes the topological structure of the mapping from the Brillouin zone to the magnetic Bloch states [22, 24, 130].

Inspired by a possible experimental connection [173], several authors also studied the three-dimensional problem [95, 131, 172]. In a three-dimensional crystal, the Hall conductivity is still given by (13.30) except that the $\mathbf{k}$ vector is three-dimensional and that one has to divide the expression by 2π and integrate k_z over the prism of the magnetic Brillouin zone (see the previous subsection). Assuming an isolated magnetic band, one then finds that the k_x, k_y integral over the base of the prism is quantized similarly to the two-dimensional case, and that it is a continuous and therefore constant function of k_z. Therefore the Hall conductivity equals the right-hand side of (13.30), i.e., the result of Thouless et al. [247], multiplied by the height of the prism and divided by 2π,

$$\sigma_{yx} = \frac{e^2}{h}\sigma/c, \qquad (13.31)$$

where c is the length of $\mathbf{c}$.

In the above analysis, we have assumed that the electric field is perpendicular to the magnetic field. In general, the current density of a filled magnetic band is given by

$$\mathbf{J} = \frac{e^2}{\hbar}\int \frac{d^3k}{(2\pi)^3}\boldsymbol{\Omega}_n(\mathbf{k})\times\mathbf{E}, \qquad (13.32)$$

where $\boldsymbol{\Omega}_n(\mathbf{k})$ is the Berry curvature vector defined by

$$\boldsymbol{\Omega}_n(\mathbf{k}) = i\left\langle\frac{\partial u_n}{\partial \mathbf{k}}\bigg|\times\bigg|\frac{\partial u_n}{\partial \mathbf{k}}\right\rangle. \qquad (13.33)$$

Let $\mathbf{W}$ represent the integral of $\boldsymbol{\Omega}_n(\mathbf{k})/(2\pi)$ over the magnetic Brillouin zone. Then the following arguments show that it is a reciprocal lattice vector $\mathbf{G}$. First, the result of the last paragraph shows that $\mathbf{W}\cdot\mathbf{c} = 2\pi\sigma$. Second, $\mathbf{W}\cdot\mathbf{b}$ may be written as 2π times the Chern number in the basal plane spanned by $\mathbf{G}_c$ and $\mathbf{G}_a/q$ which are perpendicular to $\mathbf{b}$. Similarly, $\mathbf{W}\cdot\mathbf{a}$ is $2\pi/q$ times the Chern number in the basal plane of $\mathbf{G}_b$ and $\mathbf{G}_c$ where the extra factor q comes from the fact that the thickness of the prism in the direction of $\mathbf{a}$ is q times thinner than in an ordinary Brillouin zone. Finally, because one could have used $T(\mathbf{a})$, $T(q\mathbf{b})$, and $T(\mathbf{c})$ to define the magnetic band, the number $\mathbf{W}\cdot\mathbf{a}$ must actually be an integer times 2π. In summary, the Hall current is generally of the form

$$\mathbf{J} = \frac{e^2}{2\pi\hbar}\mathbf{G}\times\mathbf{E}. \qquad (13.34)$$

13.3.5 Evaluation of the Chern Number

The Hall conductivity can also be calculated from a different point of view [26, 70, 201]. In a frame that moves with velocity $\mathbf{v} = -\frac{E_x}{B}\mathbf{e}_y$, the electric field is transformed away. The electron current in such a frame is purely due to the lattice potential which now moves in the opposite direction. Based on the theory of adiabatic particle transport, the electron current in a filled magnetic band is an integer $-c$ divided by $T = b/v$ times the number of transverse modes, where T stands for the time it takes for the lattice to move one lattice constant. Note that here we assume that the unit cell of the lattice is rectangular for simplicity. Since there is one transverse mode per lattice constant a, we obtain an electric current density ecv/ab. When we transform back to the original frame, this current density becomes $ecv/ab - e\rho v$, where $\rho = 1/qab$ is the electron density per magnetic subband. This is also the Hall current density given by $\frac{\sigma E_x e^2}{h}$ where σ is the Chern number appearing in the expression for the quantized Hall conductivity. A comparison of these expressions shows that the topological quantum numbers σ and c are related by the following simple Diophantine equation.

$$p\sigma + qc = 1. \tag{13.35}$$

The Diophantine equation (13.35) does not determine the Chern number completely. In general, one has to resort to numerical calculations to find their values. In the anisotropic limit of the tight-binding model (13.14), one can use perturbation theory to calculate the Berry curvature and the Chern number. Consider for instance the case $\Delta_2 \ll \Delta_1$. Then in the zeroth-order approximation we have a k_y-independent band with energy $E = 2\Delta_1 \cos(k_x a)$ and $\Omega(\mathbf{k}) = 0$ everywhere. For a non-zero "potential" $\Delta_2 \cos(2\pi pm/q + k_y b)$, the degenerate states at $k_x = \frac{\pi r}{qa}$ and $-k_x$ are coupled, openning a small energy gap.

Let us assume that Δ_1 is negative, so that the integer r labels the gaps from the lower to higher energies. Near such a degenerate point, we express the wave function in the form of $\psi(\mathbf{R}) = e^{ik_x X + ik_y Y}(A + Be^{-i\frac{2\pi rm}{q}})$. This leads to the following 2×2 matrix eigenvalue problem

$$\begin{pmatrix} 2\Delta_1 \cos(k_x a) & V_t e^{ik_y bt} \\ V_t^* e^{-ik_y bt} & 2\Delta_1 \cos(k_x a - 2\pi r/q) \end{pmatrix} \begin{pmatrix} A \\ B \end{pmatrix} = E \begin{pmatrix} A \\ B \end{pmatrix}, \tag{13.36}$$

where t is an integer and the coupling strength V_t is a constant independent of $\mathbf{k}$.

Some explanations are in order.

(i) If $r = p$, the two states are coupled to first order in the "potential", and we have $t = 1$ and $V_t = \Delta_2$.
(ii) If $r \neq p$, the two states are coupled in higher order in Δ_2. The integer t represents the number of quasi-momentum transfers $\frac{2\pi p}{qa}$ from the "potential" in order to make up the difference $\frac{2\pi r}{qa}$ between the two states. Because of the discreteness of the lattice, the quasi-momentum is conserved only up to an integer multiple s of $2\pi/a$, and we have the the Diophantine equation

$$tp + sq = r. \tag{13.37}$$

(iii) The coupling is dominated by the lowest order perturbation that can satisfy the above requirement. We should therefore choose the solution of the preceding equation with the smallest possible value of $|t|$. This will determine the solution uniquely except for the case of q being even and $r = q/2$ where the gap is closed. In the following, we will show that t is the Chern number for the states filled below the r-th gap. Therefore, the Diophantine equation (13.37) is completely equivalent to (13.35).

We can use the analogy between the eigenvalue equation (13.36) and the time-independent Schrödinger equation for a spin $1/2$ particle in a magnetic field (see Chap. 3) to evaluate the Berry curvature and the Chern number.

The components of the effective "magnetic field" are $X = V_t \cos(k_y bt)$, $Y = -V_t \sin(k_y bt)$, and $Z = -2\Delta_1 \sin(\pi r/q)(k_x a - \pi r/q)$. When k_y changes by $2\pi/b$, the "field" rotates about the Z-axis $-t$ rounds. As k_x passes $\pi r/qa$, the field switches from south to north ($\Delta_1 < 0$). The field configuration is a long cylinder about the Z axis because of the smallness of V_t/Δ_1. Since the cylinder is wrapped t times when (k_x, k_y) varies within a magnetic Brillouin zone, we find that the Chern number of the lower state is t and that of the upper state is $-t$. The Berry curvature in the lower state is simply half the solid angle spanned by a unit area of the parameter space (k_x, k_y). It is independent of k_y and concentrated near $k_x = \pi r/q$:

$$\Omega(\mathbf{k}) = \frac{abt \sin(\pi r/q)\Delta_1/V_t}{[1 + (2\sin(\pi r/q)\delta k_x a \Delta_1/V_t)^2]^{3/2}}, \tag{13.38}$$

where δk_x is the deviation of k_x from the degenerate point. The width of the distribution in k_x is proportional to V_t/Δ_1. The Berry curvature in the upper state is exactly the opposite. Since Berry curvatures are created in equal and opposite amount on the two sides of each gap, the above result (13.38) also stands for the total Berry curvature of all the states filled up to the r-th gap. Obviously, the same is true for the Chern number.

When the perturbation is enhanced, the distribution of the Berry curvature will be wider in k_x and non-uniform in k_y. As we will see in the next section, it will eventually become symmetrical in the two directions in the isotropic case. It is a property of the tight-binding model that no gap becomes closed when the anisotropy is changed. Therefore the Chern numbers calculated in the extremely anisotropic case remain valid in general.

13.3.6 Semiclassical Dynamics and Quantization

We wish to consider a wave packet constructed out of states in a magnetic Bloch band associated with a magnetic field $\mathbf{B}_0$ and study its dynamics in a weak electric field $\mathbf{E}$ and an extra magnetic field $\delta \mathbf{B}$. In order for the magnetic band to be well defined, the original magnetic field $\mathbf{B}_0$ should be rational with respect to the lattice (magnetic fluxes through the cell faces are rational multiple of the flux quantum). In addition, $\delta \mathbf{B}$ should be small enough to prevent transition between the magnetic bands. The semiclassical dynamics provides the basis of a Boltzmann transport theory in such a band. Moreover, with semiclassical quantization of the classical orbits, one can study how the magnetic band splits, and considering the behavior of the underlying semiclassical motion one can understand the distribution of the first Chern numbers among different subbands.

In what follows, we will consider the case of two-dimensional electrons with the magnetic field in the perpendicular direction, for simplicity. Using the method of Sect. 12.3.3 we can obtain the following Lagrangian for the position $\mathbf{r}$ and wave vector $\mathbf{k}$ of the wave packet [60, 61],

13.3 Magnetic Bands in Periodic Potentials

$$L(\mathbf{r}, \mathbf{k}, \dot{\mathbf{r}}, \dot{\mathbf{k}}) = \hbar \mathbf{k} \cdot \dot{\mathbf{r}} - e\dot{\mathbf{r}} \cdot \delta\mathbf{A}(\mathbf{r}, t) + \hbar \dot{\mathbf{k}} \cdot \mathcal{A}_n(\mathbf{k}) + e\phi(\mathbf{r}) - E_n(\mathbf{k}), \quad (13.39)$$

where ϕ and $\delta\mathbf{A}$ are the scalar and vector potentials associated with the electric field and the extra magnetic field. The other quantities in the Lagrangian describe various properties of the magnetic band;

$$\mathcal{A}_n(\mathbf{k}) = i\langle u_n(\mathbf{k})| \frac{\partial}{\partial \mathbf{k}} |u_n(\mathbf{k})\rangle \quad (13.40)$$

is the Berry connection of the magnetic Bloch state, E_n is the sum of the band energy $\epsilon_n(\mathbf{k})$ and the magnetization energy $\frac{e}{2m}\delta B L_n(\mathbf{k})$ due to the extra magnetic field, and

$$L_n(\mathbf{k}) := i\frac{m}{\hbar}\left[\left\langle \frac{\partial u_n}{\partial k_x} \middle| (H(\mathbf{k}) - \epsilon_n) \middle| \frac{\partial u_n}{\partial k_y} \right\rangle - \left\langle \frac{\partial u_n}{\partial k_y} \middle| (H(\mathbf{k}) - \epsilon_n) \middle| \frac{\partial u_n}{\partial k_x} \right\rangle\right]. \quad (13.41)$$

We will not reproduce the derivations here; the interested reader can either refer to the original paper [61] or consult the preceding chapter where a derivation of a similar Lagrangian for the case of an ordinary Bloch band was given.

The equations of motion are given by the Euler–Lagrange equations

$$\hbar \dot{\mathbf{k}} = -e\mathbf{E} - e\dot{\mathbf{r}} \times \mathbf{e}_z \delta B \quad (13.42)$$

$$\hbar \dot{\mathbf{r}} = \frac{\partial E_n(\mathbf{k})}{\partial \mathbf{k}} - \hbar \dot{\mathbf{k}} \times \mathbf{e}_z \Omega_n(\mathbf{k}), \quad (13.43)$$

where we have used the relation $\mathbf{e}_z \Omega_n(\mathbf{k}) = \frac{\partial}{\partial \mathbf{k}} \times \mathcal{A}_n(\mathbf{k})$. The equation for $\dot{\mathbf{k}}$ is the same as for an ordinary Bloch band, except that only $\delta\mathbf{B}$ appears in the Lorentz force. This is because $\mathbf{B}_0$ has already been taken into account in the band structure. In the absence of $\mathbf{E}$ and $\delta\mathbf{B}$, the wave vector $\mathbf{k}$ is a good quantum number and thus should not depend on time. The velocity formula for an ordinary Bloch band is usually given as $\hbar \dot{\mathbf{r}} = \frac{\partial \epsilon_n(\mathbf{k})}{\partial \mathbf{k}}$ in standard textbooks [20]. In the preceding chapter (see Sect. 12.3.1), we showed that this formula needs to be modified by a couple of additional correction terms which are first order in the fields. These are a magnetization energy in addition to ϵ and a Berry curvature term. There is no further modification of these terms in the case of a magnetic Bloch band. Note however that only the extra magnetic field appears in the magnetization energy. Unlike for the case of an ordinary Bloch band, here it is a rule rather than an exception that these correction terms are non-zero. This is a consequence of the fact that $\mathbf{B}_0$ breaks the time-reversal symmetry of the magnetic band structure.

We next examine the equations of motion in some important special situations. In the absence of δB, $\mathbf{k}$ drifts in the direction of the electric field giving rise to Bloch oscillations in the magnetic band. The velocity formula is now simplified as

$$\hbar \dot{\mathbf{r}} = \frac{\partial \epsilon_n(\mathbf{k})}{\partial \mathbf{k}} + e\mathbf{E} \times \mathbf{e}_z \Omega_n(\mathbf{k}). \quad (13.44)$$

In the presence of scattering, the Bloch oscillations are interrupted. If the relaxation time τ is shorter than the Bloch period, one has ohmic conductivity. In the relaxation time approximation, one obtains a current in the form

$$\mathbf{J} = -\mathbf{E} \times \mathbf{e}_z \frac{e^2}{h} \int \frac{d^3k}{2\pi^3} f(\epsilon_n) \Omega_n(\mathbf{k}) - e^2 \tau \int \frac{d^3k}{2\pi^3} \frac{\partial \epsilon_n(\mathbf{k})}{\partial \mathbf{k}} \left(\mathbf{E} \cdot \frac{\partial \epsilon_n(\mathbf{k})}{\partial \mathbf{k}} \right) \frac{\partial f}{\partial \epsilon}, \tag{13.45}$$

where f is the Fermi–Dirac distribution function. The second term gives a symmetric conductivity tensor which may be diagonalized by a suitable choice of the orientation of the reference frame. The first term gives an antisymmetric conductivity tensor corresponding to the Hall effect. In the two-dimensional problem, with $\mathbf{E} = E_x \mathbf{e}_x$ and the magnetic field in the direction normal to the plane, we obtain for the Hall conductivity

$$\sigma_{yx} = \frac{e^2}{h} \int dk^2 f(\epsilon_n(\mathbf{k})) \Omega_n(\mathbf{k}). \tag{13.46}$$

For a filled band this is exactly the result derived earlier (13.32).

In the presence of δB (but with a non-vanishing $\mathbf{E}$), the equations of motion may be combined to yield

$$\hbar \dot{\mathbf{k}} = -\frac{e/\hbar}{1 - \frac{e}{\hbar}\Omega_n(\mathbf{k})\delta B} \frac{\partial E_n(\mathbf{k})}{\partial \mathbf{k}} \times \mathbf{e}_z \delta B. \tag{13.47}$$

Therefore, the wave vector moves along a constant energy contour of $E_n(\mathbf{k})$. Following Pippard [212, 213], we call such a trajectory a *hyperorbit*. We have seen that the magnetic bands in the skeleton of the Hofstadter spectrum are due to quantized cyclotron orbits in the parent Bloch band. The hyperorbits in such a magnetic band may thus be thought of as the trajectories of the center of the quantized cyclotron orbit.

The quantization of the hyperorbit follows the same procedure as for the cyclotron orbit in an ordinary Bloch band. The $\mathbf{k}$ space area enclosed by the j-th orbit C_j is given by

$$A_j = [(2j+1)\pi - \Gamma_n(C_j)] \frac{e\delta B}{\hbar}, \tag{13.48}$$

where $\Gamma_n(C_j)$ is the Berry phase of the magnetic Bloch state along the orbit C_j. Therefore, the area enclosed between two adjacent orbits is expressed as

$$\delta A = \left[2\pi - \int d^2k \, \Omega_n(\mathbf{k}) \right] \frac{e\delta B}{\hbar}, \tag{13.49}$$

where the integral is over the area between the two orbits. It coincides with the difference between the Berry phases on the two orbits according to Stokes' theorem.

The number N of the orbits that can be fitted in the magnetic Brillouin zone is therefore given by the relation

13.3 Magnetic Bands in Periodic Potentials

$$2\pi(N-\sigma)\frac{e\delta B}{\hbar} = \frac{(2\pi)^2}{qab}, \qquad (13.50)$$

where σ is the Chern number of the magnetic band, ab is the area of the unit cell, and q is the denominator of the magnetic flux of $\mathbf{B}_0$ through this cell divided by the flux quantum. Due to the q-fold degeneracy of the magnetic Brillouin zone, N should be divided by q to give the number of distinct energy levels as

$$D = [\sigma + 1/(q\delta\phi)]/q, \qquad (13.51)$$

where $\delta\phi := e\delta Bab/h$. It is understood that this number should be rounded to the nearest integer, and that a negative value of D which occurs for negative $\delta\phi$ corresponds to hyperorbits in the reversed direction.

Finally, we wish to understand how a hyperorbit responds to a perturbation of an infinitesimal electric field. Following the method given in the last chapter, we find that a hyperorbit which is closed in $\mathbf{k}$ space will acquire a drift velocity given by $\frac{\mathbf{E}\times\mathbf{e}_z}{\delta B}$. Therefore, if a subband of closed hyperorbits has an electron density of ρ, then it will contribute to the Hall conductivity as $e\rho/\delta B$. In particular, in two dimensions, if the density is $\rho = 1/\tilde{q}ab$, then one has a Hall conductivity given by

$$\tilde{\sigma} = 1/\tilde{q}\delta\phi, \qquad (13.52)$$

in units of e^2/h.

13.3.7 Structure of Magnetic Bands and Hyperorbit Levels

In order to employ semiclassical dynamics, we need the band structures of the energy $\epsilon_n(\mathbf{k})$, the Berry curvature $\Omega_n(\mathbf{k})$, and the angular momentum $L_n(\mathbf{k})$. We take the case of the tight-binding model in (13.14) with $\frac{eBab}{\hbar} = p/q = 1/3$ and assume square symmetry with $b = a$ and $\Delta_2 = \Delta_1 = \Delta$. The detailed calculations of the three magnetic Bloch bands were carried out in [61]. Here we will give the resulting plots of the corresponding band structures. For convenience we measure energy and length in units of Δ and a, respectively.

Figure 13.5 shows a plot of the three energy bands of the Harper equation for the case $\phi = 1/3$ in the basic zone. The 3×3 matrix eigenvalue equation (13.15) leads to the characteristic equation

$$-\epsilon^3 + 6\epsilon = 2(\cos 3k_x + \cos 3k_y) \qquad (13.53)$$

whose solution describes how the three magnetic Bloch bands split from the parent band. Simple solutions are found for special points of the basic zone. For example at $(k_x, k_y) = (0, 0)$, the energies of the three bands are $-(\sqrt{3}+1)$, $(\sqrt{3}-1)$, and 2; at $(\frac{\pi}{3}, \frac{\pi}{3})$, these energies take the values -2, $-\sqrt{3}+1$, and $\sqrt{3}+1$; and at $(0, \frac{\pi}{3})$, they become $-\sqrt{6}$, 0, and $\sqrt{6}$, respectively.

Figure 13.6 depicts the $\mathbf{k}$ dispersion of the Berry curvatures for the three bands. Performing the integrals of these Berry curvatures over the magnetic

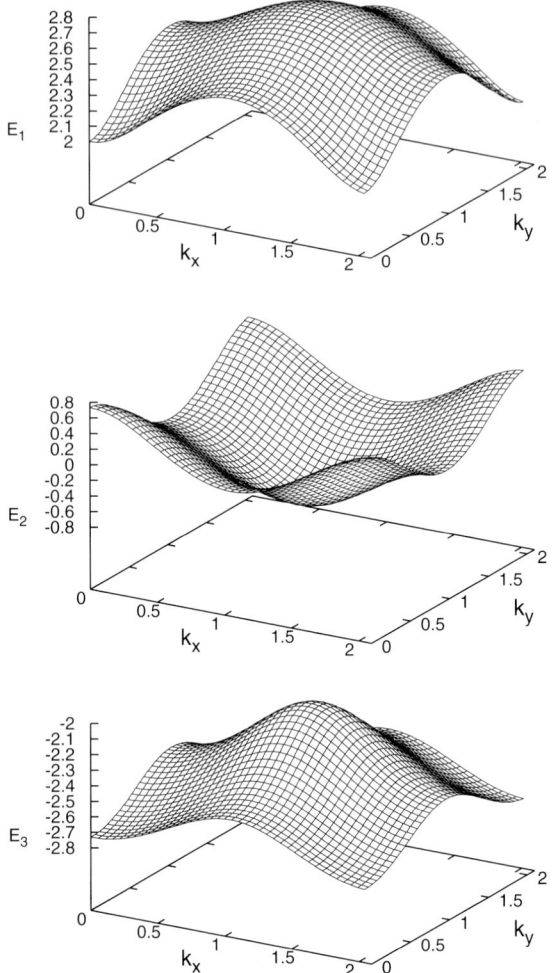

Fig. 13.5. Band structure of the energy $\epsilon_n(\mathbf{k})$ in the basic zone at $\phi = 1/3$.

Brillouin zone and dividing the result by 2π, we obtain the Chern numbers $(1, -2, 1)$ for the three magnetic bands. These Chern numbers sum to zero which is the value of the Chern number for the parent band. This is an example of a general sum rule: the sum of the Chern numbers associated with all the states in a spectral range is unchanged as long as the latter is isolated by finite energy gaps. The Berry curvatures themselves satisfy a stronger sum rule, namely $\sum_n \Omega_n(\mathbf{k}) = 0$. Note that this is a local result valid for each point in the magnetic Brillouin zone. It applies for arbitrary quantum systems with a finite number of bands [31]. Finally, one can see that the Berry curvatures are concentrated at edges of the inner band. The bump

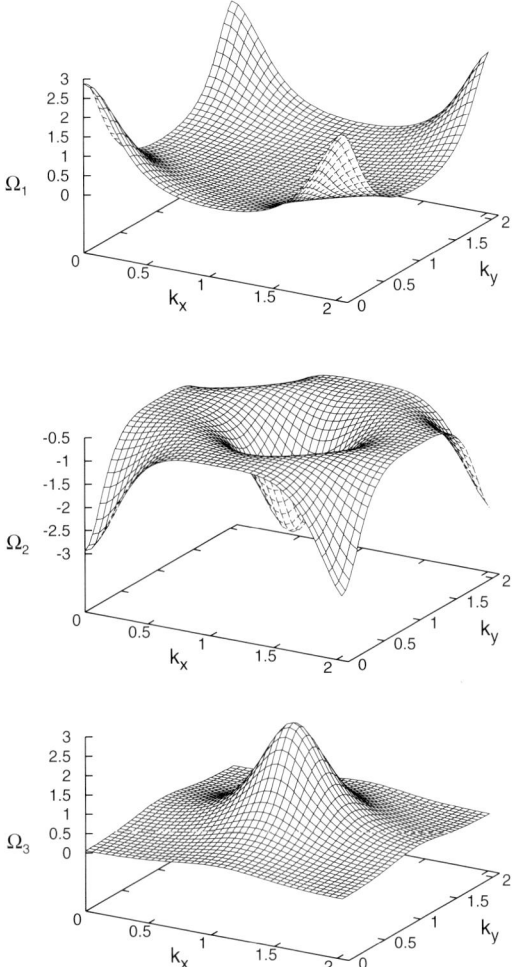

Fig. 13.6. Band structure of the Berry curvature $\Omega_n(\mathbf{k})$ in the basic zone at $\phi = 1/3$.

in Ω_3 at the top edge of its energy band at $\mathbf{k} = (\frac{\pi}{3}, \frac{\pi}{3})$ is opposed by a dip in Ω_2 at the bottom edge of the middle energy band. Ω_2 and Ω_1 display similar behavior at $\mathbf{k} = (0,0)$. The Berry curvature seems to also satisfy a sum rule which is local in energy: when the parent band is split into three bands, the Berry curvature is created in equal and opposite amounts on the two sides of each gap.

In Fig. 13.7, we present the band structures for the angular momentum $L_n(\mathbf{k})$ which is aligned along the z-axis for the two-dimensional problem at hand. Similar to the case of the Berry curvature, L_n tends to concentrate

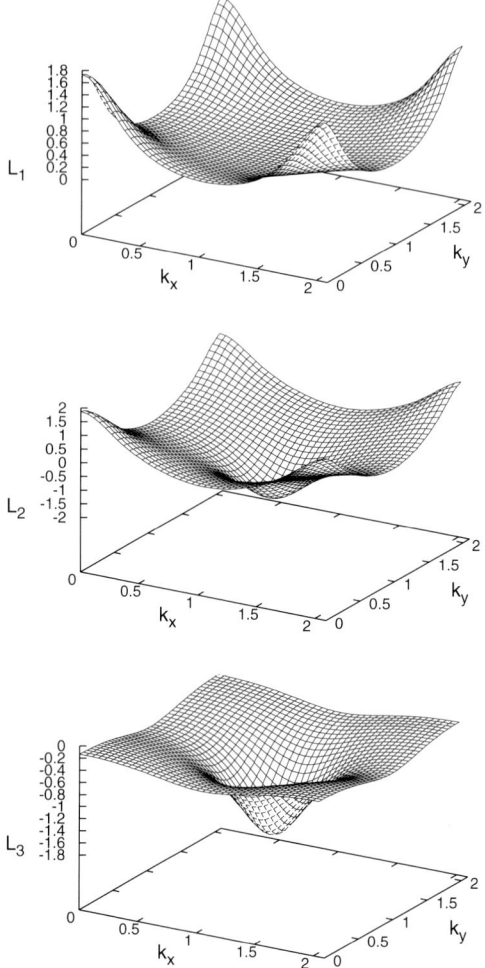

Fig. 13.7. Band structure of the angular momentum $L_n(\mathbf{k})$ in the basic zone at $\phi = 1/3$.

near the band edges. In addition the integrated angular momenta from the three bands sum to zero. However, this does not happen locally for $L_n(\mathbf{k})$ at each point in the magnetic Brillouin zone.

We are now in a position to determine the levels of quantized hyperorbits in these three magnetic bands. We consider the case of changing the flux per lattice cell from $\phi = 1/3$ to $\phi' = 22/67 = \frac{1}{3+\frac{1}{22}}$. This corresponds to a net change of $\delta\phi = -1/201$. Using the formula (13.51) and the Chern numbers for the bands, we find that the numbers of semiclassical levels in the three bands are (22, 23, 22) which are exactly what we found numerically.

Table 13.1. Magnetic subbands (for $\phi = 22/67$) and the cyclotron energies calculated from the quantization formula. Only the top ten subbands for the middle parent band are shown. The last column is the subband closest to the parent band edge $\epsilon_2(\mathbf{0}) = 0.7321$.

E_Hofst	0.0618	0.1067	0.1566	0.2124	0.2747	0.3443	0.4221	0.5098	0.6098	0.7266
	0.0678	0.1085	0.1570	0.2125						
E_cyclo	0.0632	0.1063	0.1558	0.2115	0.2738	0.3435	0.4212	0.5086	0.6082	0.7240

The energies of these levels are determined by the quantization formula (13.48). In Table 13.1, we list the results for these levels (E_cyclo) in the upper half of the middle band. The upper row (E_Hofst) includes the exact numerical results for the subbands; only the band edges for the four subbands close to the band center are distinct within four significant figures. The almost perfect agreement between the lower row and the energies of the subband centers shows that the subbands indeed originate from the levels of quantized hyperorbits. The width of the subbands derives from tunneling between hyperorbits in different magnetic Brillouin zones.

13.3.8 Hierarchical Structure of the Butterfly

Azbel [27] analyzed the splitting pattern by writing the flux $\phi \in (0,1)$ in a continued-fraction expansion

$$\phi = \cfrac{1}{f_1 + \cfrac{1}{f_2 + \cfrac{1}{f_3 + \cdots}}}, \qquad (13.54)$$

where $f_1, f_2, \cdots$ are positive integers [149]. He then argued that the spectrum should consist of f_1 subbands, each of which is split into f_2 finer subbands, each of which is split into f_3 subbands of even finer scales, and so on. For a rational ϕ this process stops at some stage, yielding a finite number of subbands. For an irrational ϕ the process never stops and eventually it yields something like a Cantor set. Before we proceed to ramify this picture, we give a brief review of the properties of the continued fraction.

The expansion coefficients in the continued fraction can be easily generated using a typical calculator: compute $1/\phi$ and call its integer part f_1, then take the inverse of the fractional part, that is find $(1/\phi - f_1)^{-1}$, and call its integer part f_2, now take the inverse of the fractional part in the preceding calculation and call its integer part f_3, and repeat this procedure. For example, the irrational number $\phi = 1/(2 + \sqrt{2})$ has the expansion with $f_1 = 3$, $f_2 = 2$, $f_3 = 2$, and so on. Truncating the expansion at the r-th stage (by formally setting $f_{r+1} = \infty$) gives the r-th order rational approximant p_r/q_r to ϕ. If ϕ itself is a rational, there is only a finite sequence of approximants. But if ϕ is irrational, then the sequence continues indefinitely. The first two approximants are given by $p_1/q_1 = 1/f_1$, $p_2/q_2 = f_1/(f_1 f_2 + 1)$. In general, one has $q_{r+1} = f_{r+1}q_r + q_{r-1}$ and $p_{r+1} = f_{r+1}p_r + p_{r-1}$. For the above example, the sequence of approximants is $1/3, 2/7, 5/17, \cdots$.

One might ask why we take the approximants using a continued-fraction expansion. To answer this question, consider a poor man's sequence of rational approximants for the above irrational number, namely

$$0/1,\ 1/2,\ 1/3,\ 1/4,\ 1/5,\ 1/6,\ 2/7,\ 3/8,\ 2/9,\ 3/10,\ 3/11,\ 5/12,\ 4/13, \cdots$$

whose denominators run through all the natural numbers. Each numerator in this sequence is chosen to give the best approximation for the given denominator and yet has no common divisors with the denominator. In Fig. 13.8 we show the values of the poor man's approximants as a function of the denominator q. The convergence to the number ϕ (dashed line) is rather slow ($\sim 1/q$), and the errors oscillate not only in sign but also in magnitude.

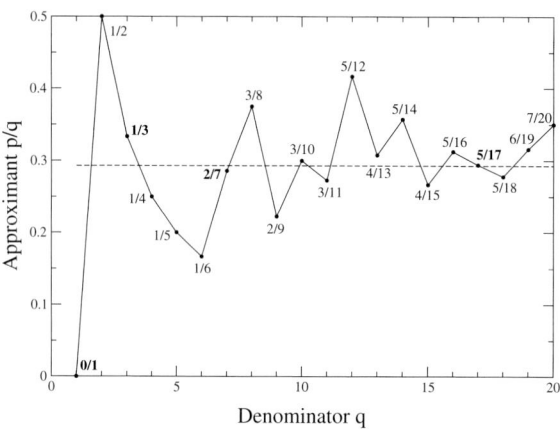

Fig. 13.8. The poor man's sequence of rational approximants to $\phi = 1/(2+\sqrt{2})$. The bold numbers are from the continued fraction expansion, which converges much faster.

On the other hand, the sequence from the continued fraction which is a subset of the poor man's sequence converges much faster ($\sim 1/q^2$), and the errors oscillate in sign but decrease monotonically in magnitude with q. This can be seen from the simple and general formula for the difference between two successive approximants

$$\frac{p_{r+1}}{q_{r+1}} - \frac{p_r}{q_r} = \frac{(-1)^r}{q_r q_{r+1}}. \tag{13.55}$$

Moreover, it is known that each approximant p_r/q_r generated from the continued fraction gives the best approximation to ϕ among all rational approximants with $q \leq q_r$ [149]. This is the reason why the approximants obtained from the continued fraction are preferred.

13.3.9 Quantization of Hyperorbits and Rule of Band Splitting

We are now in a position to explain the hierarchical structure of the Hofstadter spectrum. We will consider the case $\phi = 1/(2 + \sqrt{2})$ as an example. Starting with the zeroth approximant $p_0/q_0 = 0/1$, we have the $q_0 = 1$ band which is the parent band with a zero Chern number. We then treat the first approximant $p_1/q_1 = 1/3$ as a perturbation $\delta\phi_0$. According to (13.51), this leads to $D_0 = (\sigma_0 + \frac{1}{q_0 \delta\phi_0}) = 3$ cyclotron levels. Two of these levels on the sides correspond to closed orbits, so in view of (13.52) their Chern numbers are $\sigma_1 = 1/q_1 \delta\phi_0 = 1$. The middle level corresponds to an open orbit. We can employ the sum rule for the Chern numbers to obtain the Chern number associated with this level. This yields $\sigma_1' = -2$. The above three levels broaden into three magnetic bands for the case of $\phi = 1/3$.

Next, we treat the difference between the first and second approximants as a perturbation $\delta\phi_1 = 2/7 - 1/3 = -1/21$. In future discussions, it will be important to distinguish two kinds of magnetic bands: the closed and open magnetic bands correspond to the closed and open orbits respectively. For each of the two closed bands on the sides, we have $D_1 = (\sigma_1 + 1/q_1 \delta\phi_1)/q_1 = -2$ cyclotron levels, where the "$-$" sign refers to hyperorbits being clockwise. For the open band in the middle, we replace σ_2 by σ_2' in the above formula. This yields a different number of levels namely $D_1' = -3$. From Fig. 13.9, we clearly see that these levels correspond to the seven subbands for the case of $\phi = 2/7$.

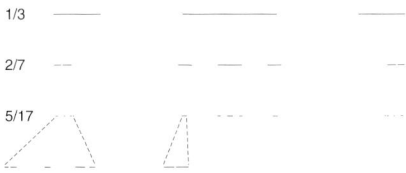

Fig. 13.9. Pattern of band splitting as ϕ goes through the sequence of rational approximants for $1/(2 + \sqrt{2})$.

The Chern numbers of the closed hyperorbits in the three magnetic bands are given by $\sigma_2 = 1/q_2 \delta\phi_1 = -3$. Assuming that there is one open orbit in each of the three magnetic bands, the Chern numbers of these orbits can be found using the sum rule for each magnetic band. The latter turn out to be the same, $\sigma_2' = 4$. Now, we note that the square symmetry allows no more than one open orbit, and that the sum rule requires the existence of at least one open orbit in each of the three magnetic bands.

We can continue the above process indefinitely and find higher order approximants. In general, in the r-th stage we have q_r magnetic subbands. These are generated from the hyperorbits in the q_{r-1} parent subbands. The closed subbands have Chern numbers equal to

$$\sigma_r = 1/q_r \delta\phi_{r-1} = (-1)^{r-1} q_{r-1}, \tag{13.56}$$

where $\delta\phi_{r-1} = p_r/q_r - p_{r-1}/q_{r-1} = (-1)^{r-1}/q_r q_{r-1}$. Also, the Chern numbers of the open subbands satisfy the sum rule $\sigma'_r q_{r-1} + \sigma_r(q_r - q_{r-1}) = 0$. This means that the sum of the first Chern numbers for all the orbitals of the original tight-binding band is zero. This in turn implies

$$\sigma'_r = (-1)^{r-1} q_{r-1} + (-1)^r q_r. \tag{13.57}$$

Here we implicitly used the facts that there is one open daughter subband per parent subband, and the Chern numbers for the open daughter subbands corresponding to all the parent subbands of the same generation are identical.[3]

In the next stage, the perturbation $\delta\phi_r = (-1)^r/q_r q_{r+1}$ gives rise to q_{r+1} levels of hyperorbits in the q_r subbands. A closed subband splits into

$$D_r = \frac{\sigma_r + (-1)^r q_{r+1}}{q_r} = (-1)^r f_{r+1} \tag{13.58}$$

levels, and an open subband splits into

$$D'_r = \frac{\sigma'_r + (-1)^r q_{r+1}}{q_r} = (-1)^r (f_{r+1} + 1) \tag{13.59}$$

levels.

The above analysis provides a renormalization group procedure for describing the hierarchical structure of the Hofstadter spectrum: from the behavior of the hyperorbits of a quantized level we can predict the Chern number of the corresponding magnetic subband and from the Chern number we can predict how the subband is going to split in the next stage. Several remarks are in order:

(i) The results can be immediately extended to systems beyond the nearest neighbor tight-binding model and to lattices with hexagonal symmetry. The square or hexagonal symmetry allows only one open hyperorbit in each magnetic band and enables us to obtain the Chern numbers from the sum rule. The only difference is that the position of the level of the open orbit tends not to lie at the center of the subband. This results in skewed splitting patterns of the spectrum.

---

[3] These statements can be established by induction. They are obviously true for $r \leq 2$. Assuming that they are true for a general r, we can use the following argument to show that they hold for $r + 1$. First, we note that the square symmetry allows no more than one open orbit. Furthermore, the Chern number $\sigma_{r+1} = (-1)^r q_r$ for a closed subband in the $r + 1$ stage cannot divide both σ_r of (13.56) and σ'_r of (13.57). Therefore, there must be at least one open daughter subband in each parent subband. Finally, the Chern number σ'_{r+1} for such an open subband is either $\sigma_r - (|D_r| - 1)\sigma_{r+1}$ or $\sigma'_r - (|D'_r| - 1)\sigma_{r+1}$ depending on whether the parent subband is closed or open. Indeed, the result is the same. It is given by the expression (13.57) with r replaced by $r + 1$.

(ii) For lattices without these symmetries, it is possible to have more than one open orbit per magnetic band. In this case one can still determine the Chern numbers from the semiclassical equations of motion of the open orbits. But the result will be sensitive to the details.

(iii) We have not discussed in any detail the width of the magnetic bands. They are narrow or wide depending on whether the corresponding hyperorbits are closed or open, respectively.

(iv) When two bands touch at a single **k** point the band structure develops a conical intersection. The hyperorbits in the immediate neighborhood of the intersection point have a Berry phase angle of π because of the degeneracy (see Chap. 9). The quantization condition (13.48) then implies that there is an energy level at the degenerate point. This result is of a general topological nature and is evident from the Hofstadter spectrum where one can see that there is indeed a level at $E = 0$ near each ϕ with an even denominator.

(v) One may use our scheme to find out how a Landau level splits into subbands. For concreteness, consider a closed subband at $\phi = 1/f_1 = p_1/q_1$ which has Chern number $\sigma_1 = 1$ and resembles a Landau level. In the next stage, it will split into $|D_1| = f_2 = p_2$ daughter subbands. One of these is open and the others are closed. Therefore, in the following stage, we have $D_2' + (f_2 - 1)D_2 = f_2 f_3 + 1 = p_3$. In general, the number of subbands splitting from a Landau level is always given by the numerator of ϕ in each stage of the rational approximation.

13.4 Quantization of Hall Conductance in Disordered Systems

13.4.1 Spectrum and Wave Functions

In the presence of impurities, the Landau levels are broadened into bands called *Landau bands*. This is reflected in the *density of states*, $n(E)$, which is defined as the number of states per unit energy interval per unit area. The density of states $n(E)$ has been calculated for the *white noise potential*,

$$\langle V(\boldsymbol{r}) \rangle = 0, \tag{13.60}$$
$$\langle V(\boldsymbol{r})V(\boldsymbol{r}') \rangle = W^2 \delta(\boldsymbol{r} - \boldsymbol{r}'), \tag{13.61}$$

with a self-consistent Born approximation [13]. For each band, $n(E)$ is a semi-elliptic curve with a width given by

$$\Gamma = \frac{2W}{\ell} \qquad \ell = \sqrt{\frac{\hbar}{eB}} = \text{magnetic length}.$$

Later numerical calculations showed that the band edges are smeared into Gaussian tails. An exact result exists for the lowest band in the limit of large

330   13. The Quantum Hall Effect

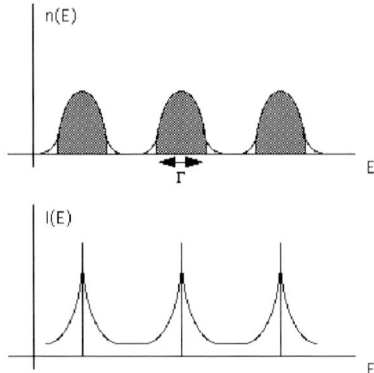

Fig. 13.10. Landau bands shown in the plots of the density of states and localization length as functions of the energy.

magnetic field [266], showing such features. The case of smoothly varying random potentials also leads to similar results for $n(E)$ [11, 252].

The precise nature of the wave functions was found only after the discovery of the quantum Hall effect (QHE). According to the theory of Anderson localization [2], in two dimensions all states are localized by any finite amount of disorder and therefore do not contribute to conductivity. One can also show that the same holds in the presence of a weak magnetic field. If this were also true for strong magnetic fields, then the Hall conductivity would vanish. This is certainly not the case in the QHE experiments, which show finite (and quantized) Hall conductivities. The conclusion is that there must be some extended states in the Landau bands.

Numerical calculations for the short-ranged disordered potentials (white noise limit) show that all the states except at the center E_c of a Landau band are localized (see Fig. 13.10). The *localization length* (defined as the width of the wave function) for these states is given by

$$\ell(E) \sim \mid E - E_c \mid^{-\nu},$$

where $\nu \simeq 2$ for the lowest Landau band [10]. For smooth random potentials one can make an analogy with two-dimensional classical continuum percolation theory [252]. In the limit where the *magnetic length*, $\ell = \sqrt{\frac{\hbar}{eB}}$, is much smaller than the length scale of the variation of the potential, an eigenfunction of energy E is concentrated along the contour of constant potential energy

$$V(\boldsymbol{r}) + \hbar\omega_c\left(\frac{1}{2} + n\right) = E,$$

where n labels the Landau bands.

Consider a non-flat region on the surface of the Earth. Let $V(\boldsymbol{r})$ denote the height at the point $\boldsymbol{r}$ and imagine that this region is covered with water up to the level $E - \hbar\omega_c\left(\frac{1}{2} + n\right)$. This leads to the formation of a water–land boundary which is the contour along which an eigenfunction of energy E is concentrated. At low water levels the water is in the form of lakes. At high water levels, the water becomes a sea while the land becomes islands. In both cases, the water–land boundaries are finite contours meaning that the eigenfunctions are localized. There is a unique height at which both the land and water become infinite; the wave function becomes extended along the infinite water–land boundary.

This analysis gives a qualitatively similar result as in the case of a short-ranged potential. The localization length diverges as

$$\ell(E) \sim |E - E_c|^{-\nu}$$

with $\nu = \frac{4}{3}$ [252]. When the tunneling correction is taken into account one finds $\nu = \frac{7}{3}$ [171].

13.4.2 Perturbation and Scattering Theory

Ando *et al.* [12] were the first to provide a systematic treatment of impurities in an otherwise ideal two-dimensional electron gas subject to a strong magnetic field. Using a self-consistent Born approximation, they calculated the Hall conductivity σ_H for the case of both short- and long-ranged scatterers. This led to the important result that if the Fermi level lies in an energy gap, the deviation of the Hall conductivity from the classical formula,

$$\Delta\sigma_H = \sigma_H + \frac{ne}{B},$$

vanishes within the domain of the validity of the approximation. Because a broadened Landau band contains the same electron density as in an unperturbed Landau level, the Hall conductivity is quantized as in the ideal case.

This seems to be the only theoretical prediction of the QHE in a non-ideal system before the experiment of von Klitzing *et al.* [260]. It has however not been considered as an explanation for the quantum Hall effect, because it does not provide a mechanism to pin the Fermi level in an energy gap for a finite interval of the gate voltage or the magnetic field in order to account for the Hall plateaus. Moreover, it is too crude to explain the high accuracy of the experiment.

The finding of von Klitzing and his collaborators led to the demand for establishing the quantization of the Hall resistance when the Fermi level lies in the energy range of localized states. Such a quantization would also explain the finite width of the Hall plateaus, because the localized states could "hold" the Fermi level.

The first important breakthrough was the work of Aoki and Ando [16]. Based on the Kubo formula written in terms of the cyclotron center coordinates, which was developed by Ando *et al.* [12] in the "pre-history time" of the QHE, they showed that the Hall conductivity does not change as the Fermi level moves within the mobility gap. Then they considered the limit of strong magnetic field to arrive at the quantization of σ_H. An important result of this investigation is that the correction due to the finiteness of the magnetic field vanishes to the order of $(\Gamma/\hbar\omega_c)^2$ where Γ is the broadening of the Landau levels due to impurities. Later, Usov and Ulinich [255] showed that the third-order correction vanishes as well.

Prange [214], on the other hand, considered the problem with a single δ-function impurity potential. By an exact calculation he showed that a localized state exists at the impurity, but the non-localized states carry an extra Hall current which precisely compensates for the loss of current due to the localized state. Later, Prange and Joynt [215] offered a more extensive and general treatment by considering various kinds of impurity potentials. Joynt's numerical results, reported in [118], provide a very clear demonstration of the compensating effect; the electrons in the extended states just move faster when they pass by their unfortunate friends trapped in the localized states.

Thouless [248] also studied the influence of impurities using a general Green's function analysis. He noticed that so long as the energy lies in an energy gap or mobility gap (no extended states), the Green's function decays exponentially at large space separations. Using this property of the Green's function in the Kubo formula for the Hall conductance, he proved that at zero temperature and with a Fermi energy not in the spectrum of the extended states, the correction to the Hall conductance vanishes to all orders of the strength of the impurity potential. This is an extension of the results of Aoki and Ando, and Prange that we mentioned above.

The work by Streda [234] has also shed much light on the problem. He proved that as the Fermi level lies in an energy gap, the Hall conductivity is given by

$$\sigma_H = e \frac{\partial \rho(E_F)}{\partial B} \tag{13.62}$$

where $\rho(E_F)$ is the density of electron states with energy lying below E_F. If the gap is between two Landau levels, which may be broadened by the impurities, one derives the quantization of σ_H from the conservation of the number of states below an energy gap. In contrast to the approximate results of Ando et al., this is an exact result. The Streda formula (13.62) is also valid for other kinds of energy gaps, but the quantization of σ_H is not generally obvious. Nevertheless, in the case of a periodic potential, Streda [235] was able to show that his formula gives the same quantized Hall conductances as the topological invariant expression of the theory of Thouless et al. [247].

13.4.3 Laughlin's Gauge Argument

Based on the fact that the QHE appears to be independent of the details of the experimental conditions, Laughlin [150] proposed that this phenomenon might be explained by a fundamental principle. He considered a two-dimensional electron gas on a cylinder of finite height with a magnetic field perpendicular to the surface and a current I in the azimuthal direction. The current I and the potential drop ΔV across the two edges are connected in the following manner. Suppose that a magnetic flux through the center of the cylinder is increased. Then the total energy of the electron system changes correspondingly. According to Faraday's law, the current I is given by

$$I = \frac{\partial U}{\partial \phi}.$$

If the flux change is equal to a flux quantum h/e, then every localized state returns to itself while the extended states can go from one to another. If there is no extended state at the Fermi energy in the interior of the cylinder, the occupation of the electron states in the interior must be the same before and after the flux change (at zero temperature). The net effect can only be a transfer of an integral number (say, p) of electrons from one edge to the other.

This gives rise to a change in energy:

$$\Delta U = pe\Delta V.$$

Replacing the derivative by $\Delta U/\Delta \phi$, we obtain a quantized Hall conductance of

$$\sigma_{\mathrm{H}} = \frac{I}{\Delta V} = p\frac{e^2}{h}.$$

The integer p would be zero if there were no extended states below the Fermi surface.

Laughlin's derivation, as described above, does not assume any detailed structure of the system. The triumph of his theory lies in the use of the gauge symmetry principle: *an adiabatic change of flux by a quantum can only leave a state invariant or exchange it with another state* [55]. This has been illustrated in the numerical work of Aoki [14, 15], and Aers and MacDonald [4].

Following the above developments, Halperin [94] gave a more detailed analysis of Laughlin's theory by clarifying the role of the edges and the disordered potentials. In particular, he pointed out that a chemical potential difference between the two edges can produce an imbalance of the diamagnetic currents localized along the two edges; and the ratio of the resultant net current to the chemical potential difference is still quantized in units of e^2/h.

13.4.4 Hall Conductance as a Topological Invariant

In this subsection we will show that the Hall conductance characterizes the topological structure of the wave function on a parameter space. We will allow for the presence of disorder in the system and interactions between the electrons. Therefore, we will establish the quantization of Hall conductance in the most general setting.

The Hall conductivity can be obtained perturbatively by treating the electron system in the absence of an external electric field as the unperturbed system and the external electric field as a small perturbation. One then calculates the current to first order in the electric field. The coefficients of the field give the conductivity tensor. The Hall conductivity, whose spatial average gives the Hall conductance, is the off-diagonal element of the conductivity tensor. In terms of the eigenstates of the electron system (labeled by n), the Hall conductance can be written as

$$\sigma_H = \frac{ie^2\hbar}{L_1 L_2} \sum_{n>0} \frac{(v_1)_{0n}(v_2)_{n0} - (v_2)_{0n}(v_1)_{n0}}{(E_0 - E_n)^2}, \qquad (13.63)$$

where the system is assumed to have the shape of a rectangle with side lengths L_1 and L_2, $v_i = \frac{1}{e}\frac{\partial H}{\partial A_i}$ are the velocity operators in the two directions, and (A_1, A_2) is the vector potential. Differentiating both sides of the eigenvalue equation,

$$H \mid \psi_n \rangle = E_n \mid \psi_n \rangle,$$

and applying $\langle \psi_0 \mid$ to both sides of the resulting expression, we obtain

$$\langle \psi_0 \mid \frac{\partial H}{\partial A_i} \mid \psi_n \rangle + E_0 \langle \psi_0 \mid \frac{\partial}{\partial A_i} \mid \psi_n \rangle = E_n \langle \psi_0 \mid \frac{\partial}{\partial A_i} \mid \psi_n \rangle. \qquad (13.64)$$

This enables us to express the velocity matrix elements in terms of ψ_0, and (13.63) becomes

$$\sigma_H = \frac{i\hbar}{L_1 L_2} \left\{ \left\langle \frac{\partial \psi_0}{\partial A_1} \bigg| \frac{\partial \psi_0}{\partial A_2} \right\rangle - \left\langle \frac{\partial \psi_0}{\partial A_2} \bigg| \frac{\partial \psi_0}{\partial A_1} \right\rangle \right\}.$$

Now, imagine that the two-dimensional electron gas is located on the surface of a torus, which can be formally obtained by identifying the parallel sides of the rectangle, i.e., by applying periodic boundary conditions on the rectangle.[4] The change in the vector potential (A_1, A_2) can then be regarded as the effect of inserting fluxes, $\phi_1 := \delta A_1 L_1$ and $\phi_2 := \delta A_2 L_2$, through the holes of the torus (see Fig. 13.11). We therefore have

$$\sigma_H = i\hbar \left\{ \left\langle \frac{\partial \psi_0}{\partial \phi_1} \bigg| \frac{\partial \psi_0}{\partial \phi_2} \right\rangle - \left\langle \frac{\partial \psi_0}{\partial \phi_2} \bigg| \frac{\partial \psi_0}{\partial \phi_1} \right\rangle \right\}.$$

Niu and Thouless [199] and Niu, Thouless, and Wu [201] have shown that σ_H is actually independent of ϕ_1 and ϕ_2 in the thermodynamic limit

---

[4] Strictly speaking, one cannot demand the wave function to be periodic in both directions of the rectangle. Instead, one can demand that the magnetic translation of the wave function in the two directions be periodic. See [201] for details.

13.4 Quantization of Hall Conductance in Disordered Systems

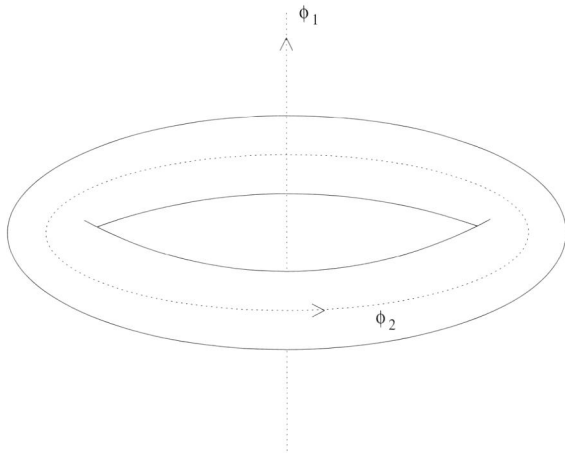

Fig. 13.11. Magnetic fluxes going through the holes of the torus.

$L_1, L_2 \to +\infty$, provided that there is a finite energy gap or mobility gap above the ground state of the system. We can therefore evaluate σ_H by averaging it over ϕ_1 and ϕ_2:

$$\sigma_H = \frac{i\hbar}{(h/e)^2} \int_0^{h/e} \int_0^{h/e} d\phi_1 d\phi_2 \left[\langle \frac{\partial \psi_0}{\partial \phi_1} | \frac{\partial \psi_0}{\partial \phi_2} \rangle - \langle \frac{\partial \psi_0}{\partial \phi_2} | \frac{\partial \psi_0}{\partial \phi_1} \rangle \right] \quad (13.65)$$

$$=: \frac{e^2}{h} \cdot C, \quad (13.66)$$

where the equation defines C, the Chern number which is known to take integer values. As we discussed in our study of $U(1)$ bundles associated with a spinning quantum particle (Sect. 6.5), the Chern number is a topological invariant. Here it describes the topological phase structure of the wave function ψ_0 on the torus of the parameter space (ϕ_1, ϕ_2). More specifically, $2\pi C$ actually gives the phase change of ψ_0 around the boundary of the rectangle $(0 < \phi_1 < h/e, \ 0 < \phi_2 < h/e)$. Consequently C is an integer.

Finally, we would like to remark on the boundary conditions. Strictly speaking, the high precision achieved in the experimental study of the quantum Hall effect leaves no room for the theorist to enjoy the luxury of choosing the most convenient boundary conditions. We have to show the quantization of Hall conductance in the geometry of the Hall bar which has a different topology than a cylinder or a torus. This issue has been addressed by Niu and Thouless [200] where it is shown that the quantum Hall effect also has a remarkable locality property: the Hall current at a given location only depends on the property of the system in a microscopic neighborhood of that location. Therefore, when we establish the relationship between the Hall voltage and current in the middle of the Hall bar we can change the boundary conditions at the two ends. A periodic boundary condition makes the sample equivalent to a cylinder. One can then apply Laughlin's argument to relate the Hall current with an imaginary process of adiabatic charge transport from one edge to the other by the insertion of a flux through the hole of the cylinder. This charge transport can be calculated from the transient current through the middle section of the cylinder. This current turns out to also have a locality property; it does not depend on the conditions on the edges which are far away on microscopic scales. We can thus also change the boundary conditions on the two edges; a periodic boundary condition makes the sample finally into a torus.

14. Geometric Phase in Condensed Matter III: Many-Body Systems

14.1 Introduction

In previous chapters, we have considered condensed matter systems that can be described effectively as single-particle problems. In this chapter, we will consider condensed matter systems where interactions between particles are very important. This includes fractional quantum Hall systems, magnetic systems, and Jahn–Teller crystals. A recurrent and central topic in the study of such many-body systems is the dynamical properties of quasi-particles and collective modes. We will show how the concept of geometric phase is used in this context. Due to the complexity caused by the presence of many degrees of freedom, an exact solution of the Schrödinger equation is usually impossible. Nevertheless, there are certain general arguments based on the concept of the geometric phase that allow for useful and sometimes exact results.

14.2 Fractional Quantum Hall Systems

Soon after the discovery of the integer quantum Hall effect (IQHE) [260], Tsui *et al.* [253] repeated the experiment on GaAs-(AlGa)As heterostructures. Due to the small effective mass of the two-dimensional electrons in this type of material, they found the quantum Hall plateaus at higher temperatures and lower magnetic fields. To their surprise, the Hall conductance also exhibited a plateau at $\frac{1}{3}\frac{e^2}{h}$ to a very high precision (10^{-5}). Later, more refined experiments revealed quantization of the Hall conductance at other simple fractions: $\frac{2}{3}, \frac{1}{5}, \frac{2}{5}, \frac{3}{5}, \frac{4}{5}, \frac{1}{7}, \ldots$. These observations led to the belief that the system of electrons should behave quite extrordinarily at densities corresponding to fractional occupation of the degenerate Landau level.

14.2.1 Laughlin Wave Function

The fractional quantum Hall effect was observed on samples of high mobility. This makes it reasonable to exclude impurities as the cause of this effect. The solution of the Schrödinger equation in the absence of many-body interactions is completely known, and there is no clue of any peculiarity at fractional fillings of the Landau level. This led the researchers to suspect that the Coulomb interactions among the electrons must play a crucial role.

The typical interaction energy per particle is $\frac{e^2}{\varepsilon d}$ where d is the average distance between electrons and ε is the dielectric constant of the semiconductor housing the two-dimensional electron gas. At high magnetic fields, this is small compared to the spacing $\hbar\omega_c$ between two Landau levels. Therefore, one can ignore the coupling between two Landau levels due to the Coulomb interaction, and all the electrons should stay in the lowest Landau level for filling fractions less than 1.

In the circular gauge, $A_x = -\frac{By}{2}$, $A_y = \frac{Bx}{2}$, the wave functions associated with the lowest Landau level may be written as

$$z^n e^{-|z|^2/4\ell^2}, \quad n = 0, 1, \ldots \infty,$$

where ℓ is the magnetic length introduced in the last chapter and $z = (x+iy)$ is the complex notation corresponding to the coordinates (x, y) in two dimensions. Therefore, any state in the lowest Landau level may be represented by a wave function of the form

$$f(z) e^{-|z|^2/4\ell^2}$$

where $f(z)$ is an analytic function of z. If there are N electrons, the N-body wave function is generally a sum of the products of single-particle wave functions. Therefore, it may be expressed as

$$g(z_1, z_2, \ldots z_N) e^{-\Sigma_i |z_i|^2/4\ell^2},$$

where g is analytic in all its variables. The Fermi statistics of the electrons dictates that this function must be antisymmetrical under the exchange of any pair of electrons.

In [151], Laughlin invoked the so-called *Jastrow ansatz* to obtain a good approximation for the ground state wave function. A Jastrow type wave function,

$$\psi(\mathbf{r}_1, \mathbf{r}_2, \ldots \mathbf{r}_N) = \prod_{i=1}^{N} f_1(\mathbf{r}_i) \prod_{i<j}^{N} f_2(\mathbf{r}_i - \mathbf{r}_j),$$

involves a pair of functions f_1 and f_2 that respectively represent the response of the system to an external field and the (pair) correlations among the particles.[1] The particular form f_1 and f_2 is determined by variational means so that the energy is minimized. Jastrow type wave functions provide a very good description of the ground state of liquid ^{3}He and ^{4}He. A Jastrow type wave function for the two-dimensional electron system has the form

$$\psi(z_1, \ldots z_N) = \prod_{i<j} f(z_i - z_j) e^{-\Sigma_i |z_i|^2/4\ell^2}.$$

---

[1] Obviously, one can consider higher order correlations by including another factor involving three particles at a time. But this is rarely necessary.

14.2 Fractional Quantum Hall Systems

Now, consider the fact that the constituent single-particle states are all in the lowest Landau level, and $f(z)$ must be analytic. Further, $f(z)$ should be an odd function in order to be antisymmetric under particle exchanges. Therefore,

$$f(z) = a_1 z + a_3 z^3 + a_5 z^5 + \ldots$$

Next, we note that the total angular momentum,

$$J := \sum_j -i\hbar \frac{\partial}{\partial \theta_j},$$

commutes with the Hamiltonian of the system,

$$H = \sum_{i=1}^{N} \frac{1}{2m}\left[\left(-i\hbar\frac{\partial}{\partial x_i} - \frac{1}{2}eBy_i\right)^2 + \left(-i\hbar\frac{\partial}{\partial y_i} + \frac{1}{2}eBx_i\right)^2\right] + \sum_{i<j} \frac{e^2}{\varepsilon\,|\,\mathbf{r}_i - \mathbf{r}_j\,|}.$$

We may therefore choose the eigenstates of H from among those of J. Since each power of z_i in ψ contributes to J an angular momentum of $\hbar$ (up to an unimportant normalization constant) the $f(z)$ appearing in the expression for ψ must be a power (m) of z,

$$f(z) = z^m.$$

In summary, the ground state wave function should have the form:

$$\psi(z_1, \ldots z_N) = \prod_{i<j}(z_i - z_j)^m e^{-\sum_i |z_i|^2 / 4\ell_0^2},$$

where m is an odd integer as required by the antisymmetry condition.

Having obtained the wave function, we can evaluate its various physical properties. For example, the density of particles at z is given by

$$\rho(z) = \frac{\int d^2 z_1 d^2 z_2 \ldots d^2 z_N \mid \psi(z_1, \ldots z_N)\mid^2 \sum_i \delta(z - z_i)}{\int d^2 z_1 d^2 z_2 \ldots d^2 z_N \mid \psi(z_1, \ldots z_N)\mid^2},$$

where $d^2 z_i := dx_i dy_i$ and $\delta(z - z_i) := \delta(x - x_i)\delta(y - y_i)$. The actual calculation of the multiple integrals appearing in this formula is itself a non-trivial task when N is very large. But Laughlin was able to get the result based on a formal analogy to a two-dimensional classical plasma. It turns out that the density profile is essentially a constant within a certain radius and drops off as a Gaussian outside this radius. The constant density is just $1/m$ times the density for a fully occupied Landau level.

One can also calculate two-particle correlations in the Laughlin state and obtain the interaction energy. The Jastrow factors make the probability of finding two particles at a small distance r very small (vanishing like r^{2m}). This helps to reduce the Coulomb repulsion between the electrons and consequently suggests that the system has a particularly low energy only for the electron densities of a simple odd-denominator fraction, $\nu = 1/m$, of the Landau level capacity. This is because in this case, as Laughlin found out, good Jastrow type wave functions are available. Indeed, small-size exact numerical calculations show that the interaction energy is particularly low at filling fractions $\nu = \frac{1}{3}, \frac{1}{5}, ...$ of the Landau level. Conventional theories such as the Hartree–Fock approximation give energies higher than the numerical results and show no pecularity at these fractional fillings. Laughlin's wave function for $m = 3$ reproduces the numerical result of the interaction energy at $\nu = \frac{1}{3}$ within a relative error of 0.05%. Moreover, the overlap (inner product) of Laughlin's state to the true ground state is nearly unity.

One can write down the Laughlin state only for densities of $\frac{1}{m}$ of that of a Landau level. Studies show that the ground states at other densities actually consist of a Laughlin state together with finite densities of quasi-particles or quasi-holes (to be discussed below). Under certain conditions, these quasi-particles or quasi-holes form a Laughlin state themselves. This gives rise to stability at filling fractions such as $\nu = \frac{2}{5}$.

14.2.2 Fractional Charged Excitations

A vortex can be created in the Laughlin state by adiabatically inserting a magnetic flux, a concentrated magnetic field, through the two-dimensional electron system. If the flux is infinitesimally thin and has a total strength of one flux quantum h/e, the resulting wave function may be written as

$$|\psi(z_0)\rangle = \prod_{i=1}^{N}(z_i - z_0) \prod_{i<j}(z_i - z_j)^m e^{-\sum_i |z_i|^2/4\ell^2},$$

where z_0 is the position of the flux. This is a quantized vortex, because the many-body state changes by a phase of 2π when each electron moves around z_0 once, or in other words, every electron acquires an angular momentum of unity ($\hbar$) around z_0. Because an infinitesimally thin flux quantum can be gauged away from the electron Hamiltonian without affecting the single-valuedness of the wave function, the above wave function may also be regarded as an excited eigenstate of the system in the absence of the flux.

Laughlin called such an excitation a quasi-hole for the reason that there is a zero of the electron density at z_0 as reflected in the fact that the wave function vanishes when any electron coordinates z_j coincide with z_0. Indeed, a plot of the electron density calculated using this new wave function reveals a hole in a background of constant electron density of the Laughlin ground state. Because the electrons carry negative charges, this quasi-hole carries a positive charge. Integrating the deficiency in the electron density, Laughlin found the value of this charge as e/m, a fraction of the proton charge for $m > 1$.

It turns out that this fractional charge can also be determined by evaluating a Berry phase [18]. If we take a particle of charge e^* around a loop in a magnetic field, the Aharonov–Bohm phase is given by $e^*\phi/\hbar$ (see Sect. 2.4). Similarly, we can also take this quasi-hole around a loop, and calculate the Berry phase of the above wave function in the parameter space of z_0. This yields

$$\oint_c d\mathbf{r}_0 \frac{\langle \psi(z_0)|i\frac{\partial}{\partial \mathbf{r}_0}|\psi(z_0)\rangle}{\langle \psi(z_0)|\psi(z_0)\rangle}, \qquad (14.1)$$

where we have taken into account the normalization of the wave function. The vector $\mathbf{r}_0$ denotes the position of the quasi-hole with coordinates given by the real and imaginary parts of z_0. Because the wave function depends on $\mathbf{r}_0$ only through the factor $\prod(z_i - z_0)$, we can write $d\mathbf{r}_0 \frac{\partial}{\partial \mathbf{r}_0}$ in (14.1) as $dz_0 \sum_i (z_0 - z_i)^{-1}$. Now, using the definition of the particle density, we find the Berry phase as

$$i\oint_c dz_0 \iint d^2z \frac{\rho(z)}{z_0 - z},$$

where d^2z denotes the area element $dxdy$.

The particle density is the sum of the constant eB/hm for the Laughlin ground state and a deficiency $\delta\rho(z)$ due to the presence of the quasi-hole. The latter is circularly symmetric about z_0 and its contribution vanishes after integrating over the phase angle of $z - z_0$. The former yields the Berry phase as $-2\pi eBS/mh$ where S is the area enclosed by the contour loop. Comparing to the Arharonov–Bohm phase of a charged particle, we find that the quasi-hole has a charge of $e^* = e/m$. One can also construct quasi-particles with negative fractional charges. The wave function has the form of

$$\prod_{i=1}^{N}\left(\frac{\partial}{\partial z_i} - z_0/\ell^2\right) \prod_{i<j}(z_i - z_j)^m e^{-\sum_i |z_i|^2/4\ell^2}.$$

The fractional charges in the quantum Hall system have been observed by Saminadayar et al. [223] in a shot noise experiment. The *shot noise* in a steady current is associated with the fact that the charge carriers are discrete particles passing through the conductor randomly. The strength of the shot noise as measured by the variance of the current is proportional to the charge of the particles. The experiment shows that the noise in the 1/3 fractional Hall system is three times smaller than normal.

14.2.3 Fractional Statistics

From elementary quantum mechanics we know that particles are either bosons or fermions, i.e., under exchange of two identical particles the corresponding many-body wave function is either symmetrical or antisymmetrical, respectively. These are the only possibilities in the three-dimensional world,[2]

---

[2] This can be proven using the basic axioms of standard local quantum field theories in (3+1)-dimensional Minkowski space [91].

14. Geometric Phase in Condensed Matter III: Many-Body Systems

but more exotic quantum statistics may arise in two dimensions owing to its peculiar topology [154, 271]. Particles with such exotic statistics are called *anyons*. Quasi-holes and quasi-particles in the fractional Hall states provide examples of anyons.

Anyons are identical particles whose many-body wave function may change by a phase factor ($\neq \pm 1$) when any two particles exchange their positions by moving along a path; the phase change only depends on the topology of the path. In this definition, two paths are regarded as having the same topology if they can be continuously deformed into each other without ever colliding the two particles.

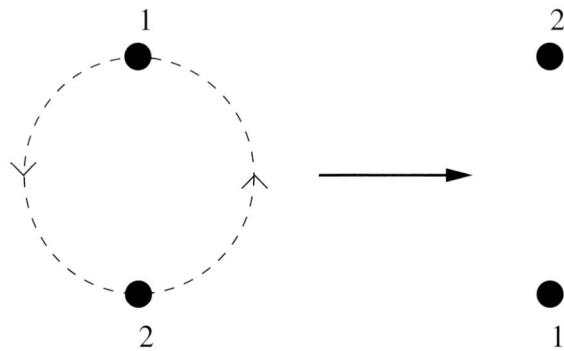

Fig. 14.1. Exchange of particles 1 and 2.

To be more concrete, consider the case of two anyons. They may be exchanged as in Fig. 14.1. Exchanging twice, we have both particles coming back to their original positions. The total path is topologically equivalent to the one in which particle 2 is fixed but particle 1 goes around particle 2 once counterclockwise. The reason is that the change of the reference frame does not involve the collision of two particles, and therefore does not change the topology of the path. However, in three or more dimensions, a path going around a fixed point is topologically equivalent to no motion at all. Therefore, two exchanges are equivalent to no exchange. If $e^{i\theta}$ is the phase factor for one exchange, we must have

$$(e^{i\theta})^2 = 1$$

which implies that $e^{i\theta} = \pm 1$. In other words, there are only fermions and bosons in three or more dimensions. The situation is very different in two dimensions, in which a path going around a fixed point cannot be shrunk to a point. The above argument for more than two dimensions fails in the present case. In the following, we will show that the exchange phase for two quasi-holes is $\theta = \pi/m$.

14.2 Fractional Quantum Hall Systems

Laughlin has considered the following wave function for a two-hole state.

$$\psi_{z_a z_b} = \prod_{i=1}^{N}(z_i - z_a)(z_i - z_b)\prod_{i<j}(z_i - z_j)^m e^{-\sum_i |z_i|^2/4\ell^2},$$

where z_a and z_b are the complex coordinates of the quasi-holes. Let us fix z_a and move z_b slowly around a large loop C. This is equivalent to exchanging the two holes twice. The associated geometric phase will be regarded as twice the exchange phase [18]. It is given by

$$\gamma(C) = -i \oint dz_b \langle \psi_{z_a z_b} | \frac{\partial}{\partial z_b} | \psi_{z_a z_b}\rangle / \langle \psi_{z_a z_b} | \psi_{z_a z_b}\rangle \qquad (14.2)$$

$$= +i \oint dz_b \langle \psi_{z_a z_b} | \sum_i \frac{1}{z_b - z_i} | \psi_{z_a z_b}\rangle / \langle z_{z_a z_b} | \psi_{z_a z_b}\rangle$$

$$= i \oint dz_b \int\int d^2 z \rho(z) \frac{1}{z_b - z}$$

$$= -2\pi \int\int_c d^2 z \rho(z). \qquad (14.3)$$

If C does not enclose z_a,

$$\gamma(C) = -2\pi A_c \cdot \frac{2}{\pi m},$$

where A_c is the area inside the contour c. If C does enclose z_a,

$$\gamma(C) = -2\pi \left(A_c \frac{2}{\pi m} - \frac{1}{m}\right).$$

The difference is the phase due to the quasi-hole z_a:

$$\delta\gamma(C) = +\frac{2\pi}{m}.$$

The exhange phase of two quasi-holes is half of this value, i.e., π/m.

If $m = 1$, the extra phase change is 2π. Therefore, the phase change for the interchange of quasi-particles is π, that is an interchange of two identical particles leads to a sign change of the wave function. This corresponds to Fermi statistics. However, if $m = 3$, the phase change is not a simple sign change and the quasi-particles are neither bosons nor fermions. As we mentioned earlier, any phase change is possible for two spatial dimensions. Therefore, anyons are particles that obey statistical rules, called *fractional statistics*, that are more general than Fermi and Bose statistics. They may be considered as flux tube composites of charged particles.

For the quasi-hole case that we discussed above, we can reproduce the extra phase change by introducing the electromagnetic vector potential

$$\mathbf{A}_\phi = \frac{h}{2\pi e} \frac{\mathbf{e}_z \times (\mathbf{r} - \mathbf{r}_b)}{|\mathbf{r} - \mathbf{r}_b|^2} = \frac{h}{2\pi e} \nabla \left(\tan^{-1} \frac{y - y_b}{x - x_b} \right), \tag{14.4}$$

in the Hamiltonian. This vector potential provides a delta function like flux tube at $\mathbf{r}=\mathbf{r}_b$ where the amplitude of the wave function is zero.

In many cases, the kinetic term of the Hamiltonian for an anyon of charge e^* has the form

$$\sum_j \frac{1}{2m^*} \left(\mathbf{p}_j + e\mathbf{A}(\mathbf{r}_j) - e^* \mathbf{A}_\phi(\mathbf{r}_j) \right)^2, \tag{14.5}$$

where

$$\mathbf{A}_\phi(\mathbf{r}_j) = \frac{h}{2\pi e} \nabla_j \sum_{k \neq j} \left(\tan^{-1} \frac{y_j - y_k}{x_j - x_k} \right), \tag{14.6}$$

[19]. The vector potential $\mathbf{A}_\phi$ provides a flux tube at each anyon. Note that the classical motion is not affected by $\mathbf{A}_\phi$.

For a further discussion of anyons and their relation to field theories consult [69].

14.2.4 Degeneracy and Fractional Quantization

Laughlin's theory shows that interacting two-dimensional electrons have a particularly low energy when the filling fraction of the Landau level has the form $1/q$ with q being an odd integer. Later developments indicate that such local stability also occurs for filling fractions of the form p/q with $p \neq 1$ and q not dividing p. In other words, a plot of the ground state energy per electron as a function of the filling fraction involves cusp-like valleys at rational fillings. The biggest valleys occur for the simple rationals with small p and q, and the size of the valleys decreases quickly with increasing p and q.

One can interpret the fractional plateaus in the Hall conductance as follows. In the vicinity of a fractional filling, e.g., 1/3, that corresponds to a range of the gate voltage for a fixed magnetic field or vice versa, the system tends to stay in the ground state of the 1/3 filling with some additional quasi-holes or quasi-particles that make up for the difference in the filling fraction. These quasi-holes or quasi-particles may be localized by the impurities in the sample. They do not contribute to the Hall current, so the Hall conductance of the system remains the same as for the exact 1/3 filling, i.e., $e^2/3h$.

The above argument for the fractional quantization of the Hall conductance provides a good physical picture but fails to explain the remarkable fact that the fractional Hall plateaus are quantized to the extremely high precision of one part in 10^5. Let us recall the topological-invariant theory of the integer quantum Hall effect. There we assumed that the ground state of the electrons is non-degenerate and that it is separated from the excited states by an energy gap that does not tend to zero in the thermodynamic limit. Under these conditions the Hall conductance takes the form

$$\sigma_H = \frac{e^2}{h} \cdot \frac{i}{2\pi} \int\int_0^{\frac{h}{e}} d\phi_1 d\phi_2 \left[\left\langle \frac{\partial \psi_0}{\partial \phi_1} \bigg| \frac{\partial \psi_0}{\partial \phi_2} \right\rangle - \left\langle \frac{\partial \psi_0}{\partial \phi_2} \bigg| \frac{\partial \psi_0}{\partial \phi_1} \right\rangle\right],$$

and is necessarily quantized as integer multiples of $\frac{e^2}{h}$. To obtain a fractional Hall conductance, at least one of the above assumptions must be relaxed. According to Laughlin's theory, the ground state energy is separated from the excitations by a finite gap corresponding to finite quasi-particle and quasi-hole energies. Therefore, we must consider the possibility of a ground state degeneracy.

The ground state degeneracy was first pointed out by Tao and Wu [243] in their attempt to generalize the gauge symmetry argument of Laughlin to the fractional QHE. It was then established by group theoretical methods and numerical calculations by Haldane and Rezayi [93] and Su [236], respectively. When the electon density is p/q of the Landau level capacity, there are q independent ground states.

Consider the case of $\frac{p}{q} = \frac{1}{3}$. There are three ground states: ψ_1, ψ_2 and ψ_3. The Hall conductance is the average of the contributions of these states:

$$\sigma_H = \frac{e^2}{3h} \frac{i}{2\pi} \int\int_0^{\frac{h}{e}} \sum_{j=1}^{3} d\phi_1 d\phi_2 \left[\left\langle \frac{\partial \psi_j}{\partial \phi_1} \bigg| \frac{\partial \psi_j}{\partial \phi_2} \right\rangle - \left\langle \frac{\partial \psi_j}{\partial \phi_2} \bigg| \frac{\partial \psi_j}{\partial \phi_1} \right\rangle\right].$$

Because of the degeneracy, there is a freedom in the choice of three orthogonal state vectors to represent the degenerate subspace. One can choose these vectors such that each of them returns to itself when ϕ_1 increases by a quantum of flux $\frac{h}{e}$, but transforms into another of these vectors when ϕ_2 does the same. We can thus rewrite the above expression for the Hall conductance as

$$\sigma_H = \frac{e^2}{3h} \cdot \frac{i}{2\pi} \int_0^{\frac{h}{e}} d\phi_1 \int_0^{3\frac{h}{e}} d\phi_2 \left[\left\langle \frac{\partial \psi_1}{\partial \phi_1} \bigg| \frac{\partial \psi_1}{\partial \phi_2} \right\rangle - \left\langle \frac{\partial \psi_1}{\partial \phi_2} \bigg| \frac{\partial \psi_1}{\partial \phi_1} \right\rangle\right].$$

In view of a similar argument we used in the description of the integer QHE (see Sect. 13.4.4), this is an integer multiple of $\frac{e^2}{3h}$.

Tao and Haldane [242] and later Wen and Niu [267] showed that, in the thermodynamic limit, the ground state degeneracy is not lifted by weak impurities. Therefore, the above treatment remains valid in the presence of the latter. Moreover, when the density of the electrons is slightly off a fraction $\frac{p}{q}$ of that of a Landau level, one expects the extra density of particles to exist in the form of quasi-holes or quasi-particles that are bound to the impurities. The many-particle system still has a q-fold ground state degeneracy with an energy separated from the excited states by a finite gap. This should explain the fractional plateaus in the Hall conductance as a function of the magnetic field or the gate voltage.

Finally, we would like to make some remarks on boundary conditions. Although there is conclusive evidence for the degeneracy of the fractional Hall states in the toroidal geometry, a similar result does not hold for other geometries. For example, no such degeneracy is ever found in spherical geometry [92, 93]. This does not invalidate our arguments, because one simply cannot, in principle, make a global measurement of the Hall conductance on a sphere. Nevertheless, the dependence of degeneracy on the space topology seems to be rather puzzling. This mystery was finally resolved by Wen and Niu [267] based on a field theoretical argument. They showed that in general the degree of degeneracy is given by q^g, where g is the genus (number of holes or handles) of the surface characterizing its topology. For a sphere $g = 0$, and there is no degeneracy. For a torus $g = 1$, and we have a q-fold degeneracy.

14.3 Spin-Wave Dynamics in Itinerant Magnets

Spin waves are collective modes of motion in the local magnetic moments in a ferromagnetic or antiferromagnetic material where the spins are aligned or antialigned. One may also generalize this concept to collective excitations in non-collinearly ordered spin systems. The simplest and most intuitive treatment of spin waves is based on the Heisenberg model where the magnetic moments are tightly bound to atomic sites. This is one limiting view of magnetism in solids, and there are many materials including the most important ferromagnets such as Fe, Co, and Ni, where the so-called *itinerant* picture of Bloch and Stoner is more relevant [103]. In the itinerant picture, the spins are carried by electrons in Bloch states which move throughout the system. Powerful first-principles calculations based on this picture provide extremely accurate predictions for the ground state magnetization. Nevertheless, the calculation of spin waves for itinerant-electron systems has been a relatively undeveloped area of computational condensed matter physics until recently.

Thanks to the concept of the geometric phase, it is now possible to fomulate the dynamics of local moments even if the underlying spins are itinerant [198, 202]. This approach is based on the adiabatic assumption that the orientational motion of the spins is much slower than the other degrees of freedom in the electronic system. This is usually the case for real materials.

14.3.1 General Formulation

Before we derive the equations for the spin waves, we need to define the basic dynamical variables, i.e., the local magnetic moments. First, we partition the space into cells of volume V_j which can be as small as the atoms or even smaller depending on one's needs. The local moment or simply the spin for each cell is defined as

$$\mathbf{s}_j := \langle \psi | \hat{\mathbf{s}}_j | \psi \rangle, \quad \hat{\mathbf{s}}_j := \int_{V_j} d^3 r\, \hat{\mathbf{s}}(\mathbf{r}), \qquad (14.7)$$

where $\hat{\mathbf{s}}$ is the operator measuring the spin density.

Note that this definition does not imply that the electrons are localized in each cell. They may flow between the cells. The local moment in a particular cell at a particular time just consists of (the spins of) those electrons that happen to be in that cell at that time. The above definition is reminiscent of the Eulerean point of view in fluid dynamics where one focuses on the physical quantities associated with (fixed) spatial points (the cells), although the individual particles can travel around in the system.

We treat the spin configuration [**s**] as the slow variables, and assume that the electron system always lies in the lowest energy state for each spin configuration, called the constrained ground state. In other words, we are interested in situations where the electrons are initially in the ground state for some spin configuration and where the subsequent movement of the spin configuration does not excite the electrons. This is much like the Born–Oppenheimer approximation (see Sect. 8.3) of the electron–nuclei system with the spin configuration playing the role of the nuclei coordinates.

In the following, we denote the lowest energy of the system for a given spin configuration [**s**] by $E[\mathbf{s}]$ and the corresponding many-body wave function by $\psi[\mathbf{s}]$. We demand that $\psi[\mathbf{s}]$ should satisfy the time-dependent Schrödinger equation. Because the wave function is specified once the spin configuration is fixed, this equation determines, in principle, how the spin configuration should move. It is convenient to use the time-dependent variational principle, in which the action $\int L dt$ calculated with the Lagrangian

$$L := \langle \psi | i\hbar \frac{\partial}{\partial t} | \psi \rangle - \langle \psi | H | \psi \rangle$$
$$= \hbar \sum_j \dot{\mathbf{s}}_j \cdot \mathbf{A}_\mathbf{s}(j) - E(\mathbf{s}) \qquad (14.8)$$

is extremized, where

$$\mathbf{A}_\mathbf{s}(j) := i \langle \psi | \frac{\partial}{\partial \mathbf{s}_j} | \psi \rangle. \qquad (14.9)$$

is the Berry connection.

After the extremization with respect to the spin configuration [**s**], we obtain the equations for the spin dynamics as

$$-\sum_{j'} \hbar \Omega(j, j') \dot{\mathbf{s}}_{j'} + \frac{\partial E}{\partial \mathbf{s}_j} = 0, \qquad (14.10)$$

where the Berry curvature $\Omega(j, j')$ is a 3×3 matrix whose (α, β) element is given by

$$\Omega(j, j')_{\alpha, \beta} = \frac{\partial}{\partial s_{j\alpha}} A_{\mathbf{s}\beta}(j') - \frac{\partial}{\partial s_{j'\beta}} A_{\mathbf{s}\alpha}(j), \qquad (14.11)$$

and $\alpha, \beta, = x, y, z$. Therefore, with a knowledge of the ground state energy of each spin configuration and the corresponding Berry curvature, both of which can be calculated using practical methods based on density functional theory [56, 85] one can study the collective dynamics of the spins for real materials.

14.3.2 Tight-Binding Limit and Beyond

In the special situation of the tight-binding limit, our general equations of motion (14.10) can be shown to reduce to the classical Landau–Lifshitz equations

$$\hbar \dot{\mathbf{s}}_j = \mathbf{s}_j \times \frac{\partial E}{\partial \mathbf{s}_j}. \qquad (14.12)$$

In this limit, one obtains the constrained ground state from the true ground state by rigidly rotating the spins within each cell, i.e.,

$$|\psi[\mathbf{s}_j]\rangle = \prod_j e^{i\boldsymbol{\theta}_j \cdot \hat{\mathbf{s}}_j} |\psi_0\rangle, \qquad (14.13)$$

where $\boldsymbol{\theta}_j$ is in the direction of $\mathbf{s}_j^0 \times \mathbf{s}_j$ and has a magnitude given by the angle between $\mathbf{s}_j$ and $\mathbf{s}_j^0$ (see Chap. 3). The spin operators $\hat{\mathbf{s}}_j$ correspond to the total spin for each cell. They satisfy the usual commutation relations for angular momentum operators, i.e.,

$$[\hat{s}_{j\alpha}, \hat{s}_{j'\alpha'}] = \delta_{jj'} \epsilon_{\alpha\alpha'\beta} \hat{s}_{j\beta},$$

where $\epsilon_{\alpha\beta\gamma}$ is the totally antisymmetric Levi Civita symbol with $\epsilon_{1,2,3} = 1$.

Having given the expression (14.13) for the ground state vector, we can use (14.9) to compute the Berry curvatures (14.11). This yields

$$\Omega(j, j')_{\alpha\alpha'} = \delta_{jj'} \epsilon_{\alpha\alpha'\beta} s_{j\beta} / (\mathbf{s}_j)^2 \qquad (14.14)$$

which does not mix different cells and its value for a single cell is given by that of a rigid spin (see Chap. 3). With this form of the Berry curvature, our general equations of motion reduce to the Landau–Lifshitz equations (14.12).

For the itinerant spins and the cases where the local moments interact with the itinerant electrons [29], the Landau–Lifshitz equations may not hold. This is because for such a system, the moments at different sites may share the same electron and there are Berry curvatures that mix them. In the following, we consider two ways in which the inter-moment Berry curvature Ω can arise: (1) electron hopping between sites, (2) other neglected degrees of freedom.

First we examine the hopping effect. For this purpose we consider the two-site t-J-B Hamiltonian:

$$H = -\frac{t}{2} \sum_\sigma [a_{1\sigma}^\dagger a_{2\sigma} + a_{2\sigma}^\dagger a_{1\sigma}] - 4J \hat{\mathbf{s}}_1 \cdot \hat{\mathbf{s}}_2 - \sum_j \mathbf{B}_j \cdot \hat{\mathbf{s}}_j, \qquad (14.15)$$

14.3 Spin-Wave Dynamics in Itinerant Magnets

where $a_{i\sigma}$ is the annihilation operator for the electron at site i with spin σ, $\hat{\mathbf{s}}_i$ is the spin operator given by

$$\hat{\mathbf{s}}_i := \frac{1}{2} \sum_{\alpha,\beta} a_{i\alpha}^\dagger \boldsymbol{\sigma}_{\alpha\beta} a_{i\beta}, \tag{14.16}$$

with $\boldsymbol{\sigma}$ being the vector of Pauli matrices

$$\boldsymbol{\sigma} = \left(\begin{pmatrix} 0 & 1 \\ 1 & 0 \end{pmatrix}, \begin{pmatrix} 0 & -i \\ i & 0 \end{pmatrix}, \begin{pmatrix} 1 & 0 \\ 0 & -1 \end{pmatrix} \right), \tag{14.17}$$

and a uniform magnetic field $\mathbf{B}_j = B\mathbf{e}_z$ is added to select a direction of magnetization. If the hopping energy is not too large, the ground state is three-fold degenerate at $B = 0$ but is split into levels of energy $-J - B$, $-J$, and $-J + B$ for finite B.

In order to obtain Ω, we apply small transverse fields in order to deform the spin configuration. Using the perturbation method, the Berry curvature can be calculated with the unperturbed ground state wave function [182]. After straightforward but tedious calculations, one finds that the non-zero elements of the Berry curvature have the form

$$\Omega(11)_{xy} = \Omega(22)_{xy} = -2 - \frac{t^2}{(J+B)^2},$$

$$\Omega(12)_{xy} = \Omega(21)_{xy} = \frac{t^2}{(J+B)^2}. \tag{14.18}$$

The above results clearly demonstrate the hopping effect; if t is set to zero, the matrix elements of the curvature associated with different sites disappear.

In order to demonstrate the second possibility, namely, the effect of the neglected degrees of freedom, we examine the following three-spin Hamiltonian.

$$H = -J(\hat{\mathbf{s}}_1 \cdot \hat{\mathbf{s}}_2 + \hat{\mathbf{s}}_2 \cdot \hat{\mathbf{s}}_3) - \sum_{j=1}^{3} \mathbf{B}_j \cdot \hat{\mathbf{s}}_j, \tag{14.19}$$

where $\mathbf{B}_j := B\mathbf{e}_z$. The spin $\mathbf{s}_2$ is treated as a neglected degree of freedom. Diagonalizing this Hamiltonian, we find the ferromagnetic ground state as a quartet with energies $-2J \pm 2B$, $-2J \pm B$ at finite B.

Now, we identify the spin 2 as a neglected degree of freedom and apply small transverse fields on spins 1 and 3 (leaving spin 2 free as it is neglected). This creates transverse deviations of the spins. Next we calculate the Berry curvature. The non-zero matrix elements are

$$\Omega(11)_{xy} = \Omega(33)_{xy} = -2\frac{B^2 + 4BJ + 5J^2}{(2J+B)^2},$$

$$\Omega(13)_{xy} = \Omega(31)_{xy} = -2\frac{J^2}{(2J+B)^2}. \tag{14.20}$$

If the spin 2 is absent (J is zero), the matrix elements of the Berry curvature between spin 1 and spin 3 disappear. Thus, the above result clearly shows that the Berry curvature between different sites arises from the neglected degree of freedom.

14.3.3 Spin Wave Spectrum

In the calculation of the spin wave spectrum, one linearizes the equations of motion (14.10)

$$\sum_{j'}[-\hbar\Omega(j,j')\dot{\mathbf{s}}_{j'} + K(j,j')\mathbf{s}_{j'}] = 0, \tag{14.21}$$

where $K(j, j')$ stands for the second derivative of the constrained ground state energy with respect to $\mathbf{s}_j$ and $\mathbf{s}_{j'}$. Therefore, if one can determine the matrices $\Omega(j, j')$ and $K(j, j')$, one can calculate the spin-wave spectrum.

The calculation of the matrix elements of $K(j, j')$ is straightforward. The calculation of the matrix elements of $\Omega(j, j')$ can be done through the formula [219]

$$\Omega(j,j')_{\alpha,\beta} = -\frac{2}{\epsilon^2}\operatorname{Im}\left[\log\left(\langle\Psi_0|\Psi_1\rangle\langle\Psi_1|\Psi_2\rangle\langle\Psi_2|\Psi_0\rangle\right)\right], \tag{14.22}$$

where ϵ is a small constant, $\Psi_0 = \Psi[\mathbf{s}_0]$, $\Psi_1 = \Psi[\mathbf{s}_0 + \epsilon\mathbf{e}_\alpha(j)]$, $\Psi_2 = \Psi[\mathbf{s}_0 + \epsilon\mathbf{e}_\beta(j')]$, $\mathbf{s}_0$ is the ground state spin configuration, and $\mathbf{e}_\alpha(j)$ is the unit vector in the α-direction in the j-th cell. If we adopt the phase convention corresponding to the requirements that $\langle\Psi_0|\Psi_1\rangle$ and $\langle\Psi_0|\Psi_2\rangle$ are real and positive, (14.22) reduces to

$$\Omega(j,j')_{\alpha,\beta} = -\frac{2}{\epsilon^2}\operatorname{Im}\left[\log\left(\langle\Psi[\mathbf{s}_0 + \epsilon\mathbf{e}_\alpha(j)]|\Psi[\mathbf{s}_0 + \epsilon\mathbf{e}_\beta(j')]\rangle\right)\right]. \tag{14.23}$$

In the following, we consider the case of a collinear magnetic system such as a ferromagnetic or antiferromagnetic system, in which there is a symmetry of spin rotation about the magnetization axis (z-axis). We consider a spin-wave in which the x- and y-components of the spins are given by

$$\mathbf{s}_j^\perp := \operatorname{Re}[e^{-i\omega t}(\mathbf{e}_x + i\mathbf{e}_y)s_j], \tag{14.24}$$

where s_j is a complex number describing the amplitude and phase angle of the spin wave at the j-th site. Then, the normal mode equations (14.21) become

$$\sum_{j'}[i\hbar\omega\Omega_{jj'} + K_{jj'}]s_{j'} = 0, \tag{14.25}$$

where

14.3 Spin-Wave Dynamics in Itinerant Magnets

$$\Omega_{jj'} := \Omega(j,j')_{xx} - i\Omega(j,j')_{xy},$$
$$K_{jj'} := K(j,j')_{xx} - iK(j,j')_{xy}. \tag{14.26}$$

With the knowledge of the K and Ω matrices, one should be able to solve for the normal mode frequencies from (14.25).

It will be useful to have a closer examination of the physical meanings of the K and Ω matrices. First, by Taylor expansion of the constrained ground state energy to second order in the spin wave amplitudes, we find that the increase in energy from the absolute ground state is given by

$$\Delta E = \frac{1}{2} \sum_{jj'} s_j^* K_{jj'} s_{j'}. \tag{14.27}$$

Second, we will show that a similar quadratic form for Ω yields the reduction of magnetization due to the spin wave, i.e.,

$$\Delta S_z := S_z^0 - \langle \hat{S}_z \rangle = \frac{i}{2} \sum_{jj'} s_j^* \Omega_{jj'} s_{j'}, \tag{14.28}$$

where $\langle \hat{S}_z \rangle$ is the expectation value of the z-component of the total spin in the constrained ground state and S_z^0 is that in the absolute ground state.

To establish the relation (14.28), we express $s_j = s_j^x - is_j^y$ by splitting it into the overall amplitude $A = A^x + iA^y$ and the relative amplitude $R_j = R_j^x + iR_j^y$ as $s_j = AR_j$. Furthermore, we introduce

$$\Omega_A := i\frac{\partial}{\partial A^x} \langle \Psi | \frac{\partial}{\partial A^y} | \Psi \rangle - i\frac{\partial}{\partial A^y} \langle \Psi | \frac{\partial}{\partial A^x} | \Psi \rangle$$
$$= \sum_{j,j'} [\Omega(j,j')_{xx}(R_j^y R_{j'}^x - R_j^x R_{j'}^y)$$
$$+ \Omega(j,j')_{xy}(R_j^x R_{j'}^x + R_j^y R_{j'}^y)], \tag{14.29}$$

so that

$$\sum_{jj'} i\frac{1}{2} s_j^* \Omega(j,j') s_{j'} = \frac{1}{2} |A|^2 \Omega_A. \tag{14.30}$$

The right-hand side of this equation represents the geometric phase around the boundary of an area of $\frac{1}{2}|A|^2$ in the complex A plane.

Now, consider a circular wedge of radius $|A|$ and angle ϕ in the complex A plane. The geometric phase (angle) associated with the boundary of this wedge is given by $\frac{1}{2}|A|^2\phi\Omega_A$. We can obtain the same geometric phase (angle) by calculating the phase of $\langle \Psi | \Psi_\phi \rangle$ directly, where Ψ is the constrained ground state vector with $A^x = |A|$, $A^y = 0$ and Ψ_ϕ is the rotated state vector with $A^x = |A|\cos\phi$, $A^y = |A|\sin\phi$. If we adopt the phase convention in which $\langle \Psi | \Psi_\phi \rangle$ is real and positive, the rotated wave function Ψ_ϕ may be written as

$$e^{i\phi(\hat{S}_z - S_z^0)}\Psi, \tag{14.31}$$

where $\hat{S}_z$ is the z-component of the total spin operator and S_z^0 is its ground state expectation value. We thus find that the geometric phase for small ϕ is given by

$$\frac{1}{2}|A|^2 \phi \Omega_A = (\langle \hat{S}_z \rangle - S_z^0)\phi. \tag{14.32}$$

Together with (14.30), this result then leads to the relation (14.28).

Based on our deeper understanding of the physical meaning of the K and Ω matrices in the normal mode equation (14.25), we now derive a simple formula for the normal mode frequency. By multiplying s_j^* on both sides of (14.25) and summing over j, we find

$$\hbar\omega = \frac{\sum_{jj'} s_j^* K_{jj'} s_{j'}}{\sum_{jj'} s_j^* i\Omega_{jj'} s_{j'}} = \frac{\Delta E}{\Delta S_z}. \tag{14.33}$$

The left-hand side of this equation can be regarded as an energy quantum of the spin wave, called a *magnon*. In a full quantum mechanical treatment of spin waves, a magnon constitutes a single spin flip which reduces the total S_z in the ground state by one [20]. Therefore, ΔS_z may be regarded as the number of magnons in the spin wave. Then, the formula (14.33) just says that the magnon energy is given by the energy of the spin wave divided by the number of magnons in the spin wave.

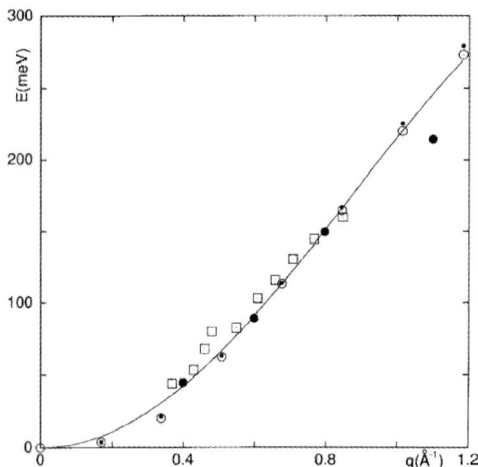

Fig. 14.2. Spin wave spectrum $\hbar\omega$ in the (100) direction of the BCC crystal of ferromagnetic Fe. The curve is a guide to the eye, the squares are experimental data of Loong *et al.* [161], and the circles are theoretical results [56, 85].

Combined with the results from density functional theory, the method described in this section has been successfully applied to calculate the spin wave spectrum for Fe [56, 85]. In Fig. 14.2, we show the spin wave spectrum for ferromagnetic Fe, where good agreement between theory and experiment is evident.

14.4 Geometric Phase in Doubly-Degenerate Electronic Bands

In this final section we consider a system of interacting doubly degenerate electronic bands where a non-trivial geometric phase appears spontaneously in the effective Hamiltonian for conduction electrons [138].

The Hamiltonian for this system is written as

$$H = \sum_{\mathbf{k}} \epsilon_n(\mathbf{k})(a_{\mathbf{k}}^\dagger a_{\mathbf{k}} + b_{\mathbf{k}}^\dagger b_{\mathbf{k}})$$
$$+ \sum_j (V_j a_j^\dagger b_j + V_j^* b_j^\dagger a_j) + \omega \sum_j |V_j|^2. \quad (14.34)$$

It may be viewed as the Hamiltonian for Jahn–Teller crystals such as manganese oxides,[3] where the first sum is the band energy for the doubly degenerate e_g electronic state of the manganese,[4] the second sum is the Jahn–Teller interaction between e_g electrons and e_g lattice distortions, and the third sum is the elastic energy of the Jahn–Teller distortions.

The same Hamiltonian may be also considered as part of a mean-field Hamiltonian for the Hubbard model given by

$$H_{\text{Hubbard}} = -t \sum_{\langle i,j \rangle} (a_i^\dagger a_j + b_i^\dagger b_j + \text{H.c.}) + U \sum_j a_j^\dagger a_j b_j^\dagger b_j, \quad (14.35)$$

where a_j and b_j denote annihilation operators for spin-up states $c_{j\uparrow}$ and spin-down states $c_{j\downarrow}$, respectively. In a mean-field approximation, the on-site Coulomb interaction term (the second sum in (14.35) is linearized as

---

[3] In an over-simplified view, a manganese oxide crystal is a periodic array of an octahedral complex MnO_6, which is an $E \otimes e$ Jahn–Teller system if the nominal manganese state is Mn^{3+} (see Appendix B for further information about the $E \otimes e$ Jahn–Teller system). In an O_h symmetry environment, $3d$ orbitals of the Mn ion split into a triply degenerate t_{2g} level and a doubly degenerate e_g level, where the former is localized at a manganese site and filled with three electrons with the same spin direction. Electrons in the e_g levels are itinerant (i.e., they move around the crystal) and form e_g conduction bands. The spin direction for the e_g electron is parallel to the underlying t_{2g} spins due to Hund's coupling, thus, the spin degrees of freedom for the e_g electrons are essentially lost.

[4] The degree of freedom associated with the local electronic state degeneracy or near-degeneracy is often called the *orbital degree of freedom*.

$$U \sum_j a_j^\dagger a_j b_j^\dagger b_j \approx -U \sum_j \langle b_j^\dagger a_j \rangle a_j^\dagger b_j - U \sum_j \langle a_j^\dagger b_j \rangle b_j^\dagger a_j$$
$$+ U \sum_j \langle a_j^\dagger b_j \rangle \langle b_j^\dagger a_j \rangle + U \sum_j (C_{aj} a_j^\dagger a_j + C_{bj} b_j^\dagger b_j),$$
(14.36)

where C_{aj} and C_{bj} are constants that satisfy $C_{aj} + C_{bj} = 1$, and $\langle \cdots \rangle$ denotes the ground state expectation value.[5] Thus, V_j in (14.34) corresponds to $-U\langle a_j^\dagger b_j \rangle$ in (14.36).

One way to obtain the energy spectrum of the Hamiltonian (14.34) is to perform so-called self-consistent calculations [109]. The self-consistency conditions are obtain as stationary conditions for the ground state energy $\langle H \rangle$ with respect to the variations of V_j^*:

$$\left\langle \frac{\partial H}{\partial V_j^*} \right\rangle = \omega V_j + \langle b_j^\dagger a_j \rangle = 0. \tag{14.37}$$

The self-consistent calculation starts with specifying initial values for V_j and repeating the following processes until the self-consistency is established:

1. Diagonalize the Hamiltonian (14.34) with the initial V_j (denoted as $V_j^{(\text{in})}$) to obtain the ground state.
2. Calcualte $\langle b_j^\dagger a_j \rangle$.
3. Update V_j (denoted as $V_j^{(\text{up})}$) using (14.37).
4. If $|V_j^{(\text{in})} - V_j^{(\text{up})}|$ are smaller than the tolerance ϵ_t (i.e., $|V_j^{(\text{in})} - V_j^{(\text{up})}| < \epsilon_t$), the self-consistency is established (end of the iteration); otherwise obtain the new initial V_j from $V_j^{(\text{in})}$ and $V_j^{(\text{up})}$ as $V_j = \alpha V_j^{(\text{in})} + (1 - \alpha) V_j^{(\text{up})}$ (typically, $\alpha \sim 0.7$), and go to 1.

Note that the final result depends on the starting V_j, therefore, in order to find the ground state, calculations must be repeated by changing the starting V_j. In the following, however, instead of performing the self-consistent calculations we simply assume that V_j can be written as $|V|e^{i\xi_j}$. We also omit the restoring energy term $\omega \sum_j |V_j|^2 = \omega \sum_j |V|^2$ since it merely gives a constant energy shift.

Now, we consider the simplest case where the spatial variation of $\xi(\mathbf{R})$ is given by $\nabla \xi(\mathbf{R}) = 1 = \text{const}$. Here, we assume that ξ is a continuous function

---

[5] The linearization (14.36) is based on the assumption that

$$0 \approx A_j(a_j^\dagger b_j - \langle a_j^\dagger b_j \rangle)(b_j^\dagger a_j - \langle b_j^\dagger a_j \rangle) + B_j(b_j^\dagger a_j - \langle b_j^\dagger a_j \rangle)(a_j^\dagger b_j - \langle a_j^\dagger b_j \rangle).$$

There are many ways to rewrite the on-site Coulomb interaction [90] and thus many ways to perform the mean-field approximation. Note that the mean field approximation for the Hubbard model is generally not reliable.

14.4 Geometric Phase in Doubly-Degenerate Electronic Bands

of the coordinate $\mathbf{R}$ and ξ_j is ξ at the j-th site $\mathbf{R}_j$ (namely, $\xi_j := \xi(\mathbf{R}_j)$). Then, we can analytically calculate the single-electron energy,

$$E(\mathbf{k})^{\pm} = \frac{\epsilon_n(\mathbf{k}) + \epsilon_n(\mathbf{k}+\mathbf{l})}{2} \pm \sqrt{\left(\frac{\epsilon_n(\mathbf{k}) - \epsilon_n(\mathbf{k}+\mathbf{l})}{2}\right)^2 + |V|^2}. \quad (14.38)$$

Under a translation $\hat{T}(\mathbf{a})$, the eigenvectors $\psi^{\pm}(\mathbf{k})$ associated with $E(\mathbf{k})^{\pm}$ transform according to

$$\begin{pmatrix} \psi^+(\mathbf{k}) \\ \psi^-(\mathbf{k}) \end{pmatrix} \to \hat{T}(\mathbf{a}) \begin{pmatrix} \psi^+(\mathbf{k}) \\ \psi^-(\mathbf{k}) \end{pmatrix} = e^{i((\mathbf{k}+\mathbf{l}/2)\cdot\mathbf{a})} U P U^{-1} \begin{pmatrix} \psi^+(\mathbf{k}) \\ \psi^-(\mathbf{k}) \end{pmatrix}, \quad (14.39)$$

where P, U are defined as

$$P := \begin{pmatrix} e^{i\mathbf{l}\cdot\mathbf{a}/2} & 0 \\ 0 & e^{-i\mathbf{l}\cdot\mathbf{a}/2} \end{pmatrix}, \quad U := \begin{pmatrix} p(\mathbf{k})^+ & p(\mathbf{k})^- \\ -p(\mathbf{k})^- & p(\mathbf{k})^+ \end{pmatrix}, \quad (14.40)$$

and

$$p(\mathbf{k})^{\pm} := \sqrt{\frac{1}{2} \pm \frac{\epsilon_n(\mathbf{k}+\mathbf{l}) - \epsilon_n(\mathbf{k})}{2\sqrt{(\epsilon_n(\mathbf{k}+\mathbf{l}) - \epsilon_n(\mathbf{k}))^2 + 4|V|^2}}}. \quad (14.41)$$

It is interesting to see that according to (14.39), the translation induces a phase $e^{i\mathbf{l}\cdot\mathbf{a}/2} P$ on the eigenvectors in addition to the ordinary Bloch phase $e^{i\mathbf{k}\cdot\mathbf{a}}$. This extra phase is nothing but a geometric phase [139]. In order to see this, we define the phase-dressed operator as

$$\tilde{a}_\mathbf{R} := e^{i\xi(\mathbf{R})} a_\mathbf{R}, \quad (14.42)$$

where $\mathbf{R}$ denotes the lattice site. Then, the connection one-form is obtained as

$$A := \begin{pmatrix} \langle \tilde{a}_\mathbf{R} \frac{d}{dt} \tilde{a}_\mathbf{R}^\dagger \rangle dt & \langle \tilde{a}_\mathbf{R} \frac{d}{dt} \tilde{b}_\mathbf{R}^\dagger \rangle dt \\ \langle \tilde{b}_\mathbf{R} \frac{d}{dt} \tilde{a}_\mathbf{R}^\dagger \rangle dt & \langle \tilde{b}_\mathbf{R} \frac{d}{dt} \tilde{b}_\mathbf{R}^\dagger \rangle dt \end{pmatrix} = \begin{pmatrix} i\nabla\xi \cdot d\mathbf{R} & 0 \\ 0 & 0 \end{pmatrix}, \quad (14.43)$$

where t is a parameter for the path $\mathbf{R}(t)$.

The geometric phase acquired after the translation by $\mathbf{a} = \mathbf{R}(1) - \mathbf{R}(0)$ is thus calculated as

$$\mathcal{P} \exp\left(\int_0^1 A\right) = \begin{pmatrix} \exp(i \int_{\mathbf{R}(0)}^{\mathbf{R}(1)} \nabla\xi \cdot d\mathbf{R}) & 0 \\ 0 & 0 \end{pmatrix}$$

$$= \begin{pmatrix} e^{i\mathbf{l}\cdot\mathbf{a}} & 0 \\ 0 & 1 \end{pmatrix}$$

$$= e^{i\mathbf{l}\cdot\mathbf{a}/2} P, \quad (14.44)$$

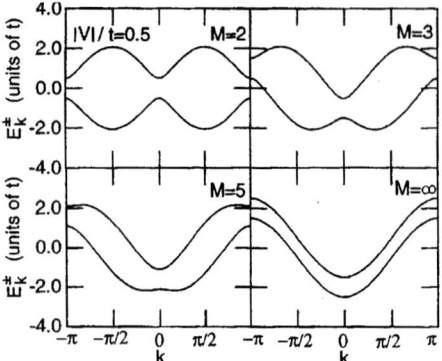

Fig. 14.3. Dispersion curves for several values of M ($l = 2\pi/M$) with $|V|/t = 0.5$. Reprinted with permission from [138].

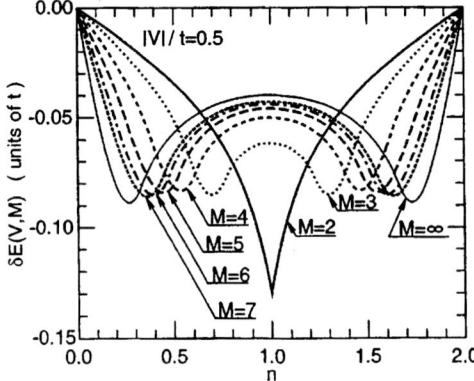

Fig. 14.4. Stabilization energy $\delta E(V, M) := E_M(V) - E_M(0)$ ($E_M(V)$ is the ground state energy with the coupling constant V and the lattice distortion period M) of the M-period Jahn–Teller distorted state as a function of the electron filling n. Reprinted with permission [138].

where $\mathcal{P}$ denotes the path-ordered integral. Therefore, the spatial variation of the phase ξ_j of the local coupling parameter V_j is shown to induce a geometric phase on the band electrons.

Next, we wish to examine the relation between the geometric phase and the band structure. Figure 14.3 shows the dispersion curves (graphs of the band energies as a function of the wave vector) for the one-dimensional band calculated with (14.38). Band dispersions are plotted for several values of M where M is related to the spatial variation of the phase, $\frac{d\xi}{dx} = l$, according to $l = 2\pi/M$. As seen from Fig. 14.4, depending on the electron filling n, the optimal value of M that minimizes the total electronic energy varies.

14.4 Geometric Phase in Doubly-Degenerate Electronic Bands

We now consider one-dimensional finite size ring systems in order to examine more closely the change of the electron-filling pattern caused by the geometric phase [135]. The results are displayed in Fig. 14.5. As seen from this figure, the optimal l value (the l value that gives the lowest energy), which depends on the number of electrons N_e in the ring, establishes the simultaneous occupancy of the time-reversal conjugate-pair states, $-K$ and K, irrespective of whether the number of electrons is even or odd. Here K is the generalized quasi-momentum given by $K = k + l/2$.

Because of this simultaneous occupancy of the time-reversal-pair states, the system will show a diamagnetic response to a magnetic field[6] [55]. In order to confirm this, we place a solenoid through the hole of the ring and calculate the electronic energy change by varying the magnitude of the magnetic flux through the solenoid. The result is displayed in Fig. 14.6. The optimal l changes depending on the magnitude of the magnetic flux. The energy minima are located at integral multiples of ch/e for the winding number $w = 5$ and half-odd integral multiples of ch/e for the winding number $w = 6$. Note that here the winding number is given by $w := \frac{N}{2\pi} l$ where N is the number of sites. As a consequence of the change in the optimal winding number (the winding number corresponds to the optimal l), local energy minima are located at integral multiples of $ch/2e$, i.e., the system shows flux quantization in units of $ch/2e$.[7] In order for this quantization to be observable, the energy barriers between the minima must be sufficiently large. This is not the case for a macroscopic one-dimensional ring, because the energy barriers between the energy minima are inversely proportional to the size of the ring. For a macroscopic ring they are negligibly small.

Next, we attempt to extend the above argument for the case where $\nabla \xi$ is not constant but position-dependent. In order to deal with this situation, we employ the effective mass formalism. We extend the single-band formalism given in Sect. 13.3.1 to that for two bands. Corrresponding to (13.10), we write the wave function ψ as

$$\psi(\mathbf{r}) = \frac{1}{\sqrt{2}} \sum_j \left[\psi_{an}(\mathbf{R}_j) W_a(\mathbf{R}_j) + e^{i\xi(\mathbf{R}_j)} \psi_{bn}(\mathbf{R}_j) W_b(\mathbf{R}_j) \right], \quad (14.45)$$

where $W_a(\mathbf{R}_j)$ and $W_b(\mathbf{R}_j)$ are Wannier functions for $a_j^\dagger |\text{vac}\rangle$ and $b_j^\dagger |\text{vac}\rangle$, respectively.

In view of the above analogies, we find a system of equations similar to (13.11) and (13.12), namely

---

[6] Here, "a diamagnetic response to a magnetic field" means that the direction of the induced current due to the application of a magnetic field is such that the induced magnetic field arising from the induced current is in the opposite direction to the applied one.

[7] At those local minima, the system shows a diamagnetic response. Consult [55] for the details.

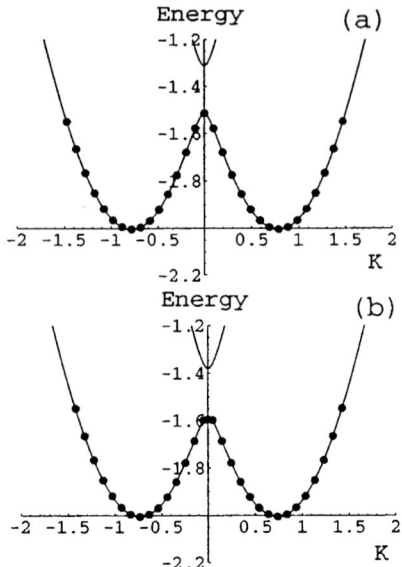

Fig. 14.5. Energy levels for the one-dimensional ring model with the optimized winding number w for the number of electrons N_e. Occupied levels are denoted by filled circles, $t = 1$, and $|V| = 0.1$, and the number of sites is 64. (a) $N_e = 31$, $w = 16$, (b) $N_e = 30$, $w = 15$ [135].

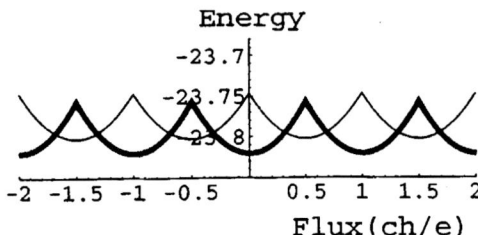

Fig. 14.6. Total energy of the one-dimensional ring model threaded with a magnetic flux. The optimal winding numbers are 5 and 6 for the cases where the magnetic flux are even (thick line) and odd (thin line) integral multiples of $ch/2e$, respectively. The numbers of sites and particles are respectively 32 and 12, $t = 1$, and $|V| = 0.3$ [135].

$$\epsilon_n(-i\nabla)\psi_{an} + |V|\psi_{bn} = E\psi_{an},$$
$$|V|\psi_{an} + \epsilon_n(-i\nabla + \nabla\xi)\psi_{bn} = E\psi_{bn}. \quad (14.46)$$

We can cast these equations in a single equation,

$$\{[\epsilon_n(-i\nabla) - E][\epsilon_n(-i\nabla + \nabla\xi) - E] - |V|^2\}\psi_{bn} = 0. \quad (14.47)$$

This equation has two energy branches. If the condition

$|V| \gg |\langle\psi_{bn}|\epsilon_n(-i\nabla) - \epsilon_n(-i\nabla + \nabla\xi)|\psi_{bn}\rangle|$

is satisfied, the eigenvalue equation for the lower branch takes the form

$$\left[\frac{\epsilon_n(-i\nabla) + \epsilon_n(-i\nabla + \nabla\xi)}{2} - |V|\right]\psi_{bn} = E\psi_{bn}. \tag{14.48}$$

Now, assume that the energy dispersion $\epsilon_n(\mathbf{k})$ is simply given by $\hbar^2 \mathbf{k}^2/(2m^*)$ where m^* is the effective mass. Then the Schrödinger equation is written as

$$\left[\frac{\hbar^2}{2m^*}(-i\nabla + \frac{1}{2}\nabla\xi)^2 + \frac{(\hbar\nabla\xi)^2}{8m^*} - |V|\right]\psi_{bn} = E\psi_{bn}. \tag{14.49}$$

It is remarkable that in this equation $\frac{\hbar}{2}\nabla\xi$ appears as a vector potential. This vector potential is equivalent to a fictitious electromagnetic vector potential $\mathbf{A}_{\text{fic}}$ given by

$$\mathbf{A}_{\text{fic}} = -\frac{\hbar c}{2e}\nabla\xi. \tag{14.50}$$

If the function ξ is single valued (and well defined all over the coordinate space), this may be simply removed by the gauge transformation

$$\psi_{bn} \to e^{-i\frac{1}{2}\xi}\psi_{bn}. \tag{14.51}$$

However, if ξ is multiple valued, this transformation is a singular-gauge transformation and not allowed.

The gauge potential $\mathbf{A}_{\text{fic}}$ leads to a non-trivial effect on the electronic motion. Namely, the conduction electrons are moving under the influence of a non-trivial fictitious magnetic flux. Note that the vector potential $\mathbf{A}_{\text{fic}}$ gives rise to a quantized flux,

$$\Phi = \oint \mathbf{A}_{fic} \cdot d\mathbf{r} = -\frac{\hbar c}{2e}\oint d\xi = -\frac{ch}{2e}n, \tag{14.52}$$

where n is an integer.

Flux states similar to the one considered here are also investigated in [5, 277]. There are arguments supporting the view that they may be relevant in explaining the anomolous transport properties observed in strongly correlated electronic systems.

Problem

14.1) Show that when $m = 1$, ψ is simply a determinant:

$$\det\{z_i^{j-1} e^{-|z_i|^2}\}.$$

A. An Elementary Introduction to Manifolds and Lie Groups

A.1 Introduction

The following is an elementary introduction to some basic concepts in modern differential geometry. The aim is to provide the reader with a clear understanding of the key ideas and to motivate the concepts rather than to promote the rigor and generality pursued in mathematical texts. Specifically, our main objective is to arrive at a comprehensive description of *smooth manifolds, Lie groups*, and their most basic properties.

In particular, we shall introduce the following specific concepts:

1. metric space;
2. open and closed subsets of a metric space;
3. equivalence relation;
4. bijection;
5. isometry;
6. continuous function between metric spaces;
7. topological space;
8. continuous function between topological spaces;
9. homeomorphism;
10. connected topological space;
11. compact topological space;
12. topological manifold;
13. differentiable and smooth manifolds;
14. diffeomorphism;
15. complex projective space;
16. smooth curve on a smooth manifold;
17. tangent and cotangent spaces;
18. vector and tensor fields;
19. associative algebra;
20. differential forms and exterior differentiation;
21. push-forward map;
22. pullback map for differential forms;
23. compact manifold;
24. group;
25. group homomorphism and isomorphism;

26. Lie group;
27. action of a Lie group on a manifold;
28. transitive and free actions;
29. Lie algebra of a Lie group;
30. abstract Lie algebra;
31. unitary groups;
32. representation of a Lie group;
33. irreducible representation;
34. representation of a Lie algebra;
35. enveloping algebra;
36. Casimir operator.

The abstract definitions are supplemented with concrete examples.

For a more thorough treatment of this material, we wish to refer the reader to the following introductory textbooks:

- B. Schutz: *Geometrical Methods of Mathematical Physics* (Cambridge University Press, Cambridge, 1980);
- C. Isham: *Modern Differential Geometry for Physicists* (World Scientific, Singapore, 1989);
- M. Nakahara: *Geometry, Topology and Physics* (Adam Hilger, Bristol, 1990);
- C. Nash and S. Sen: *Topology and Geometry for Physicists* (Academic Press, London, 1983);
- Y. Choquet-Bruhat, C. De Witt-Morette, and M. Dillard-Bleick: *Analysis, Manifolds and Physics*, Parts I and II (North-Holland, Amsterdam, 1989);
- R. Geroch: *Mathematical Physics* (The University of Chicago Press, Chicago, 1985);
- V. Guillemin and A. Pollack: *Differential Topology* (Prentice-Hall, Inc., New Jersey, 1974).

A more advanced book on the subject is:

- S. Helgason: *Differential Geometry, Lie Groups, and Symmetric Spaces* (Academic Press, New York, 1978).

Before going through the discussion of manifolds and Lie groups, we would like to introduce the reader to some more basic mathematical concepts.

Consider the real line $\mathbb{R}$ as a set of points (numbers). Let $[a,b]$ and (a,b) denote the "closed" and "open" intervals in $\mathbb{R}$, with $a < b$. The points a and b are called the "boundary" or "limit points" of both $[a,b]$ and (a,b). Alternatively, we say that the set $\{a,b\}$ is the "boundary" of $[a,b]$ and (a,b). The distinguishing feature between these two intervals is that $[a,b]$ includes its boundary whereas (a,b) does not. We can use this property to define the notions of "open" and "closed" sets in $\mathbb{R}$, namely we define a *closed set* to be a subset of $\mathbb{R}$ that includes its boundary. Similarly, we call a subset *open*, if it does not include any of its boundary points.[1]

In view of these definitions, we can make several interesting observations. First, the complement of an open set is closed and vice versa. However, this does not contradict the existence of subsets that are neither open nor closed. The half-open (half-closed) intervals such as $[a, b)$ are examples of such subsets. Next, we can easily convince ourselves that the union and the intersection of two open (closed) sets are open (respectively closed). The only trouble seems to be the case where the two open (closed) sets have an empty intersection. This is remedied by postulating that the empty set and the universal set, in this case $\mathbb{R}$, are both open and closed. This assumption is also in agreement with the statement that the complement $\mathbb{R} - O$ of an open (closed) set O is closed (open). Furthermore, we can show that the union of any infinite collection of open sets is also open whereas their intersections may or may not be open. A standard example is the infinite family of open intervals defined by

$$\left\{ \left(-\frac{n+1}{n}, \frac{n+1}{n} \right) : n \in \mathbb{Z}^+ \right\}.$$

Evidently, the union of all such intervals is the open interval

$$\bigcup_{n=1}^{\infty} \left(-\frac{n+1}{n}, \frac{n+1}{n} \right) = (-2, 2),$$

whereas their intersection,

$$\bigcap_{n=1}^{\infty} \left(-\frac{n+1}{n}, \frac{n+1}{n} \right) = [-1, 1]$$

is a closed interval.

An important fact about point sets such as $\mathbb{R}$ is that each point p can be included in some open interval. For example, for every positive number $\epsilon \in \mathbb{R}^+$, $p \in (p - \epsilon, p + \epsilon)$. The open subset

$$O_\epsilon(p) := (p - \epsilon, p + \epsilon) = \{x \in \mathbb{R} : |x - p| < \epsilon\}$$

is called an "open neighborhood" or simply a "neighborhood" of p.

So far, we have introduced several simple mathematical concepts using the example of the real line as our universal set. The set of real numbers enjoys the property of having a natural notion of distance between its elements. This is indeed the main ingredient that allows us to define limit points, open and

---
[1] These definitions are clearly based on the definition of a *boundary* or *limit point*. We do not offer the latter at this stage hoping that the reader is guided by his or her intuition. We shall present precise definitions of open and closed sets later in this appendix.

closed subsets, and the neighborhoods of points. The set $\mathbb{R}$ with its usual notion of distance is an example of a large class of mathematical structures called *metric spaces*. These are sets of points with a well-defined concept of distance between their elements. The following definition describes metric spaces in precise terms.

Definition 1: Let X be a set of points and $d : X \times X \to [0, \infty)$ be a function that assigns a non-negative real number to any two elements of X and satisfies the following conditions:

1) For every $p, q \in X$, $d(p, q) = 0$ if and only if $p = q$.
2) For every $p, q \in X$, $d(p, q) = d(q, p)$.
3) For every $p, q, r \in X$, $d(p, q) + d(q, r) \geq d(p, r)$.

Then, the pair (X, d) is said to be a **metric space** and d is called a **metric** or a **distance function**. Alternatively, it is said that X has a **metric structure** specified by d.

Once a set is endowed with a metric structure we can immediately define the *open neighborhoods* of its points. For every $p \in X$ and $\epsilon > 0$, we define the open neighborhoods of p by

$$N_\epsilon(p) := \{x \in X \,:\, d(x, p) < \epsilon\}.$$

These are also called the *open balls* centered at p. Having defined the notion of an open neighborhood, we proceed with the definitions of open and closed subsets of a metric space.

Definition 2: Let (X, d) be a metric space. A subset $O \subseteq X$ is said to be an **open subset** if every point $x \in O$ has at least an open neighborhood $N_\epsilon(x)$ that lies entirely inside O, i.e., $N_\epsilon(x) \subset O$. A subset $C \subseteq X$ is said to be a **closed subset** if its complement, $X - C$, is open.[2]

Using this definition and assuming that the empty set $\emptyset$ and the universal set X are both open and closed, we can show that the finite and infinite unions and finite intersections of open sets are open. We shall see that these properties play an important role in generalizing the concepts of openness and closedness to the structures that lack the notion of distance. This is a general pattern in the methodology of mathematics. The results or theorems that are derived as logical consequences for special structures can be employed as postulates for more general structures. In this case, these more general structures are called *topological spaces*.

In mathematics once one defines a structure such as a metric space, the next task becomes to investigate the problem of the classification of such structures. In order to pursue the classification problem, one first needs to

---

[2] We can alternatively give a definition of limit points for metric spaces and retain our definitions of open and closed sets for $\mathbb{R}$. Following this approach, a point p is said to be a *limit point* of a subset $Y \subset X$, if every open neighborhood of p intersects both Y and $X - Y$. Then, a subset is called open if it does not include any of its limit points. It is called closed if it includes all its limit points.

have a clear understanding of the notion of "equivalence" of two structures. For example, to be able to compare two metric spaces one must know under what conditions they are "identical" or "equivalent." This raises the question of what "equivalence" means in precise mathematical language. It does not take much effort to realize that an "equivalence" is a sort of "relation" between two objects. It has three intuitively sound and simple properties.

Definition 3: Let $\mathcal{S}$ be a collection of objects and $\sim$ be a relation between any two of its elements. The relation $\sim$ is said to be an **equivalence relation** if it satisfies the following conditions:

1) For every $s \in \mathcal{S}$, $s \sim s$ (reflexivity).
2) For every $s_1, s_2 \in \mathcal{S}$, $s_1 \sim s_2$ implies $s_2 \sim s_1$ (symmetry).
3) For every $s_1, s_2, s_3 \in \mathcal{S}$, if $s_1 \sim s_2$ and $s_2 \sim s_3$, then $s_1 \sim s_3$ (transitivity).

An important property of an equivalence relation is that it divides the universal collection $\mathcal{S}$ into distinct (non-intersecting) subcollections. These are called the *equivalence classes*. As is clear from the nomenclature, each equivalence class consists of objects that are equivalent, i.e., related by the equivalence relation. Whence, each member of an equivalence class can represent the whole class as equally well as any other member.

In physics, one usually uses the word *symmetry* when such a situation occurs. Physicists would say that there is a symmetry between the members of each equivalence class that allows one to represent the whole class using a particular member. Symmetry is a desirable quality because it permits the freedom of choice of the representative for each class.[3] In this sense, the notion of symmetry is associated with the notion of equivalence. In the following, we shall examine some examples of equivalence relations in the mathematical arena.

Let us consider the set of positive integers (natural numbers) $\mathbb{Z}^+$. For any pair $p, n \in \mathbb{Z}^+$, we have

$$p = mn + r,$$

where m and r are integers such that $m \geq 0$ and $0 \leq r < n$. "r" is called the *remainder* (of the division of p by n). We know from arithmetic that m and r are uniquely determined. We can use this result to set up an equivalence relation between any two integers. Let us choose a positive integer n and define any two positive integers p_1 and p_2 to be equivalent if they correspond to the same remainder r upon division by n. In mathematical symbols, we write

$$p_1 \equiv p_2 \ (\mathrm{mod} \ n).$$

This means that there are $m_1, m_2 \in \mathbb{Z}^+$, such that

$$p_1 = m_1 n + r \quad \text{and} \quad p_2 = m_2 n + r.$$

---

[3] Often, a clever choice can simplify the study of the particular problem appreciably. A concrete example of this is the gauge symmetry of electromagnetism.

We can check that, indeed, this satisfies the requirements of Def. 3. The equivalence classes of this equivalence relation are labelled by $r \in \{0, 1, \cdots, n-1\}$ and denoted by $\bar{r}$. The set of all the equivalence classes

$$\mathbb{Z}_n := \{\bar{0}, \bar{1}, \cdots, \overline{n-1}\}$$

is called *integers modulo n*. For example, let us choose $n = 2$. Then, there are two equivalence classes $\bar{0}$ and $\bar{1}$. These are known as *even* and *odd* numbers.

Next, let us consider the collection of all finite sets with no further structures on them. Two finite sets are distinguished by the "number" of their elements. In other words, two sets are said to be *equivalent* if they have the same "number" of elements. This is obviously an equivalence relation. It divides the collection of all finite sets into equivalence classes of sets with the same number of elements. The same definition is rather unsatisfactory for infinite sets since the notion of the number of elements is not well defined. A simple generalization of this notion, however, works perfectly well.

Definition 4: Let X and Y be two sets and $f : X \to Y$ be a function (map), i.e., f assigns to each element of X one and only one element of Y. The subset

$$f(X) := \{y \in Y : y = f(x) \text{ for some } x \in X\} \subseteq Y$$

is called the *image* of X under f. In some cases $f(X)$ is identical to Y but not always. If $f(X) = Y$, then f is called an *onto* or *surjective* function. Let Y_1 be a subset of Y. Then the subset

$$f^{-1}(Y_1) := \{x \in X : f(x) \in Y_1\} \subseteq X$$

is called the *preimage* or *inverse image* of Y_1 under f. Inverse images of subsets of Y are subsets of X. The inverse image of a subset Y_1 of Y which includes only a single point, i.e., $Y_1 = \{y\}$, is called the inverse image of that point, $y \in Y$. It may happen that the inverse image of a point $y \in Y$ is empty, $y \notin f(X)$, or that it consists of many elements. If the inverse images of all the points of Y have at most a single element, then f is called a *one-to-one* or *injective* function. This simply means that every $y \in f(X)$ is the image of a single point $x \in X$. The condition of one-to-oneness is the necessary and sufficient condition for the existence of the inverse function

$$f^{-1} : f(X) \subseteq Y \longrightarrow X$$

of f. The function f is said to be a **bijection** or a **one-to-one correspondence**, if it is both onto and one-to-one. Two sets X and Y are said to be *bijective* if there exists a bijection f between them.[4]

The relation of being bijective for arbitrary sets is the appropriate generalization of having the same number of elements for finite sets. We can easily

---

[4] A bijection is also called a bijective function.

show that bijective finite sets have the same number of elements. Moreover, we can check that the relation defined in this way is an equivalence relation, i.e., it satisfies all the necessary requirements of Def. 3. The utility of this equivalence relation is in the classification of all point sets. In fact, in set theory, one does not distinguish between bijective sets.

Let us return to our discussion of metric spaces where the issues of the identification and classification of mathematical structures were raised. Naturally, two "equivalent" metric spaces must be necessarily "equivalent" as sets. Therefore, the notion of equivalence is again linked to the existence of certain functions between metric spaces. Since a metric space has an additional metric structure besides the point set structure, the notion of equivalence of metric spaces is a refinement of that of point sets.

Definition 5: Let (X_1, d_1) and (X_2, d_2) be two metric spaces and $f : X_1 \to X_2$ be a bijection. f is said to be an **isometry** if f preserves the metric structures, d_1 and d_2.[5] In more precise language, for all $p_1, q_1 \in X_1$ and $p_2, q_2 \in X_2$

$$d_1(p_1, q_1) = d_2(f(p_1), f(q_1)).$$

Two metric spaces are said to be **isometric** if there exists an isometry between them.

Once more, the relation of isometry satisfies the axioms of an equivalence relation and it divides the collection of all metric spaces into distinct classes of isometric metric spaces.

We saw that the notions of open and closed subsets can be easily defined for metric spaces. Let us collect all the open sets of a metric space and then consider this collection without any reference to the metric structure on the universal set. There are important properties of this derived structure which can be defined and analyzed regardless of the details of the metric structure, i.e., the distance function. One of these properties is related to the existence of "continuous" functions.

The definition of a continuous function between two metric spaces is almost identical to the one presented in elementary calculus texts for the continuity of a function of a real variable.

Definition 6: Let (X_1, d_1) and (X_2, d_2) be two metric spaces. A function $f : X_1 \to X_2$ is said to be **continuous** at a point $p_1 \in X_1$ if for every $\epsilon > 0$, there is a $\delta > 0$ such that

$$d_1(p_1, x_1) < \delta \stackrel{\text{implies}}{\Longrightarrow} d_2(f(p_1), f(x_1)) < \epsilon,$$

for all such $x_1 \in X$. Alternatively, f is continuous at p_1, if for every neighborhood $N_\epsilon(f(p_1)) \subset X_2$ of the point $f(p_1)$ there is a neighborhood $N_\delta(p_1)$ of p_1 such that

$$f(N_\delta(p_1)) \subset N_\epsilon(f(p_1)).$$

---

[5] It is not difficult to see that if f preserves the metric structure, so does its inverse.

This definition is a little abstract. The reader is advised to draw the graph of a simple function of a real variable and examine the utility of the above definition in practice.

An important characterization of continuous functions is the following result.

Proposition 1: A function $f : X_1 \to X_2$ is continuous if and only if the inverse image of every open subset of X_2 is open in X_1.

A simple but rather instructive consequence of this result is that the notion of continuity does not directly depend on the particular metric structures of the corresponding metric spaces. For example, let us assume that (X_1, d'_1) is another metric structure on X_1 such that both distance functions, d_1 and d'_1, define the same collection of open subsets in X_1. This is to say that every open set in (X_1, d_1) is also open in (X_1, d'_1) and vice versa. Then, the continuity of a function f will not depend on which metric function we choose on X_1. The same is true for X_2. The concept of continuity, therefore, is only sensitive to the collection of open sets of the two spaces.

Let us consider the following two choices of metric function on $\mathbb{R}^2$:

$$d_1(\mathbf{x}, \mathbf{y}) := \sqrt{(x^1 - y^1)^2 + (x^2 - y^2)^2}$$
$$d'_1(\mathbf{x}, \mathbf{y}) := |x^1 - y^1| + |x^2 - y^2|.$$

It is easy to observe that both d_1 and d'_1 satisfy the axioms of a metric function. Further, they define the same notion of open subsets and consequently continuity in $\mathbb{R}^2$.

As we argued in the preceding paragraphs, the continuity of a function does not depend on the details of the metric structure. This triggers the question of identifying the minimal structure on a point set that allows for the concept of continuity to be defined. Evidently, such a structure will be more general than a metric structure and less trivial than a plain set structure. This structure is called a *topological structure* or simply a *topology* on a set.

Definition 7: Let X be a set and $\mathcal{T}$ be a family of subsets of X such that

1) The empty set $\emptyset$ and the universal set X belong to $\mathcal{T}$.
2) If O_1 and O_2 belong to $\mathcal{T}$, then so does $O_1 \cap O_2$, i.e., the intersection of a finite number of elements of $\mathcal{T}$ is also an element of $\mathcal{T}$.
3) The union of any finite or infinite number of elements of $\mathcal{T}$ is also an element of $\mathcal{T}$.

Then, the pair $(X, \mathcal{T})$ is said to be a **topological space**. The collection $\mathcal{T}$ is called a **topology** on X. The elements of $\mathcal{T}$ are called the *open subsets* of X.

In other words, a topology on a set is an assignment of the word "open" to a collection of its subsets that possess certain properties of open subsets of say metric spaces. If the set is endowed with a metric structure, then the open subsets defined by the metric satisfy the requirements of a topological space. Thus any metric space has a canonical topological structure which is called the *metric topology*. However, there is no unique topology on a given set (unless the set is empty). In particular, a metric space may be given a non-metric topology.

The notion of *continuity* can be easily defined for functions between topological spaces. This can be done either by first defining the open neighborhoods and then using a similar definition to the one presented for metric spaces, namely Def. 6, or by promoting Proposition 1 to a definition.

Definition 8: Let $(X_1, \mathcal{T}_1)$ and $(X_2, \mathcal{T}_2)$ be two topological spaces and $f : X_1 \to X_2$ be a function. Then, f is said to be continuous if the inverse image of every open subset of X_2 (every element of $\mathcal{T}_2$) is an open subset of X_1 (an element of $\mathcal{T}_1$).

Having defined topological spaces, we shall next try to define an appropriate concept of "equivalence" for topological spaces. This is done in complete analogy to the cases of point sets and metric spaces. Again the equivalence of topological spaces must reduce to that of total or universal sets. Thus, we need a bijection that preserves the topological structures. Such a function is called a *homeomorphism*.

Definition 9: A function $f : X_1 \to X_2$ between two topological spaces $(X_1, \mathcal{T}_1)$ and $(X_2, \mathcal{T}_2)$ is said to be a **homeomorphism**, if it is a continuous bijection with a continuous inverse. If there exists a homeomorphism between two topological spaces, they are called **homeomorphic**.

The existence of a homeomorphism between two topological spaces defines an equivalence relation. This equivalence relation divides the collection of all topological spaces into equivalence classes of homeomorphic topological spaces. The members of each class share the same *topological properties*. These are properties that are defined using the notion of open subsets. An intuitively simple example of a topological property is *connectedness*.

Definition 10: A topological space $(X, \mathcal{T})$ is said to be **disconnected** if there are two open subsets O_1 and O_2 such that $X = O_1 \cup O_2$ and $O_1 \cap O_2 = \emptyset$. If a topological space is not disconnected, it is said to be **connected**.

We can show that under a homeomorphism a connected topological space is mapped to another connected topological space. Alternatively, there is no homeomorphism between a connected and a disconnected topological space. Hence, connectedness is a *topological property*.

We saw that in a topological space the unions of open sets are also open. This simple property suggests a practical way of generating all the open subsets as the unions of some "more basic" ones. The collection of these basic open sets is called a *basis* of the topological space. More precisely, a subfamily $\mathcal{B}$ of a topology $\mathcal{T}$ is called a *basis* if all the elements of $\mathcal{T}$, i.e., all the open subsets, are obtained as the unions of the elements of $\mathcal{B}$.

There are other important collections of open subsets of a topological space $(X, \mathcal{T})$, subfamilies of $\mathcal{T}$. For example, let O be a subset and consider a family of open subsets $\mathcal{C} = \{O_\alpha\}$, i.e. a subfamily of $\mathcal{T}$, such that $O \subseteq \cup_\alpha O_\alpha$. Then, $\mathcal{C}$ is called an *open covering* of O. Certainly, a basis of $\mathcal{T}$ is an open covering of every subset of X. The converse is certainly not true. A subset $\mathcal{C}'$ of an open covering $\mathcal{C}$ is naturally called a subcovering.

Definition 11: A topological space $(X, \mathcal{T})$ is said to be **compact** if every (infinite) open covering of X has a finite subcovering.

Compactness is also a topological property, i.e., under a homeomorphism a compact topological space is mapped to another compact topological space. Compactness plays a very substantial role in the study of a special class of topological spaces, called manifolds, and particularly Lie groups. We shall give a more intuitive definition of compactness which is valid for manifolds in Sect. A.2.

A simple example of a topological space is the space $\mathbb{R}^n$ with a metric topology. Usually, we choose the Euclidean metric,

$$d(\mathbf{x}, \mathbf{y}) = \sqrt{\sum_{i=1}^{n}(x^i - y^i)^2},$$

to define the open neighborhoods and hence the open subsets. A simple basis for this topological space is given by the subfamily of all open balls:

$$\mathcal{B} = \{N_r(\mathbf{x}) : \mathbf{x} \in \mathbb{R}^n, r \in \mathbb{R}^+\}.$$

Now, let us consider the following family of open subsets:

$$\mathcal{C} = \{N_2(\mathbf{n}) : \mathbf{n} \in \mathbb{Z}^n \subset \mathbb{R}^n\}.$$

$\mathcal{C}$ is an open covering of $\mathbb{R}^n$. It is however not a basis. One can easily show that none of these coverings has a finite subcovering. Hence $\mathbb{R}^n$ with the Euclidean metric topology is not compact.

There are other topological structures on the set $\mathbb{R}^n$. In fact, there is an infinite number of them. Two rather trivial examples of non-metric topologies on $\mathbb{R}^n$ are

1) $\mathcal{T}_0 := \{\emptyset, \mathbb{R}^n\}$, i.e., the only open subsets are the empty set and the total space.
2) $\mathcal{T}_{\text{discrete}} := \{O : O \subseteq \mathbb{R}^n\}$, i.e., all the subsets are open.

Similar topologies can be given to any other point set X. They are known as the *trivial topology* and the *discrete topology* on X, respectively.

We shall always assume that $\mathbb{R}^n$ is endowed with the Euclidean metric topology.

Another useful fact about topological spaces is that we can induce a topology on a subset X_1 of the universal set X. This is simply done by defining the open subsets of X_1 to be the intersections of the open subsets of X and X_1. This topology is called the *subspace topology*. This allows us, for example, to speak of a homeomorphism between the subsets of two topological spaces.

As for point sets, we can define the Cartesian product of two topological spaces $(X, \mathcal{T})$ and $(X', \mathcal{T}')$. The result is called the *product topology*. It is naturally defined to be a topology on the Cartesian product $X \times X'$, whose elements (open subsets) are the Cartesian products of the elements of $\mathcal{T}$ and $\mathcal{T}'$ (open subsets of X and X'). An example of a product topology is the metric topology on $\mathbb{R}^{m+n}$. This is the product of the metric topologies on $\mathbb{R}^m$ and $\mathbb{R}^n$. Therefore, the metric topology on $\mathbb{R}$ generates $\mathbb{R}^m$, for all $m \in \mathbb{Z}^+$, as its products.

So far, we have introduced many mathematical concepts that are usually not familiar to non-mathematicians. We shall need these concepts to arrive at a fairly precise definition of a *manifold*. Our list of related topological concepts is however far from being complete. We wish to refer the interested reader to textbooks on topology, such as

- J. G. Hocking and G. S. Young: *Topology* (Dover Publications Inc., New York, 1988);
- G. F. Simmons: *Topology and Modern Analysis* (R. E. Krieger Publishing Company, Malabar, Florida, 1983).

A.2 Differentiable Manifolds

Throughout the development of geometry and topology the space $\mathbb{R}^m$ has served mathematicians as a source of key ideas and properties. Many of these ideas and properties could be generalized to define various abstract mathematical structures which are occasionally used to formulate and solve concrete physical problems. An important class of such structures that are extensively used in theoretical physics is the class of the so-called *manifolds*. Manifolds are certain topological spaces which are obtained by patching together open pieces of $\mathbb{R}^m$. In other words, a manifold is the union of a number of open subsets each of which is homeomorphic to (an open subset) of $\mathbb{R}^m$. These open subsets are called *coordinate patches* or *charts*. On each of these patches one can set up a coordinate system. This is done by mapping the points of the patch into $\mathbb{R}^m$ using the corresponding homeomorphism. In this way, one translates the local properties of a manifold into those of $\mathbb{R}^m$ and employs the knowledge of $\mathbb{R}^m$ to develop calculus, analysis, and even geometry. The main complication arises from the fact that the results obtained in one patch are only valid locally. Thus, it is necessary to check the validity of local results when applied to the whole space. On the other hand, since the choice of the coordinate charts is not unique, the physical results of any computation must be independent of this choice.

Definition 12: Let M be a topological space with a countable basis.[6] M is said to be a **topological manifold** of dimension m, if there exists an

---
[6] This means that the elements of the basis can be labelled by integers.

open covering $\{O_\alpha\}$ of M such that each O_α is homeomorphic to (some open subset of) $\mathbb{R}^m$, for a fixed m. The homeomorphisms

$$\phi_\alpha : O_\alpha \longrightarrow \phi_\alpha(O_\alpha) \subseteq \mathbb{R}^m,$$

together with the open subsets O_α are called the **charts** of the manifold. A complete collection of charts is called an **atlas**. For every pair of intersecting charts, e.g., $O_\alpha \cap O_\beta \neq \emptyset$, the functions

$$g_{\alpha\beta} := \phi_\alpha \circ \phi_\beta^{-1}\big|_{O_\alpha \cap O_\beta} : \phi_\beta(O_\alpha \cap O_\beta) \longrightarrow \phi_\alpha(O_\alpha \cap O_\beta)$$

are homeomorphisms between open subsets of $\mathbb{R}^m$. These functions are called the **transition** or **overlap functions**.

The topology of a manifold, i.e., the collection of all its open subsets, can be recovered from the (Euclidean metric) topology of $\mathbb{R}^m$. In fact, all the open subsets of a manifold are unions of images of open subsets of $\mathbb{R}^m$ under the ϕ_α^{-1}'s. Moreover, since all m-dimensional manifolds are locally "identical", the global or topological properties of a manifold depend on the way these open subsets are patched or glued together. This is done by the transition functions. Therefore, the topology of a manifold is determined by its transition functions.

Definition 13: Let $U \subseteq \mathbb{R}^{m_1}$ and $V \subseteq \mathbb{R}^{m_2}$. Then, a function $g : U \to V$ is said to be a C^N, $0 < N \leq \infty$, function if g is N-times differentiable.[7] A C^∞ function is also called a **smooth function**. g is called a **diffeomorphism** if it is a differentiable homeomorphism with a differentiable inverse. Similarly, one defines C^N diffeomorphisms by requiring a diffeomorphism and its inverse to be C^N. If the transition functions $g_{\alpha\beta}$ are C^N diffeomorphisms, then the manifold is called a C^N manifold. In particular, C^1 and C^∞ manifolds are called **differentiable** and **smooth manifolds**, respectively. Throughout this book all the manifolds are assumed to be smooth.

The homeomorphisms ϕ_α and their inverses ϕ_α^{-1} enable us to treat the points of O_α as those of a subset of $\mathbb{R}^m$. Clearly, the space $\mathbb{R}^m$ is itself a manifold. It can be covered by a single chart. If the minimum number of charts that cover a manifold is more than one, then the manifold has a different topology than $\mathbb{R}^m$. A simple example of such a manifold is the two-dimensional sphere, S^2. We shall examine the manifold structure of S^2 in detail.

As the reader must have noticed, the concept of a manifold is quite abstract. A practical way of thinking about manifolds is to think about them as the generalizations of surfaces. In fact, the simplest, and at the same time interesting, examples of a manifold are two-dimensional surfaces such as the sphere and torus. These examples are easily understood, for they can be visualized as sitting inside $\mathbb{R}^3$. In mathematical language, one says that these

---

[7] Occasionally, the symbol C^0 is used for the set of *continuous* functions.

manifolds are *submanifolds* of $\mathbb{R}^3$. A submanifold is a manifold which is a subset of another manifold. Its charts are the restrictions of those of the "bigger" manifold. If M' is a submanifold of M, then the charts $(O'_\alpha, \phi'_\alpha)$ of M' are given by

$$O'_\alpha := M' \cap O_\alpha, \quad \phi'_\alpha := \phi_\alpha|_{O'_\alpha}.$$

A substantial result of differential geometry is the *Whitney Embedding Theorem*. This theorem says that every manifold is a submanifold of $\mathbb{R}^d$ for some $d \in \mathbb{Z}^+$. Thus, we can always think of manifolds as direct generalizations of two-dimensional surfaces to higher dimensions. Nevertheless, we must keep in mind that manifolds are well-defined mathematical objects by themselves. They can be studied independently of their relation to an embedding space.

As we mentioned earlier, the sphere S^2 is an example of a "non-trivial" smooth manifold.[8] It is usually thought of as the submanifold of $\mathbb{R}^3$ defined by the set of unit vectors,

$$\{\mathbf{x} \in \mathbb{R}^3 : |\mathbf{x}| = 1\} \subset \mathbb{R}^3.$$

This is called the *round sphere*. The round sphere has more structure than a smooth manifold. Specifically, it inherits a "geometric" or "metric structure" from $\mathbb{R}^3$. This brings us to the problem of the classification and the necessity of a definition of "equivalence" for smooth manifolds. We know that a manifold is a topological space. Thus, the concept of equivalence should be defined via certain homeomorphisms between two manifolds so that the equivalent manifolds have the same topological structures. Furthermore, this homeomorphism must reflect the information about the smoothness and the chart structures. Such a function is also called a *diffeomorphism*.

Definition 14: Let M_1 and M_2 be two differentiable (C^N) manifolds whose charts are given by $(O_{\alpha_1}, \phi_{\alpha_1})$ and $(O_{\alpha_2}, \phi_{\alpha_2})$. Then, any function $f : M_1 \to M_2$, can be defined by its restrictions:

$$f_{\alpha_{1,2}} := \phi_{\alpha_2} \circ f \circ \phi_{\alpha_1}^{-1} : \phi_{\alpha_1}(O_{\alpha_1}) \longrightarrow \phi_{\alpha_2}(O_{\alpha_2}).$$

These are functions from $\mathbb{R}^{m_1}$ to $\mathbb{R}^{m_2}$, where m_1 and m_2 are the dimensions of M_1 and M_2, respectively. If all $f_{\alpha_{1,2}}$ are differentiable (C^N) functions then the function f is said to be differentiable (C^N). Similarly, a homeomorphism $f : M_1 \to M_2$ is said to be a (C^N) **diffeomorphism of manifolds**, if the corresponding functions $f_{\alpha_{1,2}}$ are (C^N) diffeomorphisms.[9]

The notion of (C^N) *diffeomorphy* defines an equivalence relation. This relation divides the collection of all (C^N) differentiable manifolds into the equivalence classes of (C^N) diffeomorphic manifolds. The elements of each class are treated as identical manifolds. In this respect, the round sphere is a

---

[8] A trivial manifold means that it is topologically equivalent (homeomorphic) to $\mathbb{R}^m$ for some $m \in \mathbb{Z}^+$. In other words, a manifold is called trivial if it can be covered by a single chart.

[9] Note that we have already defined the notion of a (C^N) diffeomorphism for $\mathbb{R}^m$.

representative of an infinite class of smooth manifolds which are diffeomorphic to one another. For example, the surface of an ellipsoid can be obtained by a smooth deformation of the round sphere and vice versa. Thus, it belongs to the same diffeomorphy class and can represent the sphere S^2 equally well.

Let us examine the manifold structure of S^2. As we mentioned, S^2 is (topologically) different from $\mathbb{R}^2$. In fact, we need at least two charts to cover S^2. A practical choice of coordinates for the points of S^2 is the spherical coordinates (θ, φ). The fact that the global (θ, φ) coordinate system is ill defined at $\theta = 0$ and $\theta = \pi$, is an indication of the necessity for at least two different coordinate charts. A rather standard set of coordinate charts for S^2 is obtained by what is known as the *stereographic projection* of S^2 on $\mathbb{R}^2$. The corresponding open covering consists of two open subsets:

$$O_1 := S^2 - \{\mathcal{S}\} \text{ and } O_2 := S^2 - \{\mathcal{N}\}.$$

Here $\mathcal{S}$ and $\mathcal{N}$ denote the south and the north poles of S^2, respectively. The homeomorphisms

$$\phi_i : O_i \longrightarrow \mathbb{R}^2, \quad i = 1, 2,$$

are defined by the following projective procedure.

Consider the tangent plane P_1 to S^2 at the north pole, $\mathcal{N}$. Let R be an arbitrary point of O_1 and draw a line through R and the south pole, $\mathcal{S}$ (Fig. A.1). This line intersects P_1 at a point p_1. The map ϕ_1 is defined by

$$\phi_1(R) = p_1.$$

By interchanging the roles of the north and the south poles we define ϕ_2 similarly. It is clear that ϕ_1 and ϕ_2 are homeomorphisms of a punctured sphere to $\mathbb{R}^2$.

Using ϕ_1 and ϕ_2, we can treat the points of S^2 as those of P_1 or P_2. On each of these planes, we can set up a coordinate system and use our knowledge

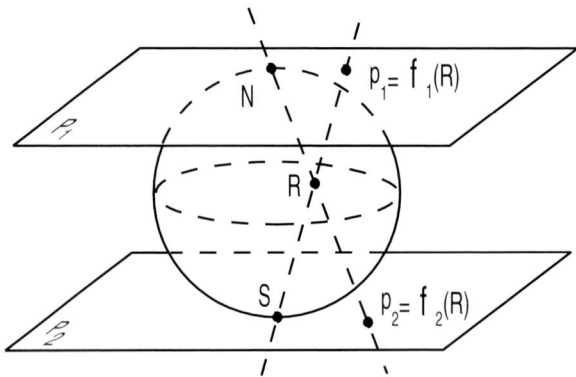

Fig. A.1. Stereographic projection of sphere.

of $\mathbb{R}^2$ to do calculations. Let us choose the usual Cartesian coordinates on P_1 and P_2,

$$p_1 \in P_1: \quad p_1 \equiv (x_1, y_1) \tag{A.1}$$
$$p_2 \in P_2: \quad p_2 \equiv (x_2, y_2).$$

We can use the Euclidean geometry of $\mathbb{R}^3$ to obtain an explicit formula for the transition function,

$$g_{21} := \phi_2 \circ \phi_1^{-1}.$$

Let us choose the x_1 and y_1 axes in P_1 to be parallel to the x_2 and y_2 axes in P_2. On the intersection of O_1 and O_2, i.e., for any R different from the north and the south poles, we have

$$\begin{aligned}
(x_2, y_2) \equiv p_2 &= \phi_2(R) \\
&= \phi_2\left(\phi_1^{-1}(p_1)\right) \\
&= \left(\phi_2 \circ \phi_1^{-1}\right)(x_1, y_1) \\
&=: g_{21}(x_1, y_1).
\end{aligned}$$

Then, a simple calculation shows that

$$g_{21}(x_1, y_1) = \left(\frac{x_1}{x_1^2 + y_1^2}, \frac{-y_1}{x_1^2 + y_1^2}\right). \tag{A.2}$$

Note that on $O_1 \cap O_2$, $x_1^2 + y_1^2 \neq 0$. Thus, the transition function $g_{21} : \mathbb{R}^2 - \{0\} \to \mathbb{R}^2 - \{0\}$ is a well-defined smooth diffeomorphism, and consequently S^2 is a smooth manifold.

The Cartesian coordinates (x_i, y_i), $i = 1, 2$, have the disadvantage that they do not reflect the desirable symmetries of S^2. The natural choice of a coordinate system that inherits these symmetries is the spherical coordinates (θ, φ). We pointed out that this coordinate system cannot be used globally as one uses the spherical coordinates (r, θ, φ) on $\mathbb{R}^3$. A compromise can be reached, however, by adopting two sets of spherical coordinates to represent the points of O_1 and O_2. Denoting these by (θ_1, φ_1) and (θ_2, φ_2), respectively, we find the following simple expression for the transition function:

$$(\theta_2, \varphi_2) = \tilde{g}_{21}(\theta_1, \varphi_1) = (\pi - \theta_1, \varphi_1).$$

This choice has its ambiguities at the poles. Nevertheless, these turn out to be unimportant. The situation is analogous to the use of the polar coordinates (r, φ) on $\mathbb{R}^2$. At $r = 0$, φ is not well defined. However, this does not prevent us from using polar coordinates on $\mathbb{R}^2$.

If we use complex coordinates in (A.1), namely

$$p_i \equiv w_i := x_i + iy_i, \quad i = 1, 2,$$

then the expression (A.2) for the transition function takes the particularly simple form

$$w_2 = g_{21}(w_1) = \frac{1}{w_1}. \tag{A.3}$$

As a function of a complex variable $g_{21} : \mathbb{C} - \{0\} \to \mathbb{C} - \{0\}$ is a complex analytic (holomorphic) function. This makes S^2 an example of a *complex (holomorphic) manifold*. Briefly, a complex manifold is an even-dimensional manifold that is locally homeomorphic to $\mathbb{C}^m$ and whose transition functions are analytic functions from $\mathbb{C}^m$ to $\mathbb{C}^m$. m is called the complex dimension of the manifold. Its real dimension is $2m$.

S^2 is a member of an important class of smooth (complex) manifolds called *complex projective spaces*.

Definition 15: Consider the space of $(N+1)$-tuple complex numbers, $\mathbb{C}^{N+1}$. As a point set the **complex projective space** $\mathbb{C}P(N)$ is the set of all complex lines in $\mathbb{C}^{N+1}$ that pass through the origin. These lines are also called "rays". Each ray is represented by a non-zero complex vector, $\mathbf{z} \in \mathbb{C}^{N+1}$, via

$$l = [\mathbf{z}] := \{\lambda \mathbf{z} \, : \, \lambda \in \mathbb{C} - \{0\}\}. \tag{A.4}$$

In fact, as seen in (A.4), $\mathbf{z}$ can be chosen to have unit length (norm). The length or the norm of $\mathbf{z}$ is defined by

$$\| \mathbf{z} \| := \sqrt{|z^1|^2 + \cdots + |z^{N+1}|^2},$$

where z^i are the complex components of $\mathbf{z}$ and $|z^i|$ are their moduli.[10] The set of all unit vectors in $\mathbb{C}^{N+1}$ defines the unit sphere S^{2N+1},

$$S^{2N+1} := \left\{ \mathbf{z} \in \mathbb{C}^{N+1} \, : \, \| \mathbf{z} \| = 1 \right\} \subset \mathbb{R}^{2N+2} = \mathbb{C}^{N+1}.$$

However, the rays are still not in one-to-one correspondence with the points of S^{2N+1}. This is because we can represent l with another unit vector $\mathbf{z}' := w\mathbf{z}$, where $w \in \mathbb{C}$ is of unit modulus. Clearly, $l = [\mathbf{z}] = [\mathbf{z}']$ and $\mathbf{z}' \in S^{2N+1}$, but $\mathbf{z} \neq \mathbf{z}'$. This suggests that although $\mathbb{C}P(N) \neq S^{2N+1}$, it may be viewed as a set of equivalence classes of points of S^{2N+1}. The desired equivalence relation is given by

$$\mathbf{z} \sim \mathbf{z}' \text{ iff there is } w \in \mathbb{C} \text{ with } \mathbf{z}' = w\mathbf{z}.$$

The equivalence class of $\mathbf{z} \in S^{2N+1}$ is denoted by $[\mathbf{z}]$ as defined in (A.4). These equivalence classes are precisely the points of $\mathbb{C}P(N)$.

There is a standard choice of coordinate charts for $\mathbb{C}P(N)$. These are called *homogeneous coordinate* charts. They are naturally induced from $\mathbb{C}^{N+1} - \{0\}$. Let us use the notation z for the coordinates of $\mathbf{z}$ in $\mathbb{C}^{N+1}$, i.e., $z := (z^1, \cdots, z^{N+1})$. Consider the $N+1$ open sets in $\mathbb{C}P(N)$ defined by

$$O_i := \{[\mathbf{z}] \in \mathbb{C}P(N) \, : \, z^i \neq 0\}, \quad i = 1, 2, \cdots N+1.$$

---

[10] If $z^i = x^i + iy^i$, the modulus of z^i is given by $|z^i| = \sqrt{(x^i)^2 + (y^i)^2}$.

A.2 Differentiable Manifolds 377

Clearly every point in $\mathbb{C}P(N)$ is included in at least one of these open subsets. Thus, $\{O_i\}_{i=1,\cdots N+1}$ forms an open covering of $\mathbb{C}P(N)$. The homogeneous coordinate charts are given by (O_i, ϕ_i), where for every $[z] \in O_i$

$$\phi_i([z]) = \left(\frac{z^1}{z^i}, \cdots, \frac{z^{i-1}}{z^i}, 1, \frac{z^{i+1}}{z^i}, \cdots, \frac{z^{N+1}}{z^i}\right),$$

or more correctly

$$\phi_i([z]) := \left(\frac{z^1}{z^i}, \cdots, \frac{z^{i-1}}{z^i}, \frac{z^{i+1}}{z^i}, \cdots, \frac{z^{N+1}}{z^i}\right) \in \mathbb{C}^N.$$

We shall next obtain the transition functions. In order to do this, we compare the coordinates of a point associated with two different coordinate charts. Let $[z] \in \mathbb{C}P(N)$ such that $z^i \neq 0$ and $z^j \neq 0$ for some $i \neq j$. Then $[z] \in O_i \cap O_j$, and we have

$$\phi_j([z]) = g_{ji}\left(\phi_i([z])\right).$$

Since z^i and z^j are both non-zero, we have

$$\left(\frac{z^1}{z^i}, \cdots, \frac{z^{i-1}}{z^i}, 1, \frac{z^{i+1}}{z^i}, \cdots, \frac{z^{N+1}}{z^i}\right) \equiv [z]$$

$$\equiv \left(\frac{z^1}{z^j}, \cdots, \frac{z^{j-1}}{z^j}, 1, \frac{z^{j+1}}{z^j}, \cdots, \frac{z^{N+1}}{z^j}\right).$$

The action of the transition function g_{ji} on an arbitrary element $w = (w^1, \cdots, w^N) = \phi_i([z]) \in \mathbb{C}^N$ is described by the following steps:

1) Identify $(w^1, \cdots, w^N)$ with $(w^1, \cdots, w^{i-1}, 1, w^i, \cdots, w^N)$.
2) Multiply this $(N+1)$-tuple by $\frac{z^i}{z^j}$, to obtain

$$\left(\frac{w^1 z^i}{z^j}, \cdots, \frac{w^{i-1} z^i}{z^j}, \frac{z^i}{z^j}, \frac{w^i z^i}{z^j}, \cdots, \frac{w^N z^i}{z^j}\right). \tag{A.5}$$

3) Since $w = \phi_i([z])$,

$$w^k = \begin{cases} \frac{z^k}{z^i} & \text{for } k = 1, \cdots, i-1 \\ \\ \frac{z^{k+1}}{z^i} & \text{for } k = i, \cdots, N. \end{cases}$$

This implies that one of the components displayed in (A.5) is identically 1. In fact, if we assume that $i < j$, $w^{j-1} = \frac{z^j}{z^i}$, that is, $\frac{w^{j-1} z^i}{z^j} = 1$. This allows us to undo the first step by dropping this 1 from (A.5) and obtain

$$\left(\frac{w^1 z^i}{z^j}, \cdots, \frac{w^{i-1} z^i}{z^j}, \frac{z^i}{z^j}, \frac{w^i z^i}{z^j}, \cdots, \frac{w^{j-2} z^i}{z^j}, \frac{w^j z^i}{z^j}, \cdots, \frac{w^N z^i}{z^j}\right) := \tilde{w}.$$

This is an N-tuple of complex numbers that we denote by $\tilde{w}$. For $i > j$, we obtain another N-tuple similarly.

4) The transition function g_{ji} is given by

$$g_{ji}(w) := \tilde{w}.$$

We could have associated each step of this list with a function and define the transition functions as the composition of these functions. Clearly, the transition functions are (complex) analytic.

Let us look at the case of $N = 1$. There are two coordinate charts:

$$O_1 = \{[(z^1, z^2)] : z^1 \neq 0\} \text{ and } O_2 = \{[(z^1, z^2)] : z^2 \neq 0\}.$$

On $O_1 \cap O_2$, both z^1 and z^2 are non-zero and

$$\phi_1\left([(z^1, z^2)]\right) = \frac{z^2}{z^1} \text{ and } \phi_2\left([(z^1, z^2)]\right) = \frac{z^1}{z^2}.$$

Following the above procedure, we denote $\frac{z^2}{z^1}$ by w. Then,

$$g_{21}(w) = \frac{1}{w}.$$

This is exactly the expression given in (A.3). Thus, in view of the fact that the structure of a manifold is determined by its transition functions, we have the following identity

$$\mathbb{C}P(1) \simeq S^2.$$

The symbol $\simeq$ stands for the word "diffeomorphic."

We have given many examples of smooth manifolds of dimension two or higher. As for one-dimensional manifolds, there are two possibilities. These are the real line $\mathbb{R}$ and the circle S^1. In fact, a more correct statement is that these are the only *connected* one-dimensional manifolds that do not have a boundary. The adjective "connected" refers to the same property as defined for topological spaces: Connectedness is a topological property.

Another basic concept that generalizes to the discussion of manifolds is that of taking products of manifolds. The resulting objects, which are themselves manifolds, are called *product manifolds*. As a topological space, a product manifold, $M = M_1 \times M_2$, has the product topology. The coordinate charts of M are obtained as the products of the coordinate charts of M_1 and M_2. In particular, if M_1 and M_2 are of dimension m_1 and m_2, then M is (m_1+m_2)-dimensional. Some of the higher-dimensional manifolds are product manifolds. A typical example of this is the torus, T^2, that is the product of two circles,

$$T^2 \simeq S^1 \times S^1.$$

More trivial examples of product manifolds are

$$\mathbb{R}^m \simeq \underbrace{\mathbb{R} \times \cdots \times \mathbb{R}}_{m\text{-times}}.$$

We have also higher dimensional tori,

$$T^m := \underbrace{S^1 \times \cdots \times S^1}_{m\text{-times}}.$$

The examples we have considered are all connected manifolds. The disconnected manifolds are usually not quite as interesting. They are essentially a collection of two or more copies of connected manifolds. An example of a one-dimensional disconnected manifold is $\{-1, 1\} \times S^1$. Here, we view the finite set $\{-1, 1\}$ as a zero-dimensional manifold. In fact, every finite set can be seen as a submanifold[11] of $\mathbb{R}$. In particular, $\{-1, 1\}$ is also called the *zero-dimensional circle*, S^0.

We have discussed the importance of the notion of a diffeomorphism in some detail. The action of a diffeomorphism can be represented by its effect on the transition functions. In fact, a diffeomorphism takes one set of smooth transition functions into another. What remains unchanged is a little more than the topological structure of the manifold. There are cases in which we can find a function between two manifolds which is a homeomorphism but not a diffeomorphism. In this case, we say that the two manifolds have identical topological structures but different *differential (smooth) structures*.[12]

The differentiability or smoothness requirement enables us to introduce the notion of a *tangent space*. This is an essential step towards defining analysis and geometry on a smooth manifold.

Tangent spaces of a manifold are direct generalizations of the tangent spaces of a two-dimensional surface in $\mathbb{R}^3$. They can be defined without any reference to the embedding of the manifold into $\mathbb{R}^d$. However, we would like to give a "geometric" definition of the tangent space at a point of a manifold. This approach makes use of such an embedding. We shall first give the definition of a *curve* on a manifold.

Definition 16: Let M be an m-dimensional smooth manifold. Then, any smooth function $C : [0, T] \to M$ is called a smooth **curve** in M.[13]

A smooth curve C on M can be viewed as a smooth curve in a Euclidean space $\mathbb{R}^d$ by embedding M into $\mathbb{R}^d$, where d is a sufficiently large positive integer.

Definition 17: Let M be a smooth manifold and $C : [0, T] \to M$ be a smooth curve on M with $C(0) = p \in M$. Choose a Euclidean space $\mathbb{R}^d$ such that M is a submanifold of $\mathbb{R}^d$. The vector $v := \frac{d}{dt}C(t)\big|_{t=0} \in \mathbb{R}^d$ is called the *tangent vector* to $C \subset \mathbb{R}^d$ at $t = 0$. It is also called the **tangent vector** to $C \subset M$ at $p \in M$. The set of all tangent vectors to all curves in M that

---

[11] The topology given to a finite set is the subset topology which is, in this case, the same as the discrete topology. This means that every subset of a finite set is postulated to be open.

[12] For a precise definition of a differential structure of a manifold see [186].

[13] Consider $[0, T]$ as a submanifold of $\mathbb{R}$. Then, C is a function between two smooth manifolds. The notion of smoothness for such a function is defined earlier.

originate at $p \in M$ forms an m-dimensional vector space ($\cong \mathbb{R}^m$).[14] This vector space is called the **tangent space** of M at p and denoted by T_pM. The dual (vector) space of T_pM – the space of all linear real-valued functions on T_pM – is called the **cotangent space** of M at p. It is denoted by T_pM^*.

The elements of a cotangent space are called *cotangent vectors* or simply *covectors*. Another terminology is to call the tangent and cotangent vectors, *contravariant* and *covariant* vectors, respectively.

Let us consider the set of all the tangent spaces of a manifold. This set inherits the structure of a smooth manifold from M. It has the special property that the tangent vectors belonging to the tangent space of each point behave as the elements of a finite-dimensional vector space. In other words, this manifold consists of an infinite collection of vector spaces which are labeled by the points of the original manifold M. This is an example of a *vector bundle*. Specifically, it is called the *tangent bundle* of M and denoted by TM. Similarly, the set of all cotangent vectors is called the *cotangent bundle* and denoted by TM^*. A review of vector bundles is provided in Chap. 5. We suffice to say that a vector bundle is locally homeomorphic to the Cartesian product of an open subset of a manifold and a vector space. For example, if $\{(O_\alpha, \phi_\alpha)\}_\alpha$ is an atlas of M, then the subsets of TM consisting of the tangent vectors at the points of O_α are homeomorphic to $O_\alpha \times \mathbb{R}^m$. This allows us to represent the points of TM by pairs of the form: (p, v_p), where $p \in M$ and $v_p \in T_pM$.

Definition 18: Let $V : M \to TM$ be a (smooth) function such that for every $p \in M$, $V(p) = (p, v(p))$ for some $v(p) \in T_pM$. Then, V is called a (smooth) **vector field** on M. Because V is determined by its values, $v(p)$, we usually identify $V(p)$ and $v(p)$ and write $V(p) \in T_pM$. Similarly, a smooth function $\Omega : M \to T_pM^*$ with $\Omega(p) = (p, \omega(p))$ and $\omega(p) \in T_pM^*$, is called a **differential one-form** or simply a **one-form**. Again, we identify $\Omega(p)$ with $\omega(p)$.

Vector fields and one-forms are also called **contravariant** and **covariant** vector fields, respectively. They have many applications in theoretical physics. This is because we can write explicit local formulas for their components and use them to do calculations. Let us choose a local coordinate chart, (O_α, ϕ_α), that includes $p \in M$. The open subset O_α is represented by its image $\phi_\alpha(O_\alpha) \subset \mathbb{R}^m$. We have the following identification:

$$p \equiv \phi_\alpha(p) \equiv (x^1, \cdots, x^m) \in \mathbb{R}^m, \quad \forall p \in O_\alpha.$$

The same applies for the points of a curve $C : [0, T] \to M$. The portion of C that belongs to O_α is mapped to a curve in $\phi_\alpha(O_\alpha)$:

$$C(t) \equiv \left(x^1(t), \cdots, x^m(t)\right) =: x(t).$$

Let us now consider an arbitrary tangent vector $v_p = \frac{d}{dt}C(t)\big|_{t=0}$. The tangent curve C can be assumed to lie entirely inside O_α. v_p is given by

---
[14] We shall often use $\cong$ to denote an *isomorphism*.

$$v_p \equiv \left(\frac{dx^1(t)}{dt}\bigg|_{t=0}, \cdots, \frac{dx^m(t)}{dt}\bigg|_{t=0} \right). \tag{A.6}$$

The m numbers "$\frac{dx^i(t)}{dt}\big|_{t=0}$" are called the *components* of v_p associated with the local coordinate chart (O_α, ϕ_α).

As we emphasized in the beginning of our discussion of manifolds, an important problem in doing analysis on manifolds is the problem of coordinate transformations. In physics, we usually associate a physical quantity with the points of a manifold. These quantities are nothing but different types of functions which assign one or more numbers to each point. These numbers are also called "components" of the quantity in question. They obey different transformation rules under coordinate transformations. In fact, one way of distinguishing different quantities on a manifold is by examining their transformation properties. The simplest mathematical objects that we can associate with a manifold are the real (complex) *scalar fields*. These are real (complex) valued functions on the manifold. By definition, a scalar function has a single component that is independent of the choice of the coordinate chart. Two simple examples of quantities with non-trivial coordinate transformation rules are vector fields and one-forms, respectively, contravariant and covariant vector fields.

Let us choose two intersecting coordinate charts (O_α, ϕ_α) and (O_β, ϕ_β). For convenience, we will place a "prime" to indicate that a quantity is evaluated in the coordinate chart labeled by the index β. For example, in this chart the coordinates of a point $p \in O_\alpha \cap O_\beta$ will be denoted by

$$p \equiv \left(x'^{1}, \cdots, x'^{m} \right) =: x'.$$

If $f: M \to \mathbb{R}$ is a real scalar field, then

$$f(p) \equiv f(x) = f'(x').$$

Let us next examine the behavior of the components of a (contravariant) vector (field) under a coordinate transformation. Consider the vector v_p of (A.6). In the primed coordinate chart, we have

$$C(t) \equiv \left(x'^{1}(t), \cdots, x'^{m}(t) \right) := x'(t),$$

and consequently

$$v_p \equiv \left(\frac{dx'^{1}(t)}{dt}\bigg|_{t=0}, \cdots, \frac{dx'^{m}(t)}{dt}\bigg|_{t=0} \right).$$

The coordinates x' are related to x via the transition function

$$x'(x) = \phi_\beta(p) = \left(\phi_\beta \circ \phi_\alpha^{-1} \right)(\phi_\alpha(p)) = g_{\beta\alpha}(x).$$

Let us denote the components of v_p in the two coordinate charts by v_p^i and $v_p'^i$. Then, the chain rule

$$\frac{dx'^i}{dt} = \frac{\partial x'^i}{\partial x^j} \frac{dx^j}{dt}, \quad i = 1, \cdots, m, \tag{A.7}$$

dictates the following transformation rule:

$$v_p'^i = \frac{\partial x'^i}{\partial x^j} v_p^j. \tag{A.8}$$

All the repeated indices are understood to be summed over their ranges. This is known as the *Einstein summation convention*.

Relation (A.8) can be taken as the definition of a (contravariant) vector. According to this definition a (contravariant) vector on an m-dimensional manifold is an object with m local components which satisfy the coordinate transformation rule given by (A.8). A simple way of remembering this transformation rule is to use the symbols $\frac{\partial}{\partial x^i}$ for the local coordinate basis vectors, i.e.,

$$v_p = v_p^i \frac{\partial}{\partial x^i}. \tag{A.9}$$

Then, recognizing that v_p does not depend on the choice of coordinates,

$$v_p^i \frac{\partial}{\partial x^i} = v_p = v_p'^i \frac{\partial}{\partial x'^i},$$

and enforcing the chain rule,

$$\frac{\partial}{\partial x^i} = \frac{\partial x'^j}{\partial x^i} \frac{\partial}{\partial x'^j},$$

we recover (A.8).

The same line of reasoning applies for (contravariant) vector fields. After all, a vector field is a vector-valued function. The only additional fact is that for a vector field, the point p in (A.9) becomes an independent variable. It is represented by its coordinates $(x^1, \cdots, x^m)$ or collectively by x. Therefore, a (contravariant) vector field V on $O_\alpha \subseteq M$ is expressed by

$$V = V^i(x) \frac{\partial}{\partial x^i}, \tag{A.10}$$

where $V^i(x)$ satisfy the coordinate transformation rule listed as (A.8).

The transformation properties of covectors (covariant vectors) is obtained similarly. Let $\omega_p \in T_pM^*$ be a covector. Then, for any $v_p \in T_pM$, $\omega_p(v_p)$ is a real scalar. We know from elementary linear algebra that T_pM^* as an m-dimensional vector space is isomorphic to $\mathbb{R}^m$. Thus, in the local chart (O_α, ϕ_α), we can write ω_p in terms of its components $(\omega_p)_i$, $i = 1, \cdots, m$. The scalar obtained by the action of ω_p on v_p can then be written componentwise,

A.2 Differentiable Manifolds

$$\omega_p(v_p) = (\omega_p)_j (v_p)^j \tag{A.11}$$

However, as we pointed out earlier, a scalar is independent of the choice of coordinate charts. Hence, if we choose another coordinate chart, say, (O_β, ϕ_β) we must obtain the same result, namely

$$(\omega'_p)_j (v'_p)^j = \omega_p(v_p) = (\omega_p)_j (v_p)^j.$$

This equality together with (A.8) yield the transformation rule for the components of a covector (covariant vector):

$$(\omega'_p)_i = \frac{\partial x^j}{\partial x'^i} (\omega_p)_j. \tag{A.12}$$

A convenient notation for the local basis covectors is dx^i. In this notation, any covector is locally written as

$$\omega_p = (\omega_p)_i \, dx^i.$$

The independence of covectors from the choice of coordinate charts and the chain rule

$$dx'^i = \frac{\partial x'^i}{\partial x^j} \, dx^j,$$

reproduces (A.12) immediately.

We would like to remark that at this stage the use of (the operators) $\frac{\partial}{\partial x^i}$ for the basis vectors and similarly the use of dx^i for the basis covectors are for practical purposes. In this notation, the duality of the basis vectors and covectors takes the following form

$$dx^i \left(\frac{\partial}{\partial x^j} \right) = \delta^i_j,$$

where δ^j_i are the components of the Kronecker delta function,

$$\delta^i_j := \begin{cases} 1 & \text{if } i = j \\ 0 & \text{if } i \neq j. \end{cases}$$

Letting p be an independent variable, we obtain the local expression for a covariant vector field or a one-form,

$$\omega(x) = \omega_i(x) \, dx^i.$$

The transformation rule for the components of a one-form is given by

$$\omega'_i(x') = \frac{\partial x^j}{\partial x'^i} \omega_j(x). \tag{A.13}$$

Again, we can take (A.13) as a definition of a covariant vector field.

384   A. An Elementary Introduction to Manifolds and Lie Groups

Covariant and contravariant vector fields are simple examples of *tensor fields*. In general, a tensor is an elements of the tensor products of a bunch of vector spaces. A tensor at a point $p \in M$ is defined similarly as an element of the tensor product space

$$\underbrace{T_pM^* \otimes \cdots \otimes T_pM^*}_{r\text{-times}} \otimes \underbrace{T_pM \otimes \cdots \otimes T_pM}_{s\text{-times}}. \tag{A.14}$$

An element of this space is called an r-times covariant and s-times contravariant tensor at $p \in M$. r and s are also called the *covariant* and *contravariant ranks* of the tensor, respectively. Alternatively, we can define a tensor as a multilinear real-valued function of several vectors and covectors. Consider a function

$$T: \underbrace{T_pM \times \cdots \times T_pM}_{r\text{-times}} \times \underbrace{T_pM^* \times \cdots \times T_pM^*}_{s\text{-times}} \longrightarrow \mathbb{R},$$

such that T is linear in all its entries. Then, T is said to be a *tensor* of covariant and contravariant ranks r and s, respectively. The vector space depicted by (A.14) is the set of all such multilinear functions. This set forms a vector space under the operations of pointwise addition and multiplication by real numbers.

Tensor fields are tensor valued functions on a manifold. They play an important role in describing various physical quantities. Probably, the best example of the use of tensor fields in physics is in the theory of electromagnetism where the electromagnetic field strength[15] F is a tensor field [117, 146]. Tensor (fields) are also defined according to their coordinate transformation properties. For instance, the electromagnetic field tensor F is a covariant tensor of rank 2, because its components, $F_{\mu\nu}$, satisfy the following transformation rule

$$F'_{\mu\nu}(x') = \frac{\partial x^\sigma}{\partial x'^\mu} \frac{\partial x^\rho}{\partial x'^\nu} F_{\sigma\rho}(x). \tag{A.15}$$

The use of the Greek indices in (A.15) is to indicate that they refer to the Minkowski spacetime coordinates. Let us use the notation dx^μ for the basic covectors. Then, we have

$$F(x) = F_{\mu\nu}(x)\, dx^\mu \otimes dx^\nu.$$

In general, a (mixed) tensor of rank (r,s) is expressed locally according to

$$T(x) = T^{j_1\cdots j_s}_{i_1\cdots i_r}\, dx^{i_1} \otimes \cdots \otimes dx^{i_r} \otimes \frac{\partial}{\partial x^{j_1}} \otimes \cdots \otimes \frac{\partial}{\partial x^{j_s}}.$$

The transformation properties of the components are easily obtained by requiring that $T(x)$ is coordinate independent.

---

[15] In the text we used the symbol $F^{(\text{el})}$ to denote the electromagnetic field strength tensor to avoid any ambiguity. Here we drop the label $^{(\text{el})}$ for simplicity.

An important class of tensor fields is the *totally antisymmetric covariant tensor fields*. These are more commonly called **differential forms**. As the name indicates, differential forms are covariant tensor fields whose components are antisymmetric in all their indices. The electromagnetic field tensor F is an example of a differential form of rank 2, a two-form. We shall occasionally drop the adjective "differential" for simplicity.

The space of all tensors (tensor fields) form a vector space. This is, by definition, the tensor product of copies of the tangent and cotangent spaces. The operation of *tensor product* makes this vector space into an (associative) *algebra*. This means that not only we can add tensors and multiply them by numbers, but we can multiply tensors by other tensors as well. To obtain the tensor product of two tensors, we multiply the components and take the tensor product of the basic tensors. For example, suppose

$$T = T_i^j \, dx^i \otimes \frac{\partial}{\partial x^j}, \quad S = S_k dx^k.$$

Then, we have

$$T \otimes S := T_i^j S_k \, dx^i \otimes \frac{\partial}{\partial x^j} \otimes dx^k.$$

In the next section we shall encounter other examples of an associative algebra. Therefore we next present the definition of an associative algebra.

Definition 19: Let $(\mathcal{A}, +, \cdot)$ be a vector space and $\otimes : \mathcal{A} \times \mathcal{A} \to \mathcal{A}$ be a binary operation satisfying the following conditions:

1) $(a_1 + a_2) \otimes a_3 = a_1 \otimes a_3 + a_2 \otimes a_3$;
2) $a_3 \otimes (a_1 + a_2) = a_3 \otimes a_1 + a_3 \otimes a_2$;
3) $c \cdot (a_1 \otimes a_2) = (c \cdot a_1) \otimes a_2 = a_1 \otimes (c \cdot a_2)$;
4) $a_1 \otimes (a_2 \otimes a_3) = (a_1 \otimes a_2) \otimes a_3$,

where a_1, a_2, and a_3 are arbitrary elements of $\mathcal{A}$ and $c \in \mathbb{R}$ ($c \in \mathbb{C}$ if $\mathcal{A}$ is a complex vector space). Then $(\mathcal{A}, +, \cdot, \otimes)$ is said to be an **associative algebra**. A subset $\{\mathcal{J}_i\}$ of $\mathcal{A}$ is said to generate $\mathcal{A}$ if every element of $\mathcal{A}$ can be written as a linear combination of products of $\mathcal{J}_i$. The elements $\mathcal{J}_i$ are called the **generators** of the algebra.

Similarly to the space of tensor fields, the space of all differential forms is a vector space. The tensor product, however, does not respect the requirement of antisymmetry. By this, we mean that the tensor product of two forms is in general a tensor whose components are not antisymmetric in all its indices. Thus, we need an alternative algebra operation for differential forms that preserves the antisymmetry. This is called the "antisymmetric tensor product" or the *wedge product*, $\wedge$. For example, we have

$$dx^\mu \wedge dx^\nu := dx^\mu \otimes dx^\nu - dx^\nu \otimes dx^\mu = -dx^\nu \wedge dx^\mu.$$

The electromagnetic field tensor can be written as

$$F = \frac{1}{2} F_{\mu\nu} \, dx^\mu \wedge dx^\nu. \tag{A.16}$$

The vector space of all differential forms together with the wedge product form an algebra known as the *exterior algebra* or the *Grassmann algebra*. A detailed introduction to differential forms can be found in H. Flanders' book: *Differential Forms with Applications to the Physical Sciences* (Dover, New York, 1989).

Nowadays, differential forms are used extensively in theoretical physics. They provide a remarkable tool for keeping the calculations short. In particular, they are useful in keeping track of the transformation properties of physical quantities. We can also develop a sort of calculus on differential forms by defining the so-called "exterior differentiation". This is an operation that takes a p-form to a $(p+1)$-form. The components of the resultant differential form are linear combinations of the first derivatives of the components of the original p-form.

Definition 20: Let Ω^p denote the space of all p-forms on a smooth manifold M. Then the **exterior derivative** is the map

$$d : \Omega^p \longrightarrow \Omega^{p+1},$$

defined locally by

$$d\omega := \left(\frac{\partial}{\partial x^j}\omega_{i_1 \cdots i_p}\right) dx^j \wedge dx^{i_1} \wedge \cdots \wedge dx^{i_p},$$

where

$$\omega = \omega_{i_1 \cdots i_p} dx^{i_1} \wedge \cdots \wedge dx^{i_p} \in \Omega^p.$$

An important observation is the following result.

Proposition 2: $d^2 := d \circ d = 0$.

A simple application of differential forms and exterior differentiation is in the theory of electromagnetism. It is not difficult to check that the components of the *vector potential* A_μ [117, 146], transform like the components of a one-form:

$$A := A_\mu dx^\mu, \quad \mu = 0, 1, 2, 3.$$

Let us calculate

$$dA = \left(\frac{\partial}{\partial x^\mu} A_\nu\right) dx^\mu \wedge dx^\nu$$

$$= \frac{1}{2}\left[\frac{\partial}{\partial x^\mu} A_\nu - \frac{\partial}{\partial x^\nu} A_\mu\right] dx^\mu \wedge dx^\nu.$$

We immediately recognize the content of the last bracket as the components of the electromagnetic field strength tensor,

$$F_{\mu\nu} = \frac{\partial}{\partial x^\mu} A_\nu - \frac{\partial}{\partial x^\nu} A_\mu.$$

Thus, we have the simple identity

$$F = dA.$$

An application of Proposition 2 is

$$dF = d^2 A = 0.$$

This simple equation is known as the "homogeneous Maxwell's equations" (in component form). These (two) equations together with the "inhomogeneous Maxwell's equations" describe all electromagnetic phenomena.

So far, we have defined different types of vector and tensor fields on smooth manifolds. Let us see how we can use a function between two manifolds to induce these fields from one manifold to another. Consider a smooth function $f : M_1 \to M_2$ between two smooth manifolds. Let v_p be a tangent vector at $p \in M_1$. v_p can be "pushed forward" to define a tangent vector at $f(p) \in M_2$.

Definition 21: Let M_1 and M_2 be smooth manifolds, $p \in M_1$, and $f : M_1 \to M_2$ be a smooth function. Then f induces a *linear* map

$$f_* : T_p M_1 \longrightarrow T_{f(p)} M_2$$

called the **push-forward** or the **differential map**. To describe this map let us consider an arbitrary tangent vector $v_p \in T_p M$ and choose a curve $C_1 : [0, T] \to M_1$ such that $v_p = \frac{dC_1}{dt}\big|_{t=o}$. The image of C_1 under f is a smooth curve

$$C_2 := f \circ C_1 : [0, T] \longrightarrow M_2$$

in M_2. The push-forward map is then defined by

$$f_*(v_p) := \frac{dC_2}{dt}\bigg|_{t=0}. \tag{A.17}$$

A useful exercise is to show that this definition is independent of the choice of the curve C_1.

The same definition applies for a vector field by taking p to be an independent variable. However, we should point out that the push-forward map is defined locally. This means that, in general, the push-forward map can be used to induce a vector field only on an open neighborhood in M_2. There are cases in which a global vector field on the whole of M_1 cannot be pushed forward to define a smooth global vector field on $f(M_1) \subseteq M_2$.

Similarly, we can induce cotangent vectors and differential forms using a smooth map between two manifolds. However, this time it is the cotangent vectors and differential forms of M_2 that are "pulled back" on M_1.

Definition 22: Let M_1, M_2, and f be as in Def. 21. Then f defines a linear map

$$f^* : T_{f(p)} M_2^* \longrightarrow T_p M_1^*,$$

called the **pullback** map. To describe the pullback map we consider an arbitrary cotangent vector $\omega_{f(p)}$ at $f(p) \in M_2$. As an element of $T_p M_1^*$, the **pullback** of $\omega_{f(p)}$ is a linear map acting on $T_p M_1$. Thus, it can be defined through its action on arbitrary elements u_p of $T_p M_1$. We have

$$f^*(\omega_{f(p)})[u_p] := \omega_{f(p)}[f_*(u_p)], \qquad (A.18)$$

where $f_*(u_p)$ is the push-forward of u_p.

The pullback map is also generalized to differential forms and general covariant tensor fields. Interestingly, it is a global mapping. That is, every global differential form on M_2 can be pulled back to define a global differential form on $f^{-1}(M_2) \subseteq M_1$. We make extensive use of the pullback operation in our discussion of universal connections and their realization in the phenomenon of geometric phase in Chaps. 6 and 7.

We will end this section by giving the definition of *compactness* for manifolds. As we mentioned in our discussion of general topological spaces, compactness is a topological property. It is defined for arbitrary topological spaces. However, the general definition is rather abstract. If a topological space is a subspace of $\mathbb{R}^d$ for some $d \in \mathbb{Z}^+$, as is a manifold, we can use a result of real analysis to arrive at a more intuitive definition.

Definition 23: Let M be an arbitrary (topological) manifold. Embed M into some $\mathbb{R}^d$. M is said to be a **compact manifold**, if it is a *closed* and *bounded* subset of $\mathbb{R}^d$. As usual, a closed subset is a subset whose complement is open. Moreover, a subset M of $\mathbb{R}^d$ is called bounded if there is an open ball

$$B_r^d = \{\mathbf{x} \in \mathbb{R}^d \ : \ |\mathbf{x}| < r\}$$

in $\mathbb{R}^d$, such that $M \subset B_r^d$.

Clearly, $\mathbb{R}^m$ or an open ball in $\mathbb{R}^m$ is not compact, whereas spheres, tori, and (finite-dimensional) projective spaces are all compact manifolds.

A.3 Lie Groups

In the preceding section, we introduced manifolds and gave several examples of smooth manifolds. In this section, we shall discuss another important class of smooth manifolds called *Lie groups*. The mathematical theory of Lie groups is one of the most well-established and substantial achievements of modern mathematics. It is also of great practical and theoretical use in physics. Our aim will be, therefore, not to attempt to present a complete review of Lie groups. Instead, we shall try to point out the basic concepts and emphasize a few of the most important and useful facts about Lie groups. More detailed discussions of the theory of Lie groups and its applications in physics can be found in

- M. Hamermash: *Group Theory and Its Application to Physical Problems* (Dover, New York, 1989);

A.3 Lie Groups

- J. P. Elliot and P. G. Dawer: *Symmetry in Physics*, Vol. 1 and 2 (Oxford University Press, New York, 1990);
- R. Gilmore: *Lie Groups, Lie Algebras, and Some of Their Applications* (Wiley, New York, 1974);
- H. Georgi: *Lie Algebras in Particle Physics* (Addison-Wesley, New York, 1982); and
- S. Sternberg: *Group Theory and Physics* (Cambridge University Press, Cambridge, 1994).

Some of the more advanced textbooks on Lie groups and their representations are

- S. Helgason: *Differential Geometry, Lie Groups, and Symmetric Spaces* (Academic Press, New York, 1978);
- T. Bröcker and T. tom Dieck: *Representations of Compact Lie Groups* (Springer-Verlag, New York, 1985);
- W. Fulton and J. Harris: *Representation Theory* (Springer-Verlag, New York, 1991).

Definition 24: Let G be a set of points, and

$$\bullet : G \times G \longrightarrow G$$

be a binary operation. Then the pair $(G, \bullet)$ is said to be a **group** if the following conditions are fulfilled.

1) $\bullet$ is associative:

$$(g_1 \bullet g_2) \bullet g_3 = g_1 \bullet (g_2 \bullet g_3), \quad \forall g_1, g_2, g_3 \in G;$$

2) there is an identity element $e \in G$ such that

$$e \bullet g = g \bullet e = g, \quad \forall g \in G;$$

3) every element $g \in G$ has an inverse, g^{-1} such that

$$g \bullet g^{-1} = g^{-1} \bullet g = e.$$

A subset H of a group G which has the structure of a group with the same group multiplication is called a *subgroup* of G. A necessary and sufficient condition for a subset H to be a subgroup of G is that for every h_1 and h_2 belonging to H, the element $h_1 \bullet h_2^{-1}$ also belongs to H.

Given a subgroup H of a group G, one can define a canonical equivalence relation on G. This equivalence relation is defined according to the requirement:

$$\forall g_1, g_2 \in G, \; g_1 \sim g_2, \quad \text{if and only if} \quad g_1 \bullet H = g_2 \bullet H.$$

The last equality means that there exist $h_1, h_2 \in H$ with $g_1 \bullet h_1 = g_2 \bullet h_2$.

It is a simple exercise to show that this relation is indeed an equivalence relation. The equivalence class including an element $g \in G$ is denoted by gH

and called a *left coset*. The set of all equivalence classes is called a *left coset space* or a *quotient space* of G and denoted by G/H. Similarly one defines the right cosets by defining a similar equivalence which involves the requirement: $H \bullet g_1 = H \bullet g_2$ for the equivalent elements g_1 and g_2.

Groups have proven to be extremely important mathematical constructions. They have been used in other areas of mathematics as well as different branches of natural sciences. Probably, the main reason for the utility of groups in natural sciences is the existence of "symmetry" in nature.

In our discussion of equivalence classes, we pointed out that once there is an equivalence relation, the universal set (class) of all points (objects) is divided into subsets of equivalence classes. The elements of each equivalence class could be treated equally. Hence, we associated the word "symmetry" with this situation. Groups, as we shall see, "quantify" symmetries. In this sense, "different" groups correspond to different types of symmetries. In the following, we shall first try to explain the meaning of the last couple of sentences. We will start with a familiar example.

In some cases, we have two different equivalence relations on a single universal set. In fact, we have encountered some examples of this. Let us consider the collection of all topological spaces. The elements are topological spaces $(X, \mathcal{T})$ that are distributed among the distinct equivalence classes of homeomorphic topological spaces. Also, we have the collection of all the universal sets X of the above collection. This is the collection of all point sets. We have the equivalence relation defined by the existence of bijections. Obviously, topological equivalence is a stronger condition than set theoretic equivalence. This is simply due to the fact that every homeomorphism is a bijection, but not every bijection is a homeomorphism. Therefore, each equivalence class of bijective sets is further subdivided into equivalence classes of homeomorphic topological spaces. So as we see in this example, we have different types of "symmetries."

Let us concentrate on topological symmetry and see how it is related to groups. Consider a point set X and let us study the set of possible topological structures on X. Any two topological structures on X, equivalent or not, are related to one another through a bijection $f : X \to X$. Hence we study the set of all such bijections which we denote by S_X.

Consider a mapping X by two consecutive bijections,

$$X \xrightarrow{f_1} X \xrightarrow{f_2} X.$$

This is done by composing the bijections f_1 and f_2. We have

$$X \xrightarrow{f_2 \circ f_1} X.$$

It is easy to check that the set S_X with the operation of composition forms a group. The operation of composition of functions is associative. The identity function

$$id: X \ni p \longrightarrow p \in X$$

is clearly a bijection. It provides the identity element of the group of all bijections. This is called the *permutation group* of the set X and denoted by S_X. Finally, the inverses exist and they are also bijections by definition.

Now consider a particular topological structure $\mathcal{T}$ on X. Under the action of a bijection f and its inverse f^{-1}, the open subsets of X are either mapped onto the same collection $\mathcal{T}$ of open subsets or some of them do not. In the first case, the bijection f is by definition a homeomorphism mapping the topological space $(X, \mathcal{T})$ to itself. Let us denote the set of all such homeomorphism by $\mathcal{H}(X, \mathcal{T})$. It is not difficult to show that $\mathcal{H}(X, \mathcal{T})$ forms a subgroup of the permutation group S_X. $\mathcal{H}(X, \mathcal{T})$ is called the *homeomorphism group* of $(X, \mathcal{T})$. Furthermore, the set of all topological spaces which are not homeomorphic to $(X, \mathcal{T})$ corresponds to the quotient set $S_X/\mathcal{H}(X, \mathcal{T})$.

Similarly to the permutation group of a set and the homeomorphism group of a topological space, we can introduce the *diffeomorphism group* $\text{Diff}(M)$, of a differentiable manifold M. This is the set of all diffeomorphisms of M together with the operation of composition.

The examples of groups that we have mentioned are indeed quite complicated mathematical structures. There are much simpler examples of groups with a finite number of elements. These are called *finite groups*. The group which has a single element is called the *trivial group*. The simplest non-trivial group is the group $\mathbb{Z}_2$. It has two elements. There are two defining "representations" of $\mathbb{Z}_2$. These are known as the additive and the multiplicative representations. The additive representation is obtained by viewing $\mathbb{Z}_2$ as the set of all integers modulo 2. The group operation is the simple addition of integers modulo 2, i.e., $\bar{a} + \bar{b} := \overline{a+b}$. The additive representation is, therefore, given by

$$\mathbb{Z}_2 = (\{\bar{0}, \bar{1}\}, +).$$

The multiplicative representation is

$$\mathbb{Z}_2 = (\{-1, 1\}, \times),$$

where "$\times$" means the ordinary multiplication of numbers. Other simple examples of finite groups are the groups $\mathbb{Z}_n$, integers modulo n with addition, and the permutation groups S_n. The latter consists of all the bijections from a finite set I_n of n elements onto I_n, i.e., $S_n = S_{I_n}$. Clearly, S_n has $n!$ elements.

We can divide all groups into two large classes based on whether the group operation is *commutative* or not. A group $(G, \bullet)$ is called a *commutative* or an *Abelian group*, if for every g_1 and g_2 in G,

$$g_1 \bullet g_2 = g_2 \bullet g_1.$$

Evidently, $\mathbb{Z}_n$ are Abelian groups, whereas S_n $(n > 2)$ are not Abelian. An instructive exercise would be to show that S_n are *non-Abelian* for $n > 2$.[16]

---

[16] Hint: Show that S_3 is non-Abelian and use the fact that S_3 is a *subgroup* of S_n, for $n > 2$.

As for every class of mathematical structures, we will need to define a concept of *equivalence* for groups. This is also defined by the existence of certain functions between two groups that preserve the *group structure*. Such a function is called a *group isomorphism*.

Let us consider two groups G and H. We will use the same notation for both group operations, "•", but denote the elements of G and H by g and h, respectively.

Definition 25: A function $f : G \to H$ is said to be a **group homomorphism** if for every $g_1, g_2 \in G$

$$f(g_1 \bullet g_2) = f(g_1) \bullet f(g_2).$$

A bijective homomorphism is called a **group isomorphism**.

We can easily check that the existence of isomorphisms defines an equivalence relation on the class of all groups. The elements of the isomorphism (equivalence) classes are not distinguished in group theory. Probably, the simplest example of a group isomorphism is

$$\mathbb{Z}_2 \cong S_2.$$

We use the notation "$\cong$" for the word "isomorphic". The classification of all finite groups (up to isomorphy) is one of the most difficult tasks in mathematics.

A natural concept in group theory is the concept of a *product group*. The reader should be able to define this concept by himself or herself.

A standard elementary textbook on group theory is *Topics in Algebra* by I. N. Herstein (Blaisdell Publishing Company, New York, 1964). A more advanced textbook is *Algebra* by T. W. Hungerford (Springer, New York, 1987).

We briefly mentioned that groups are associated with symmetries. A typical and rather instructive example of this is the group of the geometric operations on a plane that leave an equilateral triangle unchanged. These are three (0, 120, and 240 degrees) rotations about the center, and three reflections about the symmetry axes. These operations form a group which is also called *the symmetry group of a triangle*. This group is isomorphic to S_3. It is a non-Abelian group that contains the subset of the three rotations as an Abelian subgroup. The next example of a geometric symmetry group is the group of symmetries of a square. This includes four rotations and four reflections. It is not isomorphic to any of the groups that we have discussed so far, neither is it a product group of some "smaller" groups. We can proceed to introduce the symmetry groups of other geometric objects such as equilateral polygons. Evidently, the number of elements of the symmetry group – this is called the *order* of the finite group – depends on the number of sides of the polygon. The limiting object is the *round circle*. Its symmetry group has an infinite number of elements. These are the rotations about the center by arbitrary angles $\varphi \in [0, 2\pi)$ and reflections about any axis through the center. The latter can also be parameterized by an angle, namely the angle that the axis of reflection makes with, say, the x-axis in $\mathbb{R}^2$. Alternatively, one can perform an arbitrary reflection by a combination of two rotations and a fixed reflection, say about the x-axis. In order to do this, first one performs a suitable rotation so that the axis of reflection is rotated to the x-axis. This is followed by a reflection about the x-axis and finally a rotation by the same angle but in the opposite sense. This shows that the symmetry group of the circle is not parameterized by two independent continuous variables.

Let us consider the subgroup of the symmetry group of the circle consisting of rotations. This is an Abelian group. Its elements are parameterized by the angles or the points of another round circle. Let g be such a rotation, and let us view the original circle S^1 as the set of vectors of unit length in $\mathbb{R}^2$,

$$S^1 := \left\{ \mathbf{v} \in \mathbb{R}^2 \ : \ \mathbf{v} = \begin{pmatrix} x \\ y \end{pmatrix}, \ x^2 + y^2 = 1 \right\}.$$

Then, g is represented by a matrix that "multiplies" the points of S^1 from the left:

$$\begin{pmatrix} x \\ y \end{pmatrix} \xrightarrow{g} \begin{pmatrix} x' \\ y' \end{pmatrix} = \begin{pmatrix} \cos\varphi & \sin\varphi \\ -\sin\varphi & \cos\varphi \end{pmatrix} \begin{pmatrix} x \\ y \end{pmatrix}.$$

In this way, we have an explicit representation of an infinite group, namely

$$SO(2) := \left\{ \begin{pmatrix} \cos\varphi & \sin\varphi \\ -\sin\varphi & \cos\varphi \end{pmatrix} \ : \ \varphi \in [0, 2\phi) \right\}.$$

The notation $SO(2)$ means the set of all *special* (determinant = 1), orthogonal (inverse = transpose), two-dimensional real matrices. In $\mathbb{R}^3$, it corresponds to the rotations about a fixed axis.

We can also use the identity $\mathbb{R}^2 = \mathbb{C}$, to view S^1 as the set of complex numbers with unit modulus,

$$\{z \in \mathbb{C} \ : \ |z| = 1\}.$$

If $|z| = 1$, then z is a phase,

$$z = e^{i\theta}, \quad \theta \in [0, 2\pi).$$

In this representation, the rotation by an angle φ is performed by multiplication by the phase $e^{i\varphi}$,

$$e^{i\theta} \xrightarrow{g} e^{i\theta'} = e^{i\varphi} e^{i\theta} = e^{i(\theta + \varphi)}.$$

The group of rotations is then parameterized by the set of all phases,

$$U(1) := \left\{ e^{i\varphi} \ : \ \varphi \in [0, 2\pi) \right\},$$

where the group multiplication is the ordinary multiplication of complex numbers. The letter "U" stands for "unitary", since every phase is indeed a unitary (inverse = Hermitian conjugate) one-by-one matrix.

The two groups $SO(2)$ and $U(1)$ are in fact the same, i.e., they are isomorphic. It is quite easy to see that the group elements, in this case, describe the points of a smooth manifold, namely the circle, S^1. $U(1)$ is a typical example of a (compact, connected) *Lie group*.

Another example of a compact (but disconnected) Lie group is the full symmetry group of the round circle. This group consists of the (subgroup of) rotations and arbitrary reflections. We can construct this group algebraically. For this purpose we consider the round circle S^1 as the set of all unit vectors in $\mathbb{R}^2$ originating at $0 \in \mathbb{R}^2$, and find all the linear transformations which map S^1 onto itself, i.e., preserve the magnitude of the vectors. It is not difficult to see that these transformations correspond to 2×2 orthogonal matrices. The set of all such matrices is denoted by $O(2)$. We know from elementary linear algebra that the determinant of any orthogonal matrix is either 1 or -1. As we showed above the rotations correspond to the special orthogonal matrices which have unit determinant. Thus the reflections are identified by the orthogonal matrices of determinant -1. Alternatively, we can view S^1 as the set of complex numbers of unit norm (modulus), and try to find the symmetry group of S^1 as the group of linear transformations of the complex numbers $\mathbb{C}$ which preserve the norm. In this picture, the relation between reflections and the operation of complex conjugation is most interesting.

Having examined some simple examples of Lie groups, we pursue our review by presenting a precise definition of a Lie group.

Definition 26: Let $(G, \bullet)$ be a group that has, in addition, the structure of a smooth manifold. Then, G is said to be a **Lie group** if the functions defined by the group multiplication, $\bullet : G \times G \to G$,

$$\bullet(g_1, g_2) := g_1 \bullet g_2,$$

and inversion, $i : G \to G$

$$i(g) := g^{-1},$$

are smooth functions.

Other examples of Lie groups are the spaces $\mathbb{R}^m$, where the group multiplication is simply the addition of vectors. We also have the multiplicative group $(\mathbb{R} - \{0\}, .)$ which is a disconnected Lie group. Similarly, the multiplicative group $(\mathbb{C} - \{0\}, .)$ has the structure of a Lie group with the group space being the punctured complex plane. Other, more interesting groups are the so-called *classical groups*. These are different types of matrix groups. We will discuss the *unitary groups* $U(n)$ in some detail. The other classical groups are discussed in most of the textbooks on Lie group theory.

Lie groups enter into physical problems as the transformation groups of physical systems. Often, a physical system is defined on a smooth manifold M. Depending on the specific geometry of the problem there may exist quantities that are invariant under certain transformations of the manifold. These quantities are called *conserved quantities*. In fact, the set of all manageable transformations is the diffeomorphism group of the manifold. Here one can view the manifold as the submanifold of some Euclidean space $\mathbb{R}^d$ and interpret a diffeomorphism as a smooth and smoothly reversible deformation of the manifold in the embedding Euclidean space.

In field theories, physical quantities are usually represented by tensor fields $T = T(x)$.[17] In general, under a diffeomorphism a point $x \in M$ is mapped to another point $x' \in M$. As regards the (tensor) fields T, there are two possibilities. Either one keeps using the original fields evaluated at the new point, i.e., under the diffeomorphism $T(x) \to T(x')$, or one also transforms the tensor fields according to the diffeomorphism $T(x) \to T'(x')$. The latter is performed by pushing-forward or pulling-back the field T (or a combination of these) using the diffeomorphism and its inverse. Adopting this latter point of view, one may interpret the effect of a diffeomorphism as a reparameterization of the physical quantities. Thus the corresponding symmetries are related to the reparameterization invariance of the physical quantities. These symmetries are known as the internal symmetries. In a sense they are indispensable qualities of every sensible field theory. The *diffeomorphism invariance* of Einstein's general theory of relativity and the gauge symmetries of other field theories are examples of internal symmetries. The former point of view, in contrast, corresponds to the external symmetries which are related to the specifics of a physical system. They may or may not be present.

Usually, the (tensor) fields used to represent physical quantities are given by their local components. The use of local components is often necessary to perform computations. These local components, however, depend on the choice of a local coordinate chart and a local basis of the tangent and cotangent spaces. The choice of coordinates is completely subjective. Hence, the physical quantities must be independent of such a choice. A local coordinate transformation on a manifold is identical with a coordinate transformation on a copy of $\mathbb{R}^m$, where m is the dimension of the manifold. Coordinate transformations on $\mathbb{R}^m$ also form a group. The elements of this group may be used to parameterize the local coordinate transformations of the manifold. The physical quantities are however invariant under the "action" of this group. In fact, a coordinate transformation does not move the points of the manifold. It is merely a relabeling of the points of the corresponding local chart and thus corresponds to the identity element of the diffeomorphism group. Consequently we do not associate the coordinate invariance (covariance) of the physical quantities with a symmetry of a physical system.[18]

Consider a free particle moving on a plane. The manifold M is $\mathbb{R}^2$. The dynamics of the free particle must certainly be independent of the choice of the coordinate axes. Let us fix a Cartesian coordinate system. Then, any

---

[17] This includes the scalar, covariant, and contravariant vector fields.
[18] The local representation of a diffeomorphism, i.e., when it is represented in local coordinates, resembles a local coordinate transformation. This is the basis of the terminology according to which diffeomorphism symmetry is also called symmetry under general coordinate transformations.

other such system is obtained from the first one through (a combination of) rotations, translations, and interchange of the coordinate axes. The set of all such transformations in $\mathbb{R}^2$ form a group called the *Euclidean group*, $E(2)$. Note that the arbitrariness of the choice of a coordinate system is different from say an $E(2)$ symmetry of a problem.

Let us consider the longitudinal symmetry of the electrostatic properties of a long homogeneous charged metal bar of fixed cross-sectional geometry. The symmetry group is clearly $\mathbb{R}$. The presence of this symmetry allows us to reduce the three-dimensional problem to a two-dimensional one. The full solution is then obtained via the symmetry argument that the results must be independent of the longitudinal coordinate. If further we suppose that the metal bar is cylindrical, the symmetry group is even larger. It is the product group $\mathbb{R} \times O(2)$. Since the symmetry group is two-dimensional the problem is further reducible to a one-dimensional one. The final result will only depend on the distance of the observer from the metal bar.

We have seen how symmetries are related to groups. Specially, we examined the finite and infinite (Lie) groups of transformations. In general, the space which undergoes a transformation is an arbitrary smooth manifold. A transformation of a manifold by a group element is called the *action of the group element on the manifold*. As any transformation, the action of a group element is defined through a function acting on the manifold. More generally we have the following definition.

Definition 27: Let $(G, \bullet)$ and M be a Lie group and a smooth manifold, respectively. A smooth function $f : G \times M \to M$ is said to be a **left action** of G on M, if

1) for all $p \in M$, $f(e, p) = p$, where e is the identity element of G;
2) for all $g_1, g_2 \in G$ and $p \in M$,

$$f(g_1, f(g_2, p)) = f(g_1 \bullet g_2, p).$$

Usually, one abuses the notation and writes "$g \bullet p$" or even "gp" for $f(g, p)$. In this notation, the requirements of Def. 27 become $ep = p$ and $g_1(g_2 p) = (g_1 g_2)p$. The function f is called a **right action** if instead of the second requirement, we have

$$f(g_1, f(g_2, p)) = f(g_2 \bullet g_1, p).$$

Similarly, one denotes a right action by

$$f(g, p) \equiv p \bullet g \quad \text{or} \quad f(g, p) \equiv p g.$$

Every Lie group G has a natural left action on itself. This is simply given by group multiplication from the left:

$$f(g_1, g_2) = \bullet(g_1, g_2) = g_1 \bullet g_2 \equiv g_1 g_2.$$

Similarly, one can define the right action of G on itself. For each group element $g \in G$, the left and the right actions of G on itself define two *canonical* smooth functions $L_g : G \to G$ and $R_g : G \to G$, respectively. They are given by

$$L_g(h) := g \bullet h \tag{A.19}$$
$$R_g(h) := h \bullet g. \tag{A.20}$$

These functions and their push-forward and pullback maps have many applications in the theories of Lie groups and fiber bundles.

Definition 28: An action f is called **transitive** if for every two points p_1 and p_2 of M, there is some $g \in G$ such that $p_2 = f(g, p_1)$. f is called a **free action**, if for every $p \in M$ and $g \in G - \{e\}$, $f(g, p) \neq p$.

For example, let us consider the left action of $SO(2)$, rotations about the z-axis, on the unit (round) sphere S^2 inside $\mathbb{R}^3$. This is neither transitive nor free. To show this, first we need two points on S^2 that are not linked via a rotation about the z-axis. This is easily done by choosing two points with different values of z-coordinate in $\mathbb{R}^3$. This means that the action is not transitive. Next, consider the poles. They are left unchanged by all such rotations. Thus, the action is not free either.

An important example of a Lie group is the full rotation group in $\mathbb{R}^3$. An arbitrary rotation in $\mathbb{R}^3$ is specified by three numbers; these are known as the *Euler angles* [147]. An equivalent specification of an arbitrary element of this group is obtained by choosing a unit vector centered at the origin and a point of the unit sphere centered at the tip of this unit vector. One can show that this group is isomorphic to $SO(3)$. Geometrically, it is the manifold obtained by identifying the opposite points of the three sphere S^3 (with respect to the center). In fact, the whole sphere S^3 is another interesting example of a Lie group, namely the group $SU(2)$. $SU(2)$ is defined as the subgroup of the unitary group $U(2)$ that consists of the matrices of unit determinant.

We can study the action of $SO(3)$ on $\mathbb{R}^3$ by identifying the points of $\mathbb{R}^3$ by column vectors and multiplying them by elements of $SO(3)$, i.e., orthogonal 3×3 matrices with unit determinant. We can also restrict this action onto the submanifold S^2 of $\mathbb{R}^3$. Since every rotation preserves the magnitude of a vector, the action of $SO(3)$ maps S^2 to itself. In fact, we can easily see that this action is transitive but not free. An example of an action of a Lie group G that is both transitive and free is the left (right) action of G on itself.

Probably the most important concept in the theory of Lie groups is the concept of the *Lie algebra* of a Lie group. To present a definition of the Lie algebra of a Lie group we need to recall some basic facts about vector fields on a general manifold.

In our discussion of vector fields on smooth manifolds, we introduced a practical local expression for arbitrary vector fields. We labeled this formula as (A.10). In this expression, we denoted every vector field by a differential operator. In fact, it turns out that there is a one-to-one correspondence between the set of all (contravariant) vector fields and the set of all the differential operators of the form given by (A.10). Under this correspondence every vector field is viewed as a differential operator acting on the space of all scalar fields. An interesting property of operators is that they can be composed. Let us choose two vector fields $V = V^i(x)\frac{\partial}{\partial x^i}$ and $W = W^j(x)\frac{\partial}{\partial x^j}$ and consider their *commutator*:

$$\begin{aligned}[V, W] &:= V \circ W - W \circ V \\ &= V^i \frac{\partial}{\partial x^i}\left(W^j \frac{\partial}{\partial x^j}\right) - W^j \frac{\partial}{\partial x^j}\left(V^i \frac{\partial}{\partial x^j}\right) \\ &= \underbrace{\left(V^i \frac{\partial W^k}{\partial x^i} - W^j \frac{\partial V^k}{\partial x^j}\right)}_{U^k} \frac{\partial}{\partial x^k}.\end{aligned} \quad (A.21)$$

We can readily check that the components U^k of $[V, W]$ transform like the components of a (contravariant) vector field, i.e., according to (A.8). Let us denote the set of all vector fields of a smooth manifold M by $\mathcal{X}(M)$. The operation defined by (A.21) is a binary operation,

$$[\cdot, \cdot] : \mathcal{X}(M) \times \mathcal{X}(M) \longrightarrow \mathcal{X}(M),$$

that promotes $\mathcal{X}(M)$ into a *non-associative algebra*.[19] This operation is called the *Lie bracket* of two vector fields.

Definition 29: Let G be a Lie group and $\mathcal{X}(G)$ be the algebra of vector fields on G. A vector field $X \in \mathcal{X}(G)$ is said to be a **left-invariant vector field**, if for every $g, h \in G$, it satisfies

$$L_{g*}(X(h)) = X(g \bullet h),$$

where $X(h)$ and $X(g \bullet h)$ are the values of the vector field at the argument points and L_{g*} is the push-forward map induced by the left action of G on itself (A.19). It can be shown that the Lie bracket of two left-invariant vector fields is also left-invariant. Thus, the set of all left-invariant vector fields form a subalgebra of $\mathcal{X}(G)$. The algebra operation is obviously the Lie bracket. This algebra is called the **Lie algebra** of G. It is denoted by $\mathcal{G}$ or LG.

A "geometrical" interpretation of the Lie algebra $\mathcal{G}$ is offered by the following simple result.

Proposition 3: As vector spaces $\mathcal{G}$ and $T_e G$ are isomorphic.

Although this result may seem rather mysterious, it is shown quite straightforwardly. The key point is to recognize that every left-invariant vector field X can be constructed from its value at the identity, $X(e) \in T_e G$. This is done via the left action map L_g of (A.19). We have, for any $g \in G$,

---

[19] A non-associative algebra satisfies all the requirements of an associative algebra except the condition of associativity, i.e., condition 4 of Def. 19.

$$X(g) = L_{g*}(X(e)). \tag{A.22}$$

Using this equation, we can push-forward a basis of T_eG to define a basis for $\mathcal{G}$. This is sufficient to prove the isomorphy of these two vector spaces. Since $\dim(T_eG) = \dim(G)$, $\mathcal{G}$ is finite dimensional if and only if G is. Another important implication of (A.22) is that every basis of T_eG induces a set of global basis vector fields in TG. In the language of vector bundles this means that the tangent bundle of every Lie group is a trivial bundle,[20] i.e., it is a product manifold,

$$TG = G \times \mathbb{R}^m$$

where $m = \dim(G)$.

Furthermore, since the Lie algebra $\mathcal{G}$ is isomorphic to T_eG, we can write down the defining (commutation) relations in terms of the tangent vectors at the identity. This in turn indicates that if we compute the Lie bracket of the elements of the Lie algebra via (A.17), the coefficients of the right-hand side of (A.21) will be constant. For example, let us choose a basis $\{J_i\}$ of $\mathcal{G}$. Then, we have

$$[J_i, J_j] = c_{ij}^k \, J_k. \tag{A.23}$$

The basis elements J_i of the Lie algebra are also called the *generators* of the Lie group. The coefficients c_{ij}^k are called the *structure constants*. They determine the Lie algebra. This means that a complete set of commutation relations such as (A.23) specifies the Lie algebra without any reference to the structure of the Lie group. In fact, there is a way to find a Lie group associated to a given Lie algebra. The association is however not unique, i.e., we can find several Lie groups for a given Lie algebra.

In order to define a Lie algebra independently of a Lie group, we shall first try to determine the important properties of the Lie algebra of a Lie group. These are simply the antisymmetry of the Lie bracket:

$$[V, W] = -[W, V],$$

and the so-called *Jacobi identity*:

$$[[U, V], W] + [[W, U], V] + [[V, W], U] = 0,$$

where U, V, W are arbitrary Lie algebra elements. We can easily verify these identities for the Lie algebra of a Lie group using (A.21). For an *abstract Lie algebra* however, they serve as the defining postulates or axioms.

Definition 30: Let $(\mathcal{G}, +, \cdot, [\cdot, \cdot])$ be a non-associative algebra with the algebra multiplication denoted by

$$[\cdot, \cdot] : \mathcal{G} \times \mathcal{G} \longrightarrow \mathcal{G}.$$

---

[20] A manifold whose tangent bundle is trivial is called a *parallelizable manifold*, so every Lie group is parallelizable.

Then, $\mathcal{G}$ is called an **abstract Lie algebra** if the algebra operation is anti-symmetric and it satisfies the Jacobi identity.

A familiar example of an abstract Lie algebra is the algebra of operators in one-dimensional quantum mechanics. The basic elements are the coordinate operator $\hat{x}$, the momentum operator $\hat{p}$, and the identity operator $\hat{1}$. The algebra operation is the commutator of the operators. This operation satisfies all the requirements of Def. 30. The corresponding Lie algebra is called the Weyl–Heisenberg algebra. It is defined by the following commutation relations:

$$[\hat{x}, \hat{p}] = i\hat{1}$$
$$[\hat{x}, \hat{x}] = [\hat{p}, \hat{p}] = [\hat{1}, \hat{1}] = 0$$
$$[\hat{x}, \hat{1}] = [\hat{p}, \hat{1}] = 0.$$

Another simple example is the Lie algebra of all $(N \times N)$-matrices with the algebra operation being the commutator of two matrices. This Lie algebra is denoted usually by $\mathcal{G}\ell(N, \mathbb{R})$ or $\mathcal{G}\ell(N, \mathbb{C})$ depending on whether the entries of the matrices are real or complex.[21]

Lie algebras of Lie groups are used extensively in almost all aspects of Lie group theory. A particularly important tool that makes the applications of the Lie algebra possible is the so-called *exponential map*. This is a smooth map from the Lie algebra to the Lie group. In fact, if the Lie group is compact and connected then the exponential map is onto. That is, every group element can be obtained as the *exponential* of some element of the Lie algebra (a tangent vector at the identity). This is extremely important because in practice it is much easier to study a Lie algebra than a Lie group.

Let us consider a smooth manifold M and suppose that $V = V^i(x)\frac{\partial}{\partial x^i}$ is a vector field on M. At each point $x \in M$, we can obtain the value of V as the tangent vector to some curve in M. Let us choose $x_0 \in M$ and denote $V(x_0)$ by v_0. Then, we can find a curve C that starts at x_0 and is tangent to the vector field V at all its points,

$$V(C(t)) = \frac{dC(t)}{dt}, \quad \forall t \in [0, T].$$

The curve C is called an *integral curve* of the vector field V. Every integral curve is uniquely defined up to the starting point x_0. This is a consequence of the existence and uniqueness theorem for ordinary differential equations. For, an integral curve is the solution of the following first-order differential equation:

$$\frac{d}{dt} x^i(t) = V^i(x(t)) \tag{A.24}$$
$$x^i(0) = x_0^i,$$

---

[21] Note that the set of such matrices also form an associative algebra under matrix multiplication.

where $x(t) = (x^1(t), \cdots, x^m(t))$ are the coordinates of the points $C(t)$ of C, and x_0^i are the components of the initial point x_0. The integral curve C is obtained by integrating (A.24).

We can use the results of the previous paragraph to define a smooth function on M that maps an arbitrary point x_0 to the end point of the corresponding integral curve. If we set $T = 1$, i.e., $t \in [0, 1]$, this function maps $x_0 = C(0)$ to $C(1)$, where C is the solution of (A.24). There are manifolds with pathological problems that render this construction inapplicable. Lie groups and, as a matter of fact, all the manifolds we will encounter in this book do not have such problems and the above construction is valid. We denote the "end point" function by

$$\exp_M : \mathcal{X}(M) \times M \longrightarrow M.$$

It is defined by

$$\exp_M(V, x_0) := C(1). \tag{A.25}$$

In words, the function $\exp_M$ yields the end point of the integral curve defined by the vector field $V \in \mathcal{X}(M)$ and the initial condition $x_0 \in M$.

Let us return to our discussion of Lie groups and their Lie algebras. Every element of the Lie algebra of a Lie group is a left-invariant vector field X. We can obtain a function from the Lie algebra into the Lie group by restricting (A.25) to the left-invariant vector fields on G and choosing x_0 to be the identity element of the Lie group, $e \in G$. The resulting function is called the *exponential map*, $\exp : \mathcal{G} \to G$:

$$\exp(X) := \exp_G(X, e).$$

An examination of matrix groups justifies the name "exponential map". For example, the Lie algebra of $U(1)$ is the purely imaginary numbers, $u(1) = i\mathbb{R}$. Then for any $i\varphi \in i\mathbb{R}$, we have

$$\exp(i\varphi) = e^{i\varphi}$$

The same is true for all other matrix groups. An element of the Lie algebra of every matrix group is itself a matrix of the same dimension. The exponential map for these groups reduces to

$$\exp(X) = e^X := \sum_{k=0}^{\infty} \frac{1}{k!} X^k.$$

We shall next review an important class of matrix groups, called the unitary groups, $U(N)$.

Definition 31: Let $GL(N, \mathbb{C})$ denote the set of all invertible $(N \times N)$ complex matrices. $GL(N, \mathbb{C})$ form a Lie group under the operation of matrix multiplication. It is called the *(complex) general linear group*. The Lie algebra

$\mathcal{Gl}(N,\mathbb{C})$ of $GL(N,\mathbb{C})$ consists of all $(N \times N)$ complex matrices. The **unitary group** $U(N)$ is a compact Lie subgroup of $GL(N,\mathbb{C})$ defined by

$$U(N) := \{U \in GL(N,\mathbb{C}) : U^{-1} = U^\dagger\},$$

where "$\dagger$" stands for Hermitian conjugation. The unitary matrices of unit determinant form a Lie subgroup of $U(N)$ called the **special unitary group**:

$$SU(N) := \{U \in U(N) : \det(U) = 1\}.$$

The Lie algebras of $U(N)$ and $SU(N)$ are denoted usually by $u(N)$ and $su(N)$, respectively. They are Lie subalgebras of $\mathcal{Gl}(N,\mathbb{C})$ given by

$$u(N) := \{X \in \mathcal{Gl}(N,\mathbb{C}) : X^\dagger = -X\}$$
$$su(N) := \{X \in \mathcal{Gl}(N,\mathbb{C}) : X^\dagger = -X,\ \text{tr}(X) = 0\},$$

where "tr" stands for "trace". Therefore, elements of $u(N)$ are *anti-Hermitian matrices*.

Both $U(N)$ and $SU(N)$ are compact connected Lie groups and their exponential maps are onto. Hence, every unitary and special unitary matrix is obtained as the exponential of some element of the Lie algebra. Specifically, we have

$$U(N) = \{\exp(X) : X \in \mathcal{Gl}(N,\mathbb{C}),\ X^\dagger = -X\} \quad (A.26)$$
$$SU(N) = \{\exp(X) : X \in \mathcal{Gl}(N,\mathbb{C}),\ X^\dagger = -X,\ \text{tr}(X) = 0\}.$$

In physics, we are accustomed to work with Hermitian matrices rather than the anti-Hermitian matrices. This is the reason for the extra factor of "$i = \sqrt{-1}$" in physicists' definition of the exponential map. Equations (A.26) are often written in the following form:

$$U(N) = \{\exp(iX) : X \in \mathcal{Gl}(N,\mathbb{C}),\ X^\dagger = X\}$$
$$SU(N) = \{\exp(iX) : X \in \mathcal{Gl}(N,\mathbb{C}),\ X^\dagger = X,\ \text{tr}(X) = 0\}.$$

$U(N)$ and $SU(N)$ can also be viewed as transformation groups acting on $\mathbb{C}^N$. The situation is similar to the action of $O(2)$ on $\mathbb{R}^2$. The group elements are (represented by) the $N \times N$ complex matrices that multiply the complex column vectors on the left. Let U be a unitary matrix and $\mathbf{z} \in \mathbb{C}^N$. Then, we have

$$\|U\mathbf{z}\|^2 = \mathbf{z}^\dagger U^\dagger U \mathbf{z} = \mathbf{z}^\dagger \mathbf{z} = \|\mathbf{z}\|^2.$$

Hence, the unitary transformations preserve the (Euclidean) norm on $\mathbb{C}^N$.

Clearly, $U(N)$ is a subgroup of $U(N+1)$. This can be shown by representing elements of $U(N)$ by $(N+1) \times (N+1)$ matrices of the form

$$\begin{pmatrix} * & * & \cdots & * & 0 \\ * & * & \cdots & * & 0 \\ \vdots & \vdots & \vdots\vdots\vdots & \vdots & \vdots \\ * & * & \cdots & * & 0 \\ 0 & 0 & \cdots & 0 & 1 \end{pmatrix}, \quad (A.27)$$

where "*" are complex numbers forming an $N \times N$ unitary matrix in the upper left block of (A.27). This is a trivial example of an $(N+1)$-dimensional complex "representation" of $U(N)$.

Definition 32: Let G be a group and V be a complex (real) vector space. The space of all invertible linear transformations of V is denoted by $GL(V)$. A function

$$\rho : G \longrightarrow GL(V),$$

is called a **complex (real) representation** of G, if for every $v \in V$ it satisfies:

1) $\rho(e)(v) = v$;
2) $\rho(g_1 \bullet g_2)(v) = \rho(g_1)(\rho(g_2)(v)) = [\rho(g_1) \circ \rho(g_2)](v)$,

where "e" is the identity element of G and "$\circ$" is the composition of linear transformations (multiplication of matrices).

In practice, we usually choose V to be either $\mathbb{C}^N$ or $\mathbb{R}^N$ (for the finite-dimensional representations). In this case, a representation maps the group elements to non-singular matrices. In particular, this allows us to study an abstract group in terms of a subgroup of $GL(N,\mathbb{C})$ or $GL(N,\mathbb{R})$. A representation is said to be a *faithful representation* if it is a one-to-one function.

Next let us consider a representation (ρ, V) of G. It is possible that all the elements of the $\rho(G)$ are represented by block diagonal matrices of the same form, with each block corresponding to a proper vector subspace of V. If this happens, then the representation is said to be a reducible representation. The precise definition is as follows:

Definition 33: Let (ρ, V) be a representation of a group G. Then a vector subspace V' of V is said to be an *invariant* subspace if for every $g \in G$ and every $v' \in V'$, $\rho(g)[v'] \in V'$, alternatively $\rho(G)[V'] \subseteq V'$. The representation (ρ, V) is said to be an **irreducible representation** of G if the only invariant subspaces of V are the trivial vector subspace $\{0\}$ and V itself. A representation which is not irreducible is called a **reducible** representation. It can be decomposed into irreducible representations.

In physics, we are often interested in the so-called *unitary irreducible representations* of Lie groups. A finite-dimensional unitary representation takes the group elements to unitary matrices. It turns out that if the group is non-compact there are no finite-dimensional unitary representations. However, in many cases the groups of interest are compact and we can represent them as subgroups of some unitary group. In fact, a result of group theory, namely the *Peter–Weyl theorem*, says that every compact connected Lie group is isomorphic to a subgroup of some unitary group.

Earlier in this section we emphasized the importance of the notion of the Lie algebra of a Lie group and mentioned that it is much easier to study the Lie algebras. This extends also to the subject of the representations of the Lie groups, in particular, the irreducible unitary representations.

Given an arbitrary representation (ρ, V) of a Lie group G, one can use the identification of the Lie algebra $\mathcal{G}$ of G with the tangent space at the identity $T_e G$, to obtain a representation of the Lie algebra.

Definition 34: Let $\mathcal{G}$ be an arbitrary abstract Lie algebra, V a vector space, and $\mathcal{G}\ell(V)$ be the vector space of all the linear transformations of V. Then a linear function,

$$\lambda : \mathcal{G} \longrightarrow \mathcal{G}\ell(V),$$

is said to be a **representation of the Lie algebra** $\mathcal{G}$ if for every $X, Y \in \mathcal{G}$,

$$\lambda([X, Y]) = \lambda(X) \circ \lambda(Y) - \lambda(Y) \circ \lambda(X), \tag{A.28}$$

where "[,]" and "$\circ$" denote the Lie algebra operation of $\mathcal{G}$ and the composition of linear functions of V, respectively. In particular, if (ρ, V) is a representation of a Lie group G and $\mathcal{G}$ is the Lie algebra of G, then the push-forward map:

$$\rho_* : T_e G = \mathcal{G} \longrightarrow \mathcal{G}\ell(V) = T_1 GL(V),$$

defines a representation of $\mathcal{G}$. Here "1" stands for the identity operator on V.

Once a representation (λ, V) of a Lie algebra $\mathcal{G}$ is chosen, the elements of $\mathcal{G}$ and in particular its basis elements J_i, may be identified with some linear operators acting on V. These however may be composed. In fact the space $\mathcal{G}\ell(V)$ together with the operation of multiplication by numbers, addition of linear transformations, and their composition forms an associative algebra. This algebra which is also related with the Lie algebra structure of $\mathcal{G}\ell(V)$, with the Lie bracket defined by the right-hand side of (A.28), is an example of an *enveloping algebra*.

Definition 35: Let $(\mathcal{G}, [,])$ be an abstract Lie algebra with a basis $\{J_i\}$, $(\mathcal{A}, ., +, \otimes)$ be an associative algebra generated by $\{\mathcal{J}_j\}$ and $\tilde{\mathcal{G}}$ be the vector subspace of $\mathcal{A}$ spanned by $\{\mathcal{J}_j\}$. Then $(\mathcal{A}, ., +, \otimes)$ is said to be the **enveloping algebra** of $(\mathcal{G}, [,])$ and denoted by $\mathcal{E}(\mathcal{G})$, if there exists a vector space isomorphism

$$f : \mathcal{G} \to \tilde{\mathcal{G}},$$

and for all $X, Y \in \mathcal{G}$ the following condition is satisfied:

$$f([X, Y]) := f(X) \otimes f(Y) - f(Y) \otimes f(X).$$

More simply, one says that the enveloping algebra $\mathcal{E}(\mathcal{G})$ of a Lie algebra $\mathcal{G}$ is an associative algebra generated by a basis of $\mathcal{G}$.

Definition 36: An element of the enveloping algebra of a Lie algebra is called a **Casimir operator**, if it commutes with all the generators $\mathcal{J}_i$.

If $\mathcal{G}$ is the Lie algebra of a Lie group G which acts as a transformation group of a physical system, then the Casimir operators represent the invariant quantities. In quantum mechanics a symmetry is generated by a linear operator which commutes with the Hamiltonian. Hence, if G is the symmetry group, then the Hamiltonian must be a (representation of a) Casimir operator.

We end this appendix by emphasizing that the theory of group representations has played a substantial role in the development of quantum physics. In fact, many of the pioneering works in this subject were conducted by physicists such as Eugene Wigner.

B. A Brief Review of Point Groups of Molecules with Application to Jahn–Teller Systems

In this appendix we present a brief discussion of the point groups of molecules and their applications to Jahn–Teller systems. Our aim is to provide the reader with the minimum background necessary for following the related discussions in the text. Therefore we shall often sacrifice rigor for brevity.

For a more thorough treatment of the material in this appendix, the following books may be consulted:

– T. Inui, Y. Tanabe, and Y. Onodera: *Group Theory and Its Applications in Physics* (Springer-Verlag, Berlin, Heidelberg, 1990);
– S. K. Kim: *Group Theoretical Methods and Applications to Molecules and Crystals* (Cambridge University Press, Cambridge, 1999).

Point Groups

A transformation of the (configuration) space of a rigid body that maps a given configuration of the body to the same configuration is called a symmetry transformation or a symmetry operation. As we discussed in Appendix A symmetry operations form a group G. The product of two elements of G (two symmetry transformations) corresponds to their successive application on the rigid body.[1] In general all the symmetry transformations keep (at least) one point fixed (Brouwer fixed point) and leave the distance between any two points in the body unchanged. For this reason the group of symmetry transformations is also called a *point group*. In the following we study the point groups associated with nuclear configurations of various molecules.

Elements of point groups (symmetry operations) can be put into the following categories.

1. E: the identity operation which does not change the configuration space;[2]
2. C_n: a rotation by an angle $2\pi/n$ about an axis – this is called an n-fold proper rotation;
3. σ: a (mirror) reflection about a plane;

---

[1] For a rigid body in ordinary three-dimensional Euclidean space, G is a subgroup of the orthogonal group $O(3)$.
[2] In general a point group may include elements that do change the configuration space of a rigid body but leave all the points of the body unchanged. We shall differentiate between such an element and the identity element of the point group.

408 B. Point Groups

4. S_n: C_n followed by σ with the reflection plane of σ being perpendicular to the rotation axis of C_n – this is called an n-fold improper rotation;
5. i: inversion through the fixed point.

The axis corresponding to the largest rotational symmetry, i.e., the C_n axis (or S_n axis) with the greatest value of n is called the *principal axis* of the molecule. Conventionally, one works in a Cartesian coordinate frame in which the principal axis coincides with the z-axis. Depending on whether the reflection planes are vertical or horizontal with respect to the principal axis, the reflections are further differentiated by the labels σ_v and σ_h, respectively. A vertical plane which bisects an adjacent pair of another kind of σ_v or an adjacent pair of C_2 axes is denoted as σ_d, where d stands for "dihedral".

If a group has a finite number of elements (finite order), then it is called a *finite group*. Linear molecules with axial symmetry admit (an infinite number of) continuous symmetry operations, so the corresponding point groups are not finite. In the following we shall only consider molecules with a finite point group.

We begin our discussion of the point groups of molecules by studying the simple examples of the H_2O and H_3^+ molecules.

The equilibrium shape of H_2O is an isosceles obtuse triangle. The principal axis is the bisector of the obtuse angle. As shown in Fig. B.1, we will identify this axis with the z-axis and choose the x- and y-axes so that the molecule lies in the x-y plane. The point group for H_2O consists of the following four elements which are depicted in Fig. B.1.

1. The identity operation E;
2. C_2 about the z-axis, which is denoted by $C_2(z)$;

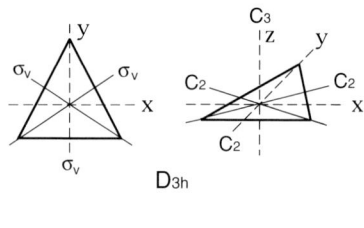

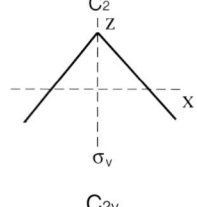

Fig. B.1. Symmetry operations of D_{3h} and C_{2v}.

3. σ through the z-y plane, which is denoted by $\sigma_v(zy)$;
4. σ through the x-z plane, which is denoted by $\sigma_v(zx)$.

The equilibrium shape of H_3^+ is an equilateral triangle. The principal axis (z-axis) is perpendicular to the plane of the triangle (this is our x-z plane) and passes through the center of the triangle where the bisectors of its angles intersect. The point group of H_3^+ has the following 12 elements.

1. The identity operation E;
2. C_3 about the z-axis, which is denote by $C_3(z)$;
3. $C_3(z)^2$ which means performing $C_3(z)$ twice;
4–6. the three C_2 rotations about the axes along the bisectors of the angles;
7. σ_h with the x-y plane as the reflection plane;
8, 9. the two improper rotations: $S_3 = \sigma_h C_3$ and $S_3^{-1} = \sigma_h C_3^2$;
10–12. the three σ_v with reflection planes defined by the z-axis and the three axes of the C_2 rotations.

In general the elements of a finite group may be obtained by multiplying a small number of more basic elements. These are called the *generators* of the group. The point groups that are generated by the proper rotations C_n are called *proper point groups*. There are five different types of proper point groups, namely

1. C_n which is the symmetry group generated by the C_n rotation;
2. D_n which is the symmetry group generated by the C_n and C_2 rotations with perpendicular axes;
3. T which is the symmetry group associated with a tetrahedron and generated by the C_2 and C_3 rotations;
4. O which is the symmetry group associated with a cube or octahedron and generated by the C_4 and C_3 rotations;
5. I which is the symmetry group associated with an icosahedron or dodecahedron and generated by the C_5 and C_3 rotations.

In general any point groups that is not proper – an *improper point group* – can be generated using the generators of a proper point subgroup, the inversion i, and the improper rotations $p = S_{2n}$ (with the axes of C_{2n} and C_n being parallel) and $v = S_2'$ (with the axes of C_2 and C_n being perpendicular). This is sometimes expressed as

$$P_a = P + aP, \qquad a = i, p, v, \qquad P = C_n, D_n, T, O, I, \qquad (B.1)$$

where P_a, P, and aP respectively denote the improper group, a proper subgroup, and the elements obtained by multiplying a by the elements of P.

Note that a given improper group P_a may be constructed using different choices for P and a. The following is a list of all improper point groups where the commonly used *Schönflies notations* are also given in parentheses.

$$\begin{aligned}
&C_{ni} \quad (C_{nh}(\text{even } n), S_n(\text{odd } n));\\
&C_{np} \quad (C_{nh}(\text{odd } n), S_n(\text{even } n));\\
&C_{nv};\\
&D_{ni} \quad (D_{nh}(\text{even } n), D_{nd}(\text{odd } n));\\
&D_{np} \quad (D_{nh}(\text{odd } n), D_{nd}(\text{even } n));\\
&T_i \quad (T_h);\\
&T_p \quad (T_d);\\
&O_i \quad (O_h);\\
&I_i \quad (I_h).
\end{aligned} \qquad (\text{B.2})$$

The Schönflies notations for the point groups of H_2O and H_3^+ are C_{2v} and D_{3h}, respectively.

Representation of a Group

A matrix representation of a group G consists of a collection M of square matrices that together with matrix multiplication form a group with the same group structure as G. This means that there is a function (a group homomorphism) $D : G \to M$ mapping the product of two group elements R and S to the product of their images. Namely

$$RS = T \qquad (\text{B.3})$$

is mapped to

$$D(R)D(S) = D(T). \qquad (\text{B.4})$$

For the matrix elements, we have

$$\sum_{j=1}^{N} D(R)_{ij} D(S)_{jk} = D(T)_{ik}. \qquad (\text{B.5})$$

The matrix group M acts in the vector space $\mathbb{R}^N$ or $\mathbb{C}^N$ depending on whether its elements are real or complex matrices. The function D together with $\mathbb{R}^N$ (respectively $\mathbb{C}^N$) is said to be an N-dimensional real (respectively complex) representation of G. $D(R)$ is said to be the representation of R in the representation D. The vector space $\mathbb{R}^N$ (respectively $\mathbb{C}^N$) in which the elements of M act is called the representation space of D.[3]

---

[3] More general representations called *ray* or *projective representations* are possible. In a ray representation $RS = T$ is mapped to

$$D(R)D(S) = \alpha_{RS}^{T} D(T), \qquad (\text{B.6})$$

where extra factors α_{RS}^{T} appear. See Section 5.4 of [113].

Representations of a point group G are usually constructed by choosing the representation space from among the subspaces of a function space. That is, in order to construct a real (respectively complex) N-dimensional representation, one chooses a set of N linearly independent real-valued (respectively complex-valued) functions $\{\phi_1, \phi_2, \cdots, \phi_N\}$ that under the application of a symmetry operation $R \in G$ transform linearly among themselves. If $R\phi_i$ denotes the application of R on ϕ_i, then

$$R\phi_i = \sum_{j=1}^{N} \phi_j D(R)_{ji}. \tag{B.7}$$

The representation space is then defined to be the span of the functions ϕ_i. This in turn means that ϕ_i form a basis of the representation space. The set $\{\phi_1, \phi_2, \cdots, \phi_N\}$ is called the basis of the representation.

In view of (B.7), the application of the product $T = SR$ of the elements $R, S \in G$ on ϕ_i is given by

$$SR\phi_i = S \sum_j \phi_j D_{ji}(R)$$

$$= \sum_{j,k=1}^{N} \phi_k D_{kj}(S) D_{ji}(R)$$

$$= T\phi_i = \sum_{k=1}^{N} \phi_k D_{ki}(T). \tag{B.8}$$

Because ϕ_i are linearly independent, this implies that the matrices $D(S)$, $D(R)$, and $D(T)$ satisfy (B.5) and (B.4); D is a representation of the group G.

In general point groups have different representations. Two representations D and D' of a point group G are said to be *equivalent* if the dimension of their representation spaces (V and V') are the same and they are related via a *similarity transformation*, i.e., there is an invertible linear operator $U : V \to V'$ such that for every $R \in G$,

$$D(R) = U^{-1} D'(R) U.$$

In the representation theory of groups one does not distinguish between equivalent representations.

We shall use the label Γ to denote non-equivalent representations. The representation of a group element $R \in G$ in the representation Γ is then denoted by $D^\Gamma(R)$. The basis functions ϕ_i are said to belong to the Γ-representation of G. A function ϕ_γ belonging to the representation space of a representation Γ is said to transform as the γ component of Γ.

The transformation property (B.7) may be extended to linear operators. Let $O : V \to V$ be a linear operator acting in the representation space V

of a Γ-representation of G. Then for all $R \in G$ and $\psi \in V$, the symmetry transformation

$$\psi \to \psi' := R\psi \tag{B.9}$$

induces a transformation $O \to O'$ on the linear operator O where $O'\psi' := RO\psi$. Hence according to (B.9), $O' = ROR^{-1}$.

An operator $\hat{O}(\Gamma)_\gamma$ is said to transform as the γ component of the representation Γ, if it satisfies

$$R\hat{O}(\Gamma)_\gamma R^{-1} = \sum_{\gamma'} \hat{O}(\Gamma)_{\gamma'} D^\Gamma_{\gamma'\gamma}(R). \tag{B.10}$$

Such an operator is called a *tensor operator*.

Consider a representation Γ of a group G and suppose that the representation space V is endowed with an inner product $\langle \ | \ \rangle$. Then Γ is said to be a *unitary representation* if for all $\psi_{\gamma_1}, \psi_{\gamma_2} \in V$ and $R \in G$,

$$\langle D^\Gamma(R)\psi_{\gamma_1} | D^\Gamma(R)\psi_{\gamma_2}\rangle = \langle \psi_{\gamma_1} | \psi_{\gamma_1} \rangle.$$

In other words, $D^\Gamma(R)$ are unitary matrices. Equivalent unitary transformations are related by unitary similarity transformations.

In the following we shall only consider the unitary representations.

Irreducible Representations

If all matrices used in a representation of a group G can be mapped to block-diagonal matrices by a single similarity transformation, i.e., if there is an invertible linear operator $C : V \to V$ acting in the representation space V such that for all $R \in G$

$$C^\dagger D(R) C = \begin{pmatrix} D_1(R) & 0 \\ 0 & D_2(R) \end{pmatrix}, \tag{B.11}$$

then the representation is said to be *reducible*. A representation which is not reducible is called *irreducible*.

The representation theory of finite groups is based on the so-called *great orthogonality relation*:

$$\sum_{R \in G} D^\Gamma_{ij}(R) D^{\Gamma'*}_{i'j'}(R) = \frac{|G|}{d_\Gamma} \delta_{\Gamma\Gamma'} \delta_{ij} \delta_{i'j'}, \tag{B.12}$$

where $|G|$ is the order of the group and d_Γ is the dimension of the representation Γ.[4] A simple application of this relation is the identity

$$\sum_\Gamma d_\Gamma^2 = |G|, \tag{B.13}$$

---

[4] This orthogonality relation also holds for ray representations.

where the sum is taken over all irreducible representations. As seen from this equation a finite group admits a finite number of non-equivalent irreducible representations.

A useful quantity in describing the representations of a finite group is the *character* of a representation:

$$\chi_\Gamma(R) := \sum_{i=1}^{d_\Gamma} D_{ii}^\Gamma(R) = \text{Tr}(D^\Gamma(R)), \tag{B.14}$$

where Γ is a representation of the group G, $R \in G$ is an arbitrary element, and Tr stands for the 'trace'.

Two group elements $S, R \in G$ are said to be related by a *conjugacy relation* or simply called *conjugate elements* if there is a group element $T \in G$ such that

$$R = T^{-1}ST. \tag{B.15}$$

The conjugacy relation is an equivalence relation.[5] The set of all elements that are conjugate to a given one, i.e., the equivalence class of the conjugacy relation, is called the *conjugacy class* (or *conjugate class*, or simply *class*) of that element. Characters of the elements of a conjugacy class are equal, for

$$\chi_\Gamma(R) = \text{Tr}(D^\Gamma(R)) = \text{Tr}((D^\Gamma(T))^{-1} D^\Gamma(S) D^\Gamma(T))$$
$$= \text{Tr}(D(S)) = \chi_\Gamma(S). \tag{B.16}$$

Here we have used (B.15) and the identity $\text{Tr}(AB) = \text{Tr}(BA)$ which holds for arbitrary square matrices A and B.

Using the orthogonality relation (B.12) we can easily derive the orthogonality relation for the characters, namely

$$\sum_{R \in G} \chi_\Gamma(R) \chi_{\Gamma'}(R)^* = |G| \delta_{\Gamma \Gamma'}. \tag{B.17}$$

Another important fact that follows from (B.12) is that *the number of inequivalent irreducible representations of a finite group is equal to the number of its conjugacy classes.*[6]

Character Tables

Tables that list characters of irreducible representations are called *character tables*.

Character tables for C_{2v}, D_{3h}, and O_h are given in Table B.1. C_{2v} and D_{3h} are the symmetry groups of H_2O and H_3^+ in their equilibrium shapes,

---

[5] See Appendix A for the definition of an equivalence relation.
[6] In ray representations, the values of characters are not necessarily equal within a class.

Table B.1. Character Tables for C_{2v}, D_{3h}, and O_h Point Groups

C_{2v}	E	C_2	$\sigma_v(zx)$	$\sigma_v(yz)$	
A_1	1	1	1	1	z, x^2, y^2, z^2
A_2	1	1	-1	-1	L_z, xy
B_1	1	-1	1	-1	x, L_y, zx
B_2	1	-1	-1	1	y, L_x, yz

D_{3h}	E	$2C_3$	$3C_2$	σ_h	$2S_3$	$3\sigma_v$	
A_1'	1	1	1	1	1	1	x^2+y^2, z^2
A_2'	1	1	-1	1	1	-1	L_z
E'	2	-1	0	2	-1	0	$(x,y), (x^2-y^2, xy)$
A_1''	1	1	1	-1	-1	-1	
A_2''	1	1	-1	-1	-1	1	z
E''	2	-1	0	-2	1	0	$(L_x, L_y), (yz, zx)$

O_h	E	$8C_3$	$3C_2$	$6C_4$	$6C_2'$	i	$8S_3$	$3S_2$	$6S_4$	$6S_v'$	
A_{1g}	1	1	1	1	1	1	1	1	1	1	$x^2+y^2+z^2$
A_{2g}	1	1	1	-1	-1	1	1	1	-1	-1	
E_g	2	-1	2	0	0	2	-1	2	0	0	$(x^2+y^2-2z^2, x^2-y^2)$
T_{1g}	3	0	-1	1	-1	3	0	-1	1	-1	(L_x, L_y, L_z)
T_{2g}	3	0	-1	-1	1	3	0	-1	-1	1	(xy, yz, zx)
A_{1u}	1	1	1	1	1	-1	-1	-1	-1	-1	
A_{2u}	1	1	1	-1	-1	-1	-1	-1	1	1	
E_u	2	-1	2	0	0	-2	1	-2	0	0	
T_{1u}	3	0	-1	1	-1	-3	0	1	-1	1	(x, y, z)
T_{2u}	3	0	-1	-1	1	-3	0	1	1	-1	

respectively. The O_h group is the symmetry group of a cube or octahedron which often arises in the discussion of octahedrally coordinated metal ions.

The first row of each character table lists the conjugacy classes of the symmetry operations. The left-hand column gives the irreducible representations using the *Mulliken symbols*. The rules for the Mulliken symbols are the following:

1. A, B for one-, E for two-, T for three-, G for four- and H for five-dimensional representations;
2. A (respectively B) is used for representations which are symmetric (respectively antisymmetric) with respect to rotations C_n about the principal axis;
3. the subscript 1 (respectively 2) is used for representations which are symmetric (respectively antisymmetric) with respect to σ_v or a C_2 with the rotation axis being perpendicular to the principal axis;
4. a prime ′ (respectively double prime ″) is used for representations which are symmetric (respectively antisymmetric) with respect to σ_h;

5. the subscript g (respectively u) is used for representations which are symmetric (respectively antisymmetric) with respect to i.

The right-hand column contains a set of functions that transform as a basis of the irreducible representation. Here (L_x, L_y, L_z) denotes the angular momentum vector.

Often a deformation of a molecule reduces the symmetry of its atomic configuration. For example, a molecular vibration of H_3^+ (Fig. B.2(g) below) deforms its equilibrium configuration, which is an equilateral triangle with D_{3h} symmetry, to an isosceles triangle with C_{2v} symmetry. This is a reduction of a point group symmetry to a subgroup symmetry. In order to describe such a deformation, one needs to correlate the irreducible representations of the point group to those of its subgroups.

It is possible that an irreducible representation Γ of a group G_1 is also a representation of another group G_2. However, Γ as a representation of G_2 is in general reducible. For example E' of D_{3h} is reducible as a representation of C_{2v}. We can correlate representations of a point group to the representations of its subgroups using their characters.

We can write the representation E' of D_{3h} as a direct sum of the representations of C_{2v} as

$$E'(D_{3h}) = \sum_{\Gamma \text{ of } C_{2v}} a_\Gamma \Gamma, \tag{B.18}$$

where a_Γ are constant coefficients relating the representation E' of D_{3h} to the irreducible representations of C_{2v}. Taking the trace of (B.18), we have

$$\chi_{E'(D_{3h})}(R) = \sum_{\Gamma \text{ of } C_{2v}} a_\Gamma \chi_\Gamma(R), \quad R \in C_{2v}. \tag{B.19}$$

Now by virtue of the orthogonality relation (B.17) we can compute

$$a_\Gamma = |C_{2v}|^{-1} \sum_{R \in C_{2v}} \chi_\Gamma(R)^* \chi_{E'(D_{3h})}(R). \tag{B.20}$$

Finally using the character tables for D_{3h} and C_{2v}, we find that the non-zero coefficients a_Γ are $a_{A_1} = 1$ and $a_{B_2} = 1$. Therefore, the E' (of D_{3h}) electronic state of an equilateral molecule splits into A_1 (of C_{2v}) and B_2 (of C_{2v}) states by the vibrational motion of Fig. B.2(g).

Product Representations

Let Γ_1 and Γ_2 be irreducible representations of a group G. The *product representation* $\Gamma_1 \times \Gamma_2$ is defined by

$$D^{\Gamma_1 \times \Gamma_2}_{ij,kl}(R) := D^{\Gamma_1}_{ij}(R) D^{\Gamma_2}_{kl}(R). \tag{B.21}$$

$\Gamma_1 \times \Gamma_2$ is indeed a representation of G, for

$$\sum_{j,l} D_{ij,kl}^{\Gamma_1 \times \Gamma_2}(R) D_{ji',lk'}^{\Gamma_1 \times \Gamma_2}(S) = \sum_{j,l} D_{ij}^{\Gamma_1}(R) D_{kl}^{\Gamma_2}(R) D_{ji'}^{\Gamma_1}(S) D_{lk'}^{\Gamma_2}(S)$$
$$= D_{ii'}^{\Gamma_1}(RS) D_{kk'}^{\Gamma_2}(RS)$$
$$= D_{ii',kk'}^{\Gamma_1 \times \Gamma_2}(RS). \tag{B.22}$$

The character of the product representation $\Gamma_1 \times \Gamma_2$ is the product of the characters for Γ_1 and Γ_2,

$$\chi_{\Gamma_1 \times \Gamma_2}(R) = \sum_{i=1}^{d_{\Gamma_1}} \sum_{k=1}^{d_{\Gamma_2}} D_{ii}^{\Gamma_1}(R) D_{kk}^{\Gamma_2}(R) = \chi_{\Gamma_1}(R) \chi_{\Gamma_2}(R). \tag{B.23}$$

In general the product representation is reducible. For example, consider the representation $E' \times E'$ of the group D_{3h}. For all $R \in G$, we have

$$\chi_{E' \times E'}(R) = (\chi_{E'}(R))^2 = \sum_{\Gamma \text{ of } D_{3h}} a_\Gamma \chi_\Gamma(R), \tag{B.24}$$

where the coefficients a_Γ are calculated using

$$a_\Gamma = |D_{3h}|^{-1} \sum_{R \in D_{3h}} \chi_\Gamma(R)^* \chi_{E' \times E'}(R). \tag{B.25}$$

This yields $a_{A_1} = a_{A_2} = a_{E'} = 1$. The direct product $E' \times E'$ is thus the direct sum of A_1, A_2, and E'. This is expressed as

$$E' \times E' = A_1 + A_2 + E'. \tag{B.26}$$

If in the decomposition of all the product representations of a group into a direct sum of its irreducible representations the non-vanishing coefficients a_Γ are all equal to 1, then the group is called *simply reducible*. Otherwise it is called *multiply reducible*. Most of the point groups are simply reducible. In the following we only consider simply reducible groups.

Consider the product of an irreducible representation Γ by itself, i.e., $\Gamma \times \Gamma$. Let $\{\phi_i\}$ and $\{\psi_i\}$ be sets of basis functions transforming in the representation Γ. The products $\phi_i \psi_j$ form a basis for the representation $\Gamma \times \Gamma$. Now consider the symmetric products $(\phi_i \psi_j + \phi_j \psi_i)$ and the antisymmetric products $(\phi_i \psi_j - \phi_j \psi_i)$. It is not difficult to see that

$$R(\phi_j \psi_l \pm \phi_l \psi_j) = \sum_{i,k} (\phi_i \psi_k \pm \phi_k \psi_i) D_{ij}^\Gamma(R) D_{kl}^\Gamma(R). \tag{B.27}$$

Therefore, the symmetric and antisymmetric products of ϕ_i and ψ_i each form basis functions of a representation of the group. These are respectively called the *symmetric* and *antisymmetric product representations* and denoted by

$[\Gamma \times \Gamma]$ and $\{\Gamma \times \Gamma\}$. Equation (B.27) also shows that the product representation $\Gamma \times \Gamma$ splits into $[\Gamma \times \Gamma]$ and $\{\Gamma \times \Gamma\}$.

We can compute the characters of the symmetric and antisymmetric product representations using the relations

$$\sum_{i,k}(\phi_i\psi_k + \phi_k\psi_i)D^\Gamma_{ij}(R)D^\Gamma_{kl}(R) = \frac{1}{2}\sum_{i,k}\{(\phi_i\psi_k + \phi_k\phi_i)$$
$$\times [D^\Gamma_{ij}(R)D^\Gamma_{kl}(R) + D^\Gamma_{kj}(R)D^\Gamma_{il}(R)]\},$$

$$\sum_{i,k}(\phi_i\psi_k - \phi_k\psi_i)D^\Gamma_{ij}(R)D^\Gamma_{kl}(R) = \frac{1}{2}\sum_{i,k}\{(\phi_i\psi_k - \phi_k\phi_i)$$
$$\times [D^\Gamma_{ij}(R)D^\Gamma_{kl}(R) - D^\Gamma_{kj}(R)D^\Gamma_{il}(R)]\}.$$

The result is

$$\chi_{[\Gamma\times\Gamma]}(R) = \sum_{i,k}\frac{1}{2}[D^\Gamma_{ii}(R)D^\Gamma_{kk}(R) + D^\Gamma_{ik}(R)D^\Gamma_{ki}(R)]$$
$$= \frac{1}{2}[(\chi_\Gamma(R))^2 + \chi_\Gamma(R^2)], \quad (B.28)$$

$$\chi_{\{\Gamma\times\Gamma\}}(R) = \sum_{i,k}\frac{1}{2}[D^\Gamma_{ii}(R)D^\Gamma_{kk}(R) - D^\Gamma_{ik}(R)D^\Gamma_{ki}](R)$$
$$= \frac{1}{2}[(\chi_\Gamma(R))^2 - \chi_\Gamma(R^2)]. \quad (B.29)$$

The Clebsch–Gordan Coefficients

As we indicated in the preceding section product representations are usually reducible. The unitary matrix $U^{\Gamma_1\times\Gamma_2}$ that reduces a product representation $\Gamma_1\times\Gamma_2$ is called the *Clebsch–Gordan matrix*. By definition, we have

$$\sum_{\gamma_1,\gamma_2,\gamma_1',\gamma_2'} U^{\Gamma_1\times\Gamma_2\dagger}_{\Gamma\gamma,\gamma_1\gamma_2}D^{\Gamma_1}_{\gamma_1\gamma_1'}(R)D^{\Gamma_2}_{\gamma_2\gamma_2'}(R)U^{\Gamma_1\times\Gamma_2}_{\gamma_1'\gamma_2',\Gamma'\gamma'}$$
$$= \sum_{\gamma_1,\gamma_2,\gamma_1',\gamma_2'}\langle\Gamma\gamma|\Gamma_1\gamma_1\Gamma_2\gamma_2\rangle D^{\Gamma_1}_{\gamma_1\gamma_1'}(R)D^{\Gamma_2}_{\gamma_2\gamma_2'}(R)\langle\Gamma_1\gamma_1'\Gamma_2\gamma_2'|\Gamma'\gamma'\rangle$$
$$= D^\Gamma_{\gamma\gamma'}(R)\delta_{\Gamma\Gamma'}, \quad (B.30)$$

where we have used the following commonly used notation.

$$U^{\Gamma_1\times\Gamma_2}_{\gamma_1\gamma_2,\Gamma\gamma} := \langle\Gamma\gamma|\Gamma_1\gamma_1\Gamma_2\gamma_2\rangle^* = \langle\Gamma_1\gamma_1\Gamma_2\gamma_2|\Gamma\gamma\rangle. \quad (B.31)$$

The matrix elements of the Clebsch–Gordan matrix are called the *Clebsch–Gordan coefficients*.

Let us write down the Clebsch–Gordan matrix for the reduction of the $E'\times E'$ representation in the D_{3h} point group. From $\Gamma_1=\Gamma_2=E'$, the resulting Γ can be A_1, A_2, or E' as given in (B.26). We denote the two components of E' as $\gamma = \theta, \epsilon$. Then, the Clebsch–Gordan matrix $U^{E' \times E'}$ is given by

$$U^{E' \times E'} = \begin{pmatrix} \frac{1}{\sqrt{2}} & 0 & -\frac{1}{\sqrt{2}} & 0 \\ 0 & \frac{1}{\sqrt{2}} & 0 & \frac{1}{\sqrt{2}} \\ 0 & -\frac{1}{\sqrt{2}} & 0 & \frac{1}{\sqrt{2}} \\ \frac{1}{\sqrt{2}} & 0 & \frac{1}{\sqrt{2}} & 0 \end{pmatrix}, \quad (B.32)$$

where the values $1, 2, 3$, and 4 of the label j (respectively i) of the (i,j) matrix elements of $U^{E' \times E'}$ correspond to $\Gamma \gamma = A_1, A_2, E'\theta, E'\epsilon$ (respectively $\gamma_1 \gamma_2 = \theta\theta, \theta\epsilon, \epsilon\theta, \epsilon\epsilon$). In Table B.2, Clebsch–Gordan coefficients for the octahedral group (O) are tabulated.

If the group is multiply reducible, there is more than one set of Clebsch–Gordan coefficients and an additional label must be added to differentiate the different sets. In the following we shall only consider the simply reducible groups. Thus, the additional label is omitted.

The unitary transformation described above is actually a change of basis given by

$$\Phi_{\Gamma\gamma} = \sum_{\gamma_1, \gamma_2} \phi_{\Gamma_1 \gamma_1} \phi_{\Gamma_2 \gamma_2} \langle \Gamma_1 \gamma_1 \Gamma_2 \gamma_2 | \Gamma \gamma \rangle. \quad (B.33)$$

Because the Clebsch–Gordan matrix is a unitary matrix, the Clebsch–Gordan coefficients satisfy the orthogonality relations:

$$\sum_{\Gamma,\gamma} \langle \Gamma\gamma | \Gamma_1 \gamma_1 \Gamma_2 \gamma_2 \rangle \langle \Gamma_1 \gamma_1' \Gamma_2 \gamma_2' | \Gamma \gamma \rangle = \delta_{\gamma_1 \gamma_1'} \delta_{\gamma_2 \gamma_2'},$$

$$\sum_{\gamma_1, \gamma_2} \langle \Gamma\gamma | \Gamma_1 \gamma_1 \Gamma_2 \gamma_2 \rangle \langle \Gamma_1 \gamma_1 \Gamma_2 \gamma_2 | \Gamma' \gamma' \rangle = \delta_{\Gamma \Gamma'} \delta_{\gamma \gamma'}. \quad (B.34)$$

Using these relations, we obtain the following relation for the product of basis (wave) functions (B.33):

$$\phi_{\Gamma_1 \gamma_1} \phi_{\Gamma_2 \gamma_2} = \sum_{\Gamma,\gamma} \langle \Gamma\gamma | \Gamma_1 \gamma_1 \Gamma_2 \gamma_2 \rangle \Phi_{\Gamma\gamma} \quad (B.35)$$

Wigner–Eckart Theorem

Consider quantum mechanical wave functions $\Psi_{\alpha\Gamma\gamma}$ labeled by an irreducible representation Γ, its component γ, and some other quantum numbers α that are necessary to specify each state. The *Wigner–Eckart theorem* states that the matrix elements of a tensor operator $\hat{O}(\bar{\Gamma})_{\bar{\gamma}}$ can be expressed as

Table B.2. Clebsch–Gordan coefficients $\langle \Gamma_1\gamma_1\Gamma_2\gamma_2|\Gamma_3\gamma_3\rangle$ for the octahedral (O) group. The bases of E, T_1 and T_2 transform as $\theta \sim 3z^2-r^2$; $\epsilon \sim \sqrt{3}(x^2-y^2)$, $\alpha \sim x$; $\beta \sim y$; $\gamma \sim z$, and $\xi \sim yz$; $\eta \sim xz$; $\zeta \sim xy$, respectively. The bases of A_1 and A_2 are denoted as e_1 and e_2, respectively. The phase conventions $\langle \Gamma_1\gamma_1\Gamma_2\gamma_2|\Gamma_3\gamma_3\rangle = \langle \Gamma_2\gamma_2\Gamma_1\gamma_1|\Gamma_3\gamma_3\rangle$ for $\Gamma_1 \neq \Gamma_2$ are used. Also note that the basis for T_2, (ξ, η, ζ) are written as a product of the bases for A_2 and T_1 as $\xi = e_2\alpha$, $\eta = e_2\beta$, and $\zeta = e_2\gamma$. Then, we can relate the CG coeffcients for $T_2 \times T_2$ to those for $T_1 \times T_1$ as $\langle T_2(e_2\gamma_1)T_2(e_2\gamma_2)|\Gamma_3\gamma_3\rangle = -\langle T_1\gamma_1 T_1\gamma_2|\Gamma_3\gamma_3\rangle$.

$A_2 \times A_2$		A_1
γ_1	γ_2	e_1
e_2	e_2	-1

$A_2 \times E$		E	
γ_1	γ_2	θ	ϵ
e_2	θ	0	-1
e_2	ϵ	1	0

$A_2 \times T_1$		T_2		
γ_1	γ_2	ξ	η	ζ
e_2	α	1	0	0
e_2	β	0	1	0
e_2	γ	0	0	1

$A_2 \times T_2$		T_1		
γ_1	γ_2	α	β	γ
e_2	ξ	-1	0	0
e_2	η	0	-1	0
e_2	ζ	0	0	-1

$E \times T_1$		T_1			T_2		
γ_1	γ_2	α	β	γ	ξ	η	ζ
θ	α	$-\frac{1}{2}$	0	0	$\frac{\sqrt{3}}{2}$	0	0
θ	β	0	$-\frac{1}{2}$	0	0	$-\frac{\sqrt{3}}{2}$	0
θ	γ	0	0	1	0	0	0
ϵ	α	$\frac{\sqrt{3}}{2}$	0	0	$\frac{1}{2}$	0	0
ϵ	β	0	$-\frac{\sqrt{3}}{2}$	0	0	$\frac{1}{2}$	0
ϵ	γ	0	0	0	0	0	-1

$E \times E$		A_1	A_2	E	
γ_1	γ_2	e_1	e_2	θ	ϵ
θ	θ	$\frac{1}{\sqrt{2}}$	0	$-\frac{1}{\sqrt{2}}$	0
θ	ϵ	0	$\frac{1}{\sqrt{2}}$	0	$\frac{1}{\sqrt{2}}$
ϵ	θ	0	$-\frac{1}{\sqrt{2}}$	0	$\frac{1}{\sqrt{2}}$
ϵ	ϵ	$\frac{1}{\sqrt{2}}$	0	$\frac{1}{\sqrt{2}}$	0

$E \times T_2$		T_1			T_2		
γ_1	γ_2	α	β	γ	ξ	η	ζ
θ	ξ	$-\frac{\sqrt{3}}{2}$	0	0	$-\frac{1}{2}$	0	0
θ	η	0	$\frac{\sqrt{3}}{2}$	0	0	$-\frac{1}{2}$	0
θ	ζ	0	0	0	0	0	1
ϵ	ξ	$-\frac{1}{2}$	0	0	$\frac{\sqrt{3}}{2}$	0	0
ϵ	η	0	$-\frac{1}{2}$	0	0	$-\frac{\sqrt{3}}{2}$	0
ϵ	ζ	0	0	1	0	0	0

$T_1 \times T_1$		A_1	E		T_1			T_2		
γ_1	γ_2	e_1	θ	ϵ	α	β	γ	ξ	η	ζ
α	α	$-\frac{1}{\sqrt{3}}$	$\frac{1}{\sqrt{6}}$	$-\frac{1}{\sqrt{2}}$	0	0	0	0	0	0
α	β	0	0	0	0	0	$-\frac{1}{\sqrt{2}}$	0	0	$-\frac{1}{\sqrt{2}}$
α	γ	0	0	0	0	$\frac{1}{\sqrt{2}}$	0	0	$-\frac{1}{\sqrt{2}}$	0
β	α	0	0	0	0	0	$\frac{1}{\sqrt{2}}$	0	0	$-\frac{1}{\sqrt{2}}$
β	β	$-\frac{1}{\sqrt{3}}$	$\frac{1}{\sqrt{6}}$	$\frac{1}{\sqrt{2}}$	0	0	0	0	0	0
β	γ	0	0	0	$-\frac{1}{\sqrt{2}}$	0	0	$-\frac{1}{\sqrt{2}}$	0	0
γ	α	0	0	0	0	$-\frac{1}{\sqrt{2}}$	0	0	$-\frac{1}{\sqrt{2}}$	0
γ	β	0	0	0	$\frac{1}{\sqrt{2}}$	0	0	$-\frac{1}{\sqrt{2}}$	0	0
γ	γ	$-\frac{1}{\sqrt{3}}$	$-\frac{2}{\sqrt{6}}$	0	0	0	0	0	0	0

Table B.2. (continued).

$T_1 \times T_2$		A_2	E			T_1			T_2		
γ_1	γ_2	e_2	θ	ϵ	α	β	γ	ξ	η	ζ
α	ξ	$-\frac{1}{\sqrt{3}}$	$-\frac{1}{\sqrt{2}}$	$-\frac{1}{\sqrt{6}}$	0	0	0	0	0	0
α	η	0	0	0	0	0	$\frac{1}{\sqrt{2}}$	0	0	$-\frac{1}{\sqrt{2}}$
α	ζ	0	0	0	0	$\frac{1}{\sqrt{2}}$	0	0	$\frac{1}{\sqrt{2}}$	0
β	ξ	0	0	0	0	0	$\frac{1}{\sqrt{2}}$	0	0	$\frac{1}{\sqrt{2}}$
β	η	$-\frac{1}{\sqrt{3}}$	$\frac{1}{\sqrt{2}}$	$-\frac{1}{\sqrt{6}}$	0	0	0	0	0	0
β	ζ	0	0	0	$\frac{1}{\sqrt{2}}$	0	0	$-\frac{1}{\sqrt{2}}$	0	0
γ	ξ	0	0	0	0	$\frac{1}{\sqrt{2}}$	0	0	$-\frac{1}{\sqrt{2}}$	0
γ	η	0	0	0	$\frac{1}{\sqrt{2}}$	0	0	$\frac{1}{\sqrt{2}}$	0	0
γ	ζ	$-\frac{1}{\sqrt{3}}$	0	$\frac{2}{\sqrt{6}}$	0	0	0	0	0	0

$$\langle \Psi_{\alpha \Gamma \gamma} | \hat{O}(\bar{\Gamma})_{\bar{\gamma}} | \Psi_{\alpha' \Gamma' \gamma'} \rangle = \langle \Psi_{\alpha \Gamma} || \hat{O}(\bar{\Gamma}) || \Psi_{\alpha' \Gamma'} \rangle \langle \Gamma \gamma | \Gamma' \gamma' \bar{\Gamma} \bar{\gamma} \rangle \qquad (B.36)$$

where $\langle \Psi_{\alpha \Gamma} || \hat{O}(\bar{\Gamma}) || \Psi_{\alpha' \Gamma'} \rangle$ are the *reduced matrix elements* that do not depend on the components of the representations. A reduced matrix element represents a physical quantity of the operator that is independent of a particular choice of the basis set.

In order to prove this theorem, we first point out that the matrix element $\langle \phi_{\Gamma \gamma} | \psi_{\Gamma' \gamma'} \rangle$ is independent of γ and γ' where $\phi_{\Gamma \gamma}$ and $\psi_{\Gamma' \gamma'}$ are functions that transform as γ and γ' components of the irreducible representations Γ and Γ', respectively. This is because

$$\begin{aligned}
\langle \phi_{\Gamma \gamma} | \psi_{\Gamma' \gamma'} \rangle &= \frac{1}{|G|} \sum_{R \in G} \langle R \phi_{\Gamma \gamma} | R \psi_{\Gamma' \gamma'} \rangle \\
&= \frac{1}{|G|} \sum_{R \in G, \gamma_1, \gamma_1'} (D^\Gamma_{\gamma_1 \gamma}(R))^* D^{\Gamma'}_{\gamma_1' \gamma'}(R) \langle \phi_{\Gamma \gamma_1} | \psi_{\Gamma' \gamma_1'} \rangle \\
&= \delta_{\Gamma \Gamma'} \delta_{\gamma \gamma'} \frac{1}{d_\Gamma} \sum_{\gamma_1} \langle \phi_{\Gamma \gamma_1} | \psi_{\Gamma' \gamma_1} \rangle, \qquad (B.37)
\end{aligned}$$

where we have used (B.7) and (B.12) and the identity $\langle \phi_{\Gamma \gamma} | \psi_{\Gamma' \gamma'} \rangle = \langle R \phi_{\Gamma \gamma} | R \psi_{\Gamma' \gamma'} \rangle$.

Another important point in the proof of the theorem is that $\hat{O}(\bar{\Gamma})_{\bar{\gamma}} \Psi_{\alpha' \Gamma' \gamma'}$ transforms as the direct product basis $\phi_{\bar{\Gamma} \bar{\gamma}} \phi_{\Gamma' \gamma'}$, because

$$\begin{aligned}
R \hat{O}(\bar{\Gamma})_{\bar{\gamma}} \Psi_{\alpha' \Gamma' \gamma'} &= R \hat{O}(\bar{\Gamma})_{\bar{\gamma}} R^{-1} R \Psi_{\alpha' \Gamma' \gamma'} \\
&= \sum_{\gamma_1, \gamma_1'} D^{\bar{\Gamma}}_{\gamma_1 \bar{\gamma}}(R) D^{\Gamma'}_{\gamma_1' \gamma'}(R) \hat{O}(\bar{\Gamma})_{\gamma_1} \Psi_{\alpha' \Gamma' \gamma_1'} \\
&= \sum_{\gamma_1, \gamma_1'} D^{\bar{\Gamma} \times \Gamma'}_{\gamma_1 \gamma_1', \bar{\gamma} \gamma'}(R) \hat{O}(\bar{\Gamma})_{\gamma_1} \Psi_{\alpha' \Gamma' \gamma_1'}. \qquad (B.38)
\end{aligned}$$

Therefore using (B.35) we can express $\hat{O}(\bar{\Gamma})_{\bar{\gamma}} \Psi_{\alpha' \Gamma' \gamma'}$ as

$$\hat{O}(\bar{\Gamma})_{\bar{\gamma}}\Psi_{\alpha'\Gamma'\gamma'} = \sum_{\Gamma,\gamma}\langle\Gamma\gamma|\Gamma'\gamma'\bar{\Gamma}\bar{\gamma}\rangle\Phi_{\Gamma\gamma}. \tag{B.39}$$

Taking the inner product with $\Psi_{\alpha\Gamma\gamma}$, we finally have

$$\langle\Psi_{\alpha\Gamma\gamma}|\hat{O}(\bar{\Gamma})_{\bar{\gamma}}|\Psi_{\alpha'\Gamma'\gamma'}\rangle = \langle\Gamma\gamma|\Gamma'\gamma'\bar{\Gamma}\bar{\gamma}\rangle\langle\Psi_{\alpha\Gamma\gamma}|\Phi_{\Gamma\gamma}\rangle. \tag{B.40}$$

Since $\langle\Psi_{\alpha\Gamma\gamma}|\Phi_{\Gamma\gamma}\rangle$ is independent of γ, we can write this in the form

$$\langle\Psi_{\alpha\Gamma}||\hat{O}(\bar{\Gamma})||\Psi_{\alpha'\Gamma'}\rangle := \langle\Psi_{\alpha\Gamma\gamma}|\Phi_{\Gamma\gamma}\rangle. \tag{B.41}$$

Thus, the expression (B.40) becomes (B.36).

Jahn–Teller Effects and Vibronic Hamiltonians

If because of the symmetry of a non-linear molecule an electronic state is degenerate, then the degenerate high-symmetry configuration is unstable with respect to spontaneous distortions that lift the degeneracy. This is known as the *Jahn–Teller theorem*.

Suppose that for a non-linear nuclear configuration $\mathbf{R} = \mathbf{R}_0$, the electronic state is $\mathcal{N}$-fold degenerate, $\varepsilon_1(\mathbf{R}_0) = \cdots = \varepsilon_{\mathcal{N}}(\mathbf{R}_0)$, with $\mathcal{N} > 1$. Expand the electronic Hamiltonian $h(\mathbf{p},\mathbf{r},\mathbf{R})$ in a power series around $\mathbf{R}_0$,

$$\begin{aligned}
h(\mathbf{p},\mathbf{r},\mathbf{R}) &= h(\mathbf{p},\mathbf{r},\mathbf{R}_0) \\
&+ \sum_i \frac{\partial h(\mathbf{p},\mathbf{r},\mathbf{R}_0)}{\partial R_i}(R_i - R_{0i}) \\
&+ \frac{1}{2}\sum_{i,j}\frac{\partial^2 h(\mathbf{p},\mathbf{r},\mathbf{R}_0)}{\partial R_i \partial R_j}(R_i - R_{0i})(R_j - R_{0j}) \\
&+ \cdots,
\end{aligned} \tag{B.42}$$

where $\mathbf{r}$ and $\mathbf{p}$ (collectively) denote electronic coordinates and momenta, respectively, and $\cdots$ stands for the higher order terms in $\mathbf{R}-\mathbf{R}_0$. We assume that $|\mathbf{R}-\mathbf{R}_0|$ is small, so that these higher order terms are negligible.

Consider the Hamiltonian $H = T_{\text{nucl}} + h$ where T_{nucl} is the nuclear kinetic energy. In a basis consisting of $\mathcal{N}$ electronic state vectors $\{|n(\mathbf{R}_0)\rangle, n = 1,\cdots,\mathcal{N}\}$, the Hamiltonian H takes the form of a matrix with entries:

$$\begin{aligned}
&T_{\text{nucl}}\delta_{nm} + \langle n|h(\mathbf{p},\mathbf{r},\mathbf{R})|m\rangle \\
&= (T_{\text{nucl}} + \varepsilon_n(\mathbf{R}_0))\,\delta_{nm} \\
&+ \sum_i \left\langle n(\mathbf{R}_0)\left|\frac{\partial h(\mathbf{p},\mathbf{r},\mathbf{R}_0)}{\partial R_i}\right|m(\mathbf{R}_0)\right\rangle (R_i - R_{0i}) \\
&+ \frac{1}{2}\sum_{i,j}\left\langle n(\mathbf{R}_0)\left|\frac{\partial^2 h(\mathbf{p},\mathbf{r},\mathbf{R}_0)}{\partial R_i \partial R_j}\right|m(\mathbf{R}_0)\right\rangle (R_i - R_{0i})(R_j - R_{0j}) \\
&+ \cdots.
\end{aligned} \tag{B.43}$$

422   B. Point Groups

Because the degeneracy is due to the symmetry of the system, we can use group theory to describe it. First, we use symmetry-labeled normal coordinates for nuclear displacements $R_i - R_{0i}$. We then write the terms linear in the normal coordinates in the form

$$\sum_{\Gamma\gamma} V_{\Gamma\gamma} Q_{\Gamma\gamma}, \tag{B.44}$$

where $Q_{\Gamma\gamma}$ is the normal vibrational coordinate that transforms as the γ component of the irreducible representation Γ, and $V_{\Gamma\gamma}$ is a parameter with the same transformation property as $Q_{\Gamma\gamma}$. This is because each term in the matrix Hamiltonian (B.43) belongs to the totally symmetric irreducible representation of the point group.[7]

If $V_{\Gamma\gamma}$ is not zero, the system can lower its energy by moving to a configuration for which $Q_{\Gamma\gamma}$ is non-zero. We wish to characterize the condition under which $V_{\Gamma\gamma}$ is non-zero. We assume that the degenerate electronic wave functions can be taken real. Then, we have

$$\begin{aligned}
V_{\Gamma\gamma} &= \left\langle n(\mathbf{R}_0) \left| \frac{\partial h(\mathbf{p},\mathbf{r},\mathbf{R}_0)}{\partial Q_{\Gamma\gamma}} \right| m(\mathbf{R}_0) \right\rangle \\
&= \int dr^{3N_e} \psi_{\Gamma'\gamma_1}(\mathbf{r},\mathbf{R}_0) v_{\Gamma\gamma} \psi_{\Gamma'\gamma_2}(\mathbf{r},\mathbf{R}_0) \\
&= \int dr^{3N_e} v_{\Gamma\gamma} \frac{1}{2}[\psi_{\Gamma'\gamma_1}(\mathbf{r},\mathbf{R}_0)\psi_{\Gamma'\gamma_2}(\mathbf{r},\mathbf{R}_0) + \psi_{\Gamma'\gamma_2}(\mathbf{r},\mathbf{R}_0)\psi_{\Gamma'\gamma_1}(\mathbf{r},\mathbf{R}_0)],
\end{aligned} \tag{B.45}$$

where Γ' is the irreducible representation corresponding to the degenerate electronic states, and

---

[7] Let Γ be a representation of a group G. Then the *complex conjugate representation* Γ^* is obtained by taking the complex conjugate of the matrices belonging to Γ. A sum of operators $\sum_\gamma A_{\Gamma^*\gamma} B_{\Gamma\gamma}$ belongs to the totally symmetric representation, because it is invariant with respect to any symmetry operations in the symmetry group:

$$\begin{aligned}
R\left(\sum_\gamma A_{\Gamma^*\gamma} B_{\Gamma\gamma}\right) R^{-1} &= \sum_\gamma R A_{\Gamma^*\gamma} R^{-1} R B_{\Gamma\gamma} R^{-1} \\
&= \sum_{\gamma,\gamma',\gamma''} A_{\Gamma^*\gamma'} B_{\Gamma\gamma''} (D^{\Gamma}_{\gamma'\gamma}(R))^* D^{\Gamma}_{\gamma''\gamma}(R) \\
&= \sum_{\gamma,\gamma',\gamma''} A_{\Gamma^*\gamma'} B_{\Gamma\gamma''} (D^{\Gamma}_{\gamma\gamma'}(R))^{-1} D^{\Gamma}_{\gamma''\gamma}(R) \\
&= \sum_\gamma A_{\Gamma^*\gamma} B_{\Gamma\gamma}.
\end{aligned}$$

If Γ is a real irreducible representation, Γ^* is equal to Γ. Note that normal vibrational coordinates belong to real representations.

B. Point Groups

$$\psi_{\Gamma'\gamma_1}(\mathbf{r}, \mathbf{R}_0) := \langle \mathbf{r} | n(\mathbf{R}_0) \rangle,$$
$$\psi_{\Gamma'\gamma_2}(\mathbf{r}, \mathbf{R}_0) := \langle \mathbf{r} | m(\mathbf{R}_0) \rangle,$$
$$v_{\Gamma\gamma} := \frac{\partial h(\mathbf{p}, \mathbf{r}, \mathbf{R}_0)}{\partial Q_{\Gamma\gamma}}. \tag{B.46}$$

Equation (B.45) shows that a necessary condition for having a non-zero value for $V_{\Gamma\gamma}$ is that the symmetric product $[\Gamma' \times \Gamma']$ has the irreducible representation Γ (in its decomposition into irreducible representations). In this case the integrand belongs to the totally symmetric representation and the integral can be nonzero.

Figure B.2 shows some of the normal vibrational modes of the octahedral complex (O_h symmetry) and equilateral triangular molecules (D_{3h} symmetry). For D_{3h} and O_h groups we have $[E \times E] = A + E$. Thus, E-symmetry vibrational modes lift the degeneracy. For the O_h group, $[T \times T] = A + E + T$, and the T- and E- symmetry vibrational modes lift the degeneracy.

We can use the Wigner–Eckart theorem (B.36) to factor the parameter $V_{\Gamma\gamma}$ as

$$\begin{aligned}(V_{\Gamma\gamma})_{\gamma_1\gamma_2} &= \langle \psi_{\Gamma'\gamma_1} | v_{\Gamma\gamma} | \psi_{\Gamma'\gamma_2} \rangle \\ &= V_\Gamma \langle \Gamma'\gamma_1 | \Gamma\gamma\Gamma'\gamma_2 \rangle \\ &:= V_\Gamma (\mathbf{C}_{\Gamma\gamma})_{\gamma_1\gamma_2}\end{aligned} \tag{B.47}$$

where V_Γ is the linear vibronic constant for the Γ-normal mode and $\mathbf{C}_{\Gamma\gamma}$ is a matrix of the (unnormalized) Clebsch–Gordan coefficients.

Similarly, we express the quadratic terms in the normal coordinates as

$$\frac{1}{2} \sum_{\Gamma_1,\Gamma_2,\Gamma,\gamma} W_\Gamma(\Gamma_1 \times \Gamma_2) \{Q_{\Gamma_1} \times Q_{\Gamma_2}\}_{\Gamma\gamma} \mathbf{C}_{\Gamma\gamma}, \tag{B.48}$$

where $W_\Gamma(\Gamma_1 \times \Gamma_2)$ is the reduced quadratic vibronic constant and we have assembled the quadratic normal coordinate terms according to their symmetry:

$$\{Q_{\Gamma_1} \times Q_{\Gamma_2}\}_{\Gamma\gamma} := \sum_{\gamma_1,\gamma_2} Q_{\Gamma_1\gamma_1} Q_{\Gamma_2\gamma_2} \langle \Gamma_1\gamma_1\Gamma_2\gamma_2 | \Gamma\gamma \rangle. \tag{B.49}$$

In the typical Jahn–Teller problem, the overall rotation of the system is neglected and the nuclear kinetic energy is simply taken to be

$$T_{\text{nucl}} = -\frac{1}{2} \sum_i \frac{\partial^2}{\partial Q_i^2}. \tag{B.50}$$

The vibronic parameters V_Γ and $W_\Gamma(\Gamma_1 \times \Gamma_2)$ are obtained from experimental data or theoretical calculations.

In the following, we give the matrix Hamiltonian for the most frequently occurring Jahn–Teller systems, namely $E \otimes e$ and $T \otimes (e + t_2)$.

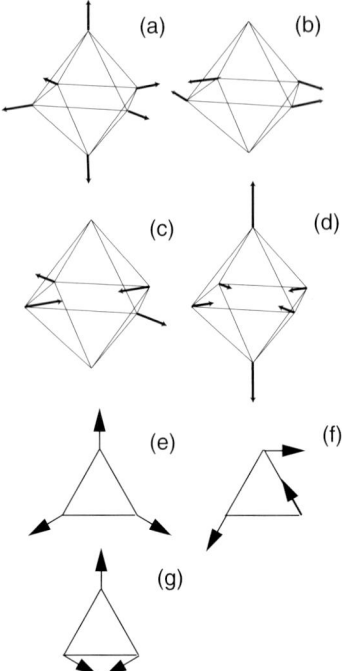

Fig. B.2. Some of the normal vibrational modes of the octahedral complex and equilateral triangular molecules: (a) Non-degenerate $A_{1g}(O_h)$ mode; (b) triply degenerate $T_{2g}(O_h)$ mode, T_ζ, that transforms as xy; two other partners T_η and T_ξ, which are not depicted here, they transform as zx and yz, respectively; (c) one of the doubly degenerate $E_g(O_h)$ modes, $E_{g\theta}$, that transforms as $x^2 - y^2$; (d) the other doubly degenerate $E_g(O_h)$ mode, $E_{g\epsilon}$, that transforms as $2z^2 - x^2 - y^2$; (e) non-degenerate $A_1(D_{3h})$ mode; (f) one of the doubly degenerate $E(D_{3h})$ modes, E_θ; (g) the other doubly degenerate $E(D_{3h})$ mode, E_ϵ.

$E \otimes e$ system. The first letter E indicates that the electronic state is doubly degenerate at the high-symmetry point and the next letter e means that the degeneracy can be lifted linearly by the doubly degenerate displacements e. The two components of E are denoted by θ and ϵ, where the former transforms as $\frac{1}{2}(3z^2 - r^2)$ and the latter as $\frac{\sqrt{3}}{2}(x^2 - y^2)$ in O_h. The E-type normal displacements are denoted as Q_θ, Q_ϵ.

The (unnormalized) Clebsch–Gordan coefficient matrices are given by

$$\mathbf{C}_{A_1} = \begin{pmatrix} 1 & 0 \\ 0 & 1 \end{pmatrix} = \sigma_0, \quad \mathbf{C}_{A_2} = \begin{pmatrix} 0 & -i \\ i & 0 \end{pmatrix} = \sigma_2,$$

$$\mathbf{C}_{E_\theta} = \begin{pmatrix} -1 & 0 \\ 0 & 1 \end{pmatrix} = -\sigma_3, \quad \mathbf{C}_{E_\epsilon} = \begin{pmatrix} 0 & 1 \\ 1 & 0 \end{pmatrix} = \sigma_1, \quad (B.51)$$

where σ_0 is the 2×2 unit matrix and σ_i with $i = 1, 2, 3$ are Pauli matrices.

Up to the second order terms in displacements, the vibronic Hamiltonian is given by

$$H = -\frac{1}{2}\left(\frac{\partial^2}{\partial Q_\theta^2} + \frac{\partial^2}{\partial Q_\epsilon^2}\right)\mathbf{C}_{A_1} + \frac{\omega^2}{2}(Q_\theta^2 + Q_\epsilon^2)\mathbf{C}_{A_1}$$
$$+ V_E(Q_\theta \mathbf{C}_{E_\theta} + Q_\epsilon \mathbf{C}_{E_\epsilon}) + W_E^{exe}[(Q_\epsilon^2 - Q_\theta^2)\mathbf{C}_{E_\theta} + 2Q_\theta Q_\epsilon \mathbf{C}_{E_\epsilon}]$$
$$+ W_A^{exe} Q_\theta (Q_\theta^2 - 3Q_\epsilon^2)\mathbf{C}_{A_1}. \tag{B.52}$$

$T \otimes (e + t_2)$ system. The first letter T indicates that the electronic state is triply degenerate at the high-symmetry point and the next letters e and t_2 mean that the degeneracy can be lifted linearly by the doubly degenerate displacements e and the triply degenerate displacements t_2. The electronic basis of the triplet T transforms as x, y, z of $T_{1u}(O_h)$, $T_2(T_d)$; and as yz, zx, xy of $T_{2g}(O_h)$, $T_2(T_d)$.

The (unnormalized) Clebsch–Gordan coefficient matrices are given by

$$\mathbf{C}_{A_1} = \begin{pmatrix} 1 & 0 & 0 \\ 0 & 1 & 0 \\ 0 & 0 & 1 \end{pmatrix} = \lambda_0, \quad \mathbf{C}_{E_\theta} = \begin{pmatrix} \frac{1}{2} & 0 & 0 \\ 0 & \frac{1}{2} & 0 \\ 0 & 0 & -1 \end{pmatrix} = \lambda_8,$$

$$\mathbf{C}_{E_\epsilon} = \begin{pmatrix} -\frac{\sqrt{3}}{2} & 0 & 0 \\ 0 & \frac{\sqrt{3}}{2} & 0 \\ 0 & 0 & 0 \end{pmatrix} = \lambda_3, \quad \mathbf{C}_{T_{2\xi}} = \begin{pmatrix} 0 & 0 & 0 \\ 0 & 0 & -1 \\ 0 & -1 & 0 \end{pmatrix} = -\lambda_6,$$

$$\mathbf{C}_{T_{2\eta}} = \begin{pmatrix} 0 & 0 & -1 \\ 0 & 0 & 0 \\ -1 & 0 & 0 \end{pmatrix} = \lambda_4, \quad \mathbf{C}_{T_{2\zeta}} = \begin{pmatrix} 0 & -1 & 0 \\ -1 & 0 & 0 \\ 0 & 0 & 0 \end{pmatrix} = -\lambda_1,$$

$$\mathbf{C}_{T_{1\alpha}} = \begin{pmatrix} 0 & 0 & 0 \\ 0 & 0 & -i \\ 0 & i & 0 \end{pmatrix} = \lambda_7, \quad \mathbf{C}_{T_{1\beta}} = \begin{pmatrix} 0 & 0 & i \\ 0 & 0 & 0 \\ -i & 0 & 0 \end{pmatrix} = -\lambda_5,$$

$$\mathbf{C}_{T_{1\gamma}} = \begin{pmatrix} 0 & -i & 0 \\ i & 0 & 0 \\ 0 & 0 & 0 \end{pmatrix} = \lambda_2, \tag{B.53}$$

where λ_0 is the 3×3 unit matrix and λ_i, with $i = 1, \cdots, 8$, are the Gell-Mann matrices.

The vibronic Hamiltonian is given, up to the second order in displacements, by

426　B. Point Groups

$$\begin{aligned}H = &-\frac{1}{2}\sum_{i=\xi,\eta,\zeta,\theta,\epsilon}\frac{\partial^2}{\partial Q_i^2}\mathbf{C}_{A_1} + \frac{\omega_T}{2}\sum_{i=\xi,\eta,\zeta}Q_i^2\mathbf{C}_{A_1} + \frac{\omega_E}{2}\sum_{i=\theta,\epsilon}Q_i^2\mathbf{C}_{A_1}\\ &+ V_E(Q_\theta\mathbf{C}_{E_\theta} + Q_\epsilon\mathbf{C}_{E_\epsilon}) + V_T(Q_\xi\mathbf{C}_{T_{2\xi}} + Q_\eta\mathbf{C}_{T_{2\eta}} + Q_\zeta\mathbf{C}_{T_{2\zeta}})\\ &+ W_E^{e\times e}[(Q_\epsilon^2 - Q_\theta^2)\mathbf{C}_{E_\theta} + 2Q_\epsilon Q_\epsilon\mathbf{C}_{E_\epsilon}]\\ &+ W_T^{t\times t}(Q_\eta Q_\zeta\mathbf{C}_{T_{2\xi}} + Q_\zeta Q_\xi\mathbf{C}_{T_{2\eta}} + Q_\xi Q_\eta\mathbf{C}_{T_{2\zeta}})\\ &+ W_E^{t\times t}[(2Q_\zeta^2 - Q_\eta^2 - Q_\xi^2)\mathbf{C}_{E_\theta} + \sqrt{3}(Q_\xi^2 - Q_\eta^2)\mathbf{C}_{E_\epsilon}]\\ &- W_T^{t\times e}\left[Q_\xi\left(\frac{1}{2}Q_\theta - \frac{\sqrt{3}}{2}Q_\epsilon\right)\mathbf{C}_{T_{2\xi}}\right.\\ &\left.+ Q_\eta\left(\frac{1}{2}Q_\theta + \frac{\sqrt{3}}{2}Q_\epsilon\right)\mathbf{C}_{T_{2\eta}} - Q_\zeta Q_\theta\mathbf{C}_{T_{2\zeta}}\right].\end{aligned} \quad \text{(B.54)}$$

Tunneling State Symmetry

Jahn–Teller systems often exhibit deep equivalent minimum. In each minimum the original symmetry G is reduced to G_L. Because the barriers between the minima have finite height, the tunneling motion of the localized states through the barriers leads to the splitting of energies. The symmetry of the tunneling-split states arising from the localized states can be calculated using group theory [34].

Suppose that the localized vibronic states are in an irreducible representation Γ_L of the point group G_L at the potential minima. Then, the number of vibronic states arising from the localized vibronic states that belong to the irreducible representation Γ of the original point group G is given by

$$n(\Gamma) = \frac{n_{\min}}{|G|}\sum_{R\in G_L}\chi_{\Gamma_L}(R)\chi_\Gamma^*(R), \quad \text{(B.55)}$$

where $n_{\min}$ is the number of equivalent minima.

As an application of the above formula, we consider $C_6H_6^+$. At the high-symmetry point, $C_6H_6^+$ has D_{6h} symmetry. Its stable configurations are speculated to be either B_{2g} or B_{3g} of D_{2h}. Using character tables for D_2 and D_6, namely Table B.3,[8] and (B.55), we have $B_1(D_6)$ and $E_1(D_6)$ tunneling states from the localized $B_2(D_2)$ states, and $B_2(D_6)$ and $E_1(D_6)$ tunneling states from the localized $B_3(D_2)$ states. In either case, the ground vibronic state is the doubly degenerate E_1, but the first excited state is B_1 for the former and B_2 for the latter case. The experimental results support the former case. For more details see Sect. 11.4.2.

---

[8] Here we use D_2 and D_6 instead of D_{2h} and D_{6h} for simplicity.

Table B.3. Character tables for the D_2 and D_6 point groups.

D_2	E	$C_2(z)$	$C_2(y)$	$C_2(x)$	
A	1	1	1	1	x^2, y^2, z^2
B_1	1	1	-1	-1	z, L_z, xy
B_2	1	-1	1	-1	y, L_y, zx
B_3	1	-1	-1	1	z, L_z, yz

D_6	E	$2C_6$	$2C_3$	C_2	$3C_2'$	$3C_2''$	
A_1	1	1	1	1	1	1	x^2+y^2, z^2
A_2	1	1	1	1	-1	-1	z, L_z
B_1	1	-1	1	-1	1	-1	
B_2	1	-1	1	-1	-1	1	
E_1	2	1	-1	-2	0	0	$(x,y), (L_x, L_y), (yz, zx)$
E_2	2	-1	-1	2	0	0	(x^2-y^2, xy)

References

1. A. Abragam: *Principles of Nuclear Magnetism* (Oxford University Press, New York, 1961)
2. E. Abrahams, P. W. Anderson, D. C. Liciardello, T. V. Ramakrishnan: Phys. Rev. Lett. 42, 673 (1979)
3. E. N. Adams, E. I. Blount: Phys. Chem. Solids 10, 286 (1959)
4. G. Aers, A. Macdonald: J. Phys. C 17, 5491 (1984)
5. I. Affleck, J. B. Marston: Phys. Rev. B 37, 3774 (1988)
6. M. A. Aguilar, M. Sokolovsky: Int. J. Th. Phys. 36, 883 (1997)
7. Y. Aharonov, A. Anandan: Phys. Rev. Lett. 58, 1593 (1987)
8. Y. Aharonov, D. Bohm: Phys. Rev. 115, 485 (1959)
9. J. Anandan, Y. Aharonov: Phys. Rev. D 38, 1863 (1988)
10. T. Ando: J. Phys. Soc. Japan 52, 1740 (1983)
11. T. Ando: J. Phys. Soc. Japan 53, 3101 (1984)
12. T. Ando, Y. Matsumoto, Y. Uemura: J. Phys. Soc. Japan 39, 279 (1975)
13. T. Ando, Y. Uemura: J. Phys. Soc. Japan 36, 959 (1974)
14. H. Aoki: J. Phys. C 15, 1227 (1982)
15. H. Aoki: J. Phys. C 16, 1893 (1983)
16. H. Aoki, T. Ando: Solid State Commun. 38, 1079 (1981)
17. I. Y. Aref'eva: Th. Math. Phys. 43, 353 (1980)
18. D. Arovas, J. R. Schrieffer, F. Wilczek: Phys. Rev. Lett. 53, 722 (1984)
19. D. Arovas, J. R. Schrieffer, F. Wilczek, A. Zee: Nucl. Phys. B 251, 117 (1985)
20. N. W. Ashcroft, N. D. Mermin: *Solid State Physics* (W. B. Saunders, Philadelphia, 1976)
21. J. E. Avron, L. Sadun, J. Segret, B. Simon: Commun. Math. Phys. 124, 595 (1989)
22. J. E. Avron, R. Seiler: J. Geom. Phys. 1, 13 (1984)
23. J. E. Avron, R. Seiler: Phys. Rev. Lett. 54, 259 (1985)
24. J. E. Avron, R. Seiler, B. Simon: Phys. Rev. Lett. 51, 51 (1983)
25. J. E. Avron, B. Simon: J. Phys. A 18, 2199 (1985)
26. J. E. Avron, L. G. Yaffe: Phys. Rev. Lett. 56, 2084 (1986)
27. M. Y. Azbel: Sov. Phys. JETP 19, 634 (1964)
28. A. B. Balachandran, G. Marmo, B. S. Skagerstam, A. Stern: *Classical Topology and Quantum States Part I* (World Scientific, Singapore, 1991)
29. Y. B. Bazaliy, B. A. Jones, S. C. Zhang: Phys. Rev. B 57, R3213 (1998)
30. P. Becker, M. Bohm, H. Joos: *Gauge Theories of the Strong and Electroweak Interactions* (John Wiley & Sons, New York, 1984)
31. M. V. Berry: Proc. Roy. Soc. London A392, 45 (1984)
32. M. V. Berry: Annals of the New York Academy of Sciences 755, 303 (1995)
33. M. V. Berry, R. Lim: J. Phys. A 23, L655 (1990)
34. I. B. Bersuker, V. Z. Polinger: *Vibronic Interactions in Molecules and Crystals* (Springer-Verlag, New York, 1989)

35. L. C. Biedenharn, J. D. Louck: *Angular Momentum in Quantum Mechanics* (Addison-Wesley, New York, 1981)
36. D. Birmingham, M. Blau, M. Rakowski, G. Thompson: Phys. Rep. 229, 129 (1991)
37. T. Bitter, D. Dubbers: Phys. Rev. Lett. 59, 251 (1987)
38. B. Bleaney, K. D. Bowers: Proc. Phys. Soc. (London) A 65, 667 (1952)
39. B. Bleaney, D. J. E. Ingram: Proc. Phys. Soc. (London) A 63, 166 (1950)
40. E. I. Blount: in: *Solid State Physics*, volume 13, p. 305 (Academic Press, New York, 1962)
41. A. Bohm: in: *Lectures on Symmetry in Science III*, edited by B. Gruber, F. Iachello (Plenum Press, New York, 1988)
42. A. Bohm: in: *Differential Geometry, Group Representations, and Quantization*, number 379 in Lecture Notes in Physics, p. 207 (Springer-Verlag, Berlin, 1991)
43. A. Bohm: *Quantum Mechanics: Foundations and Applications*, 3rd edition (Springer-Verlag, New York, 1993)
44. A. Bohm, L. J. Boya, A. Mostafazadeh, G. Rudolph: J. Geom. Phys. 12, 13 (1993)
45. A. Bohm, B. Kendrick, M. Loewe, L. J. Boya: J. Math. Phys. 33, 977 (1992)
46. M. Born, V. Fock: Z. Phys. 51, 165 (1928)
47. M. Born, K. Huang: *Dynamical Theory of Crystal Lattices* (Oxford University Press, Oxford, 1954)
48. M. Born, J. Oppenheimer: Ann. Phys. 84, 457 (1927)
49. A. D. Braisford: Proc. Phys. Soc. (London) A 70, 275 (1957)
50. B. Broda: in: *Advanced Electromagnetism: Foundations, Theory and Applications* (World Scientific, Singapore, 1995)
51. F. Brown: Phys. Rev. 133, 1038A (1964)
52. H. von Bush, M. Keil, H.-G. Krämer, and W. Demtröder: Proceedings of the XIV International Symposium on Electron–Phonon Dynamics and Jahn–Teller Effect, 311, Copyright 1999, World Scientific Publishing
53. H. von Bush, Vas Dev, H.-A. Eckel, S. Kasahara, J. Wang, W. Demtröder, P. Sebald, and W. Meyer: Phys. Rev. Lett. 82, 3560, Copyright 1999, The American Physical Society
54. F. A. Butler, F. Brown: Phys. Rev. 166, 630 (1968)
55. N. Byers, C. N. Yang: Phys. Rev. Lett. 7, 46 (1961)
56. D. Bylander, Q. Niu, L. Kleinman: Phys. Rev. B 61, R11875 (2000)
57. J. Callaway: *Quantum Theory of the Solid State* (Academic, New York, 1976)
58. M. Chaichian, N. F. Nelipa: *Introduction to Gauge Field Theories* (Springer-Verlag, Berlin, 1984)
59. W. G. Chambers: Phys. Rev. 140, A135 (1965)
60. M. C. Chang, Q. Niu: Phys. Rev. Lett. 75, 1348 (1995)
61. M. C. Chang, Q. Niu: Phys. Rev. B 53, 7010 (1996)
62. M. S. Child: *Molecular Collision Theory* (Academic Press, London, 1974)
63. Y. Choquet-Bruhat, C. DeWitt-Morette: *Analysis, Manifolds and Physics, Part II: 92 Applications* (North-Holland, Amsterdam, 1989)
64. Y. Choquet-Bruhat, C. DeWitt-Morette, M. Dillard-Bleick: *Analysis, Manifolds and Physics, Part I: Basics* (North-Holland, Amsterdam, 1989)
65. R. E. Coffman: Phys. Lett. 19, 475 (1965)
66. R. E. Coffman: Phys. Lett. 21, 381 (1966)
67. R. E. Coffman: J. Chem. Phys. 48, 609 (1968)
68. S. Coleman: in: *Unity of the Fundamental Interactions*, edited by H. Zichichi, Erice Lectures 1981 (Plenum Press, New York, 1983)

69. A. Comtet, T. Jolicœur, S. Ouvry, F. David: *1998 Les Houches Session LXIX, Topological Aspects of Low Dimensional Systems* (Springer-Verlag, Berlin, 1998)
70. I. Dana, Y. Avron, J. Zak: J. Phys. C 18, L679 (1985)
71. I. Dana, J. Zak: Phys. Rev. B 28, 811 (1983)
72. T. P. Das, E. L. Hahn: *Nuclear Quadrupole Resonance Spectroscopy*, volume Supplement 1 of *Solid State Physics* (Academic, New York, 1958)
73. G. Délacretaz, E. R. Grant, R. L. Whetten, L. Wöste, J. W. Zwanziger: Phys. Rev. Lett. 56, 2598 (1986)
74. P. A. M. Dirac: Proc. Roy. Soc. London 133, 60 (1931)
75. P. A. M. Dirac: *Principles of Quantum Mechanics* (Oxford University Press, Oxford, 1958)
76. N. A. Doughty: *Lagrangian Interactions* (Addison-Wesley, Sydney, 1990)
77. J. E. B. Wilson, J. C. Decius, P. C. Cross: *Molecular Vibrations* (Dover, New York, 1980)
78. T. Eguchi, P. B. Gilkey, A. J. Hanson: Phys. Rep. 66, 213 (1980)
79. R. R. Ernst, G. Bodenhausen, A. Wokaun: *Principles of Nuclear Magnetic Resonance in One and Two Dimensions* (Oxford University Press, New York, 1989)
80. H. J. Fischbeck: Phys. Stat. Solidi 3, 1082 (1963)
81. G. Floquet: Ann. E. N. S. 12, 47 (1883)
82. V. Fock: Z. Phys. 49, 323 (1928)
83. Y. Fukumoto, H. Koizumi, K. Makoshi: Chem. Phys. Lett. 313, 283 (1999)
84. E. Fukushima, S. B. W. Roeder: *Experimental Pulse NMR: A Nuts and Bolts Approach* (Addison-Wesley, Redwood City, 1981)
85. R. Gebauer, S. Baroni: Phys. Rev. B 61, R6459 (2000)
86. I. M. Gelfand, G. E. Shilov: *Generalized Functions*, volume 1 (Academic Press, New York, 1964)
87. P. Goddard, D. I. Olive: Rep. Prog. Phys. 41, 1357 (1978)
88. M. Goldman, P. J. Grandinetti, A. Llor, Z. Olejniczak, J. R. Sachleben, J. W. Zwanziger: J. Chem. Phys. 97, 8947 (1992)
89. H. Goldstein: *Classical Mechanics*, 2nd edition (Addison-Wesley, Reading, Massachusetts, 1980)
90. A. A. Gomes, P. Lederer: J. Phys. (Paris) 38, 231 (1977)
91. R. Hagg: *Local Quantum Physics* (Springer-Verlag, Berlin, 1992)
92. F. Haldane: Phys. Rev. Lett. 51, 605 (1983)
93. F. Haldane, E. Rezayi: Phys. Rev. Lett 54, 237 (1985)
94. B. Halperin: Phys. Rev. B 25, 2185 (1982)
95. B. I. Halperin: Jpn. J. Appl. Phys. Suppl. 26, 1913 (1987)
96. M. B. Halpern: Phys. Rev. D 19, 517 (1979)
97. F. S. Ham: Phys. Rev. A 138, 1727 (1965)
98. F. S. Ham: in: *Electron Paramagnetic Resonance*, edited by S. Geschwind, p. 1 (Plenum Press, New York, 1972)
99. F. S. Ham: Phys. Rev. Lett. 58, 725 (1987)
100. F. S. Ham: J. Phys. Condens. Matter 2, 1163 (1990)
101. J. Hamilton: *Aharonov–Bohm and Other Cyclic Phenomena* (Springer-Verlag, Berlin, 1997)
102. P. G. Harper: Proc. Phys. Soc. (London) A265, 317 (1955)
103. C. Herring: in: *Magnetism*, volume 4 (Academic, New York, 1966)
104. G. Herzberg: *Molecular Spectra and Molecular Physics*, volume 1 (D. Van Nostrand, New York, 1966)
105. G. Herzberg, H. C. Longuet-Higgins: Discuss. Faraday Soc. 35, 77 (1963)
106. U. T. Höchli, T. L. Estle: Phys. Rev. Lett. 18, 128 (1967)

107. H. Hochstadt: *Differential Equations: A Modern Approach* (Dover, New York, 1975)
108. D. R. Hofstadter: Phys. Rev. B 14, 2239 (1976)
109. T. Hotta, Y. Takada, H. Koizumi: Int. J. Mod. Phys. B 12, 3437 (1998)
110. A. C. Hurley: *Introduction to the Electron Theory of Small Molecules* (Academic Press, New York, 1976)
111. D. Husemoller: *Fiber Bundles*, 2nd edition (Springer-Verlag, New York, 1975)
112. E. L. Ince: *Ordinary Differential Equations* (Dover, New York, 1956)
113. T. Inui, Y. Tanabe, Y. Onodera: *Group Theory and Its Applications in Physics* (Springer-Verlag, New York, 1990)
114. C. J. Isham: *Modern Differential Geometry for Physicists* (World Scientific, Singapore, 1989)
115. C. Itzykson, J. B. Zuber: *Quantum Field Theory* (McGraw-Hill, New York, 1980)
116. R. Jackiw: Int. J. Mod. Phys. A3, 286 (1988)
117. J. D. Jackson: *Classical Electrodynamics* (Wiley, New York, 1975)
118. R. Joynt: J. Phys. C 17, 4807 (1984)
119. T. Jungwirth, Q. Niu, A. H. MacDonald: Phys. Rev. Lett. 88, 207208 (2002)
120. R. Karplus, J. M. Luttinger: Phys. Rev. 95, 1154 (1954)
121. T. Kato: J. Phys. Soc. Japcn. 5, 435 (1950)
122. M. Keil, H.-G. Krämer, M. Baig, J. Zhu, W. Demtröder, W. Meyer, W. B. Riesenfeld: J. Chem. Phys. 113, 7414 (2000)
123. B. Kendrick: J. Chem. Phys. 112, 5679 (2000)
124. R. D. King-Smith, D. Vanderbilt: Phys. Rev. B 47, 1651 (1993)
125. T. N. Kitsopoulos, M. A. Buntine, D. P. Baldwin, R. N. Zare, D. W. Chandler: Science 260, 1605 (1993)
126. S. Kivelson: Phys. Rev. B 26, 4269 (1982)
127. D. A. V. Kliner, D. E. Adelman, R. N. Zare: J. Chem. Phys. 95, 1648 (1991)
128. S. Kobayashi, K. Nomizu: *Foundations of Differential Geometry*, volumes 1 and 2 (Interscience, New York, 1969)
129. J. S. Koehler, D. M. Dennison: Phys. Rev. 57, 1006 (1940)
130. M. Kohmoto: Ann. Phys. 160, 343 (1985)
131. M. Kohmoto, B. I. Halperin, Y. S. Wu: Phys. Rev. B 45, 13448 (1992)
132. W. Kohn: Proc. Phys. Soc. 72, 1147 (1958)
133. W. Kohn: Phys. Rev. 115, 1460 (1959)
134. W. Kohn, J. M. Luttinger: Phys. Rev. 108, 590 (1957)
135. H. Koizumi: Phys. Rev. B 59, 8428 (1999)
136. H. Koizumi, I. B. Bersuker: Phys. Rev. Lett. 83, 3009 (1999)
137. H. Koizumi, I. B. Bersuker, J. E. Boggs, V. Z. Polinger: J. Chem. Phys. 112, 8470 (2000)
138. H. Koizumi, T. Hotta, Y. Takada: Phys. Rev. Lett. 80, 4518 (1998)
139. H. Koizumi, T. Hotta, Y. Takada: Phys. Rev. Lett. 81, 3803 (1998)
140. H. Koizumi, S. Sugano: J. Chem. Phys. 101, 4903 (1994)
141. H. Koizumi, Y. Takada: Phys. Rev. B 65, 153104 (2002)
142. I. Kovacs: *Rotational Structure in the Spectra of Diatomic Molecules* (American Elsevier, New York, 1969)
143. Z. Kunszt, A. Zee: Phys. Rev. B 44, 684 (1991)
144. A. Kuppermann, Y.-S. M. Wu: Chem. Phys. Lett. 205, 577 (1993)
145. A. Kuppermann, Y.-S. M. Wu: Chem. Phys. Lett. 241, 229 (1995)
146. L. D. Landau, E. M. Lifshitz: *The Classical Theory of Fields* (Pergamon, Oxford, 1962)
147. L. D. Landau, E. M. Lifshitz: *Mechanics* (Pergamon, Oxford, 1976)

148. L. D. Landau, E. M. Lifshitz: *Quantum Mechanics: Non-Relativistic Theory*, 3rd edition (Pergamon Press, New York, 1977)
149. S. Lang: *Introduction to Diophantine Approximations* (Springer-verlag, New York, 1995)
150. R. B. Laughlin: Phys. Rev. B 23, 5632 (1981)
151. R. B. Laughlin: in: *The Quantum Hall Effect*, edited by R. E. Prange, M. S. Girvin, p. 233 (Springer-Verlag, New York, 1987)
152. H. K. Lee, M. Rho: (1992), Hanyang University Preprint HYUPT-92-07
153. J. M. Leinaas: Physica Scripta 17, 483 (1978)
154. J. M. Leinaas, J. Myrheim: Li Nuovo Cimento 37, 1 (1977)
155. H. R. Lewis, Jr., W. B. Riesenfeld: J. Math. Phys. 10, 1458 (1969)
156. R. Lindner, K. Müller-Dethlefs, E. Wedum, K. Haber, E. R. Grant: Science 271, 1698 (1996)
157. R. G. Littlejohn, M. Reinsch: Rev. Mod. Phys. 69, 213 (1997)
158. H. C. Longuet-Higgins: Proc. Roy. Soc. London Ser. A 344, 147 (1975)
159. H. C. Longuet-Higgins, U. Öpik, M. H. L. Pryce, R. A. Sack: Proc. Roy. Soc. London Ser. A 244, 1 (1958)
160. H. C. Longuett-Higgins: in: *Advances in Spectroscopy Vol. 2*, edited by H. W. Thompson, pp. 429–72 (Interscience, New York, 1961)
161. C. K. Loong, J. M. Carpenter, J. W. Lynn, R. A. Robinson, H. A. Mook: J. Appl. Phys. 55, 1895 (1984)
162. J. M. Luttinger: Phys. Rev. 84, 814 (1951)
163. K. B. Marathe, G. Martucci: *Mathematical Foundations of Gauge Theories* (North-Holland, Amsterdam, 1992)
164. R. M. Martin: Phys. Rev. B 9, 1998 (1974)
165. C. A. Mead: Chem. Phys. 49, 23 and 33 (1980)
166. C. A. Mead: Rev. Mod. Phys. 64, 51 (1992)
167. C. A. Mead, D. Truhlar: J. Chem. Phys. 70, 2284 (1979)
168. C. A. Mead, D. G. Truhlar: J. Chem. Phys. 77, 6090 (1982)
169. R. Meiswinkel, H. Köppel: Chem. Phys. 144, 117 (1990)
170. A. Messiah: *Quantum Mechanics*, volume 2 (North-Holland, Amsterdam, 1961)
171. G. V. Milnikov: JETP Lett. 48, 536 (1988)
172. G. Montambaux, M. Kohmoto: Phys. Rev. B 41, 11417 (1990)
173. G. Montambaux, P. Littlewood: Phys. Rev. Lett. 62, 953 (1989)
174. J. Moody, A. Shapere, F. Wilczek: Phys. Rev. Lett. 56, 893 (1986)
175. D. J. Moore: Phys. Rev. A 23, L665 (1990)
176. D. J. Moore: Phys. Rep. 210, 1 (1991)
177. A. Mostafazadeh: J. Math. Phys. 37, 1218 (1996)
178. A. Mostafazadeh: J. Phys. A. 30, 7525 (1997)
179. A. Mostafazadeh: Phys. Rev. A 55, 1653 (1997)
180. A. Mostafazadeh: J. Phys. A. 31, 9975 (1998)
181. A. Mostafazadeh: J. Phys. A 32, 8175 (1999)
182. A. Mostafazadeh: J. Phys. A 32, 8325 (1999)
183. A. Mostafazadeh: *Dynamical Invariants, Adiabatic Approximation, and the Geometric Phase* (Nova, New York, 2001)
184. A. Mostafazadeh, A. Bohm: J. Phys. A 26, 5473 (1993)
185. R. S. Mulliken, A. Christy: Phys. Rev. 38, 87 (1931)
186. M. Nakahara: *Geometry, Topology and Physics* (Adam Hilger, New York, 1990)
187. H. Nakamura: *Nonadiabatic Transition* (World Scientific, Singapore, 2002)
188. M. Narasimhan, S. Ramanan: Amer. J. Math. 83, 563 (1961)

189. C. Nash: *Differential Topology and Quantum Field Theory* (Academic Press, New York, 1991)
190. C. Nash, S. Sen: *Topology and Geometry for Physicists* (Academic Press, New York, 1983)
191. J. W. Negele, H. Orland: *Quantum Many-Particle Systems* (Perseus Books, Reading MA, 1988)
192. G. Nenciu: Lett. Math. Phys. 11, 127 (1986)
193. G. Nenciu: Rev. Mod. Phys. 63, 91 (1991)
194. D. Neuheuser, R. S. Judson, D. J. Kouri, D. E. Edelman, N. E. Shafer, D. A. V. Kliner, R. N. Zare: Science 257, 519 (1992)
195. R. G. Newton: Phys. Rev. Lett. 72, 954 (1994)
196. H. H. Nielsen: Phys. Rev. 40, 445 (1932)
197. Q. Niu: Mod. Phys. Lett. B 5, 923 (1991)
198. Q. Niu, L. Kleinman: Phys. Rev. Lett. 80, 2205 (1998)
199. Q. Niu, D. J. Thouless: J. Rev. A 17, 2453 (1984)
200. Q. Niu, D. J. Thouless: Phys. Rev. B 35, 2188 (1987)
201. Q. Niu, D. J. Thouless, Y. Wu: Phys. Rev. B 31, 3372 (1985)
202. Q. Niu, X. Wang, L. Kleinman, W.-M. Liu, D. M. C. Nicholson, G. M. Stocks: Phys. Rev. Lett. 83, 207 (1999)
203. M. C. M. O'Brien: Proc. Roy. Soc. London Ser. A 281, 323 (1964)
204. M. C. M. O'Brien: J. Phys. A 22, 1779 (1989)
205. L. Onsager: Phil. Mag. Ser. 7 43, 1006 (1952)
206. J. W. Orton, P. Auzins, J. H. E. Griffiths, J. E. Wertz: Proc. Phys. Soc. (London) 78, 554 (1961)
207. T. Pacher, C. A. Mead, L. S. Cederbaum, H. Köppel: J. Chem. Phys. 91, 7057 (1989)
208. S. Pancharatnam: Proc. Ind. Acad. Sci. A 44, 247 (1956)
209. F. B. Pedersen, G. T. Einevoll, P. C. Hemmer: Phys. Rev. B 44, 5470 (1991)
210. R. E. Peierls: Z. Phys. 80, 763 (1933)
211. M. Peshkin, A. Tonomura: *The Aharonov–Bohm Effect*, number 340 in Lecture Notes in Physics (Springer-Verlag, Berlin, 1989)
212. A. B. Pippard: in: *Physics of Solids in Intense Magnetic Fields*, edited by E. D. Heidemenakis, p. 370 (Plenum, New York, 1969)
213. A. B. Pippard: *Magnetoresistance in Metals* (Cambridge University Press, Cambridge, 1989)
214. R. Prange: Phys. Rev. B 23, 4802 (1981)
215. R. Prange, R. Joynt: Phy. Rev. B 25, 2943 (1982)
216. I. I. Rabi, N. F. Ramsey, J. Schwinger: Rev. Mod. Phys. 26, 167 (1954)
217. L. E. Reichl: *A Modern Course in Statistical Physics* (University of Texas Press, Austin, Texas, 1980)
218. R. Resta: Ferroelectrics 136, 51 (1992)
219. R. Resta: Rev. Mod. Phys. 66, 899 (1994)
220. L. M. Roth: J. Phys. Chem. Solids 23, 433 (1962)
221. L. M. Roth: Phys. Rev. 145, 434 (1966)
222. J. J. Sakurai: *Modern Quantum Mechanics* (Addison-Wesley, New York, 1985)
223. L. Saminadayar, D. C. Glattli, Y. Jin, B. Etienne: Phys. Rev. Lett. 79, 2526 (1997)
224. J. Samuel, R. Bhandari: Phys. Rev. Lett. 60, 2339 (1988)
225. G. C. Schatz, A. Kuppermann: J. Chem. Phys. 65, 4642 (1976)
226. G. C. Schatz, A. Kuppermann: J. Chem. Phys. 65, 4668 (1976)
227. L. I. Schiff: *Quantum Mechanics* (McGraw-Hill, New York, 1955)
228. L. Schnieder, K. Seekamp-Rahn, J. Borkowski, E. Werde, K. H. Welge, F. J. Aoiz, L. B. Nares, M. J. D'Mello, V. J. Herrero, V. S. Rábanos, R. E. Wyatt:

Science 269, 207 (1995)
229. J. Schwinger: Science 165, 757 (1969)
230. B. Simon: Phys. Rev. Lett. 51, 2167 (1983)
231. C. P. Slichter: *Principles of Magnetic Resonance*, 3rd edition (Springer-Verlag, New York, 1990)
232. N. Steenrod: *The Topology of Fiber Bundles* (Princeton University Press, Princeton, 1951)
233. M. E. Stoll, A. Vega, R. W. Vaughan: Phys. Rev. A16, 1521 (1977)
234. P. Středa: J. Phys. C 15, L717 (1982)
235. P. Středa: J. Phys. C 15, L1299 (1982)
236. W. Su: Phys. Rev B 30, 1069 (1984)
237. G. Sundaram, Q. Niu: Phys. Rev. B 59, 14195 (1999)
238. D. Suter, G. C. Chingas, R. A. Harris, A. Pines: Mol. Phys. 61, 1327 (1987)
239. D. Suter, K. T. Mueller, A. Pines: Phys. Rev. Lett. 60, 1218 (1988)
240. A. Szabo, N. S. Ostlund: *Modern Quantum Chemistry* (McGraw-Hill, New York, 1989)
241. E. Tannenbaum, R. D. Johnson, R. J. Myers, W. D. Gwinn: J. Chem. Phys. 22, 949
242. R. Tao, F. D. M. Haldane: Phys. Rev. B 33, 3844 (1986)
243. R. Tao, Y. Wu: Phys. Rev. B 30, 1097 (1984)
244. B. N. Taylor: Phys. Today 42, 23 (1989)
245. D. Thouless: Phys. Rev. B 27, 6083 (1983)
246. D. Thouless: J. Phys. C 17, L325 (1984)
247. D. Thouless, M. Kohmoto, N. Nightingale, N. de Nijs: Phys. Rev. Lett. 49, 405 (1982)
248. D. J. Thouless: J. Phys. C 14, 3475 (1981)
249. D. J. Thouless, P. Ao, Q. Niu: Phys. Rev. Lett. 76, 3758 (1996)
250. A. Tomita, R. Chiao: Phys. Rev. Lett. 57, 937 (1986)
251. C. H. Townes, A. L. Schawlow: *Microwave Spectroscopy* (Dover, New York, 1975)
252. S. A. Trugman: Phys. Rev. B 27, 7539 (1983)
253. D. C. Tsui, H. L. Störmer, A. C. Gossard: Phys. Rev. Lett. 48, 1559 (1982)
254. R. Tycko: Phys. Rev. Lett. 58, 2281 (1987)
255. N. Usov, F. Ulinich: JETP 10, 1522 (1982)
256. D. Vanderbilt, R. D. King-Smith: Phys. Rev. B 47, 1651 (1993)
257. D. A. Varshalovich, A. N. Moskalev, V. K. Khersonskii: *Quantum Theory of Angular Momentum* (World Scientific, Singapore, 1988)
258. J. H. V. Vleck: Phys. Rev. 52, 246 (1937)
259. H. von Busch, V. Dev, H.-A. Eckel, S. Kasahara, J. Wang, W. Demtröder, P. Sebald, W. Meyer: Phys. Rev. Lett. 81, 4584 (1998)
260. K. von Klitzing, G. Doda, M. Pepper: Phys. Rev. Lett. 45, 494 (1980)
261. J. von Neumann, E. P. Wigner: Z. Phys. 30, 467 (1929)
262. S. J. Wang: Phys. Rev. A 42, 5107 (1990)
263. G. Wannier: Phys. Rev. 52, 191 (1937)
264. G. Wannier, D. Friedkin: Phys. Rev. 125, 1910 (1962)
265. J. K. G. Watson: Mol. Phys. 15, 479 (1968)
266. F. Wegner: Z. Phys. B 51, 279 (1983)
267. X. G. Wen, Q. Niu: Phys. Rev. B 41, 9377 (1990)
268. E. Werde, L. Schnieder: J. Chem. Phys. 107, 786 (1997)
269. E. Werde, L. Schnieder, K. H. Welge, F. J. Aoiz, L. B. Nares, J. F. Castillo, B. Martínez-Haya, V. J. Herrero: J. Chem. Phys. 110, 9971 (1999)
270. E. Werde, L. Schnieder, K. H. Welge, F. J. Aoiz, L. B. Nares, V. J. Herrero, B. Martínez-Haya, S. Rábanos: J. Chem. Phys. 106, 7862 (1997)

271. F. Wilczek: Phys. Rev. Lett. 49, 957 (1982)
272. F. Wilczek, A. Zee: Phys. Rev. Lett. 52, 2111 (1984), see also ref. [282]
273. N. Woodhouse: *Geometric Quantization* (Clarendon Press, Oxford, 1980)
274. Y.-S. M. Wu, A. Kupperman: Chem. Phys. Lett. 201, 178 (1993)
275. Y.-S. M. Wu, A. Kupperman: Chem. Phys. Lett. 235, 105 (1995)
276. Y.-S. M. Wu, A. Kupperman, B. Lepetit: Chem. Phys. Lett. 186, 319 (1991)
277. M. Yamanaka, W. Koshibae, S. Maekawa: Phys. Rev. Lett. 81, 5604 (1998)
278. D. R. Yarkony: Rev. Mod. Phys. 68, 985 (1996)
279. D. R. Yarkony: J. Phys. Chem. A 105, 6277 (2001)
280. J. Zak: Phys. Rev. 134, A1602 (1964)
281. J. Zak: Phys. Rev. Lett. 62, 2747 (1989)
282. A. Zee: Phys. Rev. A38, 1 (1988)
283. A. Zee: Mod. Phys. Lett. B 5, 1339 (1991)
284. G. E. Zilberman: Sov. Phys. JETP 3, 835 (1957)
285. J. W. Zwanziger, M. Koenig, A. Pines: Phys. Rev. A42, 3107 (1990)
286. J. W. Zwanziger, M. Koenig, A. Pines: Ann. Rev. Phys. Chem. 41, 601 (1990)
287. B. Zygelman: Phys. Rev. Lett. 64, 256 (1990)

Index

C_n 407
$D(R)$ 410
S_n 408
$\langle \Gamma_1\gamma_1\Gamma_2\gamma_2\Gamma\gamma\rangle$ 417
σ 407
σ_d 408
σ_h 408
σ_v 408
$|G|$ 412
d_Γ 412
i 408
Tr 413

adiabatic assumption of the local moment 346
Aharonov–Bohm effect 26
algebra 385
– associative 385
– enveloping 404
– exterior 386
– Grassmann 386
– non-associative 398
– Weyl–Heisenberg 400
antisymmetric product representation 417
anyon 342, 343
associated bundle 74

basis of a representation 411
bijection 366
bundle
– Aharonov–Anandan 59
– morphism 70
– principal 73
– reduction 74

Casimir operator 404
character of a group 413
character table 413
chart 371
class 413

Clebsch–Gordan coefficients 417
Clebsch–Gordan Matrix 417
closed form 115
compactness 370
conical intersection 198
conjugacy class of a group 413
conjugate class 413
conjugate elements 413
connectedness 369
connection 79
– Aharonov–Anandan 58
– Mead–Berry 16
– on a manifold 76
– universal 117
continuous map 367, 369
coset 390
couple-channel method 271
coupled-channel method 269
covariant
– derivative 87, 93, 98
– momentum 88
covering 370
curvature 101
– Mead–Berry 19
curve 379

de Rham cohomology 115
diabatic representation 222
diffeomorphism 372, 373
differential form 380
differential map 387
Dirac phase 27
Dirac string 195
double rotation 247
dynamical phase 49

enveloping algebra 404
equivalence relation 365
equivalent representation 411
Euclidean group 189
exact form 115

Index

experimental methods
- interferometry   248
- microwave spectroscopy   260
- nuclear magnetic resonance   231, 250
- nuclear quadrupole resonance   237, 247

experimental observations
- Abelian phase   231, 245
- Aharonov–Anandan phase   248
- non-Abelian phase   247, 260

exponential map   401
exterior derivative   386

finite group   408
Floquet theory   60
fractional statistics   343
frame bundle   76
free action   397

gauge
- field   92
- group   91
- potential   92
- theory   91, 161
- transformation   89, 100

generalized quasi-momentum   357
generator   385
generators of a group   409
Grassmann manifold   133
great orthogonality relation   412
group   389
- $U(1)$ 393
- $\mathbb{Z}_n$ 366
- action   396
- general linear   401
- Lie   394
- of gauge transformations   71, 91
- unitary   402

gyromagnetic ratio   227

holonomy   86
homeomorphism   369
homomorphism   392
homotopy   114
horizontal lift   79
Hubbard model   353

intramolecular interactions
- electronic–rotational coupling   259
- intramolecular rotational coupling   258

irreducible representation   403, 412
isometry   367

isomorphism   392

Jacobi coordinate system   153
Jahn–Teller theorem   421

Kramers degeneracy   219

Lie algebra   398, 400
Lie group   394
lift   54, 68
line bundle   72
Liouville–von Neumann equation   6
local trivialization   69

magic angle spinning   245
manifold   68, 371
- product   378
Massey parameter   200
matter field   92
metric space   364
molecular Aharonov–Bohm effect   148, 167, 195
monopole   41, 181
Mulliken symbols   414
multiply reducible   416

non-Abelian phase   240
non-adiabatic transition   198
non-crossing rule   197
NQR   237

open subset   363, 364, 368
orbital degree of freedom   353
order of a group   408

pair annihilation of conical intersections   210
parallel transport   77
parallelism   78
path-ordering operator   85
phase
- Dirac   27
potential energy surface   164
principal axis   408
principal bundle   73
projective representation   410
projective space   110, 376
proper point groups   409
proper rotation   407
pseudorotation   204
pullback
- of a form   388
push-forward map   387

Index   439

rank   67
ray representation   410
reduced matrix element   420
representation   403, 404
representation of a group   410
rotating frame   233

Schönflies notations   409
section   72
semiclassical path   218
simply reducible   416, 418
singular gauge transformation   195
smooth function   372
sphere   31, 374
structure group   70
symmetric product representation   417
symmetry operations   407

tangent   379
– space   380
– vector   379
tangent bundle   76
tensor   384
tensor operator   412
time-dependent variational principle   347
time-evolution operator   9
time-ordering operator   10
time-reversal invariance   239

topological   29, 369
– manifold   371
– phase   22, 26
– phase transition   126
– quantization   125
topological phase   202
topology   368
– basis of   369
– discrete   370
– Euclidean metric   370
– product   371
– quotient   111
– subspace   370
– trivial   370
transition function   69
transitive action   397
typical fiber   67

unitary group   402
unitary representation   412
universal
– bundle   116
– connection   117

vector potential   16
vibronic   147

wedge product   385
Wigner rotational function   268
Wigner–Eckart theorem   418

Texts and Monographs in Physics

Series Editors: R. Balian   W. Beiglböck   H. Grosse   E. H. Lieb
N. Reshetikhin   H. Spohn   W. Thirring

Essential Relativity   Special, General, and Cosmological   Revised 2nd edition
By W. Rindler

The Elements of Mechanics
By G. Gallavotti

Generalized Coherent States and Their Applications
By A. Perelomov

Quantum Mechanics II
By A. Galindo and P. Pascual

Geometry of the Standard Model of Elementary Particles
By. A. Derdzinski

From Electrostatics to Optics
A Concise Electrodynamics Course
By G. Scharf

Finite Quantum Electrodynamics
The Causal Approach   2nd edition
By G. Scharf

Path Integral Approach to Quantum Physics   An Introduction
2nd printing   By G. Roepstorff

Supersymmetric Methods in Quantum and Statistical Physics   By G. Junker

Relativistic Quantum Mechanics and Introduction to Field Theory
By F. J. Ynduráin

Local Quantum Physics Fields, Particles, Algebras   2nd revised and enlarged edition
By R. Haag

The Mechanics and Thermodynamics of Continuous Media   By M. Šilhavý

Quantum Relativity A Synthesis of the Ideas of Einstein and Heisenberg
By D. R. Finkelstein

Scattering Theory of Classical and Quantum N-Particle Systems
By. J. Derezinski and C. Gérard

Effective Lagrangians for the Standard Model   By A. Dobado, A. Gómez-Nicola, A. L. Maroto and J. R. Peláez

Quantum The Quantum Theory of Particles, Fields, and Cosmology   By E. Elbaz

Quantum Groups and Their Representations
By A. Klimyk and K. Schmüdgen

Multi-Hamiltonian Theory of Dynamical Systems   By M. B laszak

Renormalization   An Introduction
By M. Salmhofer

Fields, Symmetries, and Quarks
2nd, revised and enlarged edition   By U. Mosel

Statistical Mechanics of Lattice Systems
Volume 1: Closed-Form and Exact Solutions
2nd, revised and enlarged edition
By D. A. Lavis and G. M. Bell

Statistical Mechanics of Lattice Systems
Volume 2: Exact, Series and Renormalization Group Methods
By D. A. Lavis and G. M. Bell

Conformal Invariance and Critical Phenomena   By M. Henkel

The Theory of Quark and Gluon Interactions
3rd revised and enlarged edition
By F. J. Ynduráin

Quantum Field Theory in Condensed Matter Physics   By N. Nagaosa

Quantum Field Theory in Strongly Correlated Electronic Systems
By N. Nagaosa

Information Theory and Quantum Physics
Physical Foundations for Understanding the Conscious Process   By H.S. Green

Magnetism and Superconductivity
By L.-P. Lévy

The Nuclear Many-Body Problem
By P. Ring and P. Schuck

Perturbative Quantum Electrodynamics and Axiomatic Field Theory   By O. Steinmann

Quantum Non-linear Sigma Models
From Quantum Field Theory to Supersymmetry, Conformal Field Theory, Black Holes and Strings   By S. V. Ketov

Series homepage – http://www.springer.de/phys/books/tmp

Texts and Monographs in Physics

Series Editors: R. Balian   W. Beiglböck   H. Grosse   E. H. Lieb
N. Reshetikhin   H. Spohn   W. Thirring

The Statistical Mechanics of Financial Markets 2nd Edition By J. Voit

Statistical Mechanics A Short Treatise
By G. Gallavotti

Statistical Physics of Fluids
Basic Concepts and Applications
By V. I. Kalikmanov

Many-Body Problems and Quantum Field Theory An Introduction
By Ph. A. Martin and F. Rothen

Foundations of Fluid Dynamics
By G. Gallavotti

High-Energy Particle Diffraction
By E. Barone and V. Predazzi

Physics of Neutrinos
and Applications to Astrophysics
By M. Fukugita and T. Yanagida

Relativistic Quantum Mechanics
By H. M. Pilkuhn

The Geometric Phase in Quantum Systems
Foundations, Mathematical Concepts,
and Applications in Molecular and Condensed Matter Physics
By A. Bohm, A. Mostafazadeh, H. Koizumi,
Q. Niu and J. Zwanziger

Series homepage – http://www.springer.de/phys/books/tmp

Druck: betz-druck GmbH, D-64291 Darmstadt
Verarbeitung: Buchbinderei Schäffer, D-67269 Grünstadt